POSTHARVEST BIOLOGY AND TECHNOLOGY OF HORTICULTURAL CROPS

Principles and Practices for
Quality Maintenance

POSTHARVEST BIOLOGY AND TECHNOLOGY OF HORTICULTURAL CROPS

Principles and Practices for Quality Maintenance

Edited by
Mohammed Wasim Siddiqui, PhD

APPLE ACADEMIC PRESS

Apple Academic Press Inc.	Apple Academic Press Inc.
3333 Mistwell Crescent	9 Spinnaker Way
Oakville, ON L6L 0A2	Waretown, NJ 08758
Canada	USA

©2015 by Apple Academic Press, Inc.

First issued in paperback 2021

Exclusive worldwide distribution by CRC Press, a member of Taylor & Francis Group
No claim to original U.S. Government works

ISBN 13: 978-1-77463-226-0 (pbk)
ISBN 13: 978-1-77188-086-2 (hbk)

Library of Congress Control Number: 2015936645

Library and Archives Canada Cataloguing in Publication

Postharvest biology and technology of horticultural crops : principles and practices for quality maintenance / edited by Mohammed Wasim Siddiqui, PhD.

Includes bibliographical references and index.
ISBN 978-1-77188-086-2 (bound)
1. Horticultural crops--Postharvest technology. 2. Horticultural crops--Biotechnology.
I. Siddiqui, Mohammed Wasim, author, editor

SB319.7.P68 2015   635'.046   C2015-901597-9

Apple Academic Press also publishes its books in a variety of electronic formats. Some content that appears in print may not be available in electronic format. For information about Apple Academic Press products, visit our website at **www.appleacademicpress.com** and the CRC Press website at **www.crcpress.com**

ABOUT THE EDITOR

Mohammed Wasim Siddiqui, PhD

Dr. Mohammed Wasim Siddiqui is an Assistant Professor and Scientist in the Department of Food Science and Post-Harvest Technology at Bihar Agricultural University in Sabour, India, and is the author or co-author of 30 peer-reviewed journal articles, 18 book chapters, and 18 conference papers. He has three edited and one authored books to his credit, published by CRC Press, USA; Springer, USA; and Apple Academic Press, USA. Recently, Dr. Siddiqui has established an international peer-reviewed journal, *Journal of Postharvest Technology.* He has been honored to accept the position of Editor-in-Chief of a book series entitled Postharvest Biology and Technology, being published by Apple Academic Press. Dr. Siddiqui is also an Acquisitions Editor for Horticultural Science for AAP. He is editorial board member of several journals.

Recently, Dr Siddiqui has received the Achiever Award 2014 for outstanding research work by the Society for Advancement of Human and Nature (SADHNA), Nauni, Himachal Pradesh, India, where he is also an Honorary Board Member. He has been an active member of the organizing committees of several national and international seminars, conferences, and summits.

Dr. Siddiqui acquired a BSc (Agriculture) degree from Jawaharlal Nehru Krishi Vishwa Vidyalaya, Jabalpur, India, and received MSc (Horticulture) and PhD (Horticulture) degrees from Bidhan Chandra Krishi Viswavidyalaya, Mohanpur, Nadia, India, with specialization in postharvest technology. He was awarded a Maulana Azad National Fellowship Award from the University Grants Commission, New Delhi, India. He is a member of the Core Research Group at Bihar Agricultural University (BAU), where he helps with providing appropriate direction and assisting with prioritizing the research. He has received several grants from various funding agencies to carry out his research projects. Dr. Siddiqui has been associated with postharvest technology and processing aspects of horticultural crops, and he is dynamically involved in teaching (graduate and doctorate students) and research. He has proved himself as an active scientist in the area of postharvest technology.

DEDICATION

Dedicated to My Adored Father (Janab Md. Nasir Siddiqui) and Mother (Mrs. Sharfunnisha Siddiqui)

CONTENTS

LIST OF CONTRIBUTORS

Md. Shamsher Ahmad
Department of Food Science and Technology, Bihar Agricultural University, Sabour, Bhagalpur, Bihar 813 210, India

Shirin Akhtar
Department of Horticulture (Vegetable and Floriculture), Bihar Agricultural University, Sabour, Bhagalpur, Bihar (813210) India

Md. Arshad Anwer
Department of Plant Pathology, Bihar Agricultural University, Sabour, Bhagalpur, Bihar

Eva Arrebola
Department of Mycology, Instituto de Hortofruticultura Subtropical y Mediterranea IHSM-UMA-CSIC La Mayora

J. F. Ayala-Zavala
Centro de Investigación en Alimentación y Desarrollo, AC (CIAD, AC), Carretera a la Victoria Km 0.6, La Victoria, Hermosillo, Sonora, 83000, México

Kalyan Barman
Department of Horticulture (Fruit and Fruit Technology), Bihar Agricultural University, Sabour, Bhagalpur, Bihar 813 210, India

Donal Bhattacharjee
Department of Postharvest Technology of Horticultural Crops; Bidhan Chandra Krishi Viswavidyalaya, Mohanpur-741252, Nadia, West Bengal, India

S. Das
Department of Floriculture and Landscape Gardening, Faculty of Horticulture, Bidhan Chandra Krishi Viswavidyalaya, Mohanpur-741252, Nadia, West Bengal, India

R. S. Dhua
Department of Postharvest Technology of Horticultural Crops; Bidhan Chandra Krishi Viswavidyalaya, Mohanpur-741252, Nadia, West Bengal, India

Shailendra K. Dwivedi
College of Horticulture, RVS Krishi Vishwavidyalaya, Mandsaur Campus, Madhya Pradesh, India, PIN- 458 001

G. A. González-Aguilar
Centro de Investigación en Alimentación y Desarrollo, AC (CIAD, AC), Carretera a la Victoria Km 0.6, La Victoria, Hermosillo, Sonora, 83000, México

Norsuhada Abdul Karim
Food and Biomaterial Eng. Research Group, Bioprocess Engineering Department, Faculty of Chemical Engineering, Universiti Teknologi Malaysia, 81300 Johor Bahru, Johor, Malaysia

Nozieana Khairudin
Food and Biomaterial Eng. Research Group, Bioprocess Engineering Department, Faculty of Chemical Engineering, Universiti Teknologi Malaysia, 81300 Johor Bahru, Johor, Malaysia

Amit Kumar Khokher
Division of Fruit Science, SKUAST-K, Shalimar, Srinagar, Jammu and Kashmir (192 308) India

Vigya Mishra
Amity International Centre for Post Harvest Technology and Cold Chain Management

Ida Idayu Muhamad
Food and Biomaterial Eng. Research Group, Bioprocess Engineering Department, Faculty of Chemical Engineering, Universiti Teknologi Malaysia, 81300 Johor Bahru, Johor, Malaysia

C. Nithya
Department of Entomology, Bihar Agricultural University, Sabour, Bhagalpur-813210, Bihar

Bishun Deo Prasad
Department of Plant Breeding and Genetics, Bihar Agricultural University, Sabour, Bhagalpur, Bihar (813210) India

K. Prasad
Department of Food Engineering and Technology,S. L. I. E. T., Longowal – 148106, Punjab, India

S. N. Ray
Department of Entomology, Bihar Agricultural University, Sabour, Bhagalpur-813210, Bihar

G. R. Velderrain Rodríguez
Centro de Investigación en Alimentación y Desarrollo, AC (CIAD, AC), Carretera a la Victoria Km 0.6, La Victoria, Hermosillo, Sonora, 83000, México

Tamoghna Saha
Department of Entomology, Bihar Agricultural University, Sabour, Bhagalpur-813210, Bihar

Sangita Sahni
Department of Plant Pathology, T. C. A, Dholi, RAU, Pusa

Eraricar Salleh
Food and Biomaterial Eng. Research Group, Bioprocess Engineering Department, Faculty of Chemical Engineering, Universiti Teknologi Malaysia, 81300 Johor Bahru, Johor, Malaysia

M. L. Salmerón-Ruiz
Centro de Investigación en Alimentación y Desarrollo, AC (CIAD, AC), Carretera a la Victoria Km 0.6, La Victoria, Hermosillo, Sonora, 83000, México

A. B. Sharangi
Department of Spices and Plantation Crops, Bidhan Chandra KrishiViswavidyalaya, Mohanpur-741252, Nadia, West Bengal, INDIA

Mohammed Wasim Siddiqui
Department of Food Science and Technology, Bihar Agricultural University, Sabour, Bhagalpur, Bihar (813210) India

Jeebit Singh
Department of Postharvest Technology of Horticultural Crops; Bidhan Chandra Krishi Viswavidyalaya, Mohanpur-741252, Nadia, West Bengal, India

S. Sultana
Department of Spices and Plantation Crops, Bidhan Chandra KrishiViswavidyalaya, Mohanpur-741252, Nadia, West Bengal, India

Salehudin, Mohd Harfiz
Food and Biomaterial Eng. Research Group, Bioprocess Engineering Department, Faculty of Chemical Engineering, Universiti Teknologi Malaysia, 81300 Johor Bahru, Johor, Malaysia

Pran Krishna Thakur
Department of Postharvest Technology of Horticultural Crops; Bidhan Chandra Krishi Viswavidyalaya, Mohanpur-741252, Nadia, West Bengal, India

LIST OF ABBREVIATIONS

CA	controlled atmosphere storage
CSFRI	Citrus and Subtropical Fruit Research Institute
CT	computed tomography
EMS	ethane methyl sulfonate
FF	flesh firmness
MRI	magnetic resonance imaging
MRI	magnetic resonance imaging
MRI	magnetic resonance imaging
NAA	naphthalene acetic acid
NIR	near infrared
NMV	net magnetization vector
PAL	phenylalanine ammonia lyase
PME	pectin methylesterase
PPO	polyphenol oxidase
PPO	polyphenoloxidase
RF	radio frequency
SPI	starch pattern index
TSS	total soluble solids
ZECC	zero energy cool chamber

LIST OF SYMBOLS

λ	wavelength
f	frequency
v	velocity
E	photon energy
h	Planck's constant
μ_s	coefficient of static friction
γ	ratio of specific heat at constant pressure and specific heat at constant volume
δ_g	density of gas
P	pressure
K	bulk modulus of elasticity
δ_l	density of liquid
Y	Young's modulus
δ_s	density of solid
I_0	initial sound intensity
I_x	intensity at depth x
e	base of natural logs
x	depth in tissue
a	amplitude absorption coefficient

PREFACE

Along with the increasing production of agrihorticultural crops, particularly fruits and vegetables, postharvest loss is increasing as well, comprising about 30-35% of total production, due to the absence of proper postharvest handling/storage, preservation, and cold chain infrastructure. The ultimate goal of crop production is to provide quality produce to the consumers at reasonable prices. Most of fresh produce is highly perishable, and postharvest losses are very significant under the present scenario of postharvest management, which is insufficient to meet the requirements in the many countries. Business in the handling, packaging, and storage equipment sectors for agrihorticultural products offers good prospects worldwide. The existing postharvest technology for fresh horticultural produce is traditional, and there is an urgent need for modernization, creation, and popularization of innovations in this sector, especially in areas of fruits and vegetables to reduce large-scale wastage.

Significant achievements have been made during the last few years to curtail the postharvest losses in fresh produce and to ensure food security as well. These include advancement in understanding of breeding horticultural crops for quality improvement; postharvest physiology; postharvest pathology and entomology; postharvest management of fruits, vegetables, and flowers; nondestructive technologies to assess produce quality; minimal processing of fruits and vegetables; as well as innovations in packaging and storage technology of fresh produce. Advancements in postharvest molecular approaches are opening new ways to increase the shelf life of fresh produce.

This book, *Postharvest Biology and Technology of Horticultural Crops: Principles and Practices for Quality Maintenance,* covers the above-mentioned advancements in postharvest quality improvement of fresh horticultural produce. The book is comprised of 14 chapters. Chapter 1 describes the recent understandings of the pre- and postharvest factors responsible for the quality of fresh fruits and vegetables. The chapter provides an in-depth analysis of different aspects of postharvest practices. Chapter 2 discusses different nondestructive analysis methods being used for the quality assessment of fresh commodities. Different examples are cited in the chapter for better discussion. Pre-harvest factors are mostly responsible for affecting the postharvest quality of fresh commodities, and breeding approaches are of major importance. Chapters 3 and 4 elaborately discuss the conventional breeding efforts made to improve postharvest quality of fruits and vegetables. Chapter 5 deals with advanced storage systems of fruits and vegetables. The discussion on different parameters affecting quality of fresh products during storage is also included. Innovations such biofilm application for improved storage life are discussed in detail. Chapter 6 describes advances in packaging systems of fresh products. Active and smart packaging film for postharvest treatment of fruits and vegetables are discussed. An in-depth discussion on postharvest disease management technologies has been given in the chapter 7. Several insect-pests affect the quality of fresh commodity. The infestation begins right from the pre-harvest stage, and infestation remains a problem up to postharvest. Chapter 8 discusses integrated postharvest pest management in fruits and vegetables. Chapter 9 describes the process of flower senescence along with different factors that affect the process. In chapter 10, detailed information on postharvest management of cut flowers is given. The chapter discusses the handling process right from the harvesting stage to final marketing. Chapters 11 and 12 describe the postharvest management technologies of medicinal, aromatic, and spices crops. Biotechnological approaches have been used to increase the shelf life of fresh commodities using different technologies such antisense RNA technology. Chapter 13 discusses different biotechnological approaches to improve postharvest quality of fruits and vegetables. Chapter 14 deals with advances in postharvest disease management in vegetable crops. The chapter thoroughly discusses the mode of action and feasibility of different technologies to control the postharvest diseases in vegetables.

This book will be a standard reference work for the fresh produce industry in postharvest management to extend the shelf life by retaining the nutritional and sensory quality and increasing the safety of fresh produce.

The editor would appreciate receiving comments from readers that may assist in the development of future editions.

— Mohammed Wasim Siddiqui

ACKNOWLEDGMENT

Some rare, auspicious moments come in life when words are totally insufficient to express heartfelt emotion. It was almost impossible to reveal the deepest sense of veneration to all without whose precious exhortation this book project could not be completed. First of all, I ascribe all glory to the Gracious "Almighty Allah" from whom all blessings come. I would like to thank him for His blessing to write this book.

With a profound and unfading sense of gratitude, I wish to express our sincere thank to the Bihar Agricultural University, India, for providing me with the opportunity and facilities to execute such an exciting project, and supporting me towards research and other intellectual activities around the globe. I convey special thanks to my colleagues and other research team members for their support and encouragement for helping me in every step of the way to accomplish this venture.

I am grateful to Mr. Ashish Kumar, President, Apple Academic Press, to accomplish my dream of publishing this book series, namely *Postharvest Biology and Technology*. I would also like to appreciate Ms. Sandra Jones Sickels and Mr. Rakesh Kumar of Apple Academic Press for their continuous support to complete the project.

In omega, my vocabulary will remain insufficient to express my indebtness to my adored parents and family members for their infinitive love, cordial affection, incessant inspiration, and silent prayer to "Allah" for my well-being and confidence.

CHAPTER 1

FACTORS AFFECTING THE QUALITY OF FRUITS AND VEGETABLES: RECENT UNDERSTANDINGS

KALYAN BARMAN[1], MD. SHAMSHER AHMAD[2], and
MOHAMMED WASIM SIDDIQUI[2]

[1]Department of Horticulture (Fruit and Fruit Technology)
[2]Department of Food Science and Technology Bihar Agricultural University,
Sabour, Bhagalpur, Bihar 813 210, India
*Email: wasim_serene@yahoo.com

CONTENTS

Quality of fresh fruits and vegetables are governed by many factors. The combined effect of all, decides the rate of deterioration and spoilage. These factors if not controlled properly, lead to postharvest losses on large scale. According to Kader (2002), approximately one third of all fresh fruits and vegetables are lost before it reaches to the consumer. Another estimate suggests that about 30–40 percent of total fruit and vegetables production is lost in between harvest and final consumption (Salami et al., 2010). A considerable research priority has been given to the maturity status of fresh fruits and vegetables at harvest and the maintenance of proper storage temperature at every stage from the farm to fork during the last two decades. However, the influence of several preharvest factors and cultural practices on the postharvest physiology, quality and storage life has been given less importance. Therefore in this chapter, an attempt is made to discuss about how the postharvest quality of fresh horticultural produce gets influenced by both preharvest and postharvest factors.

1.1   WHAT IS QUALITY?

The word "quality" is derived from the Latin word "*qualitas*" meaning attribute, property or basic nature of a product. However, now-a-days it is defined as "degree of excellence or superiority" (Kader et al., 1986). The term "quality" is a complex perception felt by the consumer in different ways. It is a combination of characteristics, attributes and properties which give the commodity value to humans for food. The perception of quality varies among producer and consumer. Producers are more concerned with commodities that have good appearance and few visual defects but, for them a good cultivar must rank high in yield, disease and pest resistance, ease of harvest and shipping quality. However, consumers are interested in produce that have good appearance, firmer, good flavor and nutritive value. Although, consumers purchase produce on the basis of appearance but, their repeat purchase depends on good edible quality.

   The quality of fruits and vegetables are determined at harvest because, the produce is removed from its source of carbohydrates, water and nutrient supply. Thus, there is no possibility for further improvements in the quality of the harvested produce. In fact, the best we can do is reduction in the rate of deterioration which is progressed through the process of maturation, ripening and senescence in the produce. For this reason, it is very important to understand the both preharvest and postharvest factors that influence the quality of harvested produce and subsequently, the consumers' decision to purchase the produce.

1.2   PREHARVEST FACTORS AFFECTING QUALITY

Preharvest factors that are involved in determining the quality of fruit and vegetables can be grouped into environmental and cultural factors. Environmental factors include temperature, sunlight, wind, frost, hailstorm and pollution. Cultural factors may be mineral nutrition, organic production, irrigation, pruning, thinning, girdling, canopy position, rootstocks and growth regulators.

1.2.1 ENVIRONMENTAL FACTORS

1.2.1.1 TEMPERATURE

Temperature is an important factor in determining growth, development and post-harvest quality attributes of fresh fruit and vegetables. Maturity of fruit is highly affected by preharvest temperature in the field. For example, grapes on a particular bunch showed differential maturity time due to variation in exposure of sunlight. Fruits bear on the exposed side of sun ripen faster than those bear on the shaded side (Kliewer and Lider, 1970). These fruits also contain higher sugars and lower acidity than shaded fruits. On the other hand, avocado fruits exposed to the sun on the tree took 1.5 days longer to ripen than those which were on the shaded side (Woolf et al., 1999). Higher flesh temperature of about 35°C might have slowed down the ripening process in avocado. The same fruits when ethylene-treated exposed fruit being firmer ripened slowly than shaded fruit (Woolf et al., 2000).

Preharvest exposure of fruit and vegetables to direct sunlight leads to a number of postharvest physiological disorders among which, sunburn or solar injury is the most prevalent temperature-induced disorder in fruit and vegetables (Table 1.1, Fig. 1.1). Factors like high light intensity, high ambient temperature and moisture stress increase the propensity of sunburn. Initial symptoms of sunburn include bleaching or yellowing of the peel in apple (Bergh et al., 1980) and corky or rough fruit surface in avocado (Schroeder and Kay, 1961). Photosynthetic activity of the fruit is also hampered by high fruit temperature. In severe cases or frequent exposure lead to browning or blackening of the skin of produce due to tissue failure (Schroeder and Kay, 1961; Croft, 1995). This situation results in complete inactivation of the photosynthetic system. An important physiological disorder associated with sun exposure in the field is watercore in apple (Marlow and Loescher, 1984). Here, core as well as flesh closer to the skin is affected which are directly exposed to high temperature or sunlight. Similarly, water soaking in flesh of pineapple is also reported due to exposure of fruit to high temperature during early growing season (Paull and Reyes, 1996; Chen and Paull, 2000). In "Tahiti" lime, rupture of juicy vesicles and breakdown of stylar-end take place due to exposure of fruit to high temperatures.

TABLE 1.1 Disorders associated with preharvest exposure of fruit to high temperature or direct sunlight.

Fruit	Disorder	Symptoms	Reference
Apple	Sunburn	Skin discoloration, pigment breakdown	Bergh et al. (1980), Wünsche et al. (2000)
Apple	Watercore	Water soaking of flesh	Marlow and Loescher (1984)
Avocado	Sunburn	Skin browning	Schroeder and Kay (1961)
Pineapple	Flesh translucence	Water soaking of flesh	Paull and Reyes (1996); Chen and Paull (2000)
Lime	Stylar end breakdown	Juice vesicle rupture	Davenport and Campbell (1977)
Cranberry	Sun scald	Tissue breakdown	Croft (1995)

Source: Woolf and Ferguson (2000).

Preharvest temperature also has considerable effect on quality of fruit during and after postharvest storage. Generally, a higher temperature in the field is associated with higher sugar and lower acidity in the fruit (Kliewer and Lider, 1968, 1970). Grapes and oranges grown at high temperature contain higher level of sugar and lower level of tartaric and citric acids respectively than those grown under low temperature. Coombe (1987) reported that for every 10°C increase in temperature tartaric acid content in grapes reduce by half. The firmness of the fruit is also affected by temperature prevailing during fruit growth. Avocado fruits grown on the trees exposed to direct sunlight having temperature around 35°C showed fruits 2.5 times firmer than those grown on the shaded side (20°C) of the tree. Changes in composition of cell wall, cell number and cell turgor properties were postulated as being associated with the observed phenomenon (Woolf et al., 2000). The color development of fruit is also influenced by temperature. Generally, warm days and cool nights during growth are conducive for color development in fruit. Higher day and lower night temperatures promote synthesis of anthocyanin pigments in strawberry which lead to redder and darker fruit (Galletta and Bringhurst, 1990). Accumulation of minerals in the fruit during growth is also affected by temperature and/or sunlight. Woolf et al. (1999) reported that avocado fruits (cv. Hass) which are exposed to direct sunlight contain more calcium, magnesium and potassium compared to shaded fruits.

Bioactive compounds present in the fruits and vegetables and their antioxidant potential are also affected by high temperatures during growing period. For example, warm nights (18–22°C) and days (25°C) during growing season of strawberry (cv. Kent) facilitates to contain higher antioxidants in the fruits compared to those grown under cooler (12°C) days (Wang and Zheng, 2001). This is due to the fact that higher temperature enhanced flavonol content in the fruit which exert higher antioxidant activity. Similarly, vitamin C content in the fruit also varies depending upon growing temperature. Mandarins (cv. Frost Satsuma) grown under cool temperatures (day temperature 20–22°C, night temperature 11–13°C) accumulates more vitamin C than those grown under hot temperatures (day temperature 30–35°C, night temperature 20–25°C). Similarly, grapefruits grown in desert areas of California and Arizona contain less vitamin C than those grown in coastal areas of California.

Exposure of fruit to high temperatures at or few days before harvest may induce tolerance to high and low temperatures during subsequent postharvest storage. This is found in avocado fruits. Avocados which were exposed to direct sunlight having pulp temperature of about 40°C, stored at or above the recommended storage temperature showed lower incidence of chilling injury (at 0°C) or high temperature injury (50°C) than shaded fruits(Woolf et al., 1999). Similar result is found in high temperature exposed lemons which showlower chilling injury symptoms during storage at 2°C for 3 weeks (Houck and Joel, 1995).

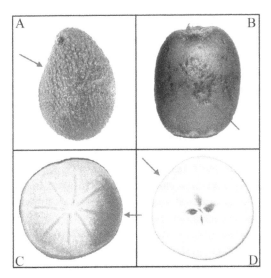

FIGURE 1.1 Effect of fruit exposure to high temperature or sun exposure on the field.(A) chilling injury of avocado fruit (cv. Hass) following four weeks of storage at 0°C, (B) skin pitting of kiwifruit (cv. Haywood) following 16 weeks of storage at 0°C, (C) internal chilling injury of persimmons (cv. Fuyu) following 6weeks storage at 0°C and3days at 20°C and (D) watercore in apples (cv. Cox's Orange Pippin) 3days after harvest (no storage).

1.2.1.2   SUNLIGHT

Exposure to direct/intermittent sunlight or shade conditions results in variations in the fruit quality as well as its postharvest ripening behavior and physiology. It also influences the incidence and severity of several physiological disorders and diseases. Citrus and mango fruits exposed to the direct sunlight were found to have lighter weight, thinner peel, higher total soluble solids, lower titratable acidity and juice percentage in comparison to those fruits growing in shade inside the canopy (Sites and Reitz, 1949, 1950). The "Fuji" apples growing in the outer side and top of the canopy receiving direct sunlight were redder and dark colored and had higher content of quercetin and cyanidin glycosides than the fruits which grew in shade in the inner parts of the tree canopy (Jakopic et al., 2009).

It is known that direct sunlight is not essentially required for the synthesis of ascorbic acid in plants. However, the total incidence, amount and intensity of sunlight on the fruits during the entire growing period have a definite influence on the content of ascorbic acid in the fruits. Ascorbic acid is synthesized from sugars generated by photosynthesis in plants. The fruits in the outer parts of the tree canopy receiving direct sunlight have a higher content of ascorbic acid in comparison to fruits growing in shaded regions of the same tree. Generally, the amount of light intensity on the fruit during the growth period is directly corelated with their total ascorbic acid content (Harris, 1975).

The "Hass" cultivar of avocado fruit on ripening at 20°C revealed that the fruits exposed to direct sunlight had higher firmness, slower ripening rates and delayed ethylene evolution peaks by about 2–5 days than the fruits which grew in shaded regions of the canopy. On inoculating these fruits with *Colletotrichumgloeosporioides,* the decay symptoms were recorded 2–3 days earlier in shaded fruits than those receiving direct exposure of sunlight. In addition, higher incidence of various symptoms of postharvest heat and chilling injury was observed in shaded fruits in comparison to fruits getting direct sun exposure when the fruits were subjected to hot water treatment (50°C) followed by storage at 0.5°C (Woolf et al., 1999, 2000). Considerably higher climacteric ethylene production peaks were recorded in shaded apple fruits over those fruits which received direct exposure of sunlight (Nilsson and Gustavsson, 2007). The synthesis of heat shock proteins in the fruits exposed to direct sunlight may have enhanced the tolerance levels of these fruits to both high as well as low temperatures.

The exposure of fruits to excessive sunlight results in the occurrence of sunscald in several fruit crops. The direct exposure of the sunlight on the fruit surface results in pigment degradation in the affected surface area of the fruit. Further, cellular death and collapse of the affected tissue occurs if the duration of exposure of sunlight or its intensity is higher than adequate or optimum levels. This influence of high sunlight exposure causing stress to the fruits is mainly thermal in nature, although some bleaching of the chlorophyll pigment can also occur. Some examples of quality degradation due to exposure to excessive or insufficient sunlight are detailed in the Table 1.2. Specific disorders like water core in apple, flesh translucence in pineapple, stylar-end breakdown in lime were also reported by various workers due to weak and insufficient light presence.

TABLE 1.2 Quality deterioration due to excess or insufficient sunlight in various fruits

Sl. No.	Name of the fruit	Researchers	Symptoms
1.	Persimmon	George et al. (1997)	Sun Scald
2.	Mandarin	Myhob et al. (1996)	Sun Scald
3.	Pomegranate	Panwar et al, 1994	Sun Scald
4.	Blueberry	Caruso (1995)	Sun Scald
5.	Pineapple	Lutchmeah (1992)	Sun Scald
6.	Apple	Sibbett et al. (1991)	Sun Scald
7.	Banana	Wade et al. (1993)	Sun Scald
8.	Strawberry	Osman and Dodd (1994)	Insufficient light typically results in smaller size fruit and decreases the surface glossiness in strawberry
9.	Apple	Campbell and Marini (1992)	Low sunlight intensity reduces the color development.
10.	Grape	Hummell and Ferree (1997)	-----do----
11.	Strawberry	Saks et al. (1996)	-----do----

1.2.1.3 WIND

During growth, wind may also cause damage to the fruit and vegetables. Damage by winds can be grouped into two categories: damage caused by less frequent severe storms; and that caused by frequent winds of intermediate strength. High velocity winds result into damage of leaves and defoliation in leafy vegetables, which cause severe damage in product appearance and market ability (Kays, 1999). In fruit crops, defoliation leads to smaller size fruit (Eckstein et al., 1996) and development of poor fruit color in citrus (Ogata et al., 1995). Mild winds may cause wind scarring disorder due to rubbing of fruits against twigs. The injured fruit developtan-to-silvery patches which increase in size with the advancement of maturity. It may also cause friction marks on kiwifruit (McAneney et al., 1984; Lizana and Stange,

1988), wind rub on persimmon (George et al., 1997) and wind scab of French prune (Michailides and Morgan, 1993). Therefore, use of windbreaks is advocated for production of fruit and vegetable in areas subjected to excessive wind (Holmes and Koekemoer, 1994).

1.2.1.4   HAIL AND FROST DAMAGE

Hailstorm during fruit growth causes direct damage to the crop by increasing percentage of deformed fruit (Hong et al., 1989) and anatomical alteration (Visai and Marro, 1986; Fogliani et al., 1985). It also causes indirect damage by increasing incidence of diseases (e.g., bacterial spot in pepper) (Kousik et al., 1994). The severity of damage depends upon factors like hail stone size, growth stage of crop and duration of exposure (Duran et al., 1994). Damage of hails can be minimized by use of hail nets over trees of fruit crop (e.g., apple, pear; Reid and Innes, 1996) or covering the fruit bunches (e.g., banana; Eckstein et al., 1996). Frost damage to the fruit and vegetables are caused at a temperature below the freezing point of water, due to the presence of solutes within the aqueous medium of the cells. The extent of damage depends upon prevailing temperature, rate of drop in temperature, duration of exposure, and susceptibility of the product (Kays, 1999). Due to exposure of fruit to frost, some fruit exhibit discoloration of the exterior (e.g., banana)while for others, internal symptoms are more pronounced (e.g., blackening of pecan kernel). In leafy vegetable like lettuce, freezing results in a limp, wilted, and unmarketable product, while crops like brussels sprout (*Brassica oleracea* L. var. *gemmifera*Zenk.) can withstand radiative freezing to some extent (Kays, 1999). In fruit crops, damage by frosts often results in smaller size misshapen fruit (e.g., strawberry) depending upon the development stage (Meesters, 1995; Goulart and Demchak, 1994). It also causes degradation in appearance such as discoloration in Kiwi fruit, frost rings in apple and frost circles in peach (Testolin et al., 1994).

1.2.1.5   POLLUTANTS

Air pollutants such as ozone, sulfur dioxide, fluoride and nitrogen oxides can cause severe damage and reduce quality of fruit and vegetables during their growth. During the summer season when high temperature and solar radiation prevails the production of ozone in the atmosphere generally increases due to increase in nitrogen species and emission of volatile organic compounds (Mauzerall and Wang, 2001). Ozone enters into the plant system through stomata and causes cellular damage by increase in membrane permeability and may cause injury (Mauzerall and Wang, 2001). Higher concentration of ozone in the atmosphere also affects the photosynthetic and respiratory processes in plants which ultimately affects postharvest quality in terms of overall appearance, color, flavor and increase turnover of antioxidant systems (Grulke and Miller, 1994; Tjoelker et al., 1995; Percy et al., 2003). Elevated atmospheric concentrations of ozone may also lead to yellowing or chlorosis in leafy vegetables, blistering in spinach, alter sugars and starch content in fruit and tuber crops and reduce fruit size by decrease in biomass production (Kays, 1999; Felzer et al., 2007). Fluoride causes discoloration in peach fruit. Similarly, higher concentration of nitrogen dioxide in the atmosphere results in marginal and interveinal collapse of lettuce leaves (Kays, 1999). Apart from air pollutants, ions of heavy metals like silver, cadmium, cobalt, magnesium, manganese, nikel, zinc, etc. which may enter into the plant system through soil amendments, runoff, or contaminated irrigation water also can cause deterioration in quality of fruits and vegetables (Kays, 1999).

1.2.2   CULTURAL FACTORS

1.2.2.1   MINERAL NUTRITION

The best possible plant performance can be obtained only if mineral nutrients are available in a balanced and timely manner. The mineral nutrition obtained from the soil influences the fruit quality significantly. However, the mineral absorption and assimilation by the plant also depends on many factors like age of the plant, type of the soil, moisture profile of the root zone and the climatic conditions. The optimal performance of the plant requires balanced availability of macro as well as micronutrients. The variations in the absorption and bio-availability in the essentially required mineral nutrition results in either a deficiency or excess of nutrients which further gets expressed as various physical/physiological disorders as well as poor quality of such fruits.

Calcium and nitrogen are two most important mineral nutrients which have profound influence on the postharvest physiology, fruit quality as well as the shelf life of the fruits. The calcium content of the fruit influences the fruit metabolism by causing alterations in the intracellular and extracellular processes. The fruit firmness, its rate of softening, fruit quality and the occurrence of different physiological disorders is also dependent on the calcium content of the fruit (Fallahi et al., 1985; Lidster and Poritt, 1978; Mason et al., 1975; Poovaiah, 1988). Calcium is also known to have a regulatory role in different processes which further affect the cell functioning and signal transduction (Marme and Dieter, 1983; Poovaiah, 1988). The lower fruit calcium content reflects in many ways like occurrence of physiological disorders like bitter pit in apples, internal breakdown, watercore, lowered poststorage disease resistance as well as accelerated rates of fruit softening thus limiting the postharvest life of the fruits considerably.

The fruit softening occurs due to cell wall degradation. Ca delays softening rates by delaying the degradation of cell wall polymers. It also has an important role in cell to cell adhesion which is important for the maintenance of the textural quality of the fruit. Fruit Ca content is an important determinant of the fruit quality. The fruits with low Ca content are susceptible to several physical, physiological and pathological disorders and have a short postharvest storage life. The development of Ca deficiency symptoms in harvested products may also be because of the manner in which calcium is transported around the plant (in the xylem only and not the phloem) and the time at which it is available to be imported into fruit (only early in the development and not during maturation) (Ferguson et al., 1999). The incidence and severity of bitter pit in apple fruit is mainly related with the calcium deficiency during fruit growth period which is detectable immediately after harvest or sometimes after protracted period of storage (Atkinson et al., 1980). The low Ca content in fruit also increases the susceptibility of apples to flesh and core browning (Rabus and Streif, 2000). The preharvest spray of Ca maintains higher firmness, SSC, titratable acidity, ascorbic acid content, lower incidence of albinism and higher resistance to gray mold rot in strawberry fruit (Singh et al., 2007). The calcium deficiency is overcome by spraying with calcium salts either during the fruit development period or by the postharvest calcium dip/drench treatments of the fruit (Hewett and Watkins, 1991).

In general, high levels of nitrogen content results in poor keeping quality. In addition to this, excessive nitrogen levels are known to delay the maturity of the stone fruits, induce poor visual red color development and also inhibit the change in ground color from green to yellow. This occurs because of the imbalance in the essential amino acids resulting due to excess soil nitrogen content. The higher nitrogen content in fruits leads to decreased calcium content and Ca/K ratio as a consequence of which the susceptibility of the fruit to abrasions, bruising injury, physiological disorders and various pathogens is enhanced. The application of high levels of nitrogen to the fruit crops also reduces the ascorbic acid content of many citrus fruits like sweet orange, lemons, mandarin and grapefruit (Nagy, 1980). It has also been found to influence and enhance the ethylene production as well as the respiration rates in the apple and mango fruits. The fruit firmness is lower in trees supplied with excessive nitrogen fertilizer with accelerated postharvest fruit softening rates. On the other hand, the deficiency of nitrogen results in small fruit size, poor fruit color and flavor, low overall production, and increased storage decay (Daane et al., 1995; Tahir et al., 2007). The balanced, judicious, timely and split application of nitrogen is known to increase the flavor synthesis in fruits.

Phosphorus have very significant role in fruit quality. A positive correlation between phosphorous content and soluble solids and acid content is found in Cox's Orange Pippin apple. A lower level of phosphorous in the soil leads to susceptibility of fruit to low temperature breakdown and senescent breakdown.Marcelle (1995) reported that a higher content of phosphorus in "Jonagold" apple increase fruit firmness and decrease dry matter content. When phosphate fertilizer is applied as foliar application, it enhances the length, diameter, firmness and soluble solids content in persimmon fruit. While it decrease the disease and pest incidence (Hossain and Ryu, 2009). In cranberry fruit, excess phosphate masks the color development.

Similar to nitrogen and phosphorous, potassium also plays an important role in postharvest fruit quality. Potassium content in fruit is highly related to pH and organic acid content in fruit. Deficiency of potassium causes poor coloration and reduced fruit size in peach and oranges, respectively. When potassium fertilizer is applied to pineapple (16 g/plant), it decreases the internal brown disorder in fruit (Soares et al., 2005). The activities of polyphenoloxidase (PPO) and phenylalanine ammonia lyase (PAL) enzymes were also found negatively correlated with potassium content. In case of citrus, it affects shape of fruit and increase acidity. While vitamin C content is reported to increase with the application of potassium fertilizer in fruit (Nagy, 1980).

Micronutrients are also useful for improving postharvest fruit quality. For example, preharvest deficiency of iron and zinc leads to reduced fruit size in citrus and both color and size reduction in peaches. Zinc deficiency also alters the shape of peaches and cherries. Similarly, deficiency of boron reduces fruit size in strawberry and causes external corking in apples. Deficiency of copper and molybdenum causes misshapen fruit in citrus and strawberry fruit respectively and affect kernel filling in walnut.

1.2.2.2   ORGANIC PRODUCTION

The consumer today is alert, health conscious and has put forth the demand for fresh fruits free of pesticide or any kind of chemical residues which has put the sustainability of conventional fruit production systems and old practices of fruit growing into question. This has resulted in a surge of interest in organic and integrated fruit production practices worldwide. Higher soil quality and significantly lower negative environmental impacts were observed in organic and integrated production systems in comparison to the conventional systems using higher fertilizers and pesticides application (Vogeler et al., 2006). The use of synthetic pesticides and fertilizers is excluded in the organic management practices and uses animal and green manures, compost, botanical insecticides, traps and other biological control methods (Peck et al., 2006).

"Bluecrop" blueberries were grown by both conventional and organic production by Wang et al. (2008). It was observed by them that organic blueberries were higher in sugars (fructose and glucose), malic acid, total phenolics, total anthocyanins (including delphinidin-3-galactoside, delphinidin-3-glucoside, delphinidin-3-arabinoside, petunidin-3-galactoside, petunidin-3-glucoside, and malvidin-3-arabinoside), and antioxidant activity than fruits obtained from the conventional culture.

"Royal Gala" and "Fuji" apple trees grown on M.7 rootstocks by both conventional and organic production systems resulted in lower concentrations of K, Mg, and N in leaves and fruit, smaller fruit size for both cultivars and lower fruit yield for "Fuji" in comparison to the conventional production system (Amarante et al., 2008). In addition to this, organically produced apple fruits had more yellowish skin background color, higher percentage of blush in the fruit skin, higher soluble solids content, higher flesh firmness and higher severity of russet than conventionally produced fruits. Weibel et al. (2000) also reported that organically produced fruits had lower titratable acidity in "Royal Gala", and higher incidence of moldy core and lower incidence of watercore in "Fuji" over the conventionally produced fruits while no difference in vitamin C content was observed. Higher vitamin E level was found in organic olive oil in one study than conventional production (Gutierrez et al., 1999). The "Robusta" bananas grown organically ripened faster at 22–25°C than conventionally grown bananasasrevealed by change in the peel color, while total soluble solids remained the same in both.

1.2.2.3   IRRIGATION

Water plays an important role in growth and development of fruit and vegetables. The amount and timing of application is very important for getting optimal quality produce. Moisture stress during fruit development stage typically result in reduced fruit size in citrus (Bielorai, 1982), peach (Veihmeyer and Hendrickson, 1949), olive (Inglese et al., 1996), muskmelon (Lester et al., 1994), etc. An appropriate water management strategy is especially important in fruit crops for optimum photosynthesis, plant growth and harvestable yield of quality produce. Sometimes obtaining higher yield is not considered as prerequisite for optimal quality fruit and vegetables. In peach, optimum irrigations during fruit growth lead to maximum sized fruit however, moderate water stress prior to harvest yield fruit of higher soluble solid content (Crisosto et al., 1994). Sorensen et al. (1995) further reported that less frequent irrigation to fruit lead to increase in concentrations of dietary fiber, protein, vitamin C, nutrients like Ca, Mg and Mn. Similarly in pear-jujube (*Zizyphus-jujuba* Mill. cv. Lizao) fruit, an increase in soluble solids, sugar/acid ratio, organic acid and ascorbic acid content is found when moderate and severe deficit irrigation is imposed to during fruit growth compared to full irrigation(Cui et al., 2008). In "Salustiano" citrus fruit, moderate and severe irrigation deficit causes higher juice percentage, soluble solids, titratable acidity and peel thickness than fruit received full irrigation (Tajero et al., 2010). When water stress is applied during early growth stage in mango, it delays the time to ripe and shows lower incidence of chilling injury during cold storage.

1.2.2.4   PRUNING, THINNING, AND GIRDLING

Several studies have reported that higher penetration of sunlight into the canopy improves the postharvest quality of fruits like apple, grapes, peach, plum, etc. When pruning operation is performed judiciously, it increases the fruit size, soluble solids content, anthocyanin accumulation, phenolics content, and reduces titratable acidity, pH, and potassium content in fruit. Fruits which are grown on the shaded portion of tree are shown to have lower flavor quality than those which are grown on the sun-light. However on the contrary, when extensive pruning is carried out during summer season by removing leaves surrounding the fruit it became harmful for the tree as these leaves supply carbohydrates to the fruit. Fruit trees which are heavily pruned found to have better color development, maximum fruit length, higher fruit weight and lower number of seeds/fruit. These fruits also have lower peel percentage, lower acidity, higher rag percentage, juice percentage and higher TSS content as compared to fruit from lightly pruned and nonpruned trees. Ahmad et al. (2006) reported that, heavy pruning of kinnow plant improved yield as well as quality of fruit.

Thinning of fruitlets during initial stages of fruit growth found to increase the fruit size however, it decreases the total yield. Therefore, a balance between fruit size and yield must be maintained to get optimum qualityfruit.Increase of fruit size by thinning is reported in apples (Batjer et al., 1957) and peaches (Westwood and Balney, 1963). Similarly, reduction of fruit size and soluble solids content is re-ported in early ripening "May Glo" nectarine and late ripening O'Henry peaches when too many fruits were left on the tree (Crisosto et al., 1995). This is due to the fact that under good production conditions, fruit crops set more fruit than desirable. Thinning when practiced, increases the leaf to fruit ration, and as a result gives larger individual fruit size (Westwood, 1993).

Apart from pruning and thinning, the balance between vegetative and fruit growth in a tree can also be altered by girdling (removal of bark). Girdling of grapes is practiced commercially in vines that have extra vigor and produced poor quality fruit. It increases the size and shape of berry and delays its ripening by lowering sugar : acid ratio and color intensity. In case of peach and nectarines, girdling 4–6 weeks before harvest increases the size of fruit and advances its maturity (Day, 1997). But, if girdling is done too early during pit hardening, it causes splits in pits of peach and nectarine. These fruits soften more quickly than intact fruits and are highly susceptible to decay (Crisosto and Costa, 2008).

1.2.2.5 ROOTSTOCK

Fruit trees are mostly produced by grafting on different rootstocks. These rootstocks provide adaptability to grow crops on different soil and climatic conditions. More-over, they also facilitate their cultivation against various diseases and pests (Dichio et al., 2004; Cinelli and Loreti, 2004). Research has revealed that growing fruit crops on different rootstocks also have profound influence on fruit quality. Most of the marketable varieties in fruit farming are produced grafted on different root-stocks. These rootstocks facilitate their culture on different soils and climatic condi-tions and provide trees against pests and diseases (Dichio et al., 2004; Cinelli and Loreti, 2004). These adaptive traits often produce increased fruit yields (Webster, 1995, 2001), but it had been assumed that rootstocks had little influence on fruit quality attributes, being considered foremost a scion cultivar-associated trait. However, in the last years different groups have reported the effect of rootstock on fruit quality for several species such as peach (Caruso et al., 1996; Iglesias et al., 2004), apple (Bielicki et al., 2004; Slowinska et al., 2004) and sweet cherry. "Allen Eureka" lemon when grafted on Cleopatra mandarin, it produces fruit of high acid content however, when it is grafted on sour orange rootstock, TSS content of fruit increases significantly (Jaleel et al., 2005). The shelf life of Jonagold apples increased when it is grafted on PB-4 and M.26 rootstocks due to lower ethylene Production and delayed ripening during storage (Tomala et al., 2008). On the other hand, Ruby Red grapefruit shows lower decay and less chilling injury incidence during storage at 4°C for 6 weeks while budded on rough lemon than *Citrus amblycarpa*. The flavor quality of fruit is also influenced by rootstock in grapefruit (Cooper and Lime, 1960) and tangerine (Hodgson, 1967). Similarly, citrus fruits grown on *Citrus volkameriana* rootstock were found inferior in flavor and juiciness and also had lower ascorbic acid content compared to the same scion grafted onto sour orange rootstock. Citrus fruits grafted on *Volkameriana* rootstock also resulted in fruit with lower titratable acidity and total soluble solids. Moreover, the bitterness of juice is also found to be influenced by rootstock (Chandler and Kefford, 1966). Apart from fruit quality parameters, composition of bioactive compounds and antioxidant capacity in fruit was found affected by rootstock. Peach fruit cv. Flavorcrest grown on Mr. S 2/5 rootstock (natural hybrid of *Prunuscerasifera*) was reported to contain highest phenolics, vitamin C, β-carotene, and antioxidant capacity than fruits grown on Ishtara, GF 677, and Barrier1 rootstocks (Remorini et al., 2008).

1.2.2.6   TREE AGE

Very limited work has been carried out on the effect of tree age on fruit quality. Tahir et al. (2007) reported that apples (cv. Aroma) produced in younger trees of below 4 years develop better color, flavor and contain higher acid/SSC ratio than those produced in older trees of more than 20 years. Furthermore, incidences of flesh and core browning found higher in fruits from young trees. Similarly in case of guava (cv. Allahabad Safeda), fruits harvested from 10-, 15-, and 20-year-old trees show considerable variation in fruit quality (Asrey et al., 2007). Higher content of total soluble solids, total sugars, ascorbic acid and lower content of acidity found to contain in fruits of 15-year-old trees. These trees also yielded copper and manganese rich fruit. However, fruits from 20-year-old trees contained higher magnesium and zinc while, fruits obtained from 10-year-old trees contained higher level of iron. The number of seed content in fruit also reported to decrease with the progressive increase in tree age.

1.2.2.7   CANOPY POSITION

When fruits are harvested at their physiological maturity stage, considerable variations in fruit quality in terms of appearance (size, shape, color) and sensory quality (texture, sugar, acid content) have been found both between as well as within trees in the orchard It has been clear from previous works that position of the fruit on the canopy has significant effect on fruit quality such as color, flavor and other quality attributes during harvest (Hoffman et al., 1995).

For example, apples (cv. Aroma) produced outside the canopy contain higher level of dry matter, soluble solids and soluble sugars than those produced inside the canopy (Nilsson and Gustavsson, 2007). These fruits also develop a darker red peel color while, those produced inside canopy remain green. This is due to higher light intensity on the outside canopy, which facilitates accumulation of soluble sugars like fructose, glucose and sucrose and also synthesis of anthocyanin pigments in the fruit. Furthermore, when Awad et al. (2001) compared the flavonoids level in apple fruit (cv. Jonagold) produced in different parts of canopy, they found that fruits in top of the canopy contain higher level of cyanidin-3-galactosideand quercetin-3-glycosides than those produced in outside as well as inside canopy. However, the levels of catechins, phloridzin, and chlorogenic acid content in fruit are independent of canopy position. The synthesis of cyaniding-3-galactosideand quercetin-3-glycoside on the fruit depends on the level of light intensity during the growing period. The fruits produced inside the tree get lower level of UV-A, blue, green, and red light which are essential for synthesis of pigments. As a result, higher levels of far-red light and higher far-red/red light ratio on the interior canopy suppress the synthesis of these pigments. Similarly, oranges (cv. Tarocco) harvested from southern side of canopy showed higher content of soluble solids and lower content of acid than fruits harvested from northern and interior parts of canopy. Moreover, these fruits also showed lower decay percentage and reduced shelf life (Agabbio et al., 1999).

Growth Regulators

Plant growth regulators (PGR) are very powerful horticultural production tools and their effects in improving postharvest fruit quality have been reported by several workers. But, except few examples, these are not widely used commercially because the concentration range between obtaining suboptimal and super optimal effects is quite narrow. For example, preharvest applications of gibberellic acid (GA_3) delay peel senescence of citrus fruit, thereby improving on-tree storage and postharvest life. In case of orange, it increases peel firmness, juice yield and delays color development (Davies et al., 1997). Shelf life of fruit is also influenced by preharvest application of gibberellin (GA_3). Eshel et al. (2000) reported that application of gibberellin to persimmon fruit extends its storage life by delaying black-spot development and fruit softening during storage at $-10°C$. It also delays fruit softening and ripening by lowering the respiration rate in fruit(Ben-Arie et al., 1986). Apart from GA_3, naphthalene acetic acid (NAA) is also used for improving fruit quality. When NAA @ 45 ppm is applied to guava cv. Red Flesh, it increases pulp/seed ratio, TSS, total sugars and ascorbic acid content (Iqbal et al., 2009).

1.3 POSTHARVEST FACTORS AFFECTING QUALITY

1.3.1 MATURITY STAGE

This is the starting point of postharvest quality management. Therefore, it must be ensured that a good quality (in terms of maturity) produce should be harvested. Immature or over matured fruits are inferior in quality and spoil more quickly even if other factors are favorable.

At which stage of maturity a fruit or vegetable should be harvested are crucial to its subsequent storage, marketable life and quality. Maturity always have a considerable influence on the quality of fresh produce as well as the storage potential and occurrence of many storage disorders (Siddiqui and Dhua, 2010). There are mainly three stages in the life span of fruits and vegetables: maturation, ripening, and senescence. Maturation is indicative of the fruit being ready for harvest. At this point, the edible part of the fruit or vegetable is fully developed in size, although it may not be ready for immediate consumption. Ripening follows or overlaps maturation, rendering the produce edible, as indicated by texture, taste, color, and flavor. Senescence is the last stage, characterized by natural degradation of the fruit or vegetable, as in loss of texture, flavor, etc. (senescence ends with the death of the tissue of the fruit). Nonclimacteric fruits should be harvested after attaining proper development of eating quality (ripening) while attached to the plant. In this case ripening treatment is not given before retailing. However, few special treatments (degreening in citrus species) are followed to enhance marketing and consumer's appeal. On the other hand, climacteric fruits are harvested at full matured stage (before onset of ripening). Here in most of the cases, ripening treatment is given before retailing. Some typical maturity indexes are described in following sections.

1.3.1.1 PEEL COLOR

This factor is commonly applied to fruits, since peel color changes as fruit ripens or matures and thus it is regarded as a quality index. Huybrechts et al. (2003) reported that peel color also acts as a maturity index for some cultivars of apple and can be used to determine the maturity of the fruit. Among vegetables, tomato quality largely depends on peel color. When choosing which tomatoes to buy, consumers use color and appearance as indicators of quality. The consumers often associate a dark red color with good quality ripe tomatoes. This color is due to the presence of high concentration of lycopene (Barrett and Anthon, 2001). Peel color is also used for grading fruits and vegetables in many countries before storage or marketing (Watkins, 2003). This is also true that few cultivars exhibit no perceptible color change during maturation (stay green character). Assessment of harvest maturity by skin color depends on the judgment and experience of the harvester. Now color charts are also available for cultivars, such as apples, tomatoes, peaches, banana, peppers, etc. Change of color in case of banana during postharvest handling also affects quality in a major way. Therefore, banana color chart is recommended (Figure 1.2, Table 1.3). There is one more reliable method to measure skin color is optical method. Here Light transmission properties can be used to measure the degree of maturity of fruits in terms of chlorophyll degradation or development of color pigments. This method is based on the principle of reduction in chlorophyll content during maturation and ripening and development of color pigments at harvest maturity. Practically fruits are allowed to pass through a camera unit on rotating conveyor belts, where camera takes twenty seven to fifty snaps to individual fruits, process it through CPU and then put in a color range say above 70 percent color, 60–70 percent color, etc. One such grader based on optical method is **GREEFA** widely used for apple grading all over the world.

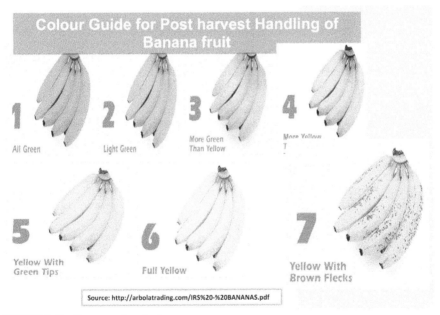

Source: http://arbolatrading.com/IRS%20-%20BANANAS.pdf

FIGURE 1.2   Banana color chart.

TABLE 1.3   Description about the color chart of banana

Indicator numbers	Quality parameters	Descriptions	Remarks
1	All green	Firm and hard fruits with very low sugar content. Before retailing, ripening treatment is suggested.	Appropriate stage for long distance transportation.
2	Light green	Fruit becomes less firm as starch begins to convert into sugar. Ripening process has begun and fruit generates heat which must be removed to control ripening	This stage confirms beginning of ripening process
3	More green than yellow	Fruit softens as starch converts into sugar continued. Heat generated in ripening chamber must be removed	Long distance retail delivery recommended
4	More yellow than green	Proper color for retail display provides. Many consumers prefer to buy at this stage	This is firm yellow stage.
5	Yellow with green tips	Proper color for retail display provides good consumer acceptance.	This is soft yellow stage.
6	Full yellow	Soft fruit with good flavor. Fruit should be on display shelf and not in storage.	On the same day consumption is recommended
7	Yellow with brown flecks	Brown flecks indicate high sugar, mealy texture and fungal infection.	Senescence started

Source: http://arbolatrading.com/IRS%20-%20BANANAS.pdf

1.3.1.2   SHAPE

In many fruits and vegetables, shape changes during maturation and thus gives an idea to determine harvest maturity. Postharvest quality depends on maturity stage of horticultural produce. For instance, a banana becomes more rounded in shape (Figure 1.3a) and less angular (Figure 1.3b) as it matures on the plant. It is the stage (Figure 1.3a) when harvesting of banana is recommended. Mangoes also change shape during maturation. This is evident by comparing the relationship between the shoulders of the fruit and the point at which the stalk is attached changes. The shoulders of immature mangoes slope away from the fruit stalk (Figure 1.4), but as maturity advances, the shoulders become level with the point of attachment and with final maturity the shoulders may be raised above this point (Figure 1.4). This stage gives maximum yield and quality attributes upon ripening.

FIGURE 1.3   Maturity **indices of banana (a)** Mature **(b)** Immature.

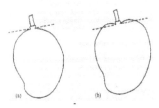

FIGURE 1.4   Maturity indices of mango.

1.3.1.3   SIZE

Change in the size of any fruit or vegetable crop while growing is frequently used to determine harvest maturity and quality. It is one of the oldest methods of maturity determination. Size increases as a fresh produce approaches toward maturity. For example, the size of a cauliflower curd or cabbage head increases up to full maturity with compactness in nature (Figure 1.5). This gives an idea for harvesting a quality curd or head.

FIGURE 1.5   Maturity indices of cauliflower (a) Immature curd (b) Mature curd (c) Over mature curd.

1.3.1.4   AROMA (FLAVOR)

Most fruits synthesize volatile compounds as they ripen. Such chemicals give fruit its characteristic odor and can be used to determine whether it is a good quality ripened fruit or not. For example in many instances, consumers bring ripe mango fruits near to nose in order to detect characteristic flavor of mango. This flavor may only be detectable by humans when a fruit is completely ripe, and therefore, this method has limited use in commercial situations. Similarly in case of litchi, quality largely depends on flavor of harvested litchi. Chyau et al. (2003) isolated and separated free and glycosidically bound volatile compounds from fresh clear litchi juice using an Amberlite XAD-2 column and concluded highest acceptability in litchi having good aroma.

1.3.1.5   LEAF/FLOWERS/INFLORESCENCE CONDITION CHANGES

In many cases, leaf quality often determines when fruits and vegetables should be harvested. In root crops, the condition of the leaves can indicate the condition of the crop below the ground. For example, if potatoes are to be harvested for storage, then the optimum harvest time (optimum maturity) is soon after the leaves and stems started yellowing and drying. If harvested earlier, (before yellowing of leaves) the skins will be less resistant to harvesting and handling damage and more prone to storage diseases. The alternative method is to kill or cut the foliage part ten to fifteen days before harvesting. This practice also hardens the skin of potato. More or less the same is true for other root and tuber crops (Figure 1.6). In case of banana, dryness of inflorescence at the tip of fingers often taken as one of the maturity index of banana.

FIGURE 1.6 Potato, onion and garlic at maturity (a) Mature potato (b) Mature onion (c) Mature Garlic.

1.3.1.6 DEVELOPMENT OF ABSCISSION LAYER

As part of the natural development of a fruit, an abscission layer is formed in the fruit stalk. For example, in melons, harvesting is done after development of abscission layer. Harvesting before the abscission layer is fully developed, results in inferior quality fruit, compared to those left on the vine till abscission layer is developed.

1.3.1.7 FIRMNESS (FLESH FIRMNESS)

The texture or firmness of any fruit changes during ripening and maturation. This is more prominent toward ripening when it losses texture more rapidly. Excessive loss of moisture may also affect the texture of the crops. These textural changes are detected by touch and the harvester presses the fruits or vegetables gently and can judge whether the crop is ready for harvest or not. This method is widely used in Indian fruit markets for apple. They also judge storable quality by simply pressing with thumb and fingers. This is an experience gained over years by producers, traders and buyers. Today sophisticated devices have been developed to measure texture in fruits and vegetables. For example, texture analyzers and pressure testers. These instruments are currently available for fruits and vegetables in various forms. A force is applied on the surface of the fruit (after peeling), allowing the probe of the Penetrometer to penetrate the fruit flesh, which then gives a reading on firmness. Hand held pressure testers are widely used in all apple and pears growing areas of the world (Figure 1.7). However, in some cases it could give variable results because the basis on which they are used to measure firmness is affected by the angle at which the force is applied and the support to hand holding the fruit. Table top-mounted pressure tester is used in order to minimize the error. Two commonly used pressure testers to measure the firmness of fruits and vegetables are the Magness-Taylor and UC Fruit Firmness. One more instrument is used for this purpose is called Instron Universal Testing Machine. It is not portable and mainly used in laboratories. Flesh firmness (FF) is the most important quality measurement of apples and pear that has been used to determine optimal storage maturity by private companies. But it varies among the cultivars (Watkins, 2003).

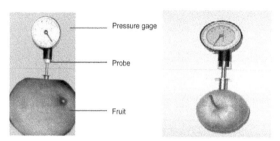

FIGURE 1.7   Hand pressure tester.

1.3.1.8 JUICE CONTENT

Juice content is an important measure of internal quality. Under or overripe fruit tend to be less juicy, which directly affects eating quality. The juice content is determined by weighing components of the whole fruit and the juice extracted. The juice content of many fruits increases as fruit matures on the tree. To measure the juice content of a fruit, a representative sample of fruit is taken and then the juice is extracted in a standard and specified manner. The juice volume is related to the original mass of juice, which is proportional to its maturity. This method is mostly used in citrus fruits.

1.3.1.9 SUGARS (TOTAL SOLUBLE SOLIDS)

In climacteric fruits, carbohydrates accumulate during maturation in the form of starch. As the fruit ripens, starch is broken down into sugars. In nonclimacteric fruits, sugar tends to accumulate during maturation. A quick method to measure the amount of sugar present in fruits is with a brix hydrometer or a refractometer. A drop of fruit juice is placed in the prism of the refractometer and a reading taken. This is equivalent to the total amount of soluble solids or Brix in the fruit juice. Brix is reported as "Degree Brix" and is equivalent to percentage. For example, a juice which is 12 degrees Brix has 12 percent total soluble solids. This method is widely used in grapes in many parts of the world to specify maturity. In the citrus industry also total soluble solids is measured in the juice. These soluble solids are primarily sugars; sucrose, fructose, and glucose. Citric acid and minerals in the juice also contribute to the soluble solids.

1.3.1.10 STARCH PATTERN INDEX (SPI)

Measurement of starch content is a reliable technique used to determine maturity in apple and to a lesser extent in pear cultivars. The method involves is cutting the fruit into two halves and dipping the pedicel end cut piece into Potassium iodide solution. The cut surfaces stain to a blue-black color in places where starch is present. This shows the pattern of starch conversion into sugar. Starch converts into sugar as maturity approaches. Harvesting begins when the samples show that 65–70 percent of the cut surfaces have turned blue-black leaving only the core (middle) or little more part unstained. The stained part gives an idea that how much starch is converted into sugar. As a fruit ripens more starch is converted into sugar, and the blue-black area becomes less prominent (Figure 1.8, Table 1.4). Ripening usually takes place from the core of the fruit toward the skin. All important stages of apple maturity in Granny Smith are shown below. This is also applicable in almost all apple cultivars.

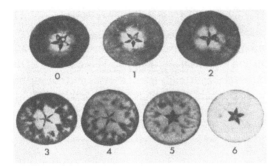

FIGURE 1.8   Starch Iodine color chart for apple (Source: Reid et al., 1982).

TABLE 1.4   Description about the color chart

Color Development	Descriptions	Remarks
0	This is immature fruit, harvesting should not be done at this stage	
1	This is borderline stage between immature to mature. Check other harvesting parameters at this stage like seed color (should turn black).	Indian apple (Red & Royal Delicious)is ready to harvest at this stage for storage in CA storage
2	Perfect mature stage, ideally suitable for both CA and Normal cold store	
3	Over mature, not suitable for CA and Normal storage or temporary storage for one Month or less.	This stage is ideal for immediate marketing.

TABLE 1.4 *(Continued)*

Color Development	Descriptions	Remarks
4	Over mature an advance stage over 3. Not suitable even for distant market sale.	Immediate sale in nearby cities or market is preferred
5	Towards senescence stage, mealy texture with no or little juice	Less commercial value
6	Advance stage of senescence, mealy texture, little pressure, not juicy at all	Not liked by consumers

1.3.1.11   BRIX: ACID RATIO

In many fruits, the acidity changes during maturation and ripening. Acidity reduces progressively as the fruit matures on the tree. Normally, acidity is not taken as a measurement of fruit maturity by itself but in relation to soluble solids, what is termed as the brix: acid ratio. The sugar-acid ratio contributes to the unique taste and flavor. At the beginning of the ripening process the sugar-acid ratio is low, because of low sugar content and high fruit acid content. This makes the fruit taste sour. During the ripening process the fruit acids are decreases and the sugar content increases as a result sugar: acid ratio achieves a higher value and taste highly acceptable.

1.3.1.12   SPECIFIC GRAVITY

Specific gravity is also one of the maturity standard used in few fresh produce. It is obtained by comparing the weight of a commodity in air with the weight of the same under water. In practice, fruit or vegetable is weighed in air and then in water (pure water). The weight in air divided by the weight in water gives the specific gravity. This will ensure a reliable measure of fruit maturity. As a fruit matures its specific gravity increases. This parameter is rarely used in practice to determine time of harvest, but could be used in cases where quality of that particular commodity is linked with other qualities. For example, specific gravity of potato is directly linked with the dry matter content of potato and dry matter content is directly linked with the yield of fried chips. Specific gravity of potatoes is commonly used by the potato processing industries as a tool for quick estimation of dry matter content. Methods like brine solution (Verma et al., 1972), hydrometer (Sukumaran and Ramdas, 1980), and weight of potatoes in air and water (Fitzpatrick et al., 1969) have been used for determining specific gravity of potatoes.

1.3.1.13 METHODS OF HARVESTING

There are basically three methods most commonly used for harvesting any fruits or vegetables (Figure 1.9).

(a) Harvesting individual fruits/vegetables with hand by pulling or twisting the fruit pedicel.

(b) Harvesting individual fruits or fruit bunch/vegetables or vegetable bunch with the help of fruit clippers/secateurs/scissors.

(c) With harvester specially designed for harvesting.

(A) HARVESTING INDIVIDUAL FRUITS/VEGETABLES WITH HAND BY PULLING OR TWISTING THE FRUIT PEDICEL

This is simple and most commonly used method of harvesting for fruits. Harvester can easily pick the optimum mature fruits. One important demerit of this method is while pulling or twisting fruit pedicel, little peel portion comes along with pedicel end. This renders fruits for quick spoilage and rapid quality deterioration takes place. Therefore, experienced harvesters are preferred.

(B) HARVESTING INDIVIDUAL FRUITS OR FRUIT BUNCH/VEGETABLES OR VEGETABLE BUNCH WITH THE HELP OF FRUIT CLIPPERS/SECATEURS/SCISSORS.

This method is an advance form of hand pulling and twisting method. Here fruit stalk is cut close to the point of attachment of fruit leaving a very small portion of pedicel attached with fruits. This is highly acceptable method and widely used in all most all citrus fruits. Postharvest quality is found better in this method compared to any others. One important precaution should be taken while harvesting is that the fruit stalk should be cut very close to the fruit leaving only a small part of fruit stalk otherwise, the part of pedicel or fruit stalk attached with fruit may puncture other healthy fruits during subsequent handling and marketing. Fruits born in bunch like grapes are harvested with scissors or secateurs.

(C) MECHANICAL HARVESTING

This method is used on commercial scale and specially designed machines are only used. Fruits and vegetables get more damage with harvester and quality deteriorates rapidly. Therefore, it is advisable to use harvester where harvested produce are intended to use for processing. This is most economic method of harvesting for processing grade fruits and vegetables. All three types of harvesting are illustrated in Figure 1.8 as mentioned below.

Harvesting by pooling   By fruit clippers   By mechanical harvester
and twisting

FIGURE 1.9   Harvesting methods.

1.3.2   TOOLS FOR HARVESTING AND ASSEMBLING

Postharvest quality also depends on the tools used for harvesting fresh produce. It is because faulty tools also affect quality. Depending on the type of fruits or vegetables, several tools are employed to harvest the produce. Commonly used tools for harvesting of fruit and vegetable are secateurs, scissors, fruit clippers, knives and handheld or pole-mounted picking shears. Few of them are shown below (Figure 1.10). When fruits or vegetables are difficult to catch with hand, such as mangoes or avocados, a pole-mounted picking shear or scissors is used. Harvested fruits are collected in a bag attached with the pole itself. Where there is no provision of collecting bags, fruits are allowed to fall on the ground directly or on a net above the ground or on a sheet of gunny bags or some cushioning material is placed on the ground just beneath the tree to prevent damage to the fruit while direct heating the ground. Harvesting bags with shoulder or waist slings can also be used for fruits with firm skins, like citrus and avocados. They are easy to carry and leave both hands free for climbing, harvesting, and assembling. Harvested produce are assembled in a container before sorting, grading, and packing. These containers too influence the quality. Plastic containers such as crates, buckets, bins are suitable containers for assembling harvested produce. These containers should be of enough strength and smooth without any sharp edges that could damage the produce. Additionally cushioning materials should be used in order to reduce bruising. Commercial growers use plastic crates and bulk bins with varying capacities and crops such as apples and cabbages are placed, and sent to packinghouses for selection, grading, and packing or directly to market. Gunny bags, plastic bags and hessian bags are used for vegetables like potato, onion, garlic, etc.

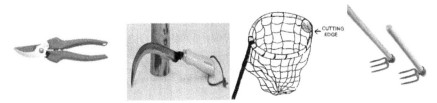

FIGURE 1.10   Different harvesting tools.

1.3.3 TIME OF HARVESTING

Harvesting time also affects quality. Fruits harvested before 10 AM in the morning and transported to pack house for sorting grading and packing yield better quality and lasts longer. Therefore, morning harvesting and within 10 AM transportation to destination pack house or market is always preferred in order to control damage due to high temperature. In case of grapes harvesting in India, it starts at 6 O'clock in the morning and harvested produce reach pack house by 10 AM. It facilitates faster precooling also and yield better quality.

1.3.4 PRECOOLING

The quality of fresh fruits and vegetable largely depends on precooling before storage and marketing. This is a compulsory postharvest treatment followed in developed countries for all most all perishable commodities. The rapid cooling of fresh produce from field temperature (pulp temperature at the time of harvesting) to its best storage temperature is called precooling. It is an important postharvest operation recommended in almost all flowers, fruits and few vegetables. The main objective of any precooling operation is to remove field temperature (field heat). This is important because it increases shelf life of the produce. Removing field heat reduces rate of respiration and all biochemical reactions from newly harvested produce. Since fruits, vegetables, and flowers are still alive after harvest. The produce continues respires. Respiration results in produce deterioration, including loss of nutritional value, changes in texture and flavor, and loss of weight. These processes cannot be stopped, but they can be slowed down significantly by precooling before storage or distribution. Generally, the higher the respiration rates of a fruit or vegetable, the greater the need for postharvest precooling. Precooling also reduces disease incidence. Wet or damp produce must be cooled, as warm, wet produce creates an environment that encourages the growth of decay organisms. Precooling facility is usually erected within pack house premises and it is the grower-shipper's pack house owner's responsibility. The four basic methods of precooling can be applied based on the texture and sale value of the product. These are forced air, hydro cooling, vacuum cooling, and icing. Each method was developed with specific crops in mind. For each crop, it is critical to know how to handle the produce at harvest, whether precooling is necessary, and which one is the best method of cooling. Forced-air cooling is the most common and widely used method of precooling. Cost, easiness and maintenance are also important consideration when you select a precooling method.

1.3.4.1   FORCED AIR

In this method, cool air with high speed moves over a product to remove the field heat. Both packed and unpacked fruits and vegetables can be precooled. Inside precooling chamber, fans pull hot air through the produce boxes and back into the cooling unit and this process continued till desired temperature not achieved. During precooling, weight loss is expected. When a room designed for precooling, provide enough refrigeration capacity and proper humidity control. These steps can prevent excess weight loss. Forced-air units are affordable for many small-scale growers and traders. An existing cold room can be converted into a precooling chamber by using portable fans, wooden or plastic pallets and tarpaulin (Figure 1.11). Cover the pallets containing fresh produce with a tarpaulin to force the system to draw air through the boxes of produce. Forced air cools most commodities effectively, but those best adapted to this method include berries, stone fruits, and mushrooms.

FIGURE 1.11   Forced-air precooling method.

1.3.4.2   HYDROCOOLING

Hydrocooling cools produce with chilled water. Hence packed fruits are difficult to cool by this method. The water usually is cooled by mechanical refrigeration, although cold well water and ice sometimes are used. The size of hydro cooling units varies depending on the size of the operation, but considerable refrigeration or large quantities of ice are required to keep the water at the desired temperature of 33° to 36°F. The produce is cooled by a water bath or sprinkler system. The produce either is dumped in the bath or under the sprinkler or is left in bins or plastic crates. Small operations might have an ice-water tank in which to "stir" the vegetables for rapid cooling. Pay special attention to water quality. Unfiltered and unsanitized water can spread undesirable microorganisms.

Most vegetables and many fruits that can withstand wetting can be hydro cooled. Asparagus, celery, cantaloupes, green peas, leaf lettuce, stone fruits, radishes, and sweet corn can be cooled successfully with this method (Figure 1.12).

FIGURE 1.12   Few images of hydro cooling.

1.3.4.3 VACUUM COOLING

This method of precooling based on the principle of "water evaporates at a very low temperature if pressure reduced" and maintained to a desired level. Vacuum cooling is one of the more rapid cooling systems and cooling is accomplished at very low pressures. At a normal pressure of 760 mm Hg, water evaporates at 100°C, but it evaporates at 1°C if pressure is reduced to 5 mmHg. Produce is placed in sealed containers where vacuum cooling is performed. This system produces about 1 percent product weight loss for each 5°C of temperature reduction. Modern vacuum coolers add water as a fine spray in the form of pressure drops. Similar to the evaporation method, this system is in general appropriate for leafy vegetables because of their high surface-to-mass ratio (Table 1.5). Produce is placed in a specially designed room, and air pressure is reduced. At lower atmospheric pressure, some water from the produce evaporates as the produce uses its own heat energy to convert water into water vapor. This results in lowering the product's temperature. Heat and moisture are removed from the vacuum tube by mechanical refrigeration. Commercial vacuum units usually cool the product to the proper storage temperature in less than 30 min. Units are available for cooling different amounts of product, from two pallets to a full truckload. Since initial investment is very and maintenance cost is also high, this method is not commercialized on large scale.

TABLE 1.5   Fresh produce suitable for vacuum cooling.

Belgian endive	Chinese cabbage	Kohlrabi	Spinach
Broccoli	Carrot	Leek	Sweet corn
Brussels sprouts	Escarole	Parsley	Swiss chard
Cantaloupe	Green onions	Pea/snow peas	Watercress

Source: Sargentet al., (2000), McGregor, (1987).

1.3.4.4 ICING

Crushed or slurry ice is placed directly into the produce box. This can be an effective way of precooling individual boxes of certain vegetables. The produce can be cooled in a short time and the temperature could be maintained in transit also. Fresh produce that can be ice are listed below (Table 1.6).

TABLE 1.6   Fresh produce suitable for Ice cooling

Belgian endive	Chinese cabbage	Kohlrabi	Spinach
Broccoli	Carrot	Leek	Sweet corn
Brussels sprouts	Escarole	Parsley	Swiss chard
Cantaloupe	Green onions	Pea/snow peas	Watercress

Source: Sargent et al. (2000), McGregor (1987).

1.3.5 SORTING AND GRADING

This is one of the most important postharvest operations after harvesting. This is done primarily for quality packing and removal of diseased and defective produce from the lot. Proper sorting and grading gives assurance of quality produce. This is either done in the farmer's field or in the pack houses. Both manual and mechanical graders are used for grading. All round shaped fruits and vegetables are easily graded by mechanical graders. Grading may be based on color, size, and extent of defects. While sorting is totally dependent on man power for removal of diseased, defected, and damaged fruits or vegetables. Grading is done by simple to highly sophisticated graders. Today many sophisticated graders are in use for fresh produce such as **GREEFA**. Both size and color grading simultaneously is possible and is being used on commercial scale in apples.

1.3.6 *PACKAGING, PACKAGING MATERIALS AND PALLATIZATION*

Both packing and packaging materials play many important roles in quality maintenance of fresh produce. Packing starts with placing the produce in the box. While placing, care must be taken to place in line, pedicel end of all fruits should be in one direction, separation layers or trays must be used where it is necessary. The box should not be under filled or over filled. Over filling is generally noticed in India where a farmer fills more in a box beyond the capacity of the box designed by the manufacturer. They do this primarily for two reasons. **(a) Demand by the commission agent (Traders)**—It was observed that those lots containing more weight in a box priced more. In general all apple boxes are designed for 20 Kg but it contains 22–23 kg and in many cases it is found 25–30 kg, where capacity is only 20 kg (Ahmad et al., 2014). This results heavy touching marks and bruising during handling.**(b) To save the cost of packing and transportation**—Farmers want to save something immediately and he/she calculates the cost of empty boxes, packing charges, handling and transport charges by saving number of boxes and to do so they prefer over filling. The practice results damage and touching (pressure marks) in almost all fruits. The type of boxes and quality in terms of strength and ventilation of each boxes play very important role in maintaining quality of any fresh produce. A number of boxes are used for packing fresh produce such as bamboo baskets, wooden boxes, CFB boxes, thermocol boxes, etc. Among them CFB boxes are most common and widely used all over world. However, quality in terms of strengths, printability and perforations varies from country to country and even region to region within the country. In India, Himachal and J&K farmers uses CFB boxes of very weak strength (3-ply) for packing apple, pear and few stone fruits. These boxes generally become very loose or even torn during transportation from Himachal or Kashmir to various wholesale markets (APMC Azadpur, Delhi, Chandigarh, etc.) of India. Just to strength these boxes buyers or forwarding agents prefer wrapping of boxes by thin plastic ropes. The idea is to give little strength to theses boxes during further transportation. Wooden boxes, the second most important packing boxes widely used in many countries including India. However, unavailability and cost of wooden box is a concern during peak season. Government regulations also discourage the use of wooden boxes for sake of trees. Bamboo baskets with gunny bags are also in use for less value crops. Repackaging of fruits and vegetables is common when the product has been packed in large containers, such as sacks, CFB boxes, plastics containers, etc. The repackaging process is often carried out by repackers who open the box, regrade, and again pack in the same box. During repacking, any damaged or rotting fruit is found, it is thrown away. It gives the product an appearance more appealing to consumers. Packages must possess few qualities before packaging any fresh produce. Otherwise, quality may deteriorate quickly. According to Wills et al. (1981), modern packaging must comply with the following requirements:

- The package must have sufficient mechanical strength to protect the contents during handling, transport, and stacking.
- The packaging material must be free of chemical substances that could transfer to the produce and become toxic to man.
- The package must meet handling and marketing requirements in terms of weight, size, and shape.
- The package should allow rapid cooling of the contents. Furthermore, the permeability of plastic films to respiratory gases could also be important.
- Mechanical strength of the package should be largely unaffected by moisture content (when wet) or high humidity conditions.
- The security of the package or ease of opening and closing might be important in some marketing situations.
- The package must either exclude light or be transparent.
- The package should be appropriate for retail presentations.
- The package should be designed for ease of disposal, reuse, or recycling.
- Cost of the package in relation to value and the extent of contents protection required should be as low as possible.

1.3.7   USE OF CUSHIONING MATERIALS

Cushioning materials are used in many stages during postharvest handling operations. But there are three main stages where it becomes compulsory in order to maintain postharvest quality. The first stage is putting harvested produce into plastic crates or any rigid container. All crates have hard surfaces and while keeping produce inside, there is a chance of dropping off form little height, causes impact bruising popularly called as touching marks. The second stage is transportation from field to pack house. In general, plastic crates are used for transportation from field to pack house and the distance may vary from few kilometers to many kilometers. Based on the condition of roads, there would be impact and vibration bruising, these bruising may not be visible immediately but after a few days, browning or blackening symptoms develops and finally produce starts rotting. Cushioning materials if used in plastic crates reduces these bruising and touching marks drastically. The third stage is transportation of packed produce from pack house to destination markets. Loading, unloading and transportation jerks causes bruising. Therefore, it is recommended to use cushioning material to preserve postharvest quality of fresh produce. There may be many types of cushioning materials such as newspaper sheets, newspaper cuttings, rice straw, bubble sheet, specially designed foam nets, molded trays, gunny bags, leaves, khaskhas, and other locally available material.

1.3.8 STORAGE

Almost all fruits are seasonal in nature. Every year, harvesting season falls during a fixed period, say 2–3 months. This period may differ from state to state for the same fruit and for different fruits also. For example, in India, apple harvesting season falls from July to October in Himachal Pradesh and from August to November in J&K every year. There may be little early or late due to prevailing weathers condition during growing periods. Demand for many fruits and vegetables are round the year. The demand of any fruit or vegetable beyond the harvesting season is called off season demand. This demand can be fulfilled only if fruits are stored in the harvesting season and sold during off season. The management of temperature, ventilation and relative humidity are three most important factors effects postharvest quality and storage life of horticultural produce. There may be many objectives of storage but the main objectives are:

- To minimize glut and distress sale in the market, thus assures good price to the farmers.
- To insure availability of food in offseason.
- Save horticultural produce from being spoiled.
- Storage in season when cost of produce are relatively low and marketing in off season at a better price. This gives higher returns to growers and traders.
- To regulate the price of the commodity during season and also in off season.
- Mostly apple, pear, grapes, potato, onion, and Chilli are stored in large quantities to feed the market round the year.

Lowering the temperature to the lowest safe level is of paramount importance for enhancing the shelf life, reducing the losses, and maintaining fresh quality of fresh produce. For example, mango needs a temperature above 8°C, banana above 12°C, apple 1–2°C, etc. The safe temperature of few important fruits and vegetables are mentioned in Table 1.7.

TABLE 1.7 Fruits, vegetables and their optimum storage temperature

SL. NO.	CROP	OPTIMUM TEMPERATURE (0°C)	RELATIVE HUMIDITY (%)
1	APPLE	1–4	90–95
2	APRICOT	−0.5 TO 0	90–95
3	ARTICHOKE	0	95–100
4	ASIAN PEAR	1	90–95
5	ASPARAGUS	0–2	95–100
6	AVOCADO	3–13	85–90
7	BANANA	13–15	90–95
8	BROCCOLI	0	90–95

TABLE 1.7 *(Continued)*

SL. NO.	CROP	OPTIMUM TEMPERATURE (0°C)	RELATIVE HUMIDITY (%)
9	BRUSSELS SPROUTS	0	90–95
10	BRINJAL	8–12	90–95
11	CABBAGE	0	98–100
12	CARROT	0	95–100
13	CASSAVA	0–5	85–96
14	CASHEW APPLE	0–2	85–90
15	CAULIFLOWER	0	95–98
16	CELERY	0	98–100
17	CHERIMOYA	13	90–95
18	CHERRIES	−1 TO 0.5	90–95
19	COCONUT	0	80–85
20	CUCUMBER	5–10	90–95
21	CUSTARD APPLE	5–7	85–90
22	DATES	−18 TO 0	75
23	FIG	−0.5 TO 0	85–90
24	GARLIC	0	65–70
25	GINGER	13	65
26	GRAPE	−0.5 TO 0	90–95
27	GRAPEFRUIT	10–15	85–90
28	GREEN ONIONS	0	95–100
29	GUAVA	5–10	90
30	JACK FRUIT	13	85–90
31	KALE	0	95–100
32	KIWIFRUIT	−0.5 TO 0	90–95
33	LEMON	10–13	85–90
34	LETTUCE	0–2	98–100
35	LIMA BEAN	3–5	95
36	LIME	9–10	85–90
37	LONGAN	1–2	90–95
38	LOQUAT	0	90
39	LYCHEE	1–2	90–95
40	MANDARIN	4–7	90–95

TABLE 1.7   *(Continued)*

SL. NO.	CROP	OPTIMUM TEMPER- ATURE (0°C)	RELATIVE HUMID- ITY (%)
41	MANGO	13	90–95
42	MANGOSTEEN	13	85–90
43	MELON (OTHERS)	7–10	90–95
44	MUSHROOMS	0–1.5	95
45	NECTARINE	−0.5 TO 0	90–95
46	OKRA	7–10	90–95
47	ONIONS (DRY)	0	65–70
48	OLIVES, FRESH	5–10	85–90
49	ORANGE	0–9	85–90
50	PAPAYA	7–13	85–90
51	PARSLEY	0	95–100
52	PARSNIP	0	95–100
53	PASSION FRUIT	7–10	85–90
54	PEACH	−0.5 TO 0	90–95
55	PEAR	−1.5 TO 0.5	90–95
56	PEAS	0	95–100
57	PEPPER (BELL)	7–13	90–95
58	PERSIMMON	−1	90
59	PINEAPPLE	7 TO 13	85–90
60	PITAYA	6–8	85–95
61	PLUM	−0.5TO 0	90–95
62	POMEGRANATE	5	90–95
63	POTATO (EARLY)	7–16	90–95
64	POTATO (LATE)	4.5 TO 13	90–95
65	PRICKLY PEAR	2–4	90–95
66	PUMPKINS	10–15	50–70
67	QUINCE	−0.5 TO 0	90
68	RADISH	0	95–100
69	RAMBUTAN	10–12	90–95
70	RASPBERRIES	−0.5 TO 0	90–95
71	RHUBARB	0	95-100
72	SAPODILLLA	15–20	85-90
73	SCORZONERA	0	95-98

TABLE 1.7 *(Continued)*

SL. NO.	CROP	OPTIMUM TEMPER-ATURE (0°C)	RELATIVE HUMID-ITY (%)
74	SNAP BEANS	4–7	95
75	SNOW PEAS	0–1	90-95
76	SPINACH	0	95-100
77	SPROUTS	0	95-100
78	STRAWBERRY	0–0.5	90-95
79	SWEET CORN	0–1.5	95-98
80	SWEET POTATO	15–20	85-90
81	SWISS CHARD	0	95-100
82	SUMMER SQUASH	5–10	95
83	TAMARIND	7	90-95
84	TARO	7–10	85-90
85	TART CHERRIES	0	90-95
86	TOMATO (MG)	12.5–15	90-95
87	TOMATO (RED)	8–10	90-95
88	TREE TOMATO	3–4	85-90
89	TURNIP	0	90-95
90	WATERMELON	10–15	90
91	WHITE SAPOTE	19–21	85-90
92	YAM	16	70-80
93	YELLOW SAPOTE	13–15	85-90

Source: Cantwell (1999); Sargent et al. (2000); McGregor (1987).

1.3.8 1   SPECIAL TREATMENTS (CURING)

Most root crops don't need curing before being placed in the storage chamber. Therefore, these crops should not be exposed to sun light. Potato, for example, turns green and become toxic if exposed to sun. However, few root crops require curing before storage for proper quality maintenance during storage and subsequent marketing. Potato also requires curing for peel hardening and wound healing (suberization) under shade. Onion and garlic require at least 1week curing process to dry out outer scaly leaves and tightening of neck portion. Pumpkin and squash need about 2weeks curing to harden their skin before storage. Don't skip curing where it is required as curing affects quality.

1.3.8.2 DON'T AND DOES FOR STORAGE OF FRESH PRODUCE

- Store only high quality produce, free of damage, decay and of proper maturity (not overripe or undermature).
- Know the requirements for the commodities you want to put into storage, and follow recommendations for proper temperature, relative humidity and ventilation. Never store carrot with apple or any fruit releases ethylene gas because carrot is very sensitive to ethylene and develops bitterness due to formation of a compound called iso-coumarin.
- Avoid lower than recommended temperatures in storage, because many commodities are susceptible to low temperature injury called freezing or chilling.
- Do not over load storage rooms or stack boxes tightly; it will hinder air movement through all boxes. Air follows the same path or easiest path if not blocked.
- Boxes should be stored on perforated wooden racks specially designed for air movement.
- Provide adequate ventilation in the storage room by keeping little space between two stack lines. Boxes should not be stored on the passage kept for the movement of staffs and labors.
- Storage rooms should be protected from rodents by keeping the immediate outdoor area clean, and free from trash and weeds.
- Containers/Boxes must be well ventilated and strong enough to with stand stacking. Do not stack boxes beyond their stacking strength.
- Monitor temperature in the storage room by placing thermometers at different locations.
- Don't store onion or garlic in high humidity environments.
- Control Inspect/Pest/rodents population inside the store.
- Check your produce at regular intervals for any sign of damage due to insect/pest/water loss, ripening, shriveling, etc.
- Remove damaged or diseased produce to prevent the spread of pathogens.
- Always, handle produce gently and never store produce unless, it is of the best quality.
- Damaged produce will lose water faster and have higher decay rates in storage as compared to undamaged produce and must be removed.

It is advisable not to store different crops together in one room of any cold store. But practically, it is very difficult to maintain and in some cases it is unavoidable, particularly at distribution or retail levels. A strategy widely practiced is to set cold chambers at an average of around 2–5°C and 90–95 percent relative humidity irrespective of specific requirement. Frequent opening and closing of cold store chamber for product loading and unloading causes an increase in temperature and decrease in relative humidity. Therefore, it is advisable for specific chambers for specific products. Thompson et al.(1999) recommended three combinations of temperature and relative humidity (RH): (1) 0–2°C and 90–98 per cent RH for leafy vegetables, crucifers, temperate fruits, and berries; (2) 7–10°C and 85–95 percent RH for citrus, subtropical fruits, and fruit vegetables; and (3) 13–18°C and 85–95 percent RH for tropical fruits, melons, pumpkins, and root vegetables. Storage of compatible groups of fruits and vegetables together (requires same temperature and RH) is advisable and necessary. Otherwise, quality of one produce affects the quality of other produce.Some fruits or vegetables can be stored together due to their common temperature and relative humidity requirements. At the same time, its reverse is also true. The below mentioned (Table 1.8) gives an over view of storage of compatible groups of fruits and vegetables.

TABLE 1.8   Compatibility groups of fruits and vegetables

Group	Temperature	Crops	Status of commodities
Group 1	0–2°C and 90–95% RH	Apple, Apricot, Asian Pear, Grapes, Litchis, Plum, Prunes, Pomegranate, Mushroom Turnip Peach.	Produce ethylene.
Group 2	0–2°C and 90–95%RH	Asparagus, Leafy greens, Broccoli, Peas, Spinach, Cabbage, Carrot, Cauliflower, Cherries.	Sensitive to ethylene.
Group 3	0–2°C and 65–70% RH	Garlic, Onions dry.	Moisture will damage these crops.

Source: Thompson et al. (1999)

1.3.9   TYPES OF STORAGE

There are many types of storage system or structure, starting from as simple as field storage to as sophisticated as controlled atmosphere (CA) and hypobaric storage. Among field storage, heap, cellar, underground tunnels or rooms, RCC rooms and evaporative cool chamber or Zero Energy Cool Chamber (ZECC) systems are important. ZECC is an important on farm storage structure based on the principle of evaporative cooling for fresh produce for short-term storage (Roy and Khurdiya, 1986;Pal et al., 1997). Among advance and technologically superior, modern cold storage, CA storage, and hypobaric storage are important. There are some basic requirement in all types of field storage are summarized below.

(i) Natural ventilation

Amongst all field storage systems, natural ventilation is required. Due to this natural airflow around the product, heat is removed regularly. Produce is placed in heaps, bags, boxes, bins, pallets, etc. Problems of pest and rodents are severe in field storage. Therefore, there must be adequate provision to keep out animals, rodents and pests. Another problem of field storage is development of hot and humid condition within the storage facility. This creates ideal conditions for the development of disease. It is possible to regulate temperature and relative humidity up to certain extent by opening and closing storage ventilation. At noon, ambient temperature increases and relative humidity decreases except rainy days. However, at night the opposite happens. To reduce the temperature of stored products, buildings ventilation should be left open at night when external air temperatures are lower.

(ii) Forced-air ventilation

Heat and gas exchange can be improved in a store room provided air is forced to pass through the stored produce. This system allows for more efficient cooling and control over temperature and relative humidity. Electric power facility is compulsory for forced air ventilation. As air follows the path of least resistance, loading patterns as well as fan capacity should be carefully calculated to ensure that there is uniform distribution of air throughout the stored produce. Inlet and exhaust fan can drag night cool air inside the chamber where difference in night a day temperature is more.

For modern cold stores, forced air ventilation is compulsory. This requirement is fulfilled by cooling fans. For smooth ventilation, perforated wooden floors for multistorey cold stores and plastic or wooden plates for a single room is necessary. For CA storage and hypobaric storage, fresh produce is stored in perforated plastic bins and crates and staked little away from the wall and door. This arrangement allows air movement through the produce.

(iii) Temperature and relative humidity

Since fruits, vegetables, and flowers are alive after harvest. All physiological process continues after harvest such as respiration, transpiration (water loss), etc. and supply of nutrient and water is not possible since produce is no more attached to the parent plant. Respiration results in produce deterioration, including loss of nutritional value, changes in texture and flavor, and loss of weight by transpiration. These processes cannot be stopped, but they can be reduced significantly by careful management of temperature and relative humidity during storage and transportation. Growth and multiplication of microorganism responsible for rotting and spoilage also associated with low temperature. At sufficiently low temperature, many disease causing microbes stop growth and multiplication. Respiration rates vary tremendously for different products and are affected by environmental conditions, mostly by temperature. As a thumb rule, lower the temperature, the slower will be its respiration rate and the growth of decay organisms. According to Van't Hoff Quotation (Q10), the rate of deteriorative reactions doubles for each 10°C rise in temperature. Generally, the higher the respiration rates of a fruit or vegetable, the greater the need for postharvest cooling.

Fruits and vegetables are composed largely of water. An important factor in maintaining postharvest quality is to ensure that there is adequate relative humidity inside the multistorey storage area. Water loss or dehydration means a loss in weight, which in turn affects the appearance, texture and, in some cases, the flavor also. Water loss also affects crispiness and firmness. Consumers tend to associate these qualities as poor with recently harvested fresh produce. For most fresh produce, relative humidity about 90–95 percent is recommended for storage and transportation. Since transportation period is only few hours to days, maintenance of RH is not of much importance except leafy vegetables. But in storage, maintenance of RH is compulsory. In modern cold stores, humidifiers are used for humidity creation. The recommended temperature and humidity for fruits and vegetables are mentioned in Table 1.1.

Controlled atmosphere storage is a system of storage fresh produce in an atmosphere that differs from normal atmosphere in respect to CO_2 and O_2 levels. At the time of loading in a CA chamber, level of CO_2 and O_2 are similar to normal air. With the passage of time, the gas mixture will constantly change due respiring fruits and vegetables in the store. Leakage of gases through doors and walls are not allowed in any CA chamber. Once the predetermined level of CO_2 and O_2 are achieved, it is constantly monitored. It is recommended that after loading, the chamber should be closed and desired level of gas composition should be established with 48 h with the help of Nitrogen generator. The gases are then measured periodically and the levels maintained by introduction of fresh air or passing the store atmosphere through a chemical to remove excess build up CO_2. Selection of the most suitable atmosphere depends on cultivars, stage of maturity, environmental and cultivation parameters. No one atmosphere is best for all produce. If the level of CO_2 increases or O_2 decreases, an anaerobic conditions can prevail with the formation of alcohol and physiological changes takes place referred as CA injury. Some examples of CA injury can be seen in Table 1.9.

TABLE 1.9   Examples of CA injury

Crop and Cultivars	CO_2 Injury Level	CO_2 Injury Symptoms	O_2 Injury Level	O_2 Injury Symptoms
Apple, red delicious	>3%	Internal Browning	<1%	Alcoholic taste
Apple, Fuji	>5%	CO2 injury	<2%	Alcoholic taint
Apple, Gala	>1.5%	CO2 injury	<1.5%	Ribbon scald
Apricot	>5%	Loss of flavor	<1%	Off-flavor
Banana	>7%	Green fruit Softening	<1%	Brown skin, discoloration

TABLE 1.9 *(Continued)*

Crop and Cultivars	CO_2 Injury Level	CO_2 Injury Symptoms	O_2 Injury Level	O_2 Injury Symptoms
Green beans	>7%	Off-flavor	<55	Off-flavor
Cabbage	>10%	Discoloration of inner leaves	<25	Off-flavor
Cherry	>30%	Brown, Discoloration	<1%	Skin pitting, offf-lavor
Mango	>10%	Softening	<2%	Skin discoloration

Source: Thompson, (1998).

Several refinements in CA storage have been made in recent years to improve quality maintenance; these include creating nitrogen by separation from compressed air using molecular sieve beds or membrane systems, rapid CA (rapid establishment of optimal levels of O2 and CO2), etc. All these refinements are for quality maintenance and to increase the length of storage period. Application of CA to all fresh produce is not found cost effective, and therefore, commercially not exploited. Commercial use of CA storage is greatest on apples and pears worldwide; less on cabbages, sweetonions, kiwifruits, avocados, persimmons, pomegranates, and nuts and dried fruits and vegetables (Kader, 1986). Classification of fresh produce according to their CA storage potential at optimum temperatures and RH is mentioned in Table 1.10.

TABLE 10 Potential storage period of few fruits and vegetable.

Sl. No.	Storage Duration (Months)	Crops
1.	>12	Almond, Brazil nut, cashew, filbert, macadamia, pecan, pistachio, walnut, dried fruits, and vegetables
2.	6–12	Some cultivars of apples and European pears
3.	3–6	Cabbage, Chinese cabbage, kiwifruit, persimmon, pomegranate, some cultivars of Asian pears
4.	1–3	Avocado, banana, cherry, grape, mango, olive, onion (sweet cultivars), some cultivars of nectarine, peach and plum, tomato (mature-green)
5.	<1	Asparagus, broccoli, cranberries, fig, lettuce, muskmelons, papaya, pineapple, strawberry, sweet corn; fresh-cut fruits and vegetables; some cut flowers

Source: Kader (1986).

1.3.10 TRANSPORTATION

Transportation is a connecting link between producers and consumers. It holds key factor in postharvest quality maintenance of all fresh produce. Most fresh produce in India and other countries of the world is transported from Farmers field to nearby market or wholesale market and from wholesale market to terminal market up to final retailers shops in open and nonrefrigerated vehicles. Only few reputed firms use refrigerated vehicles for transportation and distribution in summer months only starting from March to May/June in India. It is mainly due to the increased cost of transportation by refer van. In open-truck vehicles (nonrefer), produce is always susceptible to a loss of quality. Ambient temperature alone spoils the produce. Other means of transport include Rail transport (A/C and non-A/C), air, and ship. All imported fruits are transported in A/C containers by ships only. In every country, a dedicated port is assigned for receiving and dispatch of fresh produce containers. In India, a large number of fresh produce containers are received at Mumbai and Chennai port.

After harvest, a number of vehicles (trucks, tractors, trains, boats, ships, utility vehicles, etc.) are used to transport the product from field to either packinghouses or whole sale or retail markets. These vehicles are not equipped with refrigeration units and thus the produce decays faster, compared to that in refrigerated vehicles. If the produce is treated with edible wax or chemicals or additives after harvest, it can withstand little longer distances in open vehicles (nonrefer), without much damage. Refrigerated vehicles (trucks, trains, ships, airplanes, etc.) contain installed refrigeration units with sufficiently low temperatures to maintain fresh like quality in fresh produce. These types of vehicles are sealed with insulation material inside the walls of the container, which maintains the inside container temperature at desired level and thus preserves maximum quality. Fruits and vegetables must be classified in order to separate those susceptible to cold temperatures (mango, banana, tomato, etc.) and those not (apple, pear, cauliflower, peppers, etc.). This eliminates the possibility of product damage (chilling and freezing injury) when cooling at low temperatures during transport. Refrigeration temperatures can vary from 0°C (32°F) to 13°C (55.4°F) and RH from 70 to 95 percent. Refer van transport is an example of temporary refrigerated storage. Mixed loads cause incompatibility problems in transport also. Because packaging dimensions are different for different produce and it is not fully stackable. It is therefore, not advisable to transport mixed lots for long distance. However, for short distance, there may not be any problem.

There is usually little or no humidity control is available during transport and marketing. Thus, the packaging must be designed to provide a partial barrier against movement of water vapor from the product. Plastic liners designed with small perforations to allow some gas exchange may be an option.

(A) ROAD CONDITION AND DURATION OF TRANSPORTATION

Both road condition and duration of transportation affects quality of fresh produce. In hilly tracks and rough road surface, more touching and bruising take place as compared to smooth surface. Longer duration during transportation also affects quality. Refer van should not be hold unnecessary. It not only increases the cost of produce but also affects quality.

(B)   PATTERN OF LOADING

Pattern of loading also plays crucial role in maintaining quality of fresh produce. Here pattern of loading means number of packed boxes in one layer (stacking height). In case of fresh produce, stacking height depends on extent of perishable nature and strength of packing materials. If produce are more perishable or box strength is weak, stacking height is kept low and vice versa.

For example, height of grape boxes are kept low or it is packed in five ply corrugated boxes or thermocol boxes. This precaution must be taken to preserve postharvest quality of this highly perishable commodity. While loading, another important criteria is interlocking between the boxes. Loading and unloading fruits and vegetables directly affect quality of fresh produce. It can be done either by hand or with the aid of a forklift. Forklift is used for palletized boxes and shipping containers only. Generally, fruits and vegetables are stacked on pallets to ease the loading and unloading process and to prevent damage to the product. Exported crops arrive at the unloading port in bulk containers are unloaded directly into the storage container with the aid of conveyor belts connected from the vehicle to the container.

Another important consideration while loading is interlocking systems of loading. In this system, a little space is left in each layer on alternate basis (once in left side and once in right side). This facilitates air movement through the produce and provides strength to the boxes during transportation. Exposure to sun while awaiting loading at local mandis or transport can reduce quality drastically. The exposed portion turns black or brown and starts decaying. It is advised for nonrefer transport to move continue while under sun light and stop and park your vehicle under a tree shade especially during sunny days.

KEYWORDS

- **Preharvest**
- **Postharvest**
- **Factors**
- **Quality**
- **Fruit**
- **Vegetable**

REFERENCES

Agabbio, M.; D'Hallewin, G.; Mura, M.; Schirra, M.; Lovicu, G.; and Pala, M.; Fruit canopy position effects on quality and storage response of 'Tarocco' oranges. *Acta Hortic.* **1999**, *485*, 19–23.

Ahmad, S.; Chatha, Z. A.; Nasir, M. A.; Aziz, A.; Virk, N. A.; and Khan, A. R.; Effect of pruning on the yield and quality of kinnow fruit. *J. Agric. Soc. Sci.* **2006**, *2*(1), 51–53.

Amarante, C. V. T.; Steffens, C. A.; Mafra, A. L.; and Albuquerque, J. A.; Yield and fruit quality of apple from conventional and organic production systems. *Pesq. Agropec. Bras.* **2008**, *43*(3), 333–340.

Asrey, R., Pal, R. K.; and Sagar, V. R.; Impact of tree age and canopy height on fruit quality of guava cv. Allahabad Safeda. *Acta Hortic.* **2007**, *735*, 259–262.

Atkinson, D.; Jackson, J. E.; Sharples, R. O.; and Wallery, W. M.; *Mineral Nutrition of Fruit Trees.* Buttworths: London; **1980.**

Awad, M. A.; Wagenmakers, P. S.; and Jager, A.; Effects of light on favonoid and chlorogenic acid levels in the skin of 'Jonagold' apples. *Sci. Hortic.* **2001**, *88*, 289–298.

Barrett, D. M.; and Anthon, G. E.; Lycopene content of California-grown tomato varieties. *Acta Hortic.* **2001**, *542*(1), 65–73.

Batjer, L. P.; Billingsley, H. D.; Westwood, M. N.; and Rogers, B. L.; Predicting harvest size of apples at different times during the growing season. *Proc. Am. Soc. Hortic. Sci.* **1957**, *70*, 46–57.

Ben-Arie, R.; Bazak, H.; and Blumenfeld, A.; Gibberellin delays harvest and prolongs storage life of persimmon fruit. *Acta Hortic.* **1986**, *179*, 807–813.

Bergh, O.; Franken, J.; Van Zyl, E. J.; Kloppers, F.; and Dempers, A.; Sunburn on apples– preliminary results of an investigation conducted during the 1978:79 season. *Decid. Fruit Grower.* **1980**, *30*, 8–22.

Bielicki , P.; Czynczyk, A.; Chlebowska, D.; Effect of several new polish rootstocks and m9 subclones on growth, yield and fruit quality of two apple 'king jonagold' and 'elshof' cultivars. *Acta Hortic.* **2004**, 658, 327–332.

Bielorai, H.; The effect of partial wetting of the root zone on yield and water use efficiency in a drip- and sprinkler-irrigated mature grapefruit grove. *Irrig. Sci.* **1982**, *3*, 89–100.

Campbell, R. J.; and Marini, R. P.; Light environment and time of harvest affect 'Delicious' apple fruit quality characteristics. *J. Am. Soc. Hortic. Sci.* **1992**, *117*, 551–557.

Cantwell, M.; Características y recomendaciones para el almacenamiento de frutas y hortalizas. University of California, Davis. http://postharvest.ucdavs.edu/ Produce/Storage/spana html; **1999.**

Caruso, F. L.; *Compendium of Blueberry and Cranberry Diseases.* American Phytopathological Society: St. Paul, MN; **1995.**

Caruso, T.; Giovannini, D.; and Liverani, A.; Rootstock influences the fruit mineral, sugar and organic acid content of a very early ripening peach cultivar. *J. Hortic. Sci.* **1996**, *71*, 931–937.

Chandler, B. V.; and Kefford, J. F.; Absence of bitterness in navel oranges from rooted cuttings. *Nature.* **1966**, *210*, 868–869.

Chen, N.; and Paull, R. E.; Fruit temperature and crown removal on the occurrence of pineapple fruit translucency. *Sci. Hortic.* **2000**, *88*, 85–95.

Chyau, C. C.; Ko, P. T.; Chang, C. H.; and Mau, J. L.; Free and glycosidically bound aroma compounds in lychee. *Food Chem.* **2003**, *80*, 387–392.

Cinelli, F.; and Loreti, F.; Evaluation of some plum rootstocks in relation to lime induced chlorosis by hydroponic culture. *Acta Hortic.* **2004**, *658*, 421–428.

Coombe, B. G.; Influence of temperature on composition and quality of grapes. *Acta Hortic.* **1987**, *206*, 23–35.

Cooper, W. C.; and Lime, B. J.; Quality of red grapefruit on old-line grapefruit varieties on xyloporosis- and exocortis tolerant rootstocks. *J. Rio. Grande. Val. Hortic. Soc.* **1960**, *14*, 66–76.

Crisosto, C. H. and Costa, G.; Preharvest factors affecting peach quality. In: *The Peach: Botany, Production and Uses;* Layne, D. R., Bassi, D., Eds., **2008**; pp 536–549.

Crisosto, C. H.; Johnson, R. S.;Luza, J. G.; and Crisosto, G. M.; Irrigation regimes affect fruit soluble solids concentration and rate of water loss of 'O'Henry' peaches. *Hortic. Sci.* **1994**, *29*(10), 1169–1171.

Crisosto, C. H.; Mitchell, F. G.; and Johnson, R. S.; Factors in fresh market stone fruit quality. *Postharvest News Info.* **1995**, *6*, 217–221.

Croft, P. J.; Field conditions associated with cranberry scald. *Hortic. Sci.* **1995**, *30*, 627.

Cui, N.; Du, T.; Kang, S.; Li, F.; Zhang, J.; Wang, M.; and Li, Z.; Regulated deficit irrigation improved fruit quality and water use efficiency of pear-jujube trees. *Agric. Water Manage,* **2008**, *95*, 489–497.

Daane, K. M.; Johnson, R. S.; Michailides, T. J.; Crisosto, C. H.; Dlott, J. W.; Ramirez, H. T.; Yokota, G. T.; and Morgan, D. P.; Nitrogen fertilization affects nectarines fruit yield, storage qualities, and susceptibility to brown rot and insect damage. *Calif. Agric.* **1995**, *49*(4), 13–18.

Davenport, T. L.; and Campbell, C. W.; Stylar-end breakdown in 'Tahiti' lime: aggravation effects of field heat and fruit maturity. *J. Am. Soc. Hortic Sci.* **1977**, *102*, 484–486.

Davies, F. S.; Campbell, C. A.; and Zalman, G. R.; Gibberellic acid sprays for improving fruit peel quality and increasing juice yield of processing oranges. *Proc. Fla. State Hortic. Soc.* **1997**, *110*, 16–21.

Day, K. R.; Production practices for quality peaches. *Proc. Fla. State Hortic. Assoc.* **1997**, *77*, 59–61.

Dichio, B.; Xiloyannis, C.; Celano, G.; Vicinanza, L.; Gomez-Aparisi, J.; Esmenjaud, D.; and Salesses, G.; Performance of new selections of Prunus rootstocks resistant to Root Knot nematodes, in water logging conditions. *Acta Hortic.* *658*, 403–406.

Duran, J. M.; Retamal, N.; delHierro, J.; Rodriguez, A. E.; and Del Hierro, J.; La simulacion de danos de granizo en especies cultivadas. *Agric. Rev. Agropecu.* **1994**, *63*(740), 214–218.

Eckstein, K.; Robinson, J. C.; and Fraser, C.; Physiological responses of banana (Musa AAA; Cavendish sub-group) in the tropics. V. Influence of leaf tearing on assimilation potential and yield. *J. Hortic. Sci.* **1996**, *71*, 503–514.

Eshel, D.; Ben-Arie, R.; Dinoor, A.; and Prusky, D.; Resistance of gibberellin-treated persimmon fruit to *Alternariaalternata* arises from the reduced ability of the fungus to produce endo-1,4-β-glucanase. *Phytopathology.* **2000**, *90*(11), 1256–1262.

Fallahi, E., Richardson, D. G.; and Westwood, M. N.; Quality of apple fruit from a high density orchard as influenced by rootstocks. *J. Am. Soc. Hortic. Sci.* **1985**, *110*, 71–74.

Felzer, B. S., Cronin, T., Reilly, J. M., Melillo, J. M. and Wang, X.; Impacts of ozone on trees and crops. *Compters Rendus Geosci.* **2007**, *339*, 784–798.

Ferguson, I.; Volz, R.; and Woolf, A.; Preharvest factors affecting physiological disorders of fruit. *Postharvest Biol. Technol.* **1999**, *15*, 255–262.

Fitzpatrick, T. J.; Porter, W. L.; and Houghland, G. V. C.; Continued studies of the relationship of specific gravity to total solids of potatoes. *Am. Potato J.* **1969**, *46*, 120–127.

Fogliani, G.; Battilani, P.; and Rossi, V.; Studio dei processi riparativi dei frutti grandinati. Le lacerazioni su mele. Ponte del Concordato Italiano Grandine. **1985**, 45–51.

Galletta, G. J.; and Bringhurst, R. S.; Strawberry management. In: *Small Fruit Crop Management;* Galletta, G. J., Himelrick, D. G., Eds.; Prentice Hall: Englewood Cliffs, NJ, **1990**; pp 83–156.

George, A. P.; Collins, R. J.; Mowat, A. D.; and Subhadrabandhu, S.; Factors affecting blemishing of persimmon in New Zealand and Australia. *Acta Hortic.***1997**, *436*, 171–177.

Goulart, B. L.; and Demchak, K.; Cryoprotectants prove ineffective for frost protection on strawberries. *J. Small Fruit Vitic.* **1994**, *2*(3), 45–51.

Grulke, N. E.; and Miller, P. R.; Changes in gas exchange characteristics during the lifespan of giant sequoia: implications for response to current and future concentrations of atmospheric ozone. *Tree Physiol.* **1994**, *14*, 659–668.

Gutierrez, F.; Arnaud, T.; and Albi, M. A.; Influence of ecological cultivation on virgin olive oil quality, *JAOCS.* **1999**, *76*, 617–621.

Harris, R. S.; Effects of agricultural practices on the composition of foods. In: *Nutritional Evaluation of Food Processing,* 2nd ed.; Karmas, E., Harris, R. S., Eds.; AVI Publishing Co: Westport, CT, **1975**; pp 33–57.

Hewett, E. W.; and Watkins, C. B.; Bitter pit control by sprays and vacuum infiltration of calcium in Cox's Orange Pippin apples. *Hortic. Sci.* **1991**, *26*, 284–286.

Hodgson, R. W.; Horticultural varieties of citrus. In: *The Citrus Industry;* Reuther, W., Webber, H. J., Batchelor, L. D., Eds.; University of California Press: Berkeley, CA, **1967**; Vol. 1, pp 431–588.

Hoffman, F. O.; Thiessen, K. M.; and Rael, R. M.; Comparison of interception and initial retention of wet-deposited contaminants on leaves of different vegetation types. *Atmos. Environ.* **1995**, *29*(15), 1771–1775.

Holmes, M.; and Koekemoer, J.; Wind reduction efficiency of four types of windbreaks in the Malelane area. *Inligtingsbull. Inst. Trop. Subtrop. Gewasse.* **1994**, *263*, 16–20.

Hong, K. H.; Kim, Y. S.; Lee, K. K.; and Yiem, M. S.; An investigation of hail injury at flowering in pears. *Korean Soc. Hortic. Sci. Abstr.* **1989**, *7*(1), 150–151 (in Korean).

Hossain, M. B.; and Ryu, K. S.; Effect of foliar applied phosphatic fertilizer on absorption pathways, yield and quality of sweet persimmon. *Sci. Hortic.* **2009**, *122*, 626–632.

Houck, L. G.; and Joel, F. J.; Growth temperature influences postharvest tolerance of lemon to hot water, cold and methyl bromide. *Hortic. Sci.* **1995**, *30*, 804 (Abstract 354).

Hummell, A. K.; and Ferree, D. C.; Response of two French hybrid wine-grape cultivars to low light environments. *Fruit. Var. J.* **1997**, *51*, 101–111.

Huybrechts, C.; Deckers, T.; and Valcke, R.; Predicting fruit quality and maturity of apples by fluorescence imaging: effect of ethylene and AVG. *Acta Hortic.* **2003**, *599*, 243–247.

Iglesias, I.; Monserrat, R.; Carbo, J.; Bonany, J.; Casals, M.; Evaluation of agronomical performance of several peach root stocks in Lleida and Girona (Catalonia, NE Spain). *Acta Hortic.* **2004**, *568*, 341–348.

Inglese, P.; Barone, E.; and Gullo, G.; The effect of complementary irrigation on fruit growth, ripening pattern and oil characteristics of olive (*Oleaeuropaea*L.) cv. Carolea. *J. Hortic. Sci.* **1996**, *71*, 257–263.

Iqbal, M.; Khan, M. Q.; Jalal-ud-Din, Rehman, K.; and Munir, M.; Effect of foliar application of NAA on fruit drop, yield and physico-chemical characteristics of guava (*Psidiumguajava* L.) red flesh cultivar. *J. Agric. Res.* **2009**, *47*(3), 259–269.

Jakopic, J.; Stampar, F.; and Veberic, R.; The influence of exposure to light on the phenolic content of 'Fuji' apple. *Sci. Hortic.* **2009**, *123*, 234–239.

Jaleel, A. A.; Zekri, M.; and Hammam, Y.; Yield, fruit quality, and tree health of 'Allen Eureka' lemon on seven rootstocks in Saudi Arabia. *Sci. Hortic.* **2005**, *105*, 457–465.

Kader, A. A.; Biochemical and physiological basis for effects of controlled and modified atmosphere on fruits and vegetables. *Food Technol.* **1986**, *40*, 99–100.

Kader, A. A.; Postharvest biology and technology: an overview. In: *Postharvest Technology of Horticultural Crops;* Kader A. A., Ed.; *University of California, Division of Agriculture and Natural Resources, Special Publication 3311*, **2002**; pp 39–47.

Kays, S. J.; *Postharvest Physiology of Perishable Plant Products.* Athens: AVI; **1999**, 532 p.

Kliewer, M. W.; and Lider, L. A.; Effects of day temperature and light intensity on growth and composition of *Vitisvinifera* L. fruit. *J. Am. Soc. Hortic. Sci.* **1970**, *95*, 766–769.

Kliewer, M. W.; and Lider, L. A.; Influence of cluster exposure to the sun on the composition of Thompson seedless fruit. *Am. J. Enol. Viticul.* **1968**, *19*, 175–184.

Kousik, C. S.; Sanders, D. C.; and Ritchie, D. F.; Yield of bell peppers as impacted by the combination of bacterial spot and a single hail storm: will copper sprays help? *Hortic. Technol.* **1994**, *4*, 356–358.

Lester, G. E.; Oebker, N. F.; and Coons, J.; Preharvest furrow and drip irrigation schedule effects on postharvest muskmelon quality. *Postharvest Biol. Technol.* **1994**, *4*, 57–63.

Lidster, P. D.; and Porritt, S. W.; Some factors affecting uptake of calcium by apples dipped after harvest in calcium chloride solution. *Can. J. Plant Sci.* **1978**, *58*, 35–40.

Lizana, L. A.; and Stange, E.; Evaluacion de las causas del desecho en kiwis (*Actinidia chinensis* Planch) sele ccionados para exportacion. *Cienc. Invest. Agrar.* **1988**, *15*(3), 145–149.

Lutchmeah, R. S.; Common disorders and diseases of pineapple fruit cv. Victoria in Mauritius. *Rev. Agric. Sucr. Ile. Maurice.* **1992**, *71*, 27–31.

Marcelle, R. D.; Mineral nutrition and fruit quality. *Acta Hortic.* **1995**, *383*, 219–237.

Marlow, G. C.; and Loescher, W. H.; Watercore. *Hortic. Rev.* **1984**, *6*, 189–251.

Marme, D.; and Dieter, P.; Role of Ca and calmodulin in plants. In: *Calcium and Cell Function;* Cheung, W. Y., Ed.; Academic: New York, **1983**; pp 264–311.

Mason, J. L.; Jasmin, J. J.; and Granger, R. L.; Softening of 'McIntosh' apples reduced by a postharvest dip in calcium chloride solution plus thickener. *Hortic. Sci.* **1975**, *10*, 524–525.

Mauzerall, D. L.; and Wang, X.; Protecting agricultural crops from the effects of tropospheric ozone exposure: Reconciling science and standard setting in the United States, Europe, and Asia. *Annu. Rev. Energy Environ.* **2001**, *26*, 237–268.

McAneney, K. J.; Judd, M. J.; and Trought, M. C. T.; Wind damage to kiwifruit (*Actinidiachinensis*Planch.) in relation to windbreak performance. *New Zealand J. Agric. Res.* **1984**, *27*, 255–263.

McGregor, B. M.; Manual de ltransporte de productos tropicales. USDA, Manual de Agricultural 668, **1987**, 148 pp.

Meesters, P.; La protection d'Elsanta contre le gel: une absolue necessite. *Fruit Belg.* **1995**, *63*(456), 97–100.

Michailides, T. J.; and Morgan, D. P.; Wind scab of French prune: symptomatology and predisposition to preharvest fungal decay. *Plant Dis.* **1993**, *77*, 90–95.

Myhob, M. A.; Guindy, L. G.; and Salem, S. E.; Influence of sunburn on Balady mandrin fruit and its control. *Bull. Fac. Agric. Univ. Cairo.* **1996**, *47*, 457–469.

Nagy, S.; Vitamin C contents of citrus fruit and their products: a review. *J. Agric. Food Chem.* **1980**, *28*, 8–18.

Nilsson, T.; and Gustavsson, K.; Postharvest physiology of 'Aroma' apples in relation to position on the tree. *Postharvest Biol. Technol.* **2007**, *43*, 36–46.

Ogata, T.; Takatsuji, T.; and Muramatsu, N.; Damage caused by briny wind and measures of control in citrus. I. Effect of briny wind damage on fruit quality, rootlets and following spring growth and freezing resistance of fall shoots developed after defoliation. *Bull. Fruit Tree Res. Sta.* **1995**, *28*, 51–59.

Osman, A. B.; and Dodd, P. B.; Effects of different levels of preharvest shading on the storage quality of strawberry (*Fragaria x ananassa*Duchesne) cv. Ostara. I. Physical characteristics. *Pertanika J. Trop. Agric. Sci.* **1994**, *17*, 55–64.

Pal, R. K.; Roy, S. K.; and Srivastava, S.; Storage performance of 'Kinnow' mandarin in evaporative cool chamber and ambient condition. *J. Food Sci. Technol.* **1997**, *34*, 200–203.

Panwar, S. K.; Desai, U. T.; and Choudhari, S. M.; Effect of pruning on physiological disorders in pomegranate. *Ann. Arid Zone.***1994**, *33*, 83–84.

Paull, R. E.; and Reyes, M. E. Q.; Preharvest weather conditions and pineapple fruit translucency. *Sci. Hortic.* **1996**, *66*, 59–67.

Peck, G. M.; Andrews, P. K.; Reganold, J. P.; and Fellman, J. K.; Apple orchard productivity and fruit quality under organic, conventional, and integrated management. *Hortic. Sci.* **2006**, *41*, 99–107.

Percy, K. E.; Legge, A. H.; and Krupa, S. V.; Troposphere ozone: a continuing threat to global forests? In: *Air Pollution;* Karnosky, D. F. E. A., Ed.; Global Change and Forests in the New Millenium; Elsevier Ltd.: Amsterdam, The Netherlands, **2003**; pp 85–118.

Poovaiah, B. W.; The molecular and cellular aspects of calcium action. *Hortic. Sci.* **1988**, *23*, 267–271.

Rabus, C.; and Streif, J.; Effect of various preharvest treatments on the development of internal browning disorders in 'Braeburn' apples. *Acta. Hortic.* **2000**, *518*, 151–157.

Reid, M. S.; Padfield, C. A. S.; Watkins, C. B.; and Harman, J. E.; Starch iodine pattern as maturity index for Granny Smith apples. *New Zealand J. Agric. Res.* **1982**, *25*, 229–237.

Reid, P.; and Innes, G.; Fruit tree pollination under nets. *Australas. Beekeeper*, **1996**, 98, 229–231.

Remorini, D.; Tavarini, S.; Degl'Innocenti, E.; Loreti, F.; Massai, R.; and Guidi, L.; Effect of rootstocks and harvesting time on the nutritional quality of peel and flesh of peach fruit. *Food Chem.* **2008**, *110*, 361–367.

Roy, S. K.; and Khurdiya, D. S.; Studies on evaporative cooled zero-energy input cool chamber for the storage of horticultural produce. *Indian Food Pack.* **1986**, *40*, 26–31.

Saks, Y.; Copel, A.; and Barkai Golan, R.; Improvement of harvested strawberry quality by illumination: colour and *Botrytis* infection. *Postharvest Biol. Technol.* **1996**, *8*, 19–27.

Salami, P.; Ahmadi, H.; Keyhani, A.; Sarsaifee, M.; Strawberry post-harvest energy losses in Iran. *Researcher.* **2010**, *4*, 67–73.

Sargent, S. A.; Ritenour, M. A.; and Brecht, J. K.; Handling, cooling and sanitation techniques for maintaining postharvest quality. HS719. Horticultural Sciences Department, Florida Cooperative Extension Service, Institute of Food and Agricultural Sciences, University of Florida; **2000**.

Schroeder, C. A.; and Kay, E.; Temperature conditions and tolerance of avocado fruit tissue. *Calif. Avocado Soc. Yearbook.* **1961**, *45*, 87–92.

Sibbett, C. S.; Micke, W. C.; Mitchell, F. G.; Mayer, G.; and Yeager, J. T.; Effect of topically applied whitener on sun damage to Granny Smith apples. *Calif. Agric.* **1991**, *45*(1), 9–10.

Siddiqui, M. W.; and Dhua, R. S.; Eating artificially ripened fruits is harmful. *Curr. Sci.* **2010**, *99*(12), 1664–1668.

Singh, R.; Sharma, R. R.; and Tyagi, S. K.; Pre-harvest foliar application of calcium and boron influences physiological disorders, fruit yield and quality of strawberry (*Fragaria x ananassa*-Duch.). *Sci. Hortic.* **2007**, *112*, 215–220.

Sites, J. W.; and Reitz, H. J.; The variation in individual Valencia oranges from different locations of the tree as a guide to sampling methods and spot picking for quality. I. Soluble solids in the juice. *Proc. Am. Soc. Hortic. Sci.* **1949**, *54*, 1–10.

Sites, J. W.; and Reitz, H. J.; The variation in individual Valencia oranges from different locations of the tree as a guide to sampling methods and spot picking for quality. III. Vitamin C and juice content of the fruit. *Proc. Am. Soc. Hortic. Sci.* **1950**, *56*, 103–110.

Slowinska, I.; Tomala, K.; and Slowinski, A.; Fruit quality of "Elise" apples depending on the rootstocks. *Acta Hortic.* **2004**, *658*, 371–376.

Soares, A. G.; Trugo, L. C.; Botrela, N.; and Souza, L. F.; Reduction of internal browning of pine-apple fruit (*Ananascomusus*L.) by preharvest soil application of potassium. *Postharvest. Biol. Technol.* **2005**, *35*, 201–207.

Sorensen, J. N.; Johansen, A. S.; and Kaack, K.; Marketable and nutritional quality of leeks as affected by water and nitrogen supply and plant age at harvest. *J. Sci. Food Agric.* **1995**, *68*, 367–373.

Sukumaran, N. P.; and Ramdass, C.; A simple variable load hydrometer. *J. Indian Potato Assoc.* **1980**, *7*, 32–37.

Tahir, I. I.; Johansson, E.; and Olsson, M. E.; Improvement of quality and storability of apple cv. Aroma by adjustment of some pre-harvest conditions. *Sci. Hortic.* **2007**, *112*, 164–171.

Tajero, G.; Bocanegra, J. A.; Martinez, G.; Romero, R.; Duran-Zuazo, V. H.; and Muriel-Fernandez, J. L.; Positive impact of regulated deficit irrigation on yield and fruit quality in a commercial citrus orchard [*Citrus sinensis*(L.)Osbeck, cv. salustiano]. *Agric. Water. Manage.* **2010**, *97*(5), 614–622.

Testolin, R.; Costa, G.; Comuzzo, G.; Galliano, A.; Vittone, F.; Mescalchin, M.; Gobber, M.; Michelotti, F.; Trentini, G.; and Crivello, V.; La raccoltadell'actinidiaeipericolidigelate. *Inf. Agrar.* **1994**, *50*(38), 63–68.

Thompson, A. K.; *Controlled Atmosphere Storage of Fruit and Vegetables*. CAB International; UK: **1998**.

Thompson, J.; Kader, A.; and Sylva, K.; Compatibility chart for fruits and vegetables in short-term transport or storage. University of California, Publication 21560. http://postharvest.ucdavis.edu/Pubs/postthermo.html.; **1999**.

Tjoelker, M. G.; Volin, J. C.; Oleksyn, J.; and Reich, P. B.; Light environment alters response to ozone stress in seedlings of Acer saccharum Marsh and hybrid Populus L. I. In situ net photosynthesis, dark respiration and growth. *New Phytolog* **1993**, *124*, 627–636.

Tomala, K.; Andziak, J.; Jeziorek, K.; and Dziuban, R.; Influence of rootstock on the quality of 'Jonagold' apples at harvest and after storage. *J. Fruit Ornam. Plant Res.* **2008**, *16*, 31–38.

Veihmeyer, F. J.; and Hendrickson, A. H.; The application of some basic concepts of soil moisture to orchard management. *Proc. Wash. State Hortic. Assoc.* **1949**, *45*, 25–41.

Verma, S. C.; Joshi, K. C.; Sharma, T. R.; and Malhotra, V. P.; Relation between specific gravity and dry matter content of potato (*Solanumtuberosum* L.). *Indian J. Agric. Sci.* **1972**, *42*, 709–712.

Visai, C.; and Marro, M.; Fenomeni di spaccatura e cicatrizzazionenelmelo 'StaymanWinesap'. *Not. Ortoflorofrutticoltura*, **1986**, *12*(2), 47–53.

Vogeler, I.; Cichota, R.; Sivakumaran, S.; Deurer, M.; and McIvor, I.; Soil assessment of apple orchards under conventional and organic management. *Aus. J. Soil Res.* **2006**, *44*, 745–752.

Wade, N. L.; Kavanagh, E. E.; and Tan, S. C.; Sunscald and ultraviolet light injury of banana fruit. *J. Hortic. Sci.* **1993**, *68*, 409–419.

Wang, S. Y.; and Zheng, W.; Effect of plant growth temperature on antioxidant capacity in strawberry. *J. Agric. Food Chem.* **2001**, *49*(10), 4977–4982.

Wang, S. Y.; Chen, C.; Sciarappa, W.; Wang, C. Y.; and Camp, M. J.; Fruit quality, antioxidant capacity and flavonoid content of organically and conventionally grown blueberries. *J. Agric. Food Chem.* **2008**, *56*, 5788–5794.

Watkins , C. B.; Principles and Practices of Postharvest Handling and Stress. In: *Apple Botany, Production, and Uses;* Ferree, D., Warrington, I., Eds.; CABI Publishing: Wallingford, Oxon, UK /Cambridge, MA; **2003**; pp 585–615.

Webster, A. D.; Rootstock and interstock effects on deciduous fruit tree vigour, precocity and yield productivity. *New Zealand J. Crop. Hortic. Sci.* **1995**, *23*, 373–382.

Webster, A. D. Rootstocks for temperate fruit crops: current uses, future potential and alternative strategies. *Acta. Hortic.* **2001**, *557*, 25–34.

Weibel, F. P.; Bickel, R.; Leuthold, S.; and Alfoldi, T.; Are organically grown apples tastier and healthier? A comparative field study using conventional and alternative methods to measure fruit quality. *Acta Hortic.* **2000**, *517*, 417–426.

Westwood, M. N.; and Balney, L. T.; Nonclimatic factors affecting the shape of apple fruit. *Nature.* **1963**, *200*, 802–803.

Westwood, M. N.; *Temperate-Zone Pomology: Physiology and Culture.* Timber Press; : Portland, OR **1993**.

Wills, R. B. H.; Lee, T. H.; Grahan, D.; McGlasson, W. B.; and Hall, E. G.; *Postharvest.* An introduction to the physiology and handling of fruits and vegetables. New South Wales University Press Limited; Kensington, Australia: **1981**, 150 pp.

Woolf, A. B.; and Ferguson, I. B.; Postharvest responses to high fruit temperatures in the field. *Postharvest Biol. Technol.* **2000**, *21*, 7–20.

Woolf, A. B.; Bowen, J. H.; and Ferguson, I. B.; Preharvest exposure to the sun influences postharvest responses of 'Hass' avocado fruit. *Postharvest Biol. Technol.* **1999**, *15*, 143–153.

Woolf, A. B.; Wexler, A.; Prusky, D.; Kobiler, E.; and Lurie, S.; Direct sunlight influences postharvest temperature responses and ripening of five avocado cultivars. *J. Am. Soc. Hortic. Sci.* **2000**, *125*(3), 370–376.

Wünsche, J. N.; Greer, D. H.; Lang, A.; McGhie, T.; and Palmer, J. W.; Sunburn – the cost of a high light environment. *Acta. Hortic.* **2000**, 557.

CHAPTER 2

ADVANCES IN NONDESTRUCTIVE QUALITY MEASUREMENT OF FRUITS AND VEGETABLES

K. PRASAD[1]

[1]Department of Food Engineering and Technology, S. L. I. E. T., Longowal – 148106, Punjab, India; Email: dr_k_prasad@rediffmail.com

CONTENTS

2.1 INTRODUCTION

The term quality implies the degree of preference or excellence (Mahajan, 1995). It is a perceived perception comprising many characteristics or properties and considered as a measure of purity, fulfillment of promises, absence of defects, strength, workmanship, or its suitability for the particular use. Quality of any produce thus reflects its sensory properties (appearance, texture, flavor, and overall characteristics), chemical and nutritive attributes, microbiological characteristics, and engineering properties or its defects. High quality products never fail to prevail therefore consumer recognizes brands that maintain their quality as the standard set for that particular product. The eventual objective of the production, handling, and distribution of food materials is to at least hold the quality to meet the needs of the consumer (Kalia, 2010). Enhanced turnover of any processing unit are thus influenced by the outgrowth of quality practices (Mahajan, 1995; Harrison, 2003).

There are various methods to assess the quality attributes of any produce in order to meet out the legal, voluntary, industry, or consumer standards (Gunasekaran and Irudayaraj, 2001). Conventionally, the methods applied to measure those quality attributes may be of subjective, based on the opinion of the investigators or objective, based on the recognized standard scientific tests of evaluation, which are primarily destructive in nature. Unlike these methods, nondestructive quality evaluations are now a day's of more concern for the organizations dealing with the quality of their postharvest produce (Jha, 2010).

Nondestructive testing, measurement, evaluation, or inspection is the process of properties determination without affecting the material's ability to fulfill its intended function. Under present industrial era nondestructive evaluation must be used to ensure the quality assessment right from raw material stage through processing to intermediate or final products. Apart from ensuring the quality, integrity, and reliability of raw materials or processed products in the food processing fields, today nondestructive measurement finds extensive applications in allied fields too and utilized mainly for:
- Measurements of material's properties
- Process design for materials manufacturing
- Quality control and
- Online process control

Quality of food is the prime concern amongst the producer, processor, and consumers (Mahajan, 1995). Fresh fruits and vegetables as the protective food are often considered as healthful and nutritious. Horticultural produce are bound to change as it proceeds from producer to handler after harvesting. The quality attributes changes from handling to processing to marketing to consumption. An understanding of quality by associated handlers in postharvest processing and distribution is essential for effective utilization of postharvest produce. The cultivators, processors, transporters, distributors, and importers reevaluate the quality for their onward effectiveness as the utmost requirement (Jha, 2010; Abdullah, 2007). Moreover, variability associated with the fruits and vegetables and the quality of individual pieces may differ greatly from the representative statistical average. The sampling process does not identify the undesirable or the outstanding fruit or vegetable. However, sampling predicts the average quality, and perhaps quality distribution of the lot. Thus, the large number of per piece measurements is essential to precisely achieve the measure of statistical central tendencies with the associated dispersions. This may be effective in sorting, an important unit operation necessary to segregate the undesirable with acceptable and outstanding individual pieces to ensure the quality uniformity. To maintain at higher turnover of plantshigh-speed sorting is required. Moreover, the advancement in the nondestructive sensors for measuring several attributes at a time on every piece of fruit or vegetable serve the purpose. Further, with the application of artificial intelligence through soft computing techniques such as Fuzzy logic, artificial neural network, and multivariate statistical classification procedures decision may be applied with an appropriate control mechanism to physically place the individual piece into its proper category by a mechanical system (Abdullah, 2007, Prasad, 2012).

The available objective and empirical method of quality measures assess the particular quality attribute with a high accuracy and reproducibility. Whereas, the subjective and sensory method of quality evaluation techniques simultaneously reveal several quality attributes with less precision as well as reproducibility. Thus the possibility of clear cut quality assessment for superior or inferior fruits or vegetables within lot may not be possible through the sensory analysis. Again, the subjective analysis may not be fully reliable, which is a result of individual perception. The accuracy and precision of measurement may be affected and become unreliable through sensory evaluation in a situation of unknown postharvest treatment if subjected. So, it is crucial to distinguish parameters to be measured through indirect measurement techniques. Even the possible limitations of visual inspections with the constraints may thus be recorded for the development of soft computing database. Also, it is assumed that there is a valid unidirectional, consistent linear or nonlinear relationship prevailed between the sensor response and the magnitude of a particular constituent or attribute. To establish the relationships among instrumental measurements, sensory responses, and perceived quality or acceptability may thus be predicted as the results of nondestructive testing.

2.2 QUALITY ASSESSMENT TECHNIQUES

Techniques for assessing quality of postharvest produce could broadly be classified into two methods such as subjective or sensory methods and analytical or objective methods (Ranganna, 2000). Both the methods have their own merits and demerits. Subjective methods have limitations in quality evaluation of hazardous materials (Jha, 2010). Whereas, objective methods of quality assessment may either be destructive or nondestructive in nature. Generally small amount of samples are analyzed under destructive technique using chemical or nonchemical analysis. In nondestructive methods small as well as bulk samples may be analyzed and even remain untouched, intact as well for future use.

Postharvest sorting and grading of agricultural products is difficult and labor-intensive. Manual sorting is costly, unreliable, subjective, and slow. Many of such measurements have traditionally been 'off-line'. Traditional methods are time taking, cumbersome, labor–intensive, and costly. Moreover, food exhibits dynamic and complex behavior with respect to time and varietal dependency. High volume production of foods involves rapid, online, nondestructive analysis of product characteristics for quality assurance.

Nondestructive testing method is classified according to its underlying physical principle are visual inspection, X-ray imaging and computed tomography, near-infrared spectroscopy, Ultrasonic testing, thermographic testing, electromagnetic testing, liquid penetrant testing, magnetic particle testing, acoustic emission testing, infrared and thermal testing, magnetic resonance imaging, electronic nose, etc. Concerted efforts are required to have an accelerated development in developing countries to have their own commodity based precision postharvest technology to save huge postharvest losses. Important nondestructive techniques related to applications of electromagnetic radiations through visual inspections, imaging, ultrasonic measurements, and spectroscopy with associated theories and applications are dealt in this chapter.

Investigative imaging employs waves of electromagnetic radiations. Nondestructive quality evaluation through imaging thus engages the properties of X-ray, γ-ray, and radiofrequency for the food materials, which are partly but not completely transparent. These rays are extreme radiations of the electromagnetic spectrum and attenuated on passing through the biological materials (Figure 2.1). Thus X-ray and radio waves are mostly used for quality characterizations. In order to understand the production and application of electromagnetic waves, it is important to deal with the structure of atom.

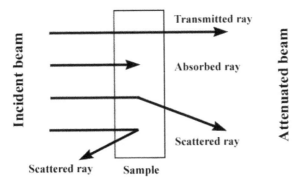

FIGURE 2.1   Interaction of electromagnetic waves with matter.

2.3   STRUCTURE OF ATOM

An atom consists of central nucleus containing "A" number of nucleons called as mass number. The nucleons contain "Z" number of proton, which is termed as atomic number and thus having (A–Z) number of neutrons. The electron in "Z" number orbiting in specific shells designated as K, L, and M around the positively charged central nucleus represented for the sodium atom (Figure 2.2). The outermost shell known as valance shell can occupy a maximum of 8 numbers of electrons and decides the chemical, thermal, optical, and electrical properties of elements. Metals have very few free electron(s) in their valence shell and easy to detach therefore, more conductive to heat and electricity. The removal of electron from the atom ionizes the atom and is energy dependent process. Being the detached negatively charged electron to unoccupied shell make the remnant atom a positively charged. When the electron falls back, the energy is released in form of packet or photon of visible or ultraviolet light (Farr and Roberts, 2006).

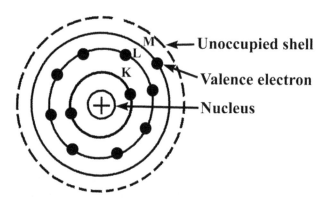

FIGURE 2.2   Electron of sodium atom in different shells.

2.4   ELECTROMAGNETIC SPECTRUM

Electromagnetic radiation is sinusoidal wave, propagate through space in combination of oscillating electric and magnetic fields mutually perpendicular to each other. The velocity of electromagnetic wave is near 3×10^8m/s in vacuum. Electromagnetic waves are formed when an electron vibrates, causing distortion in the electric field exerted by a positively charged particle (Abdullah, 2007). Electromagnetic radiation in both visible and nonvisible regions is arranged from shorter to longer wavelengths with the information regarding resolutions and energy distribution (Figure 2.3). Thus the nature and applicability of electromagnetic radiation varies with the change in wavelength.

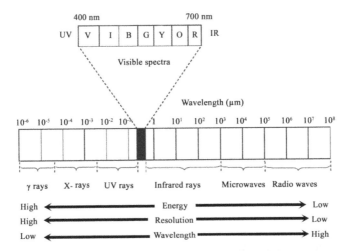

FIGURE 2.3   Electromagnetic spectrum comprising of visible and invisible regions.

Propagation of electromagnetic radiation has been classically explained in two different and partly opposing theories as wave theory and particle theory or Planck's quantum theory. Both can comfortably explain properties like reflection, refraction, and diffraction. However, limitation of particle theory prevailed toward describing the polarization phenomenon(Farr and Roberts, 2006).

Wave theory describes the wavelength (λ) and frequency (f) is inversely proportional to each other and their products are equal to the velocity (v) of all kinds of wave motions including the radio wave.

$$v = f.\lambda$$

Whereas, particle theory states that the photon energy (E) is proportional to the frequency and the proportionality constant is known as Planck's constant (h).

$$E = h.f$$

Properties of electromagnetic radiation
- travel with the same velocity in free space ,
- transfer energy from place to place in quanta ,
- travel in straight lines ,
- obey inverse square law and
- intensity of the radiation is reduced due to absorption and scattering

2.5   X-RAY

X-ray is a form of electromagnetic radiation with a wavelength broadly in the range of 10 to 0.01 nanometers having frequency in the range of 30×10^{15} to 30×10^{18} Hz. The wavelength of X-rays is longer than γ rays but shorter than UV rays. X-ray concerns with the inner shells of the atom, whereas, the radioactive γ –ray involves with the radioactive nuclei. Artificially controlled production of these ionizing radiations is used in medical diagnosis primarily as diagnostic radiography and also in computed tomography (CT).

The electron in an atom may be removed completely from the shell by applying the binding energy. The extent of binding energy increases on increasing the atomic numbers of the same K shell $\left[\left\{E_{(74^*)}=70\text{keV}\right\}>\left\{E_{(53')}=33\text{keV}\right\}>\left\{E_{(42^{kb})}=20\text{keV}\right\}>\left\{E_{(29^{Cu})}=29\text{keV}\right\}\right]$ but in any case not greater than 100keV. Increase in the shell number in tungsten reduces the binding energies of the same atom[$(E_{k}=70$ keV$) >(E_{L}=11$keV$) >(E_{M}=2$keV$)]$ (Farr and Roberts, 2006).

2.6   PRODUCTION OF X-RAY

X-rays are produced in an X-ray tube, which consists of a vacuum glass enclosure fitted with filament as cathode and a target as anode (Figure2.4). On heating the filament electrons are emitted further increases in temperature (2,200°C) and releases more electrons(Farr and Roberts, 2006). Vacuum inside the enclosure helps to move the emitted electrons freely to the target. As a large number of electrons arrives at the surface of the target with the kinetic energy interacts with the inner shells (Figure 2.5) of the atom or the field of nucleus (Figure2.6) and produce enormous heat with the X-rays.

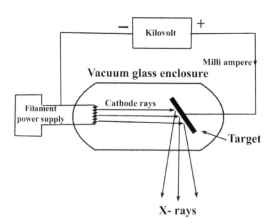

FIGURE 2.4   X-ray production unit.

Electron generated from the filament when colloid with the K shell of the target atom with the energy higher than the binding energy eject the electron b and the hole thus created in position is most likely to be filled by the electron of L shell with the production of X-ray (K_α radiation) of the energy equals to E_K-E_L (Figure 2.5).

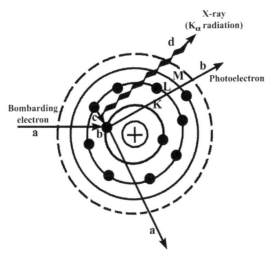

FIGURE 2.5   Production of X-ray.

As the bombarded electron approaches fast close to the nucleus, deflected and leaves less quickly with losing some of its energy and release as soft X-ray, braking radiation or bremsstrahlung (Figure 2.6), which form a continuous X-ray spectrum with low photon energy (Figure 2.7).

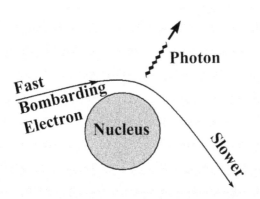

FIGURE 2.6   Production of bremsstrahlung.

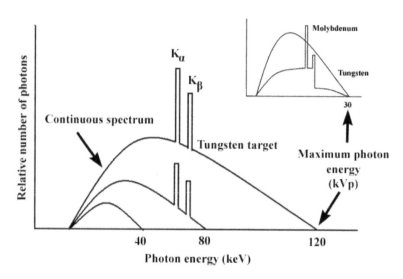

FIGURE 2.7   Effect of potential difference on X-ray spectra and inset as effect of target.

In medical applications, usually tungsten or a more crack-resistant alloy of rhenium (5%) and tungsten (95%) are used but use of molybdenum have also been observed in the specialized applications for generating the soft X-rays for specialized purposes to diagnose the soft tissues. Whereas, for the characterization of organic and inorganic, micro-, ornanoparticulate powdered materials through X-ray diffraction crystallography, a copper or cobalt target is used (Jha et at., 2009a; 2009b; Prasad et al., 2010b, 2012).

2.7 COMPUTED TOMOGRAPHY (CT)

Internal structure of any object can be reconstructed from multiple projections. The projections are formed by scanning a thin cross-section of the object with a thin X-ray beam and computing the transmitted radiation with a precise detector. The detector then adds up energy of all transmitted photons and the data is processed in a computer with the help of specialized algorithm to generate an image. The developed image is the direct reflection of the number of projections used to reconstruct the image in computed tomography.

2.8 APPLICATIONS OF X-RAY AND COMPUTER TOMOGRAPHY

The suitability of X-ray and gamma-rays are reported for quality assessment of various food products. Both the investigative radiations penetrate the exposed food materials depending on the variable attenuation properties, which primarily dependent on the density (Tollner et al., 1992). The application of X-ray work in three different ways as 2-dimensional radiography, which is mostly used in conventional diagnosis, line scanning used for assessing the products passing in a particular plane mostly used in baggage checking or through computed tomography (CT) for getting both 3-dimensional as well as 2-dimensional images (Lim and Barigou, 2004). Although, a total quality evaluation would be comprehensive when X-ray transmission and CT information is clubbed with other investigations. An exhaustive information in available with the image analysis of 3-dimensional cellular microstructure of various food products using CT and quantitative information on spatial cell size distribution, cell wall thickness distribution, connectivity, and porosity (Lim and Barigou, 2004).

Physiological changes are associated with the growth, maturation, ripening, and senescence of fruits and vegetables. Even the cell breakdown decay on physical damage has negative effects on quality, which are detectable by X-ray. Zaltzman et al., (1987) presented a comprehensive overview of the studies related to quality evaluation of agricultural products based on density differences. Since long, X-ray has been successfully used in the quality inspection of fruits and vegetables. Detection of hollow heart in potato was reported (Nylund and Lutz, 1950; Finney and Norris, 1978). X-ray CT was used to detect the changes in the densities of tomato during maturation (Brecht et al., 1991), peaches (Barcelon et al., 1999), and lettuce (Lenker and Adrian, 1971; Schatzki et al. 1981). This technique is also used to detect the split pit in peaches, olive, and cherries with the changes in the internal structure of peach during the process of ripening (Han et al., 1992; Tollner et al. 1992; Keagy et al. 1996; Schatzki et al., 1997). Tollner et al., (1992) related the X-ray absorption with the density and moisture content of apples.

Application of this technique has also been reported for the determination of water content in apples (Tollner et al., 1992), bruises and internal defects in apples (Diener et al. 1970; Schatzki et al., 1997), injury in papaya (Suzuki et al., 1994), and granulation in oranges (Johnson, 1985). Possibility of assessing the internal injuries of various fruit using digitized X-ray imaging analysis is reported (Yang et al. 2006). The pore-size distribution in apple of different cultivars has been reported based on X-ray imaging (Mendoza et al., 2010). Development and application of new automatic and effective quarantine system for the detection of pest infestation in agricultural products was presented (Chuang et al., 2011).

With limited success, internal sprouting and ring separations due to microbial rot in onion were demonstrated with X-ray line scan (Tollner et al., 1995). Freeze damaged citrus fruits could be separated using x-ray even in packaged conditions (Abbott et al., 1997). Ogawa et al. (1998) used X-rays to identify the foreign matters in food based on density difference.

2.9   VISUAL INSPECTION AND IMAGE PROCESSING

Since time immemorial visual inspections are often used in the quality inspection and evaluation of fresh as well as the processed produce. Visual inspections based on the experience of the trained human inspectors were also used in fixation of the price of the items, suitability of the materials for the desired purposes, prediction of tentative shelf life, etc. Although, visual inspection of quality evaluation is non-destructive in nature but this method is highly variable with inconsistent result and costly too. Physical assessments of fruits and vegetables through representations in form of images are undoubtedly the preferred method in reproducing the quality characteristics of the product as analyzed in direct visual inspection using the human brain.

Various limitations associated with manual visual inspection could be overcome by the use of electronic imaging systems. Rapid decrease in the cost of computer, peripheral, and image capturing device the applicability of image analysis as nondestructive quality evaluation is gaining popularity. Now a day image analysis systems include real-time imaging device and analysis using notebook computers. The inspection through image acquisition of foodstuffs for various quality factors is reproducible. This machine vision systems are ideally suited for routine inspection and quality assurance tasks. This system is also clubbed with other soft computing techniques to make a powerful artificial intelligence system (Jha, 2010; Abdullah, 2007).

Images could be obtained in both visible and nonvisible ranges. The image is acquired in the visible range extending from 10^{-4} to 10^{-7} m in wavelength. The invention of image acquisition system in larger part of the electromagnetic spectrum to be applied in the advanced inspection devices those are computed tomography (CT), magnetic resonance imaging (MRI), nuclear magnetic resonance (NMR), single photon emission computed tomography (SPECT) and positron emission tomography (PET), which operate in the shorter wavelengths of 10^{-8} to 10^{-13} m (Abdullah, 2007). On the other side infrared and radio cameras, which enable visualization to be performed at wavelengths greater than 10^{-6} m and 10^{-4} m, respectively are also available.

2.10   BENEFITS OF IMAGE ANALYSIS

Sensory evaluation performed by the customers or trained professional is the proven age old technique and also used presently in various field of interests. Although this technique of food evaluation is rapid but not suitable for the large number of samples due to limitation in the analysis of the sample by a particular observer. At the same time change in the observer affects the quality and variability prevailed in the subjected analysis of products. Thus the results are subjective. The objective methods of quality evaluations have different types of limitations. To overcome the disadvantages and limitations of sensory, chemical, and advanced instrumental method, new economical, fast, and environment friendly techniques are need of the present days. As a result, the nondestructive technique based on the image analysis in recent days becoming more and more popular in food evaluation as this method is suitable for online monitoring to make the inspection and monitoring process even in real-time systems.

2.11 STEPS OF IMAGE ANALYSIS

Human vision is tried to simulate by incorporating the artificial intelligence in the image analysis system. The analysis system covers the process of image acquisition; processing and interpretation to extract useful information, Extracted features through image analysis are further processed with the help of soft computing or multivariate technique for quality classification. Various components of image analysis system consists of image acquisition system a camera, which is analogous to the sense organ eye, computer with the software, which is the representation of human and human brain, illumination system, analogous to facilitate the image in presence of light and the monitor or printer to represent the captured and processed image for further processing (Gunasekaran, 2001; Abdullah, 2007).

Thus the image analysis system following the steps of:

a. Image acquisition
b. Image processing
c. Image interpretation.

The raw image of fruits and vegetables under suitable level of illumination is acquired using any digital image acquisition device. Further, the contrast of the raw image is enhanced using histogram equalization on the acquired raw image. Thresholding is further performed on the histogram equalized image to allow segmentation of pixels in the image into multilevel classes. Using this technique each pixel is classified either based on object of interest or the background of an image. The processing of the images could also be done using software NI vision assistance for Labview environment or other image processing software (Prasad et al., 2010a, 2012). Using the different functions the image can be analyzed for physical and optical characteristics directly and other characteristics through correlation and regression analysis.

2.12 APPLICATION OF IMAGE ANALYSIS

Nondestructive methods for measuring different quality parameters based on optical method are based on either direct imaging or analyzing the interaction of the product with the light. The reflected, transmitted, absorbed, or scattered light is recorded to act as fingerprint in characterization of fruits and vegetables having any kind of pigments such as chlorophylls, carotenoids, and anthocyanins. During growth, maturation and ripening and senescence of fruits and vegetable the characteristics such as size; shape and color alters. This physical parameter has often been referenced by chemical analysis.

2.13   DIMENSIONAL CHARACTERIZATION

The macroscopic food materials for the dimensional and dimension dependent parameter through direct image acquisition and analysis have successfully been performed (Prasad et al., 2010a). Apple (*Mauls domestica*) cultivars Red delicious, Golden delicious, Maharaji, Ambri, and American trel obtained from Shopian (33.72°N, 74.83°E) located at an average elevation of 2,057 m above the sea level in Kashmir Valley (Figure 2.8). All the harvested apples of five cultivars were determined for the dimensional and optical characteristics (Sheikh, 2012). For each fruit, three linear dimensions length (L); equivalent distance of the stem (top) to the calyx (bottom), width (W); the longest dimension perpendicular to L, and thickness (T); the longest dimension perpendicular to L and W were measured as reported (Figure 2.9) elsewhere and verified using a digital caliper with accuracy of 0.01 mm (Mitutoyo Corporation, Japan). The derived parameters such as geometric mean diameter, aspect ratio, sphericity, and packing coefficient were also determined.

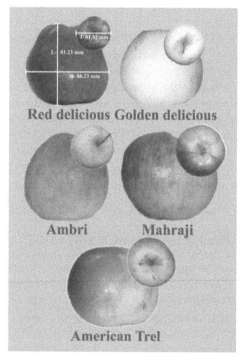

FIGURE 2.8   Images of apple cultivars.

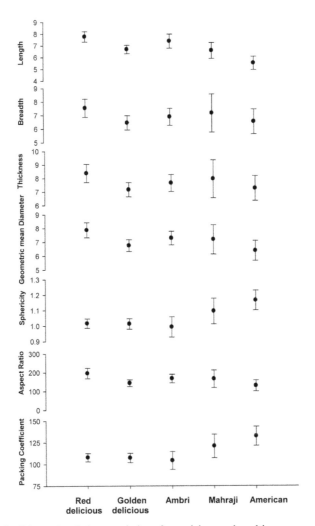

FIGURE 2.9   Dimensional characteristics of promising apple cultivars.

On the other hand if the particle size is small then the use of microscope is employed to visualize the enlarged image for the acquisition, processing, and characterization of the microscopic materials (Prasad et al., 2012). The dehydrated moringa leaves is crushed to mixed powder and fractionated using sieve, 150 BSS further to course and fine dehydrated moringa leaf powder (Singh and Prasad, 2013). Using image analysis technique the particle size and particle size distribution characteristics have been analyzed and reported (Figure 2.10).

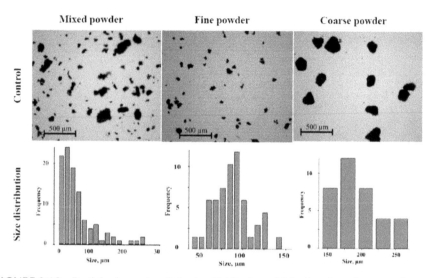

FIGURE 2.10   Particle size and particle size distribution of dehydrated moringa powder.

2.14   FRICTIONAL CHARACTERIZATION

The tangent of the angle at which the granular or powdered material moves on an inclined surface as coefficient of static friction (μ_s) and the angle at which the heap formed as angle of repose are the important frictional characteristics. Coefficient of friction have successfully been determined accurately by determining the either by angle between the base plate and inclined surface or by analyzing the vertical and horizontal height through image analysis technique. Similarly, the base and height could be measured to get the angle of repose (Figure 2.11).

FIGURE 2.11   Angle of repose through image analysis.

2.15 OPTICAL CHARACTERIZATION

Optical properties of apple cultivars were determined in terms of RGB value using image analysis technique. The RGB values for the apple cultivars varied from 171.3±18.1 to 209.8±40.1, 71.5±33.4 to 189.8±43.5 and 56.3±8.17 to 85.2±28.7, respectively. Among five cultivars Golden delicious was found to have the significant difference in color attributes (Figure 2.12). The acceptability of the apple depends on the uniformity of external color, repeatability of fruit color, with the physical defects, dents, browning, bruising, and stage of maturity. Thus, the color attributes may play critical role in quality characterization of apple on online system based on nondestructive evaluation based on optical methods.

FIGURE 2.12   Optical characteristics of promising apple cultivars.

2.16 MISCELLANEOUS APPLICATIONS

Fruits and vegetable of specific variety at particular level of maturity reflects somewhat similar characteristics. Efforts have been made by various researchers to reflect level of ripeness, firmness, soluble solids and even shelf life based on optical parameters judged through image acquisition, processing, and characterization.

2.17 NIR SPECTROSCOPY

Near infrared (NIR) region is the part of electromagnetic spectrum having the wavelengths ranged between 700 and 2,500 nm (Nicolai et al., 2007). The wavelength of light is inversely related to its energy level (Sun, 2007). Molecules have discrete energy states and the electromagnetic wave can cause a molecule to change from one energy state to another if the energy in the photon matches the energy required to elevate the energy state of molecule. When light comes in contact with a biological material, the photons of light can interact with the material at the molecular level.

The energy is said to be attenuated as it is absorbed, reflected, or transmitted through the matter (Gunasekaran, 2001). The absorbed portion of radiation can be converted to heat, chemical changes or luminescence as other forms of energy. Either wave length or wave number (cm⁻¹) is used to measure the position of a given infrared absorption (Silverstein et al., 1981). This property is wave frequency and types of molecule reflect the unique interactions between the energy of the light and the energy states of the molecule and become a part of vibrational spectroscopy. The wave number is directly proportional to the absorbed energy, whereas the wavelength is inversely proportional to the absorbed energy. Different functional groups in foods absorb energy on exposing with IR radiation at different wavelengths and the amplitude of that vibration increases. When the molecule reverts from the exited state to the original ground state, the absorbed energy is released as heat (Dyer, 1965).

Near infrared (NIR) spectroscopy deals with exposing food products with light of near infra red for collecting and analyzing the obtained absorbance spectrum (Jha, 2010). NIR radiation is highly penetrative and applied directly to the sample without any preparation, to get the spectra which give quantitative information of major organic components of food. The technique is affordable and robust. NIR spectrometers consist of light source, wavelength selector, and detector. NIR methods are applied in evaluation of fruits and vegetables in two distinct ways:

- Reflectance measurements
- Transmittance measurements

Level of absorption of electromagnetic wave at a particular depth or thickness of the sample is used in the transmittance measurement. The extent of light reflected from the surface of the fruit provides the characteristic information of the subjected materials in reflectance measurement technique.

2.18 APPLICATION OF NIR SPECTROSCOPY

Near Infrared Spectroscopy (NIR) is probably the most explored nondestructive technique for agricultural produce. Most of the study focuses on mango, apple, banana, and tomato. NIR spectroscopy is a rapid, noninvasive, less time taking, chemical-free technique for continuous fruit quality evaluation (Huang et al., 2011). The inherent advantages with the suitability of NIR spectroscopy in nondestructive quality evaluation has been widely used in the field of fruits and vegetables (Nicolai et al., 2007). The reflectance measurements has been accurately used in predicting the fruit maturity and also well correlated to the determination of internal composition of whole fruit. The use of NIR has been reported for the detection of internal fruit disorders (McGlone and Kawano, 1998; Schaare and Fraser, 2000; McGlone et al., 2002; Clark et al., 2003) and classification of apple during storage (Camps et al., 2007).

Light transmission technique is used for nondestructive evaluation of quality characteristics of small translucent fruits like grapes. Nondestructive optical methods based on NIR spectrometry have been used for nondestructive estimation of internal starch, soluble solids content, oil contents, water content, dry matter content, acidity, firmness, stiffness factor and other physiological properties of batches of apple, citruses, mandarins, fresh tomatoes, mangos, melons, kiwifruits, peaches, pineapples. The determination of total soluble solids, acidity, sugar acid ratio and firmness in fruit using NIR a spectroscopy and multivariate calibration technique is demonstrated (Dull et al., 1989; Osborne and Kunnemeyer, 1999; Jha and Matsuoka, 2004; Xie et al. 2011; Liu et al., 2010; McGlone et al., 2002; Moons and Sinnaeve, 2000; Ventura et al., 1998; Zude et al., 2006). Using visual and NIR spectrometry system in appropriate spectral range, the quality of mango and banana were nondestructively assessed (Jha et al., 2005, 2006, 2007, 2010, 2012a, b; Schmilovitch et al., 2000). Jha et al., (2014) proposed a formula for prediction of maturity index using physicochemical characteristics and overall acceptability for mangoes. Insect infestation in fruits is also an indicator of internal quality for fruits. NIR spectroscopy has been proved to have abilities to detect insects in fruits such as cherries, mangoes, and blueberries (Sun et al., 2010). As promising technique it is found to have the use in predicting the reducing sugar content of grape ripening as well as fermentation and aging of white and red wines (Novales et al., 2009).

2.19   ULTRASONIC METHODS

Ultrasound is a longitudinal wave beyond the upper frequency limit (20,000 Hz) of human audibility, where as waves beyond the lower frequency limit (20 Hz) of human audibility is infra sound. Ultrasound has come to play an important role and becoming a versatile tool used for studying the characterization of solid and liquid food (Award et al., 2012). Ultrasonic waves as travel through biomaterials cause ultrasonic attenuation mainly due absorption and scattering of wave length and reflecting the presence of lattice imperfections, phonon-phonon interactions, thermoelastic losses and electron-phonon interactions. Absorption of energy of these waves arises because of chemical relaxation effect, structural viscosity and volume viscosity effects, scattering due to nonlinear effects and rotational isomerism (Singh and Prasad, 1997). High power ultrasound somewhat affects the foods substantially. The large energy levels of high power ultrasound cause small transient air bubbles to form thus produces enhanced cavitation effect. The ultrasound is highly attenuated by the formed or present gas cells and therefore restricts its application in processing purposes (). Whereas, low power ultrasound does not alter the properties of foods significantly and thus applied for analytical applications in revealing the useful information about macroscopic and microscopic characterization of foods. Further, growing interest of low power ultrasound is due to its application as nondestructive, nonintrusive, and noninvasive technique. Recent advances have largely been reflected in the field of phased array ultrasonic. The phased firing of arrays of ultrasonic elements in a single transducer allows for precise couture of the resulting ultrasonic waves introduced into the test piece. Ultrasound is proved to have great potential as a probe for food materials both in online and laboratory. Most of the available analytical methods are either cumbersome, time taking or costly. Its potential in determination of acoustically measurable properties such as adiabatic compressibility, rigidity, and acoustic impedance with simplicity in data acquisition and interpretation enable the velocity measurement an acceptable method for assessing properties of solid and liquid foods.

2.20 CHARACTERISTICS OF ULTRASOUND

- Medium is necessary for its propagation
- Velocity is different in different medium
- Vibrating body produces waves in the surrounding medium
- Transmission is a wave motion
  - Takes time to travel
  - It is reflected
  - It is refracted
  - It exhibits interference
  - It shows diffraction

2.21 PRODUCTION OF ULTRASOUND WAVES

Magnetostriction and piezoelectric methods are the main ultrasonic wave production method (Sen, 1990). Steady current magnetizes a rod of ferromagnetic material, and then an alternating current of variable frequency is super imposed on it. The frequency of alternating current on matching with the natural frequency of the rod and thus the rod vibrates with the maximum amplitude, giving rise to a very strong beam of ultrasonic wave.

Piezoelectric method is the most commonly and widely used method for the production of ultrasound. On applying the mechanical pressure in a particular direction (mechanical axis) then a maximum potential difference is developed in a direction perpendicular (electrical axis) to it, with is known as piezoelectric effect. Most transducers for converting electrical energy to acoustical energy and vice-versa are based on this effect. Quartz, tourmaline, rochells salt are commonly used material for transducer. Ceramic transducers are also sometimes used.

2.22 ULTRASOUND VELOCITY AND VELOCITY MEASUREMENTS

The particle displacement is parallel to the direction of wave propagation in the longitudinal wave (Figure 2.13). Therefore, ultrasonic testing employs an extremely diverse set of methods based upon the generation and detection of mechanical vibrations or waves within test objects, which are not restricted to solid, liquid, metal, ornonmetal. The velocity of ultrasound traveling through a material is a simple function of the material's modulus and the density (Krasilnikov, 1963; Schmidt, 2007). Thus the analytical method based on ultrasound is uniquely suited and material specific in its characterization. Ultrasound as mechanical waves propagate through a medium (solid, liquid, or gas) at different speed depends on the elastic and inertial properties of that medium. The differential speed in different food products has paved the way to explore the analytical techniques depends on the ultrasonic velocity measurement.

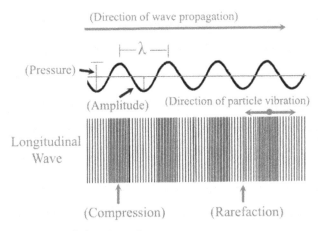

FIGURE 2.13   Characteristics of sound wave.

The cause of these effects is the vibration of particles produced by ultrasound. The ultrasonic velocity in the liquid system is found to be a function of certain variables such as frequency of wave n, temperature T, external pressure P, amplitude of wave a and concentration of constituents C_1, C_2, C_3 .... besides the nature and properties of liquid and solution. It may be expressed as:

$$v = f(C_1, C_2, C_3, \ldots \ldots \ldots P, T, n, a)$$

If the experimentation is conducted at a constant pressure with a fixed frequency of wave of considerably smaller amplitude, the other variables on which the ultrasonic waves depends may be expressed as:

$$v = f(C, T)$$

The velocity of sound wave through matter depends on the material itself and not the sound (Table 2.1). The velocity of sound (v) in any medium is given by

$$v = n.\lambda$$

n = frequency
λ = wave length
  The velocity of ultrasound in gas (v_g) is given by:

$$v_g = \sqrt{\frac{\gamma P}{\delta_g}}$$

γ = ratio of specific heat at constant pressure and specific heat at constant volume
P = Pressure
δ_g = density of gas
  The velocity of sound in liquid (v_l) is given by:

$$v_l = \sqrt{\frac{K}{\delta_l}}$$

K = Bulk modulus of elasticity
δ_l = density of liquid
  The velocity of sound in solid (v_s) is given by

$$v_s = \sqrt{\frac{Y}{\delta_s}}$$

Y = Young's modulus
δ_s = density of solid

TABLE 2.1   Ultrasonic velocity of some foods and related materials.

Materials	Ultrasonic velocity (m/s)	Materials	Ultrasonic velocity (m/s)
Water (20⁰C)	1,482.3	Oils (40°C)	1383 1462
Water (100⁰C)	478	Fats (40°C)	1,359–1,405
Ice (-20⁰C)	3,840	Ethyl alcohol (25°C)	1,145
Sugar solution (20–60°C)	1480–1884	Egg white (20°C)	1,505–1,565
Water +70% Glucose (20°C)	1,940	Egg yolk (20°C)	1,510–1,570
Water +5% Starch (25°C)	1,514	Skimmed milk (28°C)	1,522
Water +20% NaCl (25°C)	1,730	Milk chocolate (25°C)	1,020–1,740
Fruit juices (25°C)	1521–1576	Aerated milk chocolate (25°C)	900–1,000

Techniques used for the measurement of ultrasonic velocity are described below.

2.23   CONTINUOUS WAVE INTERFEROMETER

Principle used in the measurement of velocity is based on the precise determination of the wavelength in the experimental medium. Ultrasonic standing waves of known frequency are setup and produced by the quartz crystal fixed at the bottom of the cell containing the sample (Krasilnikov, 1963). The distance between probe and reflector is varied, a fixed ultrasonic frequency, and the distance corresponding to successive maxima in the standing wave pattern is the wavelength (Figure 2.14). The acoustic resonance gives rise to an electrical reaction on the generator driving the quartz crystal and the anode current of the generator becomes maximum (Figure 2.15). On changing the position of the reflector vary the crystal current in determination of wavelength for known input frequency in order to get the ultrasonic velocity (Singh and Prasad, 1997).

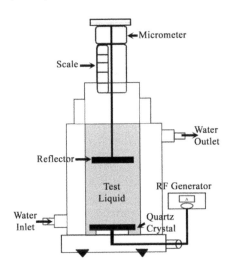

FIGURE 2.14   Experimental arrangements of Interferometer.

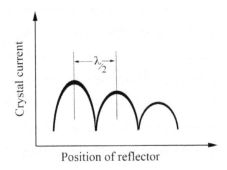

FIGURE 2.15   Position of reflector vs crystal current plot.

2.24   PULSE ECHO TECHNIQUE

The original descriptions of this technique contain both velocity and attenuation is measured (Figure 2.16). This can be done visually on an oscilloscope or automatically with a timer with a variable trigger level. A continuous reading of velocity can be obtained, provided the sample length is known (Povey and McClements, 1988).

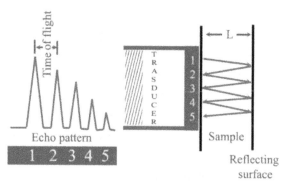

FIGURE 2.16   Pulse echo technique.

2.25   ACOUSTIC GRATING

When alone ultrasonic wave is setup in a liquid the pressure varies with the distance from the source periodically (Figure 2.17). The distance between any two nearest points in the same phase is the wavelength. The density and refractive index of the medium will therefore show a periodic variation with distance from the source along the direction of propagation of the wave and if light is passed through the liquid at right angles to the waves the liquid behaves as a diffraction grating (Povey and McClements, 1988). Such grating is called as acoustic grating. The grating element is equal to the wavelength of the ultrasonic waves.

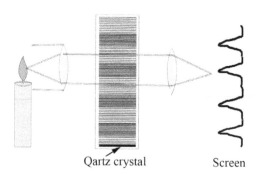

FIGURE 2.17   Acoustic grating.

ULTRASOUND IN IMAGING TECHNIQUE

Ultrasonic imaging techniques have been applied extensively in a situation of spatial heterogeneity of food materials. The property of ultrasonic absorption by the materials and ultrasonic reflection at boundaries on change of material properties are basically used in the quality characterization. As sound travels through a medium, it loses sound intensity or is attenuated, Attenuation may occur owing to the absorption of sound in the material and also to the scattering of sound.

The attenuation can be expressed by the equation:

$$I_x = I_0 e^{-2ax}$$

I_0 = Initial sound intensity
I_x = Intensity at depth x
e = Base of natural logs
x = Depth in tissue
a = Amplitude absorption coefficient

The absorption coefficient "a" of a given material increases as the frequency of sound increases. Total depth possible is inversely related to the frequency (Figure 2.18). Thus for a deep penetration a lower frequency must be used, and for a very small depth, a higher frequency should be used (Schmidt, 2007).

Sound absorption is an important aspect of attenuation and expressed as

$$S = \frac{20 \log \dfrac{A_0}{A_x}}{x}$$

S = Absorption
A_0 = Amplitude at zero depth
A_x = Amplitude at depth x
x = Depth in tissue

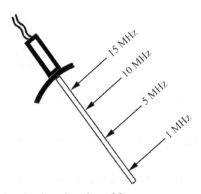

FIGURE 2.18 Penetration depth as function of frequency.

According to the equation sound absorption in tissue is approximately 1dB loss/cm/MHz. Accoustic impedence (Z) is determined by density (δ) of the material and the velocity (v) of the sound through the material.

$$Z = \delta v$$

It determines the amount of sound energy reflected from or transmitted through different materials (Table 2.2). Sound traveling through material continues until it comes to another material of different impedance (Table 2.2). Two adjacent materials of different acoustic impedance establish an acoustic boundary (Coupland and McClements, 2001). Depending on the impedance of each material, a reflection occurs.

TABLE 2.2   Acoustic impedance of biological material.

Materials	Impedance ($\times 10^{-2}$ kg/m²s)
Water	1.48
Fat	1.38
Blood	1.61
Muscle	1.70
Kidney	1.62
Spleen	1.64
Liver	1.65
Bone	7.80

When the incident sound beam is perpendicular to this acoustic boundary or interface, an echo results based on the following equation:

$$R = \left[\frac{Z_2 - Z_1}{Z_2 + Z_1} \right]^2 \times 100$$

R = Percent of sound beam reflected
Z_1 = Acoustic impedance of first material
Z_2 = Acoustic impedance of second material
     The following equation gives the amount of sound transmitted through boundary

$$T = \frac{4 Z_2 Z_1}{\left(Z_2 + Z_1 \right)^2} \times 100$$

T = Percent of sound beam transmitted

The total reflected and transmitted ultrasound at the boundary must be equal to 100 percent.

$$R + T = 100$$

Resolution is one of the important characteristics in the ultrasonic imaging techniques. It is the capability of a system to demonstrate acoustic boundary that lie parallel to the beam (axial resolution) or that lie perpendicular to beam (lateral resolution). The closer these two points can be resolved, the better the resolution of the system. Higher frequencies result in shorter wave lengths and increased resolution (Figure 2.19).

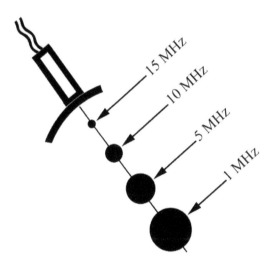

FIGURE 2.19   Resolution as function of frequency.

2.26   APPLICATION OF ULTRASOUND

The features of noninvasive, precise, rapid, and potential for real-time application provide low power ultrasound a versatile nondestructive evaluation tool in quality assessment of raw and processed fruits and vegetables (Povey and McClements, 1988, Sohale, 1994; Mittal, 1997). Optically opaque materials could be evaluated for the quality evaluation with the limitation of its use in the presence of aerated sample due to attenuation of ultrasonic energy to be acquired adequately. The effect of the porosity substantially increases if the sample is irregular shaped or defective in nature. Also the noncontact ultrasonic is not practically preferable for very high acoustic impedance materials such as fruits and vegetables. However, presently ultrasonic technique has been found to be having good potential for the nonde-structive evaluation of pre and postharvest quality assessment of fresh fruit and vegetables.

2.27   QUALITY CHARACTERIZATION OF FRUITS AND VEGETABLES

The process of fruit development, maturation, ripening, and senescence are associated with changes in chemical composition and mechanical properties. This change could be observed with the use of changes in the ultrasonic properties of the materials. However, the results should be correlated with the multivariate statistical methods or with the soft computing techniques to incorporate the artificial intelligence in precise prediction of subjected quality parameters of fruits and vegetables in order to link the theoretical, acoustic propagation and structure.

Ultrasonic techniques as analytical method closely imitate the quality attributes are applied in the assessment of characterization of fruits and vegetables. Sarkar and Wolfe (1983) have attempted this ultrasonic imaging technique in quality evaluation. Vegetables being irregular in shape and high intercellular air cells restrict the versatility of the result with high reproducibility. However, quality characterization of whole potatoes with hollow heart could successfully been used applying the sound energy in nondestructive ways (Yanling and Haugh, 1994). Texture measurement during the growth and development of apple and during ripening in banana and avocado have successful example of ultrasonic technique (Mizrach, 2008). Feasibility of ultrasonic parameters for the measurement of mealiness levels and distinguishing the bruised from unbruised tissue in apple was explored (Upchurch, 1986).

Nondestructive monitoring of physicochemical changes in whole avocado during maturation is evaluated. Ultrasonic attenuation could be good quality predictor for quality evaluation as ultrasonic attenuation increased with the decrease of firmness regardless of storage temperatures. The physicochemical properties of mango were evaluated with the ultrasonic technique. The quality parameters of melon for firmness and sugar contents were evaluated. Ultrasonic attenuation was linearly related to the firmness of the tomato during storage was demonstrated.

2.28   CONCENTRATION DETERMINATION

Ultrasonic technique was successfully been used for the estimation of sugar content in juice and beverages (Contreras, et al., 1992; Prasad et al., 2000). Linear relationship between ultrasonic velocity and solute concentration was revealed. For concentration determination across the pipeline, calibration is done empirically for the transit time of an ultrasonic echo. The advantages of such a technique are that it can be noninvasive, nonintrusive accurate and relatively cheap. This technique is also used for the measurement of suspended solids, haze, and dissolved solids (Hammond, 1992).

2.29 LEVEL MEASUREMENT

Ultrasonics is an established liquid level measurement. A variety of probe arrangements is available and using variety of sound paths, level detection may be combined with concentration determination. Ultrasonic level sensor can be used for level measurement and control of liquid, slurry, and solid levels as there is no contact with the material, and is unaffected by the physical and chemical properties of most materials. This method is easy to install and the measurement accuracy of these units was claimed to be ±0.1% with the temperature compensation.

2.30 FLOW MEASUREMENT

Ultrasonic technique could be used for the flow measurement (Lynnworth, 1979). Careful choice of the appropriate probe arrangement can improve the situation and time gating of the signal and selecting out the signal from a given portion of the flow offers the prospects of determining flow profiles.

2.31 TEMPERATURE MEASUREMENT

Ultrasonic technique is found as temperature measurement and its potential for food systems. The technique depends upon the relationship of the velocity or ultrasound with the temperature. It is having potential uses for the temperature measurement in food processing systems were investigated using invasive sensors (Richardson and Povey, (1990).

2.32 MISCELLANEOUS APPLICATIONS

Ultrasonic has been used for can head space measurement, using a technique similar to the level detector (Grau and Ors, 1978). The hydrolysis of starch particles was monitored using detection of ultrasonic attenuation (Povey and Rosenthal, 1984). Experimental validation for the change in convective heat transfer coefficient with sonic energy in natural convection is reported (Sastry et al., 1989). Measurement of specific gravity at specific temperature could be derivedfor the liquid (Skragatic et al., 1983).

2.33   MAGNETIC RESONANCE IMAGING (MRI)

Nuclei of $_1$H, $_{13}$C, $_{23}$Na, and $_{31}$P exhibit magnetic moment and align along the applied strong static magnetic field (Faust et al. 1997). The hydrogen nuclei are naturally abundant in living materials and contain a proton, which has a charge and the property of nuclear spin (Figure 2.20). Thus the magnetic resonance imaging (MRI) applications are aimed at imaging hydrogen nuclei in animal as well as most of the horticultural products. This imparts magnetic properties to the nucleus in the image formation.Normally random orientation of protons present in the materials resulting no net magnetic field but on placing under strong external magnetic field (B_0), the proton spins and align themselves in a particular directions either parallel or antiparallel to the applied magnetic field.

Since, the parallel direction is at lower steady-state energy, a small excess of protons is seen aligned in this direction. The microscopic magnetizations of these excess protons sum up into a net magnetization vector (NMV) or longitudinal magnetization(M_o).

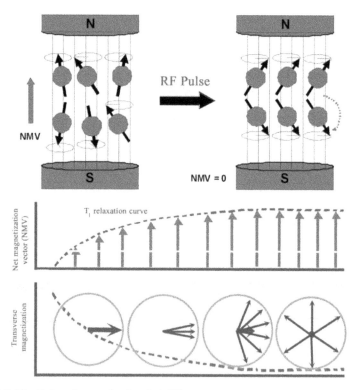

FIGURE 2.20   Relaxation mechanism in MRI.

The strength of M_o determines the maximum available signal strength. It increases linearly with the strength of B_o, providing the primary reason for the use of higher field strength.

The application of the radio frequency (RF) pulse results in absorption of the energy and this pulse is also called as excitation pulse. The absorption of applied RF energy results in an increases number of high energy spin up nuclei. On applying right amount of energy, the number of nuclei in the spin up position equals down position. The transverse magnetization vector processes in the transverse plane, producing in a time varying magnetic field. This induces a current signal, which forms the basis for image formation in MRI.

Immediately after ceasing of applied RF pulse, the system starts to relax back to its equilibrium state by releasing the absorbed energy. The process repeats and takes less than milliseconds up to several seconds depending on the proton's environment and the magnetic field. Since, this precession involves the transfer of energy from the proton spin to their environment or lattice, which termed as spin-lattice relaxation or T_1 relaxation. If the effect of external magnetic field inhomogeneities is removed then the relaxation of transverse magnetization purely through spin spin energy transfer or T_2 relaxation. The two relaxation rates, T_1 and T_2 relaxation rates are intrinsic features of the underlying tissue and vary with tissue type and serve as contrast mechanisms in exquisite depiction of MRI. Signal components from all protons are simultaneously detected in the receiver coil and the magnetic field gradients create a mapping of frequency to position in image formation.

2.34   APPLICATIONS OF MAGNETIC RESONANCE IMAGING

Magnetic resonance imaging (MRI) is a nondestructive and noninvasive technique that can be used to acquire two or three dimensional images. MRI quality analysis can be divided into two categories. Extraction of a morphological characteristic is the main goal in this case first category where phase contrast is not desirable. Nuclear magnetic resonance imaging (MRI) provides high resolution images of internal structures of intact fruit but not practically suitable presently for routine quality testing as per the cost of equipment, operating cost and speed of measurement (Wang and Wang, 1989; Clark et al. 1997; Clark and Burmeister, 1999). This technique has great potential for evaluating the internal quality of fruits and vegetables. Applications of nuclear magnetic resonance imaging in horticultural produce and various foods have recently been reviewed (Clark et al. 1997; Faust et al. 1997 and Abbott et al. 1997; Mariette, 2004). On the other hand, in the determination of moisture or fat content for the extraction of information related to changes in microstructure during heat and mass transfer within food materials. This provides the information at both cell and macroscopic levels for the physiological changes in fruit tissues during development, ripening, storage, and processing for the internal quality evaluation. MRI has been used to show the morphology, growth, ripening, seeds, pits, voids, damages, bruises, and changes due to various food processing operations (Chen et al. 1989; Saltveit, 1991; MacFall and Johnson, 1994; Clark et al. 1997; Faust et al. 1997). Wang et al. (1988) demonstrated the internal structure of apple with the presence of water core. Hinshaw et al. (1979) reported the magnetic resonance image of orange showing the segments with the membranes.

As hydrogen nuclei show high response to magnetic fields therefore, NMR to measure water content and water distribution is demonstrated. Variations in concentration and state of water and fat with the concentration variations was assessed under ripe, defected, and decayed fruits and vegetables to evaluate the maturity and quality parameters for understanding the physiological processes (Faust et al.1997).

Mealiness in apples was assessed by Barreiro et al. (2000) using MRI techniques. Behaviour of apple under controlled atmospheric storage compared and correlated mealiness as obtained through the use of Magness-Taylor firmness test. Magnetic resonance imaging was also applied to identify the seed of oranges and effect on quality during the freeze injury under online inspections (Hernandez et al. 2005).

MRI is effective in determining the distribution of apparent micro porosity in fruits and vegetables. The microscopic quality characterization could be performed for the unit operation where changes in the moisture. Ice formation during food freezing of fruits and vegetables can be examined. Packaged foods could successfully be examined.

KEYWORDS

- **Noninvasive**
- **Internal quality**
- **Spectroscopy**
- **Near infrared**
- **Magnetic resonance**
- **Imaging**

REFERENCES

Abbott, J. A.; Lu, R.; Upchurch, B. L.; and Stroshine, R. L.; Technologies for nondestructive quality evaluation of fruits and vegetables. *Hortic. Rev.* **1997**, *20*, 1–120.

Awad, T. S.; Moharram, H. A.; Shaltout, O. E.; Asker, D.; and Youssef, M. M.; Applications of ultrasound in analysis, processing and quality control of food: A review. *Food Res. Int.* **2012**, *48*(2), 410–427.

Barcelon, E. G.; Tojo, S.; and Watanabe, K.; X-ray CT imaging and quality detection of peach at different physiological maturity. *Trans. Am. Soc. Agric. Eng.* **1999**, *42*, 435–441.

Brecht, J. K.; Shewfelt, R. L.; Garner, J. C.; and Tollner, E. W.; Using X-ray computed tomography (X-ray CT) to nondestructively determine maturity of green tomatoes. *Hortic. Sci.* **1991**, *26*, 45–47.

Camps, C.; Guillermin, P.; Mauget, J. C.; and Bertrand, D.; Discrimination of storage duration of apples stored in a cooled room and shelf-life by visible-near infrared spectroscopy. *J. Near Infrared. Spec.* **2007**, *15*(3), 169–177.

Chen, P.; McCarthy, M. J.; and Kauten, R.; NMR for internal quality evaluation of fruits and vegetables. *Trans. Am. Soc. Agric. Eng.* **1989**, *32*, 1747–1753.

Chuang, Cheng-Long; Ouyang, Cheng-Shiou; Lin, Ta-Te; Yang, Man-Miao; Yang En-Cheng; Huang, Tze Wei; Kuei, Chia-Feng; Luke, Angela and Jiang, Joe-Air; Automatic X-ray quarantine scanner and pest infestation detector for agricultural products. *Comput. Electron. Agric.* **2011**, *77*, 41–59.

Clark C. J.; and Burmeister D. M.; Magnetic resonance imaging of browning development in 'Braeburn' apple during controlled-atmosphere storage under high CO2. *Hortic. Sci.* **1999**, *34*, 915–919.

Clark, C. J.; Hockings, P. D.; Joyce, D. C.; and Mazucco, R. A.; Application of magnetic resonance imaging to pre and post-harvest studies of fruit and vegetables. *Postharvest Biol. Technol.* **1997**, *11*(1), 1–21.

Clark, C. J.; McGlone, V. A.; and Jordan, R. B.; Detection of brown heart in 'braeburn' apple by transmission NIR spectroscopy. *Postharvest Biol. Technol.* **2003**, *28*, 87–96.

Contreras, N. I.; Fairley, P.; McClements, D. J.; and Povey, M. J. W.; Analysis of the sugar content of fruit juices and drinks using ultrasonic velocity measurements, *Int. J. Food Sci. Technol.* **1992**, *27*(5), 515–529.

Coupland, J.; and McClements. D. J.; Ultrasonics. In *Nondestructive Food Evaluation: Techniques to Analyze Properties and Quality,* Ed. Gunaskaran, S.; New York: Marcel Dekkaer, Inc.; **2001**.

Diener, R. G.; Mitchell, J. P.; and Rhoten, M. L.; Using an X-ray image scan to sort bruised apples. *Agric. Eng.* **1970**, *51*, 356–361.

Dyer, J. R.; *Application of Absorption Spectroscopy of Organic Compounds.* London: Prentice-Hall International, Inc.; **1965**.

Farr, R. F.; and Roberts, P. J. A.; *Physics for Medical Imaging.* New Delhi: Elsevier India Pvt. Ltd.; **2006**.

Faust, M.; Wang, P. C.; and Maas, J.; The use of magnetic resonance imaging in plant science. *Hortic. Rev.* **1997**, *20*, 225–266.

Finney, E. E.; and Norris, K. H.; X-ray scans for detecting hollow hearths in potatoes. *Am. Potato. J.* **1978**, *55*, 95–105.

Grau, J.; and Ors, J.; The headspace control test in canned food stuff using ultrasonic methods. In: *International Institute of Welding Symposium*. New York; International Institute of Welding; **1978**, pp 12–28.

Gunaskaran, S.; *Nondestructive Food Evaluation: Techniques to Analyze Properties and Quality.* New York: Marcel Dekkaer, Inc.; **2001**.

Gunasekaran, S.; and Irudayaraj, J.; Optical methods: Visible, NIR, and FTIR Spectroscopy. In: *Nondestructive Food Evaluation: Techniques to Analyze Properties and Quality;* Gunasekaran, S. Ed.; New York, Marcel Dekker, Inc.; **2001**.

Hammond, R.; Measurement and control techniques, post- fermentation. *Brew. Distill. Int.* **1992**, *23*(6), 30–32.

Han, Y. J.; Bowers, S. V.; and Dodd, R. B.; Nondestructive detection of split-pit peaches. *Trans. Am. Soc. Agric. Eng.* **1992**, *35*, 2063–2067.

Harrison, I.; Non-destructive testing for fruit quality assurance. *Innov. Food Technol.* **2003**, (No. 19), 86–87.

Hernandez, N.; Barreiro, P.; Ruiz-Altisent, M.; Ruiz-Cabello, J.; and Fernandez-Valle, M. E.; Detection of seeds in citrus using MRI under motion conditions and improvement with motion correction. *Concep. Magn. Reson. B. Magn. Reson. Eng.* **2005**, *26B*(1), 81–92.

Hinshaw, W. S.; Bottomley, P. A.; and Holland, G. N.; A demonstration of the resolution of NMR imaging in biological systems. *Experientia.* **1979**, *35*, 1268–1269.

Jaiswal, P.; Jha, S. N.; and Bharadwaj, R.; Non-destructive prediction of quality of intact banana using spectroscopy. *Sci. Hortic.* **2012**, *135*, 14–22.

Jha, S. N.; *Nondestructive Evaluation of Food Quality Theory and Practices.* New York: Springer; **2010**.

Jha, S. N.; and Garg, Ruchi.; Non-destructive prediction of quality of intact apple using near infrared spectroscopy. *J. Food Sci. Technol.* **2010**, *47*(2), 207–213.

Jha, S. N.; and Matsuoka, T.; Non-destructive determination of acid-Brix ratio of tomato juice using near infrared spectroscopy. *Int. J. Food. Sci. Technol.* **2004**, *39*(4), 425–430.

Jha, S. N.; Chopra, S.; and Kingsly, A. R. P.; Determination of sweetness of intact mango using visual spectral analysis. *Biosyst. Eng.* **2005**, *91*(2), 157–161.

Jha, S. N.; Kingsly, A. R. P.; and Chopra, S.; Nondestructive determination of firmness and yellowness of mango during growth and storage using visual spectroscopy. *Biosyst. Eng.* **2006**, *94*(3), 397–402.

Jha, S. N.; Chopra, S.; and Kingsly, A. R. P.; Modeling of color values for nondestructive evaluation of maturity of mango. *J. Food. Eng.* **2007**, *78*, 22–26.

Jha, A. K.; Prasad, K.; Prasad, K.; and Kulkarni A. R.; Plant system: nature's nanofactory. *Coll. Surf. B: Biointerf.* **2009a**, *73*(2), 219–223.

Jha, A. K.; Prasad, K.; Kumar, V.; and Prasad, K.; Biosynthesis of silver nanoparticles using *Eclipta* leaf. *Biotechnol. Prog.* **2009b**, 25(5), 1476–1479.

Jha, S. N.; Narsaiah, K.; Sharma, A. D.; Singh, M.; Bansal, S.; and Kumar, R.; Quality parameters of mango and potential of non-destructive techniques for their measurement: a review. *J. Food. Sci. Technol.* **2010**, *47*(1), 1–14

Jha, S. N.; Jaiswal, P.; Narsaiah, K.; Gupta, M.; Bhardwaj, R.; and Singh, A. K.; Non-destructive prediction of sweetness of intact mango using near infrared spectroscopy. *Sci. Hortic.* **2012**,*138*, 171–175.

Jha, S. N.; Narsaiah, K.; Jaiswal, P.; Bhardwaj, R.; Gupta, M.; Kumar, R.; and Sharma, R.; Nondestructive prediction of maturity of mango using near infrared spectroscopy. *J. Food. Eng.* **2014**, *124*, 152–157.

Johnson, M.; Automation in Citrus Sorting and Packing. In: *Agrimation Conference and Expo*, Chicago, IL; **1985**, pp 63–68.

Kalia, M.; *Food Quality Management.* Udaipur: Agrotech Publishing Academy; **2010**.

Kaur, H.; *Instrumental Methods of Chemical Analysis.* Meerut, India: Pragati Prakashan; **2010**.

Keagy, P. M.; Parvin, B.; and Schatzki, T. F.; Machine recognition of navel orange worm damage in X-ray images of pistachio nuts. *Lebensm.-Wiss. Technol.* **1996**, *29*, 140–145.

Krasilnikov, V. A.; *Sound and Ultrasound Waves in Air, Water and Solid Bodies.* Jerusalem: Israel Program for Scientific Translations; **1963**.

Lenker, D. H.; and Adrian, P. A.; Use of X-ray for selecting mature lettuce heads. *Trans. Am. Soc. Agric. Eng.* **1971**, *14*, 894–898.

Lim, K. S.; and Barigou, M.; X-ray micro-computed tomography of cellular food products. *Food Res. Int.* **2004**, *37*(10), 1001–1012.

Liu, Y.; Sun, X.; and Ouyang, A.; Nondestructive measurement of soluble solid content of navel orange fruit by visible–NIR spectrometric technique with PLSR and PCA-BPNN. *LWT – Food. Sci. Technol.* **2010**, *43*, 602–607.

Lynnworth, L. C.; Ultrasonic How meter. In *Physical Acoustics*, Ed. Mason, W. P.; and Thurston, R. N.; New York: Academic Press; **1979**, pp 408–516.

MacFall, J. S.; and Johnson, G. A.; The architecture of plant vasculature and transport as seen with magnetic resonance microscopy. *Can. J. Bot.* **1994**, *72*, 1561–1573.

Mahajan, M.; *Statistical Quality Control.* Delhi: Dhanpat Rai and Sons; **1995**.

Mariette, F.; NMR relaxometry and MRI techniques: a powerful association in food science. *Comptes Rendus Chim.* **2004**, *7*(3/4), 221–232.

McGlone, V. A.; and Kawano, S.; Firmness, dry-matter and soluble solids assessment of postharvest kiwifruit by NIR spectroscopy. *Postharvest Biol Technol.* **1998**, *13*, 131–141.

McGlone, V. A.; Jordan, R. B.; Seelye, R.; and Martinsen, P. J.; Comparing density and NIR methods for measurement of kiwifruit dry matter and soluble solids content. *Postharvest Biol. Technol.* **2002**, *26*, 191–198.

Mendoza, Fernando; Verboven, Pieter; Ho, Quang Tri; Kerckhofs, Greet; Wevers, Martin and Nicolai, Bart.; Multi fractal properties of pore-size distribution in apple tissue using X-ray imaging. *J. Food Eng.* **2010**, *99*, 206–215.

Mittal, G. S.; Computer Based Instrumentation Sensors for Inline Measurement pp 13–53. In: *Computerized Control Systems in Food Industry;* Mittal, G. S. Ed.; New York: Marcel Dekker Inc.; **1997**, 597 p.

Mizrach, A.; Ultrasonic technology for quality evaluation of fresh fruit and vegetables in pre- and postharvest processes-a Review. *Postharvest Biol. Technol.* **2008**, *48*, 315–330.

Moons, E.; and Sinnaeve, G.; Non-destructive Vis and NIR spectroscopy measurement for the determination of apple internal quality. *Acta Hortic.* **2000**, *517*, 441–448.

Nicolai, B. M.; Beullens, K.; Bobelyn, E.; Peirs, A.; Saeys, W.; Theron, K. I.; and Lammertyn, J.; Nondestructive measurement of fruit and vegetable quality by means of NIR spectroscopy: a review. *Post Harvest Biol. Technol.* **2007**, *46*, 99–118.

Novales, J. F.; López, M. I.; Sánchez, M. T.; Morales, J.; and Caballero, V. G.; Shortwave-near infrared spectroscopy for determination of reducing sugar content during grape ripening, wine-making, and aging of white and red wines. *Food Res. Int.* **2009**, *42*, 285–291.

Nylund, R. E.; and Lutz, J. M.; Separation of hollow heart potato tubers by mean of size grading, specific gravity and X-ray examination. *Am. Potato J.* **1950**, *27*, 214–22.

Ogawa, Y.; Morita, K.; Tanaka, S.; Setoguchi, M.; and Thai, C. N.; Application of X-ray CT for detection of physical foreign materials in foods. *Trans. Am. Soc. Agric. Eng.* **1998**, *41*, 157–162.

Povey, M. J. W.; and Rosenthal, A. J.; Technical note: Ultrasonic detection of the degradation of starch by α-amylase. *J. Food Technol.* **1984**, *19*, 115–119.

Povey, M. J. W.; and McClements, D. J.; Ultrasonics in food engineering. Part I: Introduction and experimental methods, *J. Food. Eng.* **1988**, *8*(4), 217–245

Prasad, K.; Statistical optimization: Response surface methodology (RSM) Approach, Compendium (SCPPO-12), 6-10 February, 2012, SLIET, Longowal, India; **2012**.

Prasad, K.; Jale, R.; Singh, M.; Kumar, R.; and Sharma. R. K.; Non-destructive evaluation of dimensional properties and physical characterization of *Carrisacarandas* fruits. *Int. J. Eng. Stud.* **2010a**, *2*(3), 321–327.

Prasad, K.; Jha, A. K.; Prasad, K.; and Kulkarni. A. R.; Can microbes mediate nanotransformation? *Indian J. Phys.* **2010b**, *84*(10), 1355–60.

Prasad, K.; Nath, N.; and Prasad, K.; Estimation of sugar content in commercially available beverages using ultrasonic velocity measurement. *Indian J. Phys.* **2000**, *74A*(4), 387–389.

Prasad, K.; Singh, Y.; and Anil, Anjali; Effects of grinding methods on the characteristics of Pusa 1121 rice flour. *J. Trop. Agric. Food. Sci.* **2012**, *40*(2), 193–201.

Ranganna, S.; *Handbook of Analysis and Quality Control for Fruit and Vegetable Products.* New Delhi: Tata McGraw-Hill Publishing Company Limited; **2007**.

Richardson, P. S.; and Povey, M. J. W.; Ultrasonic temperature measurement and its potential for food processing systems. *Food. Control.* **1990**, *1*(1), 54–57.

Saltveit, M.; Determining tomato fruit maturity with nondestructive in vivo nuclear magnetic resonance imaging. *Postharvest Biol. Technol.* **1991**, *1*, 153–159.

Sarkar, N.; and Wolfe, R. R.; Potential of ultrasonic measurement in food quality evaluation. *Trans. Am. Soc. Agric. Eng.* **1983**, *26*(2), 624–629.

Schaare, P. N.; and Fraser, D. G.; Comparison of reflectance, interactance and transmission modes of visible-near infrared spectroscopy for measuring internal properties of kiwifruit (Actinidia chinensis). *Postharvest Biol. Technol.* **2000**, *20*, 175–184.

Sastry, S. K.; Shen, G. Q.; and Blaisdell, J. L.; Effect of ultrasonic vibration on fluid to particle convective heat transfer coefficients. *J. Food Sci. Nutr.* **1989**, *18*(2), 153–159.

Schatzki, T. F.; Witt, S. C.; Wilkins, D. E.; and Lenker, D. H.; Characterization of growing lettuce from density contours— I. Head selection. *Patt. Recogn.* **1981**, *13*, 333–340.

Schatzki, T. F.; Haff, R. P.; Young, R.; Can, I.; Le, L. C. and Toyofuku, N.; Defect detection in apples by means of X-ray imaging. *Trans. Am. Soc. Agric. Eng.* **1997**,*40*, 1407–1415.

Schmidt, G.; *Ultrasound.* Stuttgart, Germany: Georg Thieme Verlag; **2007**.

Schmilovitch, Z.; Mizrach, A.; Hoffman,A.; Egozi, H.; and Fuchs, Y.; Determination of mango physiological indices by near-infrared spectrometry. *Postharvest Biol. Technol.* **2000**, *19*, 245–252.

Sen, S. N.; *Acoustics Waves and Oscillation.* New Delhi: New Age International (P) Limited, Publishers; **1990**.

Sheikh, A. S.; Standardization of Process Parameters for the Development of Apple Based Fruit Bar, M. Tech. Thesis, SLIET, Longowal; **2012**.

Silverstein, R. M.; Bassler, G. C.; and Morrill, T. C.; *Spectrometric Identification of Organic Compounds.* New York: John Wiley & Sons; **1981**.

Singh, N. P.; and Prasad, K.; *Experiments in Material Science.* Delhi: Dhanpat Rai and Co. (Pvt.) Ltd.; **1997**, 131 p.

Singh, Y.; and Prasad, K.; *Moringaoleifera* leaf as functional food powder: Characterization and uses. *Int. J. Agric. Food. Sci. Technol.* **2013**, *4*(4), 317–324

Skargatic, D. M. J.; Mitchinson, J. C.; and Graham, J. A.; Measurement- of specific gravity. PCT-international Patent-application WO87102770; **1987**.

Sohale, S.; Ultrasonic inspection methods for food products. *Lebensm-Wissen, Technol.* **1994**, *27*, 210–213.

Sun, Da-Wen; *Computer Vision Technology for Food Quality Evaluation.* Amsterdam/London; Elsevier; **2007**.

Sun, T.; Huang, K.; Xu, H.; and Ying, Y.; Research advances in nondestructive determination of internal quality in watermelon/melon: a review. *J. Food Eng.* **2010**, *100*, 569–577.

Suzuki, K.; Tajima, T.; Takano, S.; Asano, T.; and Hasegawa, T.; Nondestructive methods for identifying injury to vapor heat treated papaya. *J. Food Sci.* **1994**, *59*, 855–857, 875.

Tollner, E. W.; Hung, Y. C.; Upchurch, B. L.; and Prussia, S. E.; Relating X-ray absorption to density and water content in apples. *Trans. Am. Soc. Agric. Eng.* **1992**, *35*, 1921–1929.

Tollner, E. W.; Hung, Y. C.; Maw, B. W.; Sumner, D. R.; and Gitaitis, R. D.; Nondestructive testing for identifying poor quality onions. *Proc. Soc. Photo-Opt. Instrum. Eng.* **1995**, *2345*, 392–397.

Ventura, M.; De Jager, A.; De Putter, H.; and Roelofs, P. M. M.; Non-destructive determination of soluble solids in apple fruit by near infrared spectroscopy (NIRS). *Postharvest Biol. Technol.* **1998**, *14*, 21–27.

Wang C. Y.; and Wang P. C.; Nondestructive detection of core breakdown in 'Bartlett' pears with nuclear magnetic resonance imaging. *Hortic. Sci.* **1989**, *24*, 106–109.

Xie, L.; Ye, X.; Liu, D.; and Ying, Y.; Prediction of titratable acidity, malic acid, and citric acid in bayberry fruit by near-infrared spectroscopy. *Food. Res. Int.* **2011**, *44*, 2198–2204.

Yang, En-Cheng; Yang, Man-Miao; Liao, Ling-Hsiu; and Wu, Wen-Yen; Non-destructive quarantine technique- potential application of using x-ray images to detect early infestations caused by oriental fruit fly (*Bactrocera dorsalis*) (Diptera: Tephritidae) in fruit. *Formosan. Entomol.* **2006**, *26*, 171–186

Yanling, C.; and Haugh, C. G.; Detection hollow heart in potatoes using ultrasound. *Trans. Am. Soc. Agric. Eng.* **1994**, *37*(1), 217–222.

Zaltzman, A.; Verma, B. P.; and Schmilovitch, Z.; Potential of quality sorting of fruits and vegetables using fluidized bed medium. *Trans. Am. Soc. Agric. Eng.* **1987**, *30*, 823–831.

Zude, M.; Herold, B.; Roger, J. M.; Bellon-Maurel, V.; and Landahl, S.; Nondestructive tests on the prediction of apple fruit flesh firmness and soluble solids content on tree and in shelf life. *J. Food. Eng.* **2006**, *77*, 254–260.

CHAPTER 3

ADVANCES IN CONVENTIONAL BREEDING APPROACHES FOR POSTHARVEST QUALITY IMPROVEMENT OF FRUITS

AMIT KUMAR KHOKHER[1]

[1]Division of Fruit Science, SKUAST-K, Shalimar, Srinagar, Jammu and Kashmir (192 308) India; Email: khokherak@rediffmail.com

CONTENTS

Fruit breeding is a long-term process, which takes a minimum of about a decade from the original cross to a finished cultivar due to two challenges: long juvenile periods and large plant size. Thus, much thought needs to go into which objectives to be emphasized in the breeding. Although certain objectives, such as yield is always important, the overall lifestyle, environmental, marketing, and production trends affects the objectives that breeders emphasize in their programs as they strive to anticipate the future needs of the fruit industry. The importance of each trend varies with the crop and environment. The major trends are to develop cultivars which simplify orchard practices, have increased resistance to biotic and abiotic stress, extend the adaptation zones of the crop, create new fruit types, create fruit cultivars with enhanced health benefits, and provide consistently high quality. In spite of these difficulties, breeding programs have been developed in all important perennial fruit crops, aimed at the improved economic profitability of the crops by increasing yields, altering the harvest window, creating new fruit types, and improving fruit quality while simplifying management. The characterization of fruit quality is always difficult and often controversial. The problem is further aggravated when considering fruits and nuts due to the complexity of their physical and chemical composition and the ongoing metabolism of these components during ripening and postharvest. Differences in consumer preferences make a universal definition of fruit quality difficult, even ephemeral (Janick, 2005). Fruit cultivars can be characterized for commercial quality, industrial quality, sensory quality, and nutritional quality. Commercial quality refers to all aspects related to the external presentation of the product, including size, shape, surface texture, color, and ultimately marketable yield. Industrial quality refers to how compatible the cultivar is with the various handling, transport, processing, and storage practices typical to its preparation for different markets. Sensory or organoleptic quality refers to those factors that determine consumer preference and thus is subjective and highly variable, while nutritional quality refers to the specific nutrients provided and the overall contribution to consumer health.

Fruit quality is complex concept and it is not possible to define it by one absolute parameter. The concept could be defined as those intrinsic properties of the product that satisfy the consumer's expectations. Nowadays these expectations are diverse and not always easy to interpret. From the consumer's point of view, quality might have different aspects: the fulfillment of our hedonic expectation; tastes and aromas that frequently recall our childhood or pleasant moments in our lives; the health benefits or the ease of purchasing the product at the moment whenever we like, regardless of the season or where we live.

Forecasting consumer preference trends should be considered in planning breeding programme seeking to anticipate the development of new high-quality cultivars. Consumer's satisfaction is a basic goal of the fresh fruit industry. The first issue that should be taken into account is visual quality, given that it triggers the purchase decision (Kays, 1999). Absence of defects and homogeneity of the fruit correspond to visual quality factors, whereas size, color, and shape are characteristics of particular cultivars and requires experience to be perceived. So the external appearance, too often addressed as "quality", is not sufficient in and of itself to guarantee consumer satisfaction and repeated sales. Sensory techniques are tools to gain insight into consumer preferences, which combined with the determination of physical and chemical parameters, provide a more detailed definition of fruit quality (Infante et al., 2008b).

Sensory analysis techniques are increasingly being used to support breeding and to rest new cultivars and storage practices. Fruit quality based only on external appearance and SSC has become obsolete. A more detailed knowledge of the genetic basis of traits determining sensory quality such as sourness, astringency, and aroma could be of major importance for breeding. Sensory evaluation offers the possibility of providing a complete profile of the fruit as perceived at the point of eating. This allows the weight of a parameter to be determined in terms of its influence on the perception. The choice of sensorial parameters that judges must evaluate is a basic point for correlation with physical and chemical measurements and any molecular approach (Colaric et al., 2005).

Fruit quality is associated with different attributes that vary during ripening, providing changes in seasonal traits and consumer acceptance. Sensory evaluation in stone fruits should be consider the climacteric phase of the ripening process and a standardized flesh firmness near 1–2 kg (Infante et al., 2008a). An effective sensory evaluation must be executed during the time span that goes from commercial ripening upto the beginning of senescence. This span varies 3–5 days, depending on the cultivar (Amoros et al., 1989).

Fruit breeding programs have a wide range of specific objectives such as increased cold hardiness or disease resistance. However, all commercial scion releases must have high-quality fruit. It does not matter how disease resistant or productive the tree may be, if the fruit is not of acceptable quality it will not be a commercial success. Fruit quality is complex. There are many definitions and standards set by each industry but the simple definition of fruit quality is: Whatever the consumer desires (Barritt, 2001; Elia, 2001). Since people are different, their desires and ideas of quality are different and breeders need to provide alternatives to meet these market needs.

How does the consumer perceive quality? The consumer initially judges quality by the appearance of the fruit at the point of sale and then by the taste of the fruit. The appearance will determine the first buy but taste will determine the return buy (Kader, 2002). The challenge for breeders is to provide an attractive fruit with desired taste that will survive the process of reaching the consumer. Attractiveness is usually based on fruit size and color, and taste is generally reflected on a combination of texture, flavor, aroma, and sugar to acid ratios. The process of getting the fruit to the consumer is dependent on the rate of softening and the overall control of the ripening process.

3.1   FRUIT QUALITY TRAITS

3.1.1   APPEARANCE

The characteristics that affect the appearance are primarily size and color. Consumer surveys in peach and apple have shown that bigger is better and consumers are willing to pay enough more to make larger fruit more profitable (Parker et al., 1991; Bruhn, 1995). Bright, clear colors are preferred (Francis, 1995).

1.   **Size** Fruit size has a large genetic component thus selecting for larger fruit is relatively straightforward (Janick and Moore, 1996). Fruit size is a function of cell number, cell volume, and cell density (Coombe, 1976; Scorza et al., 1991).

2.   **Color** This trait is an important aspect of appearance. The overall color of a fruit is reflected by the color of the outer pericarp and the flesh color. Pigments responsible for the colors are various modifications of anthocyanins, lycopenes, and carotenoids. Predicting colors is difficult because small modifications of, or combinations of pigments result in unpredictable colors. There is a degreening process during ripening which exposes the colors in both the pericarp and the flesh. The degreening process is the breakdown of chlorophylls, which is usually an ethylene response. The other pigments are no longer masked by chlorophyll and the fruit "colors". One of the potential problems of some modern cultivars is that brightly colored blush in the pericarp has been selected that appears before ripening. This in itself masks the chlorophylls thereby negating the degreening as an indication of ripeness (Sistrunk and Moore, 1983; Dong et al., 1995; Janick and Moore, 1996; Winkel-Shirley, 2001).

3.1.2   TASTE

The most important aspect of fruit quality is taste. The fruit may be the most desirable looking, but if it doesn't taste good the consumer will not buy it again. Consumer preference is for higher sweetness, more intense flavors, and firmer fruit that soften prior to consumption (Bruhn, 1995; Stockwin, 1996; Baldwin, 2002; Kader, 2002).

1.   **Sweetness** Major fruits have from 9 to 20° B when ripe. Brix is highly correlated with the amount of sugar contained in the juice. The levels of sucrose, fructose, and glucose are what determine sweetness; however, the level of acidity affects the perception of sweetness such that fruit with high sugar and moderate levels of acid will be perceived to be as sweet as fruit with moderate levels of sugar and low acid. The acid levels are primarily based on the concentration of malic or citric acid. New cultivar development in peach has concentrated on high sugar with low acid to fill a niche for the Asian market. The fact that sugar accumulation occurs prior to final

ripening makes it easier to harvest at a time with high sugar (Coombe, 1976; Byrne et al. 1991; Parker, 1993; Janick and Moore, 1996; Baldwin, 2002).

2. **Flavor** Flavor and aromatics are determined by a combination of volatiles. There are three main pathways for volatile production; cleavage of lipids followed by alcohol dehydrogenase activity to yield short-chain aldehydes and alcohols, the shikimic acid pathway, and the degradation of terpenoids. Interestingly the color pigments are also derived from these pathways, anthocyanins from the shikimic acid pathway and β-carotene and lycopene from the degradation of terpenoids. As fruit ripen there are hundreds of volatiles detected but only some above threshold levels that taste panels can detect. Of those, a few have been shown to determine the characteristic aroma/flavor of particular fruits: p-hydroxyphenylbutan (raspberry), cinnamate derivatives (strawberry), cyanidin-3-rutinoside (litchi), decadienoate esters (pear), γ-decalactone, and linalool (peach) (Horvat and Chapman, 1990; Janick and Moore, 1996; Kumar and Ellis, 2001; Lewinsohn et al., 2001; Baldwin, 2002).

3. **Texture** The texture of the fruit flesh is based on how cells shear in the chewing process, mouth feel. Texture ranges from crisp to melting and all the stages in-between. In melting texture, swelling and softening of the cell wall is evident, but in crisp texture, cell wall swell is not observed during ripening. Three enzymes, polygalacturonase (PG), β-subunit of PG, and pectin methylesterase (PME) have been associated with texture determination. Their substrate is the homogalacturonans or pectin located primarily in the middle lamella of cell walls (Redgwell et al., 1997; Brummel and Harpster, 2001; Redgwell and Fischer, 2002).

3.1.3 KEEPING QUALITY

Fruit can be harvested at various times in relationship to their peak quality and that time is dependent on the desired texture, the handling process, and the shelf life of each commodity. Some fruit (nonclimacteric such as blueberry and strawberry) are harvested eating ripe and then stored. Climacteric fruit such as peach or apricot are harvested at earlier stages in order for the fruit to withstand the handling. Such type of fruit will finish ripening during storage and transport. The rate at which the fruit ripens and softens determines when it must be harvested to withstand handling and arrive to the consumer either in the process of ripening or eating ripe. These aspects can be modified postharvest but there is also a large genetic component that can be taken into account in a breeding program.

1. **Softening** Softening is attributed to the disruption of the cellulose/xyloglucan network. Numerous enzymes have been postulated to be involved including β-galactosidase, expansin (EXP), pectatelyase (PEL), endo-(2-4) β-D-glucanase (EGase) and xyloglucanendotransglycolsylase (XET).

Xyloglucan does not seem to be a substrate for EGases-not even in avocado where there is an abundance of EGase enzyme as the fruit begins to soften (Brummell and Harpster, 2001; Jimenez-Bermudez et al., 2002; Smith et al., 2002). One of the complications of these studies is that not only are many of these genes members of small related gene families with potentially different substrates and times of activation, but many of the substrates are differently modified. Observations in one species may not extend to other fruit species. Polyuronides are depolymerized to very small size during ripening in avocado but not at all in strawberry or banana. Matrix glycans become highly depolymerized in strawberry but not in avocado. Galactose loss from cell walls has been noted in numerous species but not in plum (Brummell and Harpster, 2001). Some processes may be fundamental to firmness decrease. The highly variable softening behavior and textural changes of fruit from different species may be due in part to differences in the relative extent of the various ripening-related cell wall changes that occur, coupled with differences in initial cell wall composition.

2. **Control of ripening** The expression of quality traits normally is coordinately regulated and peaks at ripening. Breeders have selected for early expression of some of these traits such as skin blush, but the texture, softening, flavor development, reduction of acid and phenolic compounds, and color development peak at the ripe stage. Sugar accumulation takes place prior to the ripest stage (Coombe, 1976; Seymour and Manning, 2002). The problem with harvesting fruit at the peak of quality and ripeness is that the fruit at that stage has practically no shelf life. The fruit needs to be picked prior to peak ripe, at a stage that combines the maximum development of desirable traits and the maximum shelf life. In climacteric fruit, the increase in the amount of ethylene synthesized triggers final ripening. Nonclimacteric fruit do not increase ethylene with ripening. Many of the genes involved in those ripening traits are under the control of ethylene. It is unclear whether or not low levels of ethylene in nonclimacteric fruit are enough to induce those ripening-related genes or if there are other mechanisms to control ripening (Seymour and Manning, 2002). The discovery of a MADS box transcription factor as the gene responsible for the Rin phenotype in tomato has clarified this issue. A deletion of this gene results in an inhibited ripening and nonresponsiveness to ethylene (Vrebalov et al., 2002). A similar gene has been found in the nonclimacteric strawberry that acts early in ripening. The idea postulated is that this MADS gene initiates the ripening program in all fruits, including the induction of ethylene in climacteric fruits (Vrebalov et al., 2002).

3.1.5   BREEDING APPROACHES FOR FRUIT QUALITY

Breeding for fruit quality can require extended periods of time, particularly for tree fruits since fruit evaluation cannot be done until the tree is mature and fruiting and the progeny will be in the field for several years before the first evaluations can be done. Secondly, a balance must be achieved to produce beautiful fruit that has desirable taste and an adequate shelf life to get that fruit to the consumer still beautiful with desirable taste. This task requires the combination of multiple complex traits and precise evaluation.

Traditional breeding is the primary approach to selecting fruit quality. Selection is by the eye and taste of the breeder and then affirmed by taste panels. Quality traits are then quantitated (brix, firmness, acid, color, blush) and evaluations are repeated in other locations under different growing systems. In many cases, parents are chosen based on their expression of the desired traits being enhanced. The progeny can then be initially screened for pleasing appearances and taste and then for keeping qualities. As the number of potential selections decreases, taste panels can be used to confirm the breeders' selections. The quality traits can then be quantified and then trees can be sent to other locations to perform similar quantitative tests and taste panels. This method has served well but involves analyzing many progeny at the fruit stage. In perennial fruit crops this necessitates having many progeny planted and maintained in the field for a number of years. However, with the advent of molecular genetics, traditional breeders would like to incorporate tools that would make the selection for fruit quality traits simpler.

3.1.6   MANGO

There are two main types of mango the Indian types with monoembryonic seeds and susceptible to anthracnose while the Indo-Chinese types with polyembryonic seeds and are tolerant to anthracnose (Lespinasse and Frederic, 1998). Breeding methods involve: selection from open pollinated seedlings occurring naturally, controlled pollinations (hand pollination of limited flowers on large number of panicles), enclosing self-incompatible female and male parents and cross pollinating with houseflies, maintaining hybrid populations by grafting scions on established plants, and preselection of mango hybrids to discriminate undesired material. Most of the hybrids arise from selection among two varieties or primary hybrids and no recurrent selection is reported. Selection from natural mutants for important agronomical traits (such as precocity, yield, regular bearing and resistance to diseases) might be improved by sport selection.

Mukherjee et al. (1968) stated that breeding results has not been encouraging and hand crossing is remarkably unrewarding. The handicaps involve long life cycle and occurrence of polyembryony. Iyer and Subramanyam (1993) carried out breeding of mango cv. Alphonso to overcome the physiological disorder of spongy tissue formation. Alphonso was crossed with seedlings from natural cross-pollination. In Florida, following intensive introduction by the end of the nineteenthcentury, some important export varieties have resulted from seedlings derived from open pollinated (or not) identified mother plants. Today, most of the new Indian hybrids are regular bearing, with good quality fruits (free of spongy tissue) and attractive skin color. All the hybrids have higher pulp yield and possessed lower peel, stone, and fiber content. The adoption of the new varieties is still fairly low. In South Africa, the outstanding new variety "Heidi" was released in 1990 and is commercialized internationally. Mango breeding in Israel has resulted in the identification of 15 hybrids. In Australia, very promising progenies were obtained from crosses between the clone "Kensington" (good flavor) and "Sensation" (bringing favorable agronomic traits). In South Africa, genetic improvement was achieved by selection and four new cultivars were released in 1990's. As a result of the implementation of controlled pollinations, good evaluation and selection procedures in the field, new outstanding hybrids of mango were released. According to Israeli breeders, mango breeding is still in its infancy and considerable genetic and varietal progress should be expected from long-termintegrated programmes. Mango breeders will have to take into account the improved knowledge of inheritance of specific characters (recessivity for polyembryony, dwarfism, regular bearing and precocity). The recent findings on heritability suggest that the additive genetic variance was small and nonsignificant; whereas, the nonadditive variance was large and significant in most of the traits, which should also be taken into consideration.

Despite the many problems associated with mango breeding for cultivar development, many useful hybrids have been released. The earliest attempts were probably made in the West Indies to combine the good qualities of the Indian mango with the indigenous types by controlled pollination (Brooks, 1912). Burns and Prayag (1921) were the first to initiate hybridization during 1911 at Pune with a view to breed varieties having regular bearing habit, good fruit quality, high yield and resistance to pests and diseases. Subsequently, planned hybridization was started at Kodur (Naik, 1948) and Sabour (Sen et al., 1946) during the early forties and later at Saharanpur (Singh, 1954), Punjab (Singh, 1960), and Krishnanagar (Mukherjee et al., 1961). Intervarietal hybridization in India has resulted in the release of many cultivars (Table 3.1).

TABLE 3.1   Mango hybrids released from different stations of India

Variety	Parentage	Characteristics	References
Fruit Research Station, Sabour			
Mahmood Bahar	Bombay x Kalapady		Roy et al, (1956)
Prabha-shanker	Bombay x Kalapady		

TABLE 3.1 *(Continued)*

Variety	Parentage	Characteristics	References
Sundar Langra	Sardar Pasand x Langra	Having Langra quality and regular bearing habit	Hoda and Ramkumar, 1993
Alfazli	Alphanso x Fazli	Fruit is medium to large in size (228.70 to 460 g) with 79% pulp, moderate TSS content (16.20 to 18.5 0 Brix) 0.24 % acidity. Keeping quality is good. Superior to "Fazli" in quality and early ripening, free from spongy tissue	
Sabri	Gulabkhas x Bombai	Having Bombai fruit shape and color of Gulabkhas with regular bearing habit. The pulp is deep orange, very sweet with a pleasant flavor, less fibrous, pulp recovery being 74 %, good keeping quality.	
Jawahar	Gulabkhas x Mahmood Bahar	Fruits are oblong, elliptical in shape, sweet to taste with a pleasant flavor. Pulp is less fibrous, pulp recovery being 79.5 %.	
Fruit Research Station, Kodur			
Swarna-jehangir	Jehangir x Chinnaswarnarekha	Have high quality and attractive color	
Neeluddin	Neelum x Himayuddin		
Neelgoa	Neelum x Yerra Mulgoa		
Neeleshan	Neelum x Baneshan		
IARI, New Delhi			
Mallika	Neelum x Dashehari	Fruit is apricot yellow, medium sized with high pulp (74 %), flesh is free from fiber and melting, High TSS (24°B) and good keeping quality.	Singh et al, (1972)
Amrapalli	Dashehari x Neelum	Fruit are green apricot yellow, medium sized, sweet in taste, TSS (22–24°B) with high pulp content (75 %), flesh is fiberless and deep orange in color, excellent quality and is also very rich in Vitamin A	

TABLE 3.1   *(Continued)*

Variety	Parentage	Characteristics	References
Pusa Arunima	Amrapalli x Sensation	Fruits size is medium having attractive red peel color. TSS (19.56 %) and is rich in vitamin C (43.6 mg /100 g pup) and β-carotene content, has a good flavor, acceptable for its attractive red peel color, fiberless pulp, mild flavor, excellent sugar: acid blend and long shelf life	
Pusa Surya	a selection from Eldon	Fruit is medium to large in size, with attractive apricot yellow peel color and TSS (18.5%), rich in vitamin C (42.6 mg/100 g pulp) & β-carotene content, good shelf life (8–10 days) at room temperature after ripening.	
Pusa Prathibha	Dashehari x Amrapali	Fruit size is medium with higher pulp content (71.1 %). TSS (19.6 %) and rich in Vitamin C (34.9 mg/100 g pulp) and β-carotene content (11,474 μg/100 g pulp). It has pleasant flavor with improved shelf life (7 to 8 days) at room temperature acceptable because of its attractive red peel color, fiberless pulp, mild flavor, excellent sugar:acid blend and good shelf life.	
Pusa Shreshth	Amrapali x Sensation	Fruit size medium with attractive red peel color and higher pulp content (71.9 %). TSS is 20.3 %, rich in vitamin C (40.3 mg/100 g pulp) and β-carotene content (10,964 μg/100 g pulp), excellent sugar:acid blend, has pleasant gustatory aroma with enhanced shelf life (7–8 days) at room temperature. Attractive red peel color, fiberless pulp, mild flavor.	
Pusa Lalima	Dashehari x Sensation	Fruit is about 209g with attractive red peel color and higher pulp content (70.1 %). TSS (19.7 %), vitamin C (34.7 mg/100 g pulp) and β-carotene content (13,028 μg/100 g pulp) and it has approving flavor with good shelf life (5 to 6 days) at room temperature after ripening making this an exportable mango	

TABLE 3.1   *(Continued)*

Variety	Parentage	Characteristics	References
Pusa Peetamber	Amrapali x Lal Sundari	Fruits weigh about 213 g with striking yellow peel and higher juicy pulp content (73.6 %). TSS (18.8 %), rich in vitamin C (39.8 mg/100 g pulp) and β-carotene content (11,737 µg/100 g pulp). It has appealing flavor with good shelf life (5 to 6 days) at room temperature. It is suitable for domestic as well as international market, because of its attractive yellow peel color, fiberless pulp, mild flavor.	
CISH, Lucknow			
Ambika	Amrapali x Janardhan Pasand	Fruits are medium sized, oblong in shape, skin is smooth, tough, and bright yellow with dark red blush. TSS 21°B.	Negi et al. (1996)
Arunika	Amrapali x Vanraj	Fruits of this variety are attractive and with red-blush. Fruit have high TSS (24°Brix) and high carotenoids content. Pulp is firm.	
IIHR, Bangalore			
Arka Aruna	Banganapalli x Alphonso	Fruit size is large, skin color attractive, pulp cream color, fiberless with good flavor, TSS (22°B), pulp recovery (78 %), free from spongy tissue	Iyer and Subramanyam, 1993
Arka Puneet	Alphonso x Banganapalli	Fruit are medium sized with attractive skin color, good aroma, fiberless, and high TSS (22°B) and pulp (74 %). Good keeping quality, free from spongy tissue, suitable for table as well as processing	
Arka Anmol	Alphonso x Janardhan Pasand	Fruit are medium sized with good aroma, golden yellow color, flesh is orange, firm, fiberless, and TSS (20°B) excellent sugar:acid blend and keeping quality, free from spongy tissue.	
Arka Neelkiran	Alphonso x Neelum	Fruit are medium sized, pulp is deep yellow in color, TSS (20°B)	

TABLE 3.1   *(Continued)*

Variety	Parentage	Characteristics	References
FRS, Vengurla, Maharashtra			
Ratna	Alphonso x Neelum	It is regular bearing, produces medium sized fruits weighing on an average about 250 g. Pulp is orange in color and free from spongy tissue and fiber.	Salvi and Gunjate, 1988
Sindhu	Ratna x Alphonso	Fruits weigh on an average about 150-220g. Pulp is deep yellow in color and it has good sugar acid blend. Fruits are almost seedless with very thin stone, though fruits above 200 g have well developed seeds.	Gunjate and Burondkar, (1993)
Fruit Research Station, Sangareddy, Andhra Pradesh			
Au-Rumani	Rumani x Mulgoa	Flesh is moderately firm, melting, fiberless, flavor excellent taste very sweet, juicy, and stone is small	Swamy et al. (1972)
Manjeera	Rumani x Neelum	Fruit are round in shape with light yellow skin, flesh in firm, yellow, fiberless, and sweet in taste	
Paria Research Station, Gujarat			
Neelphonso	Neelum x Alphonso	Fruit medium size, peel apricot on ripening, smooth thick, adhering to pulp, pulp firm fiberless, TSS (19.06 %) highly suited for table and juice purpose	Sachan *et al.*, 1988
Neeleshan	Neelum x Baneshan	Fruit medium, oval, peel yellow cadmium, thin, smooth, adhering to pulp, pulp firm fiberless, fruit quality very good (TSS 16.26 %) highly suited for HDP	
Neeleshwari	Neelum x Dashehari	Fruit medium, oval, long, sinus prominent, peel apricot, thin, smooth, adhering to pulp, pulp moderately firm fiberless, fruit quality very good (TSS 20.53 %) highly suited for HDP	
Sonpari			

TABLE 3.1 *(Continued)*

Variety	Parentage	Characteristics	References
Horticultural Research Station, Periyakulam			
PKM-1	Chinnas-warnarekha x Neelum	Regular bearing, produces good quality fruits and bears in clusters, fruits are sweet to taste	Shanmugavelu et al. (1987)
PKM-2	Neelum x Alphonso	Regular bearing and produces good quality fruits in clusters	
Govind Ballabh Pant University of Agriculture & Technology, Pantnagar			
Pant Chandra	Clonal selection of Dashehari	Fruit weight is up to 150g. Fruit pulp is reddish yellow with total soluble solids of 18% and having pleasant aroma	
Pant Sinduri	clonal selection of Dashehari	Fruit color is yellow with pink shoulder. Average fruit weight is up to 200g. Fruit pulp is yellow with pleasant aroma. Total soluble solids vary from 16-18% with average yield up to 150 kg per tree	

Mango hybridization was reported from Hawaii in the 1920s, but no outstanding problem appears to have been addressed or solved (Pope, 1929). A number of crosses have been reported in Florida (Young and Ledin, 1954; Sturrock, 1969), but all of the Florida cultivars are chance seedlings and none came from controlled pollinations.

There is an extensive breeding programme in Israel aimed at producing higher yielding cultivars with good quality, attractive fruit and with longer harvest periods. Several hundred seedlings from open and controlled pollinations have been evaluated, and 14 of them have been identified as being of interest (Lavi et al., 1993). The rootstock breeding programme is aimed at developing rootstocks resistant to or tolerant of soil stresses, i.e. calcareous soils, saline irrigation water and heavy nonaerated soils that predominate in the mango-growing regions of Israel. Several interesting mono embryonic and polyembryonic rootstocks have been selected (Lavi et al., 1993), but none has performed better than "13-1", the currently preferred rootstock in Israel (Gazit and Kadman, 1980).

A breeding programme to develop a new cultivar which retains the characteristic flavor of "Kensington", but with improved productivity, greater disease resistance, enhanced skin color and better postharvest performance, was initiated in Queensland, Australia. These features are found in many Florida cultivars (i.e. "Irwin", "Sensation" and "Tommy Atkins") which are being used as maternal parents in crosses with "Kensington" (Whiley et al., 1993). Promising hybrids have been identified in crosses involving "Sensation", for example "Calypso"™. "Calypso"™ has increased shelf life, firmer fruit, extra blush for cosmetic appeal, a higher flesh-to-seed ratio and consistent yields of high-quality fruit. The Australian mango breeding programme was strengthened since 1994 by launching a major effort involving various organizations located in different agro-climatic zones in hybrid production, as well as regional testing.

Breeding has been initiated in the tropical savannah of Brazil to develop cultivars that are dwarf and with good quality fruit. Hybridizations have involved local, Indian, and Florida cultivars. "Amrapali" and "Imperial" were good male parents to confer dwarfing in the progeny (Pinto and Byrne, 1993). Out of 2,088 seedlings in the field, 209 seedlings were selected in the first year and 42 of these were later identified as promising, from which four have been released as new cultivars (Pinto et al., 2004). These four are: "Alfa" ("Mallika"× "Van Dyke"), which is semidwarf, high yielding and regular bearing; "Beta" ("Amrapali"× "Winter"), high yielding and moderately resistant to anthracnose and Oidium; "Roxa" ("Amrapali"× "Tommy Atkins"), with excellent fruit quality; and "Lita" ("Amrapali"× "Tommy Atkins"), high yielding with excellent fruit quality.

The South African breeding programme at the Citrus and Subtropical Fruit Research Institute (CSFRI) is based on introductions, openpollination and mass selection. Four new cultivars have been released: "Heidi", "Neldawn", "Neldica" and "Ceriese". In addition, 12 promising selections have been identified for further evaluation (Marais, 1992).

3.2 MUTATIONS

3.2.1 SOMATIC MUTATIONS

Asexual propagation enables the preservation of accumulated mutations (macro and micro) which would normally be eliminated during sexual propagation. In Many fruit crops, bud mutations and chimeras occur rather frequently and can provide an additional source of variability for selection. However, such reported instances are relatively few in mango. Roy and Visweswariya (1951) observed mutants of "Puthi" in which the number of palisade cell layers differed from the original cultivar. Naik (1948) observed significant variation among trees of the same clone with respect to fruit shape, size, color, and quality, which was ascribed to bud mutations. "Davis Haden" a sport of "Haden" is larger than "Haden" and its season of maturity is about a month earlier. Rosica from Peru is a bud mutant of Rosado de Ica. Unlike its parent, Rosica is high yielding and regular bearing and does not produce seedless fruits.

Oppenheimer (1956) after a survey of many orchards in India, reported wide variability in the performance of trees of the same clone within a single orchard. Mukherjee et al, 1983 conducted a survey of mangoes in eastern India and identified some superior clones. Singh and Chadha (1981) in a study of orchards of "Dashehari" located four clones which were superior in performance. Singh et al., (1985) isolated two high yielding clones from orchards of "Langra". Within "Kensington" strains have also been identified that show improved resistance to bacterial black spot (Whiley et al., 1993).

Roy (1950) observed a mutant of "Alphonso" with respect to fruit shape and suspected it to be a mericlinal chimera. Pandey (1998) has described seven clones "Alphonso": "Alphonso Behat" and "Alphonso Bihar" from Bihar "Alphonso Batli", "Alphonso Black" and "Alphonso Bombay" from Maharashtra "Alphonso Punjab" from Punjab and "Alphonso White" from the North Canara district of Karnataka. Rajput et al., (1996) assembled several "Dashehari" variants and after 14 years of observation, reported that the clone "Dashehari 51" was superior with respect to yield and regular bearing. Other somatic mutants include: "Cardozo Mankurad" with large fruits of attractive color and high yields from Mankurad of Goa, dwarf selections from the Rumani and Bangalora (Ramaswamy, 1989), development of Paiyur, a dwarf selection from Neelum (Vijaya Kumar et al., 1991), Rati Banganapalli snf Nuzuvid from Banganapalli (Anonymous, 1999), and MA-1, regular bearing and high yielding with resistance to spongy tissue from "Alphonso" (Mukunda, 2003).

In Thailand, Chaikiattiyos et al., (2000) selected clone SKoo7 from 320 Kaew plants; SKoo7 has higher yield and superior quality. Jintanawongse et al., (1999) also made superior selections for yield and fruit quality from "Nam Dok Mai", "Khiew Sawoey", "Rad" and "Nang Klang Wau" and DNA fingerprints of all these clones were made for comparison with the parental clone.

3.2.2 INDUCED MUTATIONS

Mutation induction using ionizing radiation was attempted by Siddiqui et al. (1966). Siddiqui (1985) irradiated dormant buds of "Langra"with high doses of γ rays and grafted them onto 1 year old seedlings. A bud graft exposed to 3.0 kR bore fruits which were heavier, larger, and had a more cream yellow pulp than the control. This variability was stable over three seasons. Sharma and Majumder (1988b) irradiated bud sticks, top worked them onto 10 year old seedlings and found that dosages above 5 kR are lethal for mango and that the lethal dose required for 50 percent mortality (LD_{50}) lies between 2 and 4 kR. Effective dosages of the chemical mutagens, ethane methyl sulfonate (EMS) and N-nitroso methyl urea (NMU, were 1.5 and 0.05 per cent, respectively. The spectrum of mutations induced by physical and chemical mutagens was observed to be more or less the same, indicating the high sensitivity of certain loci. The mutants included dwarfness, changes in shape and serration of leaves and in TSS content in "Dashehari". As in other perennial crops, mutagenesis techniques that can allow useful traits to be targeted as well as isolating mutated sectors from a chimera are essential.

3.2.3 BANANA

The conventional breeding approaches in banana include selection of improved genotypes having desirable characters like better quality, higher quality, dwarfness in stature and disease and incorporating one or more desirable characters into a particular clone through crossing. Parthenocarpy and seedlessness are often inseparable barriers in using cross breeding as a tool for crop improvement. Due to serious problems like sterility, inter specific hybridity polyploidy and heterozygosity in most of the clones, it was earlier thought that the crop is intractable in terms of conventional breeding approaches. Till 1990's no man bred cultivar of banana has been grown successfully, however, Morpurgo et al., (1997) considered these difficulties as an enigma to banana breeding. Now several improved hybrids and resistant genotypes have identified (Table 3.2).

TABLE 3.2   Banana and Plantain hybrids selected for disease resistance and yield attributes

Hybrid (Genome)	Parentage	Attributes
KAU, Trichur		
H 1 (AAB)	Agniswar (AB) x Pisang Lilin (AA)	Dessert banana, resistance to leaf spot, *Fusarium wilt*, medium tall plants, bunch weight 14-16 kg, fruit slightly acidic
H 2 (AAB)	Vennan (AAB) x Pisang Lilin (AA)	Dessert banana, tolerant to leaf spot and nematode, bunch medium (15–20 kg), suitable for subsistence cultivation
FHIA, Honduras		
Goldfinger (AAAB) FHIA-1	Patra Ana (AAB) x SH-3142 (AA)	High quality dessert banana, resistant to Sigatoka and *Fusarium wilt*, very good pulp quality
FHIA-3 (AABB)	SH-3486 (ABB) x SH-3320 (AA)	Cooking banana, skin to Cardaba, drought tolerant, plant robust, bunch weight 20–25 kg, Sigatoka resistant
FHIA-6 (AAAB)	Maqueno (AAB) x SH-3486 (ABB)	Plantain resistant to black Sigatoka, higher yielder, better quality
FHIA-17 (AAAA)	Highgate (AAA) x SH-3486 (ABB)	A better quality Cavendish hybrid, resistant to black Sigatoka
FHIA-21 (AAAB)	AVP 67 (AAB) x Pisang Jari Buaya (AA)	Improved French plantain, produces larger bunches, fruit shape, color, and flavor better and excellent when fried green or baked ripe

TABLE 3.1 *(Continued)*

Hybrid (Genome)	Parentage	Attributes
IITA, Nigeria		
BITA-1 (AABB)	Bluggoe (ABB) x Calcutta-4 (AA)	Cooking banana, resistant to black leaf streak, black Sigatoka and *Fusarium wilt*
BITA-2 (ABBB)	Pisang Awak (ABB) x *Musa balbisiana*	Cooking banana, resistant to black leaf streak, black Sigatoka and *Fusarium wilt*
PITA-9 (AAAB)	Agbagha FR (AAAB) x Calcutta-4 (AA)	False Horn plantain hybrid with good horticultural traits, partially resistant to black Sigatoka, sweet, and aromatic yellow fruit pulp, good postharvest fruit quality
EMBRAPA-CNPMF, Brazil		
Caipra (AAAB)	Patra Ana (AAB) x improved diploid (AA)	Dessert banana, better quality than Patra Ana, resistant partially to Sigatoka disease
PV 03-44 (AAAB)	Poovan (AAB) x Calcutta-4 (AA)	Dessert banana, better in qualiy, resistant to Sigatoka

In spite of severe sterility problems, some commercial AAA triploids were able to set seeds when crossed with fertile diploids. The tetraploids (female triploids crossed with male diploids) produced were utilized for producing triploids through crossing 4x X 2x (Wilson, 1946; Simmonds, 1960; Rowe, 1984). Production of the superior diploids has often been the most important activity in banana breeding as these diploids are the main source of genetic variability in synthesizing new hybrids. Usually tetraploids are bred by crossing a triploid female (say Highgate AAA) with the improved diploid male (AA or AB). In French breeding programme, diploids were selected for longer green life, better fruit quality and disease resistance and were converted to tetraploids by chromosome doubling. A number of good AAAA clones exist but they are yet virtually unexploited. Although some outstanding tetraploids have been released for commercial adoption (eg. Goldfinger).

Spontaneous somatic mutants have played an essential role in the speciation and domestication of plantain and banana. All bananas and plantains that we grow and eat were selected in prehistory from spontaneous mutations. Since mutations provide a valuable source of creating variation in plant material, attempts have been made to induce it artificially by treating the corms, bits, corm-buds, suckers etc. with chemical mutagens or exposing them to ionizing radiation such as gamma rays (Satyanarayana et al., 1980; Tulmann Neto, 1989; Radha Devi and Nayar, 1992; Rowe and Rosales, 1996). Several natural mutants (sports) of well-established commercial clones have been recognized. For example, Highgate (AAA) and Cocos (AAA) are semidwarf mutants of Gros Michel (AAA); Motta Poovan (AAB) is a sport of Poovan (AAB); Ayiranka Rasthali is a sport of Rasthali; Barhari Malbhog is a sport of Malbhog; Krishna Vazhai is a natural mutant of Virupakshi and Sombrani Monthan (ABB) is a mutant of Monthan (ABB). Natural mutation thus appears to be an evolutionary course which is responsible for variations in clones of similar genetic makeup (Daniells, 1990; Gross and Simmonds, 1954; Simmonds, 1966).

3.2.4 APPLE

The overall objectives of modern breeding programs are to increase the marketability of fruit and reduce production costs. Apples are sold fresh, juiced, and processed in numerous ways, but the largest overall market involves fresh fruit. Dessert apples are sold primarily based on appearance (size, color, shape, and freedom from blemishes) and quality (taste and texture) (Janick et al., 1996; Laurens, 1999; Brown and Maloney, 2003). There is considerable regional variation in taste preferences from a desire for tartness in Europe and the U.S. Midwest, to a preference for sweetness and low acidity in Asia. Storage life is also a critical parameter, as most apples are stored for long periods of time. Resistance to apple scab and powdery mildew are common breeding goals. Niche markets are also arising for improved nutritional aspects such as higher antioxidants. The attributes needed for processed fruit depends on their final market. Some of the most important markets are for cider, sauce, and slice (Crassweller and Green, 2003). Less browning is a particularly critical parameter in the fresh cut and slices market.

Most of the characters related to fruit quality are under polygenic control, which means that when two cultivars are intercrossed, there will be a wide and continuous range of expression of all the characters among the seedlings. Precise estimates of genetic parameters for quantitative traits controlled by the polygene are not necessary for apple breeders, who over the years, have relied on information and intuition obtained from experience from successful breeding programs. However, as in the case of scab resistant cultivars, it is difficult and requires a long time to improve quantitative traits, especially fruit quality which is constituted of many traits and must be define what good quality is, and which characters are essential for good quality. An answer to these questions can be obtained from fruit that are attractive to consumers. Preference maps constructed from consumers hedonic responses and the sensory attributes of apple indicate that consumers generally prefer apples that are juicy, crisp, and sweet. Crispness is a textural attribute of fruit. Because the acid content decreases during storage, making the taste of the fruit insipid, acidity is often associated with fresh fruit. Acidity is also a desirable characteristic of fresh apples. Characters that receive a high value of liking are those that will be target for breeders.

There are few traits in which segregation in progeny is observed in relation to fruit quality. In polygenic traits, narrow sense heritability is usually estimated. The heritability expresses the degree of correspondence between phenotypic and breeding values that determines their influence on the next generation. When breeders choose individuals as parents according to their phenotypic values, the success of breeders in changing the characters of the population can be predicted only from the heritability (Falconer and MacKay, 1996).

Texture, juiciness and flavor are major components determining eating quality. The quality of fresh is closely related to texture and consumers tend to use texture as the primary limiting factor for acceptability. Excessive softening not only makes fruit less attractive to consumers but also increases shipping and storage costs because of higher pathogen susceptibility. Flavour also plays an important role in the selection and overall liking of apple cultivars. Improvement of flavor, however, has been neglected in breeding programs because the component of flavor is complicated and the direction of improvement is uncertain. The appreciation of flavor is personal and qualities considered good by one may be considered otherwise by others. Flavour can be divided into two categories: taste and aroma. Taste is primarily related to the balance between sugars and acids. Aroma depends on the components of volatiles. Wild species of apple usually have an unpleasant flavor.

Acidity is a character related to eating quality and the only one that has been found to segregate within seedling from early genetic studies. The acid in the mature fruit is almost entirely malic acid. The distribution of seedling in the malic acid concentration shows that there are two types of distribution among progeny. One is simple distribution with peak, and the other has two peaks. One peak below 0.3 % (or pH > 3.8), and the other peak appears in a higher rang of acid (Nybom, 1959; Visser et al., 1968; Brown & Harvey, 1971). Because low acid behaves as a recessive character, medium to high acid is hypothesized to be dominant in the low-acid type, and acidity is likely controlled by a major gene (*Ma*).

To produce new cultivars quickly and efficiently, breeders tend to use only a few high-quality parents. Noiton and Alspach (1996) analyzed the pedigrees of 50 apple cultivars produced from modern breeding programs around the world. They revealed that "Cox's Orange Pippin", "Golden Delicious", "Red Delicious", "Jonathan"and "McIntosh"were the most frequent progenitors and there were high levels of coancestry among many modern cultivars.

Until the mid twentieth century, most apple cultivars were selected from seedlings (Janick et al. 1996). Apple diversity is very high due to polymorphism (Pereira-Lorenzo et al. 2003, 2007), but commercial types depend on a reduced number of cultivars. Noiton and Alspach (1996) determined that 64 percent of 439 selections had their origin in among five cultivars: "McIntosh"(101 cultivars), "Golden Delicious"(87 cultivars), "Jonathan"(74 cultivars), "Cox's Orange Pippin"(59 cultivars), and "Red Delicious"(56 cultivars). Among them, 96 cultivars had two or more as parents. Other cultivars used frequently in crosses were "James Grieve", "Rome

Beauty", and "Wealthy." Estimations have shown that in the last 5 years, 43 percent of the registered cultivars in France were mutations from commercial cultivars in use at the time and six of them cannot be differentiated clearly from the originals (Le Lezec et al. 1996).

The first cultivar obtained by crossing was attributed to Thomas Andrew Knight (1759–1838). Another method to obtain new cultivars consisted in the selection of mutations and chimeras (Janick et al. 1996) these develop shoots with a stable variation when they are propagated vegetatively. The crossing of two parents is now, as it always has been, the main method in apple breeding (combination breeding).

Currently, the main characteristics of cultivated apples are (1) size over 100 g or 70 mm as a minimum for the market; (2) colors: yellow, green, red, bicolor, and brown in susceptible apples to russeting; (3) acidity: sweet apples when malic acid is lower than 4.5 g/L and bitter when it is over that limit; (4) tannins: sharp apples are those with more than 2 g/L of tannic acid; (5) sweetness: most of the cultivars contain between 12.8and 18.8°B (6) harvesting period from August to December; and (7) resistance to diseases and abiotic stress.

The eating quality is difficult to measure objectively (Hampson et al. 2000). Contribution of crispness accounts for about 90 percent of the variation in texture liking. Juiciness, aroma, sweetness, and sourness change their relative importance from year to year. They account for about 60 percent of variation in flavor liking. Sweetness and sourness are better predictors of liking than analytical measurements of soluble solids and titratable acidity. Formal sensory evaluation is a reliable way for screening breeding selections (Hampson et al. 2000). Some researchers have found poor correlation between soluble solids (% SS), titratable acidity (TA), and firmness with sensory perceptions of sweetness, sourness, and texture (Bourne, 1979a, b, c; Watada et al. 1981).

The main cultivars used for cider are differentiated on the basis of their acidity and tannin levels. Four groups of apples can be classified considering acidity and tannin contents (Downing, 1989; Lea, 1990): bittersweet apples contain more than 0.2% (w/v) tannins and less than 0.45% (w/v) acidity (calculated as malic acid). Sharp apples have less than 0.2% (w/v) tannins and more than 0.45% (w/v) acidity. A subgroup of this classification, bittersharps, has the same range of acidity but tannin content over 0.2% (w/v). Sweet apples have less than 0.2% (w/v) tannins and 0.45% (w/v) acid.

The ideal cider apple is slightly riper than the fresh market one (Downing, 1989). As apples mature, the starch turns into sugar, increasing sweetness and flavor. Unripe apples produce juice with a "starchy" or "green apple" flavor. Acidity and astringency also decrease after harvest, both with a pronounced effect on the flavor of the juice. If we compare commercial cultivars' characteristics (Iglesias et al. 2000) with some of the most frequently used cider apple cultivars in Spain (Table 3.3), we can say that the acidity of various groups, such as "Elstar"and "Reinetas,"is equivalent to some cider cultivars, such as "Raxao." Cultivars producing high levels of tannins are rarer, such as "Teorica" or "Collaos."

With a view to combine high dessert quality with good keeping quality, breeding work was initiated in Kashmir in 1956 (Chadha, 1962). Two hybrids Lal Ambri (Red Delicious x Ambri) and Sunhari (Ambri x Golden Delicious) were released (Chadha, 1987). Breeding work on similar lines was started in Himachal Pradesh in 1960 (Chadha, 1962). Combining ability of different parents was studied (Chadha and Sharma, 1975). As a result three promising hybrids viz. Ambred (Red Delicious x Ambri), Ambrich (Rich-a-Red x Ambri) and Ambstarking (Starking Delicious x Ambri) were released for mass cultivation (Chadha and Sharma, 1978).

TABLE 3.3 Main characteristics from main commercial cultivars and local ones

Cultivar	Blooming	Harvest	Colour	Caliber (mm)	Brix	Tanins	Acidity	Firmness (kg)	Origin
Commercial cultivars									
Gala Group	5 April	10–25 Aug	Bicolour	73-82	12-14		2.6-4.9	7-9	New Zealand
Elstar Group	10 April	10–30 Aug	Bicolour	74-78	13-15		7.7-9.9	6-7	Golden Delicious x Ingrid Marie
Delicious Group	5 April	1–15 Sept	Red	75-91	11-15		2.5-3.4	7-8	Seedling, USA
Golden Group	20 April	15–25 Sept	Yellow-Green	69-90	13-17		3.6-9.0	6-10	Golden Reineta x Grime Golden, USA
Reinetas Group	20 April	5–15 Sept	Yellow-Brown	74-85	12-18		11-13	8-10	Ancient cultivars, origin unknown
Jonagold Group	10 April	1–15 Sept	Bicolour	80-92	14-17		5-6	5-8	Delicious x Jonathan
Braeburn Group	1 April	25 Oct–5 Nov	Bicolour	74-84	12-14		5-7	8-9	Seedling, New Zealand
Granny Smith Group	14 April	1–25 Nov	Green	78-90	12-13		9	8	Seedling from French Crab, Australia

TABLE 3.3   *(Continued)*

Fuji Group	10 April	5–25 Nov	Bicolour	70-84	13-17		3.1-4.2	7-9	Seedling from Rallis Janet x Red Delicious, Japan
Local cultivars									
Blanquina	9 May	23 Oct	Yellow-Green	50-70	12	0.9	4.4	11	Spain
Collaos			Red			4.9	2.1		Spain
Cristalina	13 May	23 Oct	Red	56-65	12	0.9	5.7	11-12	Spain
De la Riega			Bicolour			1.4	5.2		
Marialena	5-11 May	22 Sept	Bicolour	53-59	13	1.4	4.8	8-9	Spain
Raxao	20 May	23 Oct	Bicolour	56-71	12	1.5	7.7	10-12	Spain
Teorica	12 May	22 Oct	Red	47-58	14	4.0	14.5	11-13	Spain
Mingan		25 Oct	Bicolour	60-66	12-15	2	3	10-13	Spain
Tres en Cunca	4 April	20 Sept	Yellow	53-72	13	1.2	7.2	10	Spain

Data adapted from Iglesias et al. (2000)

Subsequently hybrid Ambroyal (Royal Delicious x Ambri) was added to the list of promising selections (Sharma et al., 1986). At Horticulture Research Station, Chaubattia, breeding work was started in 1970 and two promising hybrids, Chaubattia Princess and Chaubatia Anupam (Red Delicious x Early Shanburry) was evolved (Seth and Ghildyal, 1984) (Table 3.4).

On the similar objectives hybridization programs were undertaken at SKUAST-Kashmir, Shalimar (Jammu & Kashmir) wherein some known commercial cultivars of apple *viz.* Red Delicious, Golden Delicious, American Apirouge, Cox's Orange Pippin, Maharaji (White Dotted Red) were hybridized with indigenous Ambri which was used as a common parent, which transmit a good environmental adaptation (Farooqui et al. 2005). Rigorous screening and selection among progenies have resulted in identification of some productive combinations of promising apple hybrids. Out of them Akbar and Shalimar Apple 2 have been released for cultivation so far. Of them, Lal Ambri is an ideal apple for export promotion (Bhat et al. 2005) and is providing consumers with new perceptions of quality and flavor.

Another breeding programme was initiated for disease resistance particularly against scab. It is the first disease targeted in the breeding programmes. A number of scab resistant cultivars have been introduced since the incidence of apple scab in epidemic form in 1972-1973 in Jammu and Kashmir and 1978–1979 in Himachal Pradesh. Florina, Macfree, Nova Easy Grow, Co-op-12, Co-op-13 (Red free), Nova Mac, Liberty, Freedom. Florina, an introduction from France, has shown promising performance in Himachal Pradesh and may become a good substitute for Delicious apples in scab prone areas. Keeping in view the problem of apple scab breeding programme was initiated during 1978–1980, wherein crosses of apple cultivars were made with scab resistant cultivars viz. Prima, Pricilla, Florina, carrying *Vf* resistance gene from crab apple *Malus floribunda* 821. The progenies derived were screened for scab resistance at seedling stage and after back crossing have come into bearing. These plants are being evaluated to identify genotypes which besides having resistance to scab have desirable fruit quality characters and their response to other major diseases and pests. Till date three scab resistant cultivars namely Firdous, Shireen, and Shalimar Apple-1 have been released (Farooqui et al., 2006).

Some other promising scab resistant selections viz. ALP-16, ALP-19, ALP-42, ASP-1, ASP-69 have been identified and are under evaluation.

TABLE 3.4 Apple hybrids developed at different stations

Cultivars	Parentage	Characters
Jammu & Kashmir		
Lal Ambri	Red Delicious x Ambri (1973)	High quality apple with attractive fruit size, shape with fine protrusions as in Red delicious, color stripped red covering more than 75 per cent of fruit with yellow ground color. Flesh crisp, sweet, juicy, Excellent keeping quality. Average yield 20–30 metric tons per hectare.
Sunhari	Ambri x Golden Delicious (1973)	Medium to large size apple, Fruit color yellow with golden blush. Flesh creamy white, juicy, sweet, and crisp. Excellent keeping quality. Av. Yield 15–17 metric tons per hectare.
Akbar	Ambri x Cox's Orange Pippin (2001)	Medium size apple, red stripped over yellowish white background, flesh sweet sub acidic, juicy. Average yield upto 30–35 metric tons. Prone to biennial bearing.
Firdous	Golden Delicious x Prima x Rome Beauty (1995)	Medium to large size apple. Dark red-blush colored apple with yellowish green background. Flesh sweet, juicy, crisp. Average yield 12–15 metric tons per hectare.
Shireen	Lord Lambourne x Melba x *R-12740-7A* (1995)	Medium size apple, dark red colored over yellowish green background. Lenticels prominent. Flesh sweet, richly juicy with pleasant flavor. Average yield 15–17 metric tons per hectare.

TABLE 3.4   *(Continued)*

Shalimar Apple-1	Sunhari x Prima (2009)	High quality apple, reddish orange blushed with sweet taste. Medium sized. Average yield 23–25 metric tons per hectare.
Shalimar Apple-2	Red Delicious x Ambri (2009)	Fruit medium to large in size. yield 25–28 metric tons per hectare. Ambri aroma but more juicy than Ambri. Flesh white, juicy, and sweet.
Himachal Pradesh		
Ambred	Red Delicious x Ambri	Fruits are medium in size, skin is medium in thickness, flesh is whitish, crisp, firm, aromatic, and juicy, keeping quality is good upto 3 months in air cooled storage
Ambrich	Rich-a-Red x Ambri	Fruits are medium in size, skin is thick, flesh is white, firm, crisp and subacid, aromatic and juicy with good dessert quality
Ambstarking	Starking Delicious x Ambri	Fruits are medium in size, skin is smooth, flesh is whitish firm, crisp, and juicy, keeping quality is good
Ambroyal	Royal Delicious x Ambri	Fruit is medium in size, skin is thin, flesh is white, soft, sweet, juicy with dessert quality, storage quality is comparable with Royal Delicious
Uttarakhand		
Chaubattia Princess	Red Delicious x Early Shanburry	Fruit is medium in size, skin is thin, flesh is creamywhite, crisp in texture, firm, juicy, and very sweet. TSS (14 %) and acidity (0.22 %). Fruit pressure at maturity is 14-15lbs/sq inch.
Chaubatia Anupam	Red Delicious x Early Shanburry	Fruit is medium to large in size, skin is thin, flesh is light yellow to creamywhite, fairly firm, crisp, juicy and aromatic and very sweet. TSS (14.5 %) and acidity (0.2 %). Fruit pressure at maturity is 15 to 16 lbs/sq inch.

3.2.5   *PRUNUS*

Prunus fruits are commonly called "stone fruits"because of hard pit that protects the embryo, which is characteristic of all the species of this genus. In most species, a juicy mesocarp is the edible part of the fruit, whereas for the almond the seeds are the edible portion. As results, the parameters and criteria that determine the fruit quality of the species that bear fleshy fruits are quite different from those relevant to the almond.

In these species, a major cultivar renewal process is currently taking place to satisfy consumer and industry demands. Fruit quality is fundamental to consumers accepting new cultivars, especially because of high competition in the markets with the presence of numerous recent cultivars, alternative fruits and other appealing fruits (Infante et. al., 2008b). Consumers value the appearance aroma and flavor of high-quality stone fruits, whereas other parameters such as size, resistance to manipulation and conservation aptitude are taken into greater account by the industry. The new *Prunus* cultivars have principally been generated through controlled crosses and open pollination.

If we consider the whole production chain for stone fruits, there are three main user categories: the grower, for whom quality would correspond to features regarding the plant's productivity and the farm performance; the dealer for whom quality most corresponds to the case of product handling and manipulation and nowadays most important the end consumers that buy and eat the product.

The fruit market has a tendency to split into two classes of produce: high quality and commodity (i.e. low price) (Ruiz-Altisent et al., 2006). In a sustainable breeding program, new cultivar ideotypes should be designed to reach the top class and this position with the support of consumer appreciation. Market requirements change with time and new cultivars must be introduced in order to ensure logistic requirements and consumer loyalty.

The modern breeding programme oriented toward peaches and plums should also consider the development of cultivars adapted to this kind of product to respond to a future demand in a timely fashion. Enhanced fruit quality is the universal goal of all tree fruit breeders. Fruit quality must be subdivided and specific characteristics evaluated by the breeder in order to measure genetic gain from planned hybridizations. Individual characters collectively comprise fruit quality include fruit size and the degree of flesh firmness, aroma and flavor characters, color of flesh, skin and overcolor and fruit juiciness. Each of these characters can be measured objectively with appropriate instrumentation. Couranjou (1995) demonstrated that good genetic gain is possible in apricot breeding by choosing parents based on fruit phenotype. Thus, parental apricots used in hybridizations that are markedly superior in specific aspects of fruit quality (high overcolor, strong aroma and flavor or large fruit size) generally pass along those quality characteristics to the next generation of seedlings. Breeding programs based on apricots from the European ecogeographic group cloud benefit substantially in the development of higher quality fruit by utilizing germplasm from the other ecogeographical groups.

3.2.6   BLACKBERRIES

The critical traits associated with high fruit quality in blackberry include physical characteristics such as size, shape, color, firmness, skin strength, texture, seed size and ease of harvest and chemical characteristics such as flavor, soluble solids, pH, titratable acidity and nutritional content. Obviously, whether the fruit is being grown for the fresh or processing market determine which trait rise or drop in importance. Fruit that is processed need high soluble solids, titratable acidity levels and relatively low pH in order that the products that they are made into have greater shelf stability. Similarly, since many times fruit for processing is only a small portion of a product, it is essential that they have intense flavor and color.

Breeding for fruit size has seen an interesting evolution over the past century (Moore et al., 1974a). Blackberry breeders have increased fruit size tremendously compared to wild germplasm. While large fruit size has never been that important in processed berries especially once machine harvesting became the standard practice, it has been a common objective to select for large fruit size.

3.2.7   CITRUS

Citrus fruits are usually consumed fresh or processed and the fruit quality requirements are different in some species. Mandarins, navel oranges and some common oranges, pummelos, and grapefruits are mainly for fresh markets, lemons are for both fresh and juice markets. Generally, both external and internal fruit quality are equally important in the fresh market. Suitable fruit shape, deep peel color and smooth and shiny appearance are the most attractive external quality characters for the consumers. Easy peeling, pleasant flavor, seedlessness, and fragrance are the most desired internal quality parameters.

Breeding efforts for the improvement of fruit quality have been focused on peel and pulp color, seedlessness, fragrance, and pulp taste. Hundreds of cultivars have been released to meet different market requirements. Citrus breeding methods include bud and seedling mutation selection, mutagenesis, and hybridization

3.2.8   BREEDING FOR SEEDLESS FRUITS

Seedless fruit is highly desired in the fresh market. Satsuma mandarin, navel orange, Valencia, blood orange, some grapefruit and lemon cultivars are seedless and have been dominant in fresh products. Meanwhile many seeded cultivars have gradually lost their attractiveness and disappeared from the market. Seedlessness, in fact, becomes the most important requirement for a citrus cultivar destined for the fresh market.

The cultivars that produce seedless fruit must exhibit parthenocarpy. In citrus the degree of parthenocarpy is different among genotypes, with some cultivars being highly parthenocarpy such as navel orange Satsuma mandarin, some grapefruit, and others being capable of setting seedless fruit without any external stimulation many cultivars, however, have weak parthenocarpy and generally produce fruits with fewer but variable seed numbers (Frost and Soost, 1968).

Male and female sterility as well as self-incompatibility play important roles in the production of seedless fruit, female sterile cultivars always produce seedless fruits, even after being hand pollinated with fertile cultivars. Navel Orange and Satsuma mandarin are female sterile, though not completely and in a controlled pollination they well produce a few seed. In the case of female fertility, the male sterility or self-incompatibility are necessary for setting seedless fruits, in parthenocarpic cultivars under open pollination and fruits with many seeds can be produced with hand pollination (Vardi, 1992; Yamamoto et al., 1995).

The breeding efforts for seedlessness have concentrated on the trait of parthenocarpic cultivars, selecting a parthenocarpic mutant or acquiring parthenocarpic gene (s) by conventional or biotechnological methods is the main emphasis of contemporary breeding goals. Concerning genotypes with acceptable parthenocarpy male and female sterility or self-incomatibility, mutants have been selected from seedling and bud mutations, and by irradiation. Creation of abnormal chromosome division by breeding triploid genotypes is widely accepted for seedless cultivar selection. According to Deng et al. (1996a), during the past 20 years, 119 seedless cultivars (clones) of ponkan, sweet orange, pummello, and kumquat were selected in china. These results suggest that mutagenesis with selection is an effective method for breeding seedless fruit.

Comparatively citrus flesh is the distinctive and edible tissue that contains anthocynanins and those are sometimes visible as by the red blushes on the rind of blood orange. In variety development programs more attention has been paid lately to flesh coloration improvement caused by anthocyanins to improve fruit health quality. So far, only blood orange cultivars, such as "Budd", "Moro", "Tarocco", "Sanguinello" and so on, have been reported to produce fruits colored by anthocyanins (Lee, 2002; Rapisarda et al., 2009). Major anthocyanins in blood orange fruit and juice are cyaniding-3-glucoside and cyaniding-3-(6"-malonylglucoside). Additionally minor anthocyanins and anthocyanin derived pigments such as 4-vinylcatechol adducts of cyaniding-3-glucoside, cyaniding-3-(3"-malonylglucoside, delphinidin 3-glucoside, cyaniding 3-sophoroside, cyaniding -3-(6"dioxalylglucoside), have also been identified (Hillebrand et al., 2004).

With unknown origin, blood oranges have been cultivated for several centuries in many countries including Italy, Morocco, and Spain (Tribulato and La Rosa, 1994, 2000; Saunt, 2000). Several varieties of blood oranges were selected in the field or from nucellar lines. For example, "Ippolito" and "Sant Alfio" are mutants of Tarocco (Tribulato and La Rosa, 1994, 2000), and "Spanish Sanguineli" and "Washington Sanguine"were selections from mutants of "Doble Fina"(Saunt, 2000). Novel hybridizations have also been employed. For example, it was reported in Italy that a hybrid "Omo-31" from Clementine cv. "Oroval" (*C. clementina* Hort. Ex. Tan.) and "Moro" orange (*C. sinensis*) was superior to its parents in containing notably higher amount of anthocyanin and flavanones (Rapisarda et al., 2003), and thus possessed important nutritional benefits. Two new triploid hybrids, "Taole" and "Clara", producing easy peeling fruits and containing anthocyanins in the flesh, were also obtained from a cross of "Monreal" clementine x "Tarocco" orange (Rapisarda et al., 2008). Later on, "Red Taole", a mutation from "Taole" was selected for its even higher content of anthocyanins (U.S. Patent 20080189813, 2008).

Although little attention has been paid to anthocyanins presented on citrus rind as blushes, rinds with large amount of evenly distributed anthocyanins will undoubtedly be another attractant to consumers. In addition, a novel anthocyanin pigmented peel mutant from common pummelo, named "Zipi"pummel, was found in China. It revealed light violet coloration on the outside but exhibited a normal colored flesh.

It is known that low temperature in pre and postharvest stages can induce anthocyanins accumulation in blood orange juice sacs (Lo Piero et al., 2005). Thus, it may be promising for future breeding programs if the cold signal could be enhanced in the flesh via various techniques, including perhaps genetic engineering.

The flesh of citrus contains carotenoids, though several fold higher amount of carotenoids are found in the rind than in the flesh of blond citrus. However, in citrus with red or pink flesh such as "Star Ruby"grapefruit (*C. paradisi*), "Cara Cara"navel orange (*C. sinensis*) and "Guanxi sweet pummelo mutant" (*C. grandis*), flesh contains several times higher amounts of carotenoids than rind and lyocopene and β-carotene are responsible for the pink or red color (Xu et al., 2006), whereas β-cryptoxanthin is the main carotenoid found in red flesh of satsuma mandarin and "Valencia" oranges (Kato et al., 2004). It is reported that red grapefruit has higher concentration of bioactive compounds and more antioxidant potential than the blond varieties (Gorinstein et al., 2006).

Owing to their brilliant color and nutrition value, the selection of pink or red fleshed citrus by carotenoids amounts is one of the important goals in coloration quality improvement. Pink or red fleshed fruit mutations are commonly found in grapefruits, sweet oranges and occasionally in lemon (Xu and Deng, 2002). So far due to unknown mechanisms varieties of pink or red fleshed pummelo and grapefruit have been recorded more than that of sweet orange and nevertheless no pink or red fleshed mandarin has been reported.

Among 20 recorded varieties of pink or red fleshed grapefruits many come from the selection of bud and seedling mutations and some are obtained by irradiation of buds and seeds. "Pink Marsh", "Ruby", "Rio Red" and "Star Ruby" were selected by one of the aforementioned methods. In accordance with the large cultivation area, many pummelo with colored flesh were selected from bud mutations of local cultivars in China such as "Fengdu red flesh pummel", "Guanxi sweet pummelo red mutant" (Xu and Deng, 2002) and "Shultu seedless red pummelo" (Shun and Chen, 1998).

Pink or red flesh sweet oranges are selected from bud mutation. For instance "Cara Cara" navel orange is assume to be a mutant of "Washington Navel" selected in Venezuela (Xu et al., 2006), "Honganliu" sweet orange was resulted from "Anliu" sweet orange in China (Liu et al., 2007), "Sarah" was originated in Israel from

"Shamouti" sweet orange (Saunt, 2000), "Vaniglia Sanguigno" with unknown origin in Italy is also red in flesh and accumulates carotenoids (Saunt, 2000).

Besides the aforementioned varieties from selection, a few flesh pigmented cultivars were originated by hybridization. "Hirado Butan" pummelo, is a natural hybrid of "Hirado" grapefruit and "Buntan" pummelo and is pink in juice sacs and segments (Xu et al., 2006). "Chandler" is a hybrid pummelo from "Siamese Pink"(Saunt, 2000).

New approaches such as chimera breeding is promising (Burge et al., 2002), wherein the spontaneous chimera sections have altered coloration are purified and mutants propagated having new color of fruit pulp. In fact, colored chimera fruits are common in the orchard and from these chimeras, new sweet orange cultivars with alternatively colored rind and flesh could be easily separated in the laboratory. "Zaohong" navel orange is just such new cultivar that has been recently selected from a spontaneous chimera shoot that emerged at graft union (graft chimeras) and which accumulates β-crytopxanthin as the dominant carotenoid in its flesh. It was a periclinal chimera with L-1 derived from satsuma mandarin and L-2/L-3 from the "Robertson"navel orange (Zhang et al., 2007).

In addition to flesh, new varieties with altered carotenoids composition and contents on the rind may also be important. As a mutant derived from the bright orange "Navelate" navel orange, the "Pinalate" navel orange has yellow rind, caused by accumulation of phytoene, phytofluene, and carotene in the flavedo (Rodrigo et al., 2003). Interestingly "Pinalate" has lowered production of abscisic acid.

Citrus accumulates large quantities of flavonoids in their fruit and although flavonoids are beneficial to human health, the flavanone neohesperidosides such as naringin, confer unpalatable bitterness to fruits and juices. For this reason, some breeding programs aim to decrease flavonoid components in citrus (Ufuk et al., 2009). Few reports however have concentrated on flavonoids improvements in citrus. It has been reported that rootstocks and interstocks are important factors that affect the total flavonoid content of lemon juice. It was found that interstocks could alter the contents of eriocitrin, diosmin, and hesperidin, whereas rootstock might affect 6,8-di-C-glucoside diosmetin content (Gil-Izquierdo et al., 2004). Therefore rootstock breeding programme might improve flavonoid profiles in fruits.

In addition to breeding programs focused on the improvement of flesh color, other breeding programs are focused on peel color improvement (Dou and Gmitter, 2007). Peel improvement may not benefit fruit qualities for fresh consumed varieties, but the higher content of flavonoids and carotenoids inside waste by product of citrus such as rind is of great importance in the food, pharmaceutical, or cosmetic industries.

Fruit internal quality includes flavor composition, texture, fragrance, and nutritional content. The taste is determined by total soluble solids (TSS), sugar, titratable acid contents and the ratio of TSS-to-acid. Consumers in different regions may have different flavor requirements relative to TSS/acid, but it is commonly acceptable that higher TSS produces tastes that are better accepted. In citrus, fruit quality breeding, the flavor improvement concerns mainly the increase of TSS with suitable acidity. Easy peeling and segment membrane texture are also important in fresh marketing quality. People feel more pleased with the easy peeling fruits and most of the mandarins have this desired loose skin trait. Thick or tough membranes within the fruit negatively influence pulp taste perception.

Natural mutations are a main source variation selected by breeders to create cultivars with improved internal fruit quality. A newly patented sweet orange culti-vars "Alvarina" was selected from limb mutation of "Valencia Late". This cultivar is of high flavor and is easier to peel (U.S. Patent 20080189813, 2007). Numerous cultivars of Sweet Orange, Mandarin, Clementine, and Lemon with improved fruit quality have been released (Sparta et al., 2006). "Ippolito" is a clone of "Tarocco" and has excellent flavor quality and very high content of anthocyanins (Tribulato and La Rosa, 2000).

Low acid fruit (or acidless) or sugar orange cultivars represent one of the smaller groups of citrus. The unknown origin of "Succari" in Egypt and "Vaniglia" in Italy are cultivated in limited areas (Hodgson, 1967). In China, a low acid orange "Bing-tang Cheng" was found in 1930s in a seedling population of local cultivars and two seedless cultivars of low acidity (<0.4 %) were selected in 1960s. Since then, many different clones have been selected. In the 1990s, a small tangerine called "Shat-angju" was selected from a local cultivar "Shiyueju" (Deng, 2008). "Shatangju"is easy peeling of high sugar and low acid content and becomes acceptable in Chinese and Southeast Asian fresh fruit markets.

Mutagenesis by irradiation to improve fruit quality has been attempted in many different citrus types. Recently a mandarin cultivar "Tango" was released (U.S. Pat-ent 20070056064, 2007). This cultivar, selected from an irradiated bud of the dip-loid mandarin cultivar "W. Murcott" at the University of California Riverside, has smoother peel and texture, high TSS (13.5–15°B) and low acidity (0.54–0.82 %). Satsuma fruit is usually low in TSS content because increased fruit flavor is the key breeding aim. However, "Miyagawa" Satsuma buds were irradiated by ^{60}Co γ-ray and two mutants having TSS of 15°Brix were selected (Wang and Xiang, 1998).

Citrus hybridization represents the most important tool for improved internal fruit quality, permitting the transfer of characters from one species or genus to an-other or combination of quality characters of two genotypes. The earliest hybrid-ization program was carried out by Swingle and Webber in Florida in 1893 for the purpose of improved disease tolerance (Camerson and Frost, 1968). "Sampson"and "Thornton"tangelos were obtained from crosses made between grapefruit and "Dan-cy" mandarin and later the "Orlando" and the "Minneola" tangelos were introduced. Successive crosses performed between many other parents have included diverse grapefruit, tangelo, mandarin, sweet orange, lemon, and lime genotypes. Some of the hybrids are still in cultivation. For example, "Nova" (Reece et al., 1964), a hybrid between Clementine mandarin and Orlando tangelo (*C. reticulate* x [*C. paradisi* x *C. reticulata*]) is cultivated in some areas for its red rind and good flavor.

Williams (2010) reported several newly released cultivars with high fruit quality obtained by hybridization. "Gold Nugget" is a hybrid of "Wilking" ("King" x "Willowleaf") x "Kincy" ("King" x "Dancy"). The fruit quality is high with TSS 13.5 to 16°Brix and acidity 0.80–0.60 percent. Three sibling triploid hybrids of high sugar content including "TDE2"—Shasta Gold, "TDE3"- Tahoe Gold and "TDE4"- Yosemite Gold come from the cross made in 1973 between a tetraploid female parent ("Temple" tangor × 4n "Dancy" mandarin) and a pollen parent "Encore" mandarin.

In the breeding for flavor quality improvement, a hybrid "Kiyomi" tangor has made an important contribution. This tangor of superior fruit quality was obtained from a cross of "Miyagawa" Satsuma (*C. unshiu*) and "Trovita" orange (*C. sinensis*), and is monoembryonic, male sterile and highly parthenocarpic (Nakano et al., 2001). As an ideal seed parent, "Kiyomi" has been crossed with numerous pollen parents and up to now many hybrid cultivars of "Kiyomi" pedigree have been released (Table 3.5). One of them, "Shiranuhi" has a high TSS content, in most cultivation areas reaching 13 to 16°Brix (Matsumoto, 2001). In some extreme cases even arriving 18°Brix.

TABLE 3.5   Hybrids obtained from crosses with "Kiyomi" as seed parent

Hybrid	Pollen Parent	TSS (°B)	Reference
Akemi	"Seminole" tangelo	12–13	Yoshida & Yamaka (2001)
Amaka	Encore	11–12	Matsumoto et al. (2001a)
T378	Kiyomi x Okitzu		
Amakusa	T378 x Page	11–12	US Patent PP09550
E-647	Osceola		Nakano et al. (2001)
Ehime Kashi 32	Ougonkan	11	Yukinori et al. (2008)
Harumi	"F-2342" ponkan	12–13	Yoshida et al. (2000)
Nishinokaori		11–13	Matsumoto et al. (2003)
Okitsu No. 41	Wilking		Nakano et al. (2001)
Okitsu No. 45	Wilking		Nakano et al.,(2001)
KyEn No. 5	Kiyomi x Encore		
Reikou	KyEn No. 5 x Murcott	12–13	Yoshioka et al. (2009)

TABLE 3.5   *(Continued)*

Hybrid	Pollen Parent	TSS (°B)	Reference
Kucjinotsu No. 37	Kiyonii x Encore No. 2		
Setoka	Kuchinotsu No. 37 x Murcott	12–13	Matsumoto et al. (2001b)
Shirunui	"Nakano No. 3" ponkan	13.5–15	Matsumoto (2001)
Tsunokaori	Okitsu	13–15	US Patent P085599P
Youkou	"Nakano No. 3" ponkan	14	Matsumoto et al. (1999)

In a breeding program for mandarin improvement Satsuma was crossed with various mandarins, and a series of mandarin hybrids were released (Table 3.6) and are widely cultivated in different areas in the world (Tribulato and La Rosa, 1993; 1996a: Deng et al., 2002; La Rosa and Tribulato, 2002; Tribulato et al., 2002). All the hybrids keep the seedless fruit characteristic as the female Satsuma, the fruit quality is significantly improved mainly in flavor, aroma, membrane texture, rind, and pulp color. This result shows how Satsuma can be a good seed parent for improving fruit quality by hybridization.

TABLE 3.6   Mandarin hybrids obtained from crosses using Satsuma as seed parent

Hybrid	Satsuma Seed Parent	Pollen Parent	Main Fruit Characteristics	Mature Period
Primosole	Miho	Carvalhais mandarin	Seedless, 140–150 g, rind red in color, flavor pleasant	October
Desiderio	Miho	Comune Clementine	Seedless, deep red and thin rind, juicy, and flavor very tasty	November
Simeto	Miho	Avana mandarin	Seedless, 10–180 g, flavor similar to Avana mandarin	Nov-Dec
Sirio	Miho	Comune Clementine	Seedless, 160–170 g, deep red rind, juice content 50 %, flavor delicious	December-January

TABLE 3.6 *(Continued)*

Hybrid	Satsuma Seed Parent	Pollen Parent	Main Fruit Characteristics	Mature Period
Etna	Okitsu	Comune Clementine	Seedless, 140–160 g, dark orange pulp, red rind, flavor similar to Clementine Comune	November
Bellezza	Okitsu	Carvalhais mandarin	Seedless, >150 g, deep orange rind, aromatic juicy pulp	December

3.2.9 GRAPE

At present more than 1,200 varieties have been introduced from abroad to our country which includes Thompson Seedless, Perlette, Cardinal, Beauty Seedless, Pearl of Casaba, Black Muscat, Delight, Himrod, Hur etc and found suitable for yield and fruit quality characters under our conditions (Randhawa and Chadha, 1964; Randhawa et al., 1965; Prasad, 1968; Yadav et al., 1971, 1977; Jindal, 1985). The most popular variety of South India, Anab-e-Shahi is also an introduction from Middle East made by Abdul Baquer Khan way back in year 1890 (Singh and Murthy, 1993). Several selections have been made from open pollinated seedlings of these seedlings.

The development of cv. Cheema Sahebi and Selection-94 clearly demonstrates the utility of the method (Gopala Krishna and Phadnis, 1960; Phadnis et al., 1968). Some promising seedling from the open pollinated population of Pandhari Sahebi and Kabul Monukka were selected by Gupta (1971) and Phadnis and Gupta (1972). They reported that some of them proved better than the existing commercial cultivars. Pusa seedless a popular variety of North India, is a clonal selection from Thompson seedless (Jindal, 1985). Its vines are vigorous with medium to large compact bunches. The berries are small, seedless, and greenish yellow in color. The flesh is tender and sweet with TSS up to 22 percent. Singh et al. (1974a) identified a clone (HS 37-6) from Perlette, earlier in maturity by 15 days. Tas-a-Ganesh, Rao Sahebi, Sonaka, Manik Chaman and Dilkhush are some other selection made by the enlightened grape growers (Singh and Murthy, 1993).

Cross breeding between native species, between a species and a hybrid previously produced, or crossing of two cultivars (or compatible hybrids) have produced many good varieties in different grape growing countries. In India, as many as twelve hybrids have been derived by hybridization at IARI, New Delhi and IIHR, Bangalore (Table 3.7). Critical appraisal of hybridization at different stations was focused on the objectives of earliness, seedlessness, and good fruit quality. Jindal et al. (1983) suggested that for inducing seedlessness in the progeny, the varieties Banqui-Abyad, Katta Kurghan and Hur should be selected as female parent on account of their high seed index (i.e., high ratio of berry weight to seed weight). When both the parents were seeded the percentage of seedless progeny was very low and hence seedless varieties are often used as male parent to incorporate seedlessness. The fruits of seedless cultivars are of two types: (i) those in which the seeds abort while still small and soft (stenospermocarpic) and (ii) those in which the seeds do not develop at all (parthenocarpic)-(Pratt, 1971). The pollens of such cultivars have been found to be viable and used in controlled crosses to produce seedless hybrids (Weinberger and Harmon, 1964).

TABLE 3.7   Grape hybrids developed in India

Hybrid	Parentage	Characteristics
IARI, New Delhi		
Pusa Navrang	Angenine x Rubi Red	Extremely early, uniform in ripening, basal bearer, heavy yielder, teinturier (color in skin & pulp), resistant to mildew good for juice & colored wine
Pusa Urvasi	Hur x Beauty Seedless	Early & uniform in ripening, basal bearer, more productive with loose bunches, berries uniform oval, greenish yellow and seedless with 20–22 % TSS, good as dessert grape and for raisins
IIHR, Bangalore		
Arkavati	Black Champa x Thompson Seedless	Prolific bearer with large cluster, berries yellowish green, seedless thin skin. TSS (23°B), acidity 0.7 % good as table grape and for raisins
Arka Kanchan	Anab-e-Shahi x Queen of Vineyards	Prolific bearer with medium to large clusters, golden yellow berries, seeded, and muscat flavor, TSS 20°B acidity 0.7 % good for table and wine purpose
Arka Shyam	Bangalore Blue x Black Champa	Moderate to heavy yielder, medium clusters big sized round berries, TSS 24°B, acidity 0.6 %, suitable for wine purpose
Arka Hans	Anab-e-Shahi x Bangalore Blue	Prolific bearer, bunches medium in size, yellowish green berries, seeded, TSS 21°B, acidity 0.5 %, suitable for wine making

TABLE 3.7 *(Continued)*

Arka Sweta	Anab-e-Shahi x Thompson Seedless	Prolific bearer, a hybrid for table use and raisin making, av. bunch weight 260 g, greenish yellow colored; obovid uniform seedless berries, av. berry wt. 4.08 g, TSS 18-19° B, acidity 0.5–0.6 %
Arka Majestic	Angur Kalan x Black Champa	Vigorous plant, prolific bearer, high yielding, berries deep red, obovoid, bold, and seeded suitable for table use, av. bunch wt. 370 g, 2–3 small seeds/berry, av. berry wt. 7.7 g, TSS 18–20° B, acidity 0.4–0.6 %, pedicel attachment very good, ideal for export, all buds are fruitful
Arka Neelamani	Black Champa x Thompson Seedless	Vigorous plant, well filled to slightly compact bunches weighing on an average 360 g, black colored and seedless, av. berry wt. 3.2 g, TSS 20–22° B, acidity 0.6–0.7 %, all buds on a cane are fruitful; does not require specific pruning
Arka Chitra	Angur Kalan x Anab-e-Shahi	Prolific, high yielding, berries-golden yellow with pink blush, seeded but attractive, suitable for table purpose, average bunch weight 260 g, av. berry wt 4 g, TSS 18–19° B
Arka Soma	Anab-e-Shahi x Queen of Vineyards	Heavy yielder, white berries, seeded, meaty pulp with muscat flavor, good for wine making
Arka Trishna	Bangalore Blue x Convent Large Black	Prolific bearer, berries deep tan colored, seeded very sweet pulp, male sterile, good for wine making
Arka Krishna	Black Champa x Thompson Seedless	Prolific bearer, berries-black, seedless and sweet more juicy and suitable for beverage industry

IARI, (1997), Singh and Murthy, (1993), Singh et al., (1993), Singh et al., (1998)

In India, seedless cultivars like Thomson seedless and Beauty seedless have been widely used as male parent in hybridization programmes. Use of tetraploids and diploids to obtain seedless triploids has not been successful. Because of the wide range of variability observed in the degree of development of stenospermocarpic seeds, it is believed that there are some modifying genes (Sharma and Uppal, 1978) that influence the character.

Polyploids, particularly tetraploids produce distinctly large berries. Polyploidy forms which occur spontaneously in nature may arise from a latent bud near a pruning wound or from somatic chromosome doubling in initial cells of a shoot meristem. It has been induced artificially with an aqueous solution of 0.25–0.50 percent colchicine with 5–10 percent glycerin (Das and Mukherjee, 1967) or 10 ppm GA3 (Iyer and Randhawa, 1965) (Table 3.8)

TABLE 3.8   Tetraploid grape varieties

Tetraploid variety	Source/parent variety	Nature and place of evolution
Marvel Seedless	Delight	Bud mutation cross at 4 x level between Campbell and Niagra, West Germany.
Early Niabel	Campbell (4x) x Niagra (4x)	
Lonetto		
Perle		West Germany
Early Giant	Campbell	
Muscat Common Hall	Muscat of Alexandria	
Black King	Campbell	Japan
Wallis Giant	Concord	USA
Otsubu Catawba	Catawba	Japan
Benikawachi	Delaware	Japan
Otsubu Koshu	Koshu	Japan
Ostubu Niagra	Niagra	Japan
Malbek		Argentina
Sweeney, Jones	Gords (syn. Muscat of Alexandria)	Australia
Centennial	Waltham (Dattier, Rosaki)	Australia
Baumann	Listan (syn. Palomino)	Australia
Case	Sultana	Australia

Pandey and Pandey (1990)

Chromosome doubling may sometimes be induced by frequent cutting of stems (pruning) and by heat shocks (Olmo, 1952, Wagner, 1958, Ourecky et al., 1967). But these treatments are far less reliable then colchicine treatment. Some tetraploid varieties have been evolved with colchicines but none of them has attained commercial importance except cold-hardy "Perle" which is grown commercially in Germany (Brieder, 1964). Available reports on the performance of induced tetraploids show that there is no difference in the dates of bud burst anthesis earliness, acidity, and TSS between diploids and tetraploids (Chadha and Mukherjee, 1972).

Spontaneous mutations are frequently observed especially in old cultivars and provide a valuable source of variation in grapes. The mutations reported involve all kinds of characteristics including yield, earliness, size, number, andcolor of berries, hardiness, resistance to diseases. "New Perlette" (Loose Perlette) with comparatively loose bunch has been evolved with x-rays (2.5 kR) treatment of "Perlette". The self-thinning property of "New Perlette" is a result of meiotic irregularities caused by chromosomal translocation. "Red Niagara" (Red fruited type from Niagara) and Robin Cardinal (early-maturing type from Cardinal) are some other induced grape mutants.

3.2.10 GUAVA

Guava is an open pollinated crop and seedlings are extensively used to raise new plantations (Hayes, 1957). Selection from these seedlings can be used to obtain superior strains with respect to fruit yield and quality. Several attempts (Table 3.9) have been made to select superior types from seedling population (Phadnis, 1970, Singh, 1959, Hulamani et al., 1981, Thonte and Chakrawar, 1981, Pathak and Dwivedi, 1988, Iyer and Subramanyam, 1988, Rajan et al., 1997).

TABLE 3.9 Name of guava cultivars developed through selection

Name of selected cultivar	Selection made from	Characteristics
Sardar (L-49)	Allahabad Safeda	Semidwarf tree2.5 to 3.5 m, leaves are large (12.8–13.2 cm long). Fruits are large, roundish ovat, flesh soft, white with few seeds.
Allahabad Surkha	A local red fleshed type	Profuse bearing, large fruits, soft deep pink flesh, few seeds strong flavor, very sweet, yield 120 kg after six years
CISH-G₁	A local red fleshed type	Attractive fruits wit deep red skin and uniform fruit size, seeds are soft and less in no. TSS 15 brix, shelf life better than Allahabad Safeda, suitable for export trade.
CISH-G₂	A local type	Crimson colored attractive fruits with white stripes in groves, seeds soft less in number
Lalit (CISH-G₃)	A local type	Released in 1998, fruits are attractive saffron yellow color with occasional red blush, medium in size (average in wt.185 g.). flesh is firm and pink in color and has good blend of sugar and acid and suitable for both table and processing (jelly and beverage) purposes. It gives 25% more yield than Allahabad Safeda.
CISH-G₄	Selected from open pollinated seeds of a colored guava variety	Fruits are sub globose with few soft seeds. High TSS 14 Brix and attractive pink blush. It is a good yielding potential

TABLE 3.9 *(Continued)*

Bangalore Local	A local genotype	Better in fruit quality, average fruit weight 146g. seeds soft (100 seed wt.1.7g.) flesh soft and white
Arka Mridula (Selection-8)	Seedlings of Allahabad Safeda	Fruits are smooth white flesh with soft seeds (100 seed wt.1.6g.) TSS 12 Brix) flesh firm good keeping quality.

In guava, majority of the commercial varieties are diploids (2n=22) while the seedless variety is triploid and shy bearing in nature. To evolve a variety with less seeds and better yield potential crosses were made between a triploid (seedless) and diploid (Allahabad Safeda) at IARI, New Delhi. Distinct variation in tree growth habit, leaf, and fruit characters were observed (Majumder and Mukherjee, 1972, Mukherjee, 1977). At IIHR, Bangalore several F_1 hybrids were raised by making several crosses involving cultivars like Allahabad Safeda, Red Flesh, Chittidar, Apple Colour, L-49 and Banaras. Arka Amulya, Hybrid 16-1 and Hybrid 31-1 has already been released on account of their better performance (Subramanyam and Iyer, 1993). A cross between Arka Mridula x Red Fleshed has also been isolated for deep red pulp color with soft seeds and good TSS (12°B). Similarly crosses between Arka Amulya x Apple Colour and Apple Colour x Red Flesh have produced some interesting hybrids at IIHR, Bangalore. Rama Rao and Dayanand (1977) obtained six promising hybrids at Fruit Research station Anantharajupet, Andhra Pradesh (Table 3.10). Two hybrids have been released in this context from Hissar, Haryana.

TABLE 3.10   Different hybrids released from different stations

Hybrids	Parents	Characteristics
Fruit Research Station, Anantharajupet, Andhra Pradesh		
H-1	Red flushed x Sahranpur seedless	Fruit medium to big sized, smooth with thin skin, fles pinkish with good flavor, sweet in taste and good quality with moderate seed content.
H-2	Smooth Green x Nagpur Seedless	Fruits medium sized, smooth with thin skin, fles soft, cream color with good color flavor , sweet in taste and good quality , less seeded, seed soft and easily crushed.
H-3	Allahabad x Red flushed	Fruit medium sized, smooth with thin skin, flesh white with fine flavor, sweet in taste and good quality with moderate seed content.
H-4	Smooth Green x Sahranpur Seedless	Fruit medium sized, smooth with thin skin, flesh white with good flavor, sweet in taste and good quality many seeded seed soft and easily crushed.

TABLE 3.10 *(Continued)*

H-5	Red Fleshed x Nagpur Seedless	Fruit medium sized, smooth with thin skin, flesh white with good flavor, sweet in taste and good quality, seed big in size but few and soft.
H-6	Banarsi x Allahabad	Fruits medium to big in size, smooth with thin, flesh cream, coloure with good flavor, sweet in taste, less seeded, seeds soft and easily crushed.
IIHR, Bangalore		
Arka Amulya	Seedless x Allahabad Safeda	Plant semivigorus, heavy yield, fruit size medium, pulp white with few soft seeds, keeping quality good
Hybrids 16-1	Apple color x Allahabad Safeda	Semivigorous plants, moderate yield, fruit skin bright red all over, seeds few , flesh firm, high TSS, good keeping quality
Fruit Research Station, Sangareddy		
Safed Jam	Allahabad x Kohir	Tree growth and fruit quality similar to A. Safeda, fruit size bigger, few seeds
Kohir Safeda	Kohir x Allahabad Safeda	Tree vigorous, fruits large in size, white flesh, few seeds.
CCS Haryana Agricultural University, Hissar		
Hisar Safeda	Allahabad Safeda x Seedless	Upright tree growth, compact crown, round fruits with smooth, surface, creamy yellow skin average fruit weight 92 g. creamy white flesh, seed count low, seeds soft, TSS 13.4, acidity 0.38%, average yield 114 kg/tree /year.
Hissar Surkha	Apple Colour x Banarsi Surkha	Tree crown broad to compact, round fruits with smooth surface, skin yellow with red dots in low temperature, average fruit weight 86 g, pink color flesh, seed count medium, seed texture medium, TSS 13.6 per cent, acidity 0.46 per cent, yield 86 kg/tree/year

3.2.11 PAPAYA

The main breeding objectives of papaya are cultivars which give high yield of better quality fruits. Six principal characters namely yield, flesh color, texture, fruit size, sweetness, and storage are considered. Under breeding programs emphasis must be on increased latex yield for papain production, for inducing a sex linked vegetative character for eliminating unwanted sex forms in the early seedling stage, early and bearing with short internodes, uniformity of fruit shape, texture, andflavor for export trade. Papaya is a highly cross pollinated crop, a great deal of variation exists in shape, size, quality, taste, color, andflavor of the fruit.

Breeding for gynodioecious lines involves selfing of a regular and prolific bearing hermaphrodite or sibmating the female with the hermaphrodite. Suitable hermaphrodite plants whose sex does not change with climatic variations are selected. The major advantage of developing a gynodioecious line is that all plants bear fruits.

TABLE 3.11   Some promising gynodioecious cultivars are

Varieties	Characteristics
CO_1	Selection from progenies var. Ranchi, plant dwarf, fruit round or oval
CO_2	A pure line selection from a local type, plant medium tall, fruits ovate and large in size, good for papain production
CO_3	A hybrid between CO_2 x Sunrise Solo, a tall vigorous plant, fruit medium size, sweet with good keeping quality
CO_4	A hybrid between CO_1 x Washington, plant medium fruit large, flesh thick, and yellow, with good keeping quality
CO_5	A selection from Washington variety, good from papain production yield 1500-1600 kg papain/ha
CO_6	A selection from Pusa Majesty, dioecious, dwarf in stature, fruit size large, good for papain production
CO_7	A hybrid between Coorg Honey Dew x CP85, high yielding red fleshed uniformity in shape, gynodioecious with good edible character
Pusa Delicious	A selection from progenies of var. Ranchi gynodioecious excellent fruit quality, deep orange flesh 13 per cent TSS
Pusa Majesty	A selection from Ranchi, gynodioecious, gives better papain yield than CO_2 and CO_6, resistant to nematode
Pusa Giant	A selection from Ranchi, dioecious, large size fruit, sturdy, and tolerant to strong winds
Pusa Dwarf	A selection from Ranchi, dioecious, very dwarf in stature, fruits medium in size and oval in shape
Pusa Nanha	A mutant of a local type, very dwarf plant, more suitable for high density planting, dioecious, good fruit quality
Coorg Honey	A selection from Honey dew, gynodioecious, excellent fruit quality under south Indian agro climate
Surya	A gynodioecious variety developed by crossing Sunrise Solo x Pink Flesh Sweet at IIHR, Bangalore fruits medium in size, pulp red with thickness of about 3-3.5 cm, TSS 13-15°, sweet in taste, keeping quality good
Pant 1	Selected at Pantnagar, 1-1.5 kg oblong fruits, good quality
Punjab Sweet	Selection made at PAU, Ludhiana, dioecious, fruits large, oblong, weighing more than 1.0 kg each, frost tolerant,
HPSC-3	A hybrid between Tripura Local x Honey Dew, gives high fruit yield 197.7 t/ha, papain yield 5 g/fruit, resistance to papaya mosaic potex virus

Polyploidy has received considerable attention in papaya breeding programme. Hofmeyr (1942, 1945) were able to induce polyploidy in papaya. They found that the quality of tetraploid fruit was better than diploid and the fruit was also compact, with smaller cavity, but tetraploids were observed to be less fertile than diploid. Zerpa (1957) reported colchicine induced tetraploid hermaphrodite plants which were used as male parent in a cross with a female diploid and the tetraploid produced a few seeds without endosperm.

Bankapur and Habib (1979) attempted mutation in papaya through radiation and observed that doses of gamma rays ranging from 5 to 15 kR were able to produce significant changes in characters like seed germination, survival of seedlings and ratio of male to female sexes in the population. Ram and Majumder (1981) evolved a dwarf mutant line by treating papaya seeds with 15 kR gamma rays. Initially, three dwarf plants were isolated from M_2 population. Repeated sibmating among the dwarf plants helped in establishing a homozygous dwarf line which was later named as "Pusa Nanha". It is dwarf in height and the fruiting in this strain starts at lower height (30 cm).

Fruit breeding has been in a state of flux because the traditional support by government experiment stations received throughout the world has declined as a reaction to the high costs of long-term programs that are necessary. However, the globalization of the fruit industry has underscored the economic value of improved germplasm indicating that support for fruit breeding needs to be increased. Returns from patents offer one solution to impede the decline in funding and to encourage breeding effort. Patents are essentially a direct tax on the industry to improve breeding and also offer incentives to breeders when they share in patent royalties. In some countries the costs of foreign patents has encouraged investment in local breeding programs.

The objectives of breeding efforts in the future must be twofold. The most important is to increase quality in order to increase consumption. This is important to reverse the decline in per capita fruit consumption in developed countries. The second effort must be to increase the efficiency with special emphasis on productivity, annual bearing and pest resistance. However, the future of fruit breeding is crop specific. For example the tremendous advances in seedless table grapes have intensified breeding efforts. In contrast, the breeding of wine grapes is impeded by the association of traditional cultivar name with the name of the wine. As a result the wine industry has not encouraged cultivar change except for clonal selection. Practically no advances have been made in the genetic improvement of banana but there are major projects via international efforts. The lack of diversity in export banana makes breeding essential due to disease pressure (such as sigatoka) but seedless banana is notorious difficult to hybridize.

Similarly, citrus industry has increased in efforts to produce new seedless, easy peel clones while the threat of diseases such as hanglongbing (citrus greening) is an upcoming problem that may require a genetic solution. The success of breeding in Prunus has encouraged a number of private breeding organizations especially vibrant in peach and plums. Recent studies of a partially stoneless plum described by Luther Burbank suggest that the hard stone might be eliminated in Prunus (Callahan et al., 2009). Similarly the success of breeding efforts to produce late ripening cherries in British Columbia, Canada has indicated the high economic returns made possible by sustained breeding effort. Recent success with blackberry with the introduction of thornlessness, primocane fruiting, reduced seediness and large, tasteful fruit has created a new global industry (Finn and Clark, 2011).

Finally advances in biotechnology must be attached to breeding efforts. Marker assisted selection could offer significant help in disease and pest resistance breeding but much genetic work must still be pursed to locate these quality traits. This is a conservative approach somewhat similar to backcross breeding but it is true that many high-quality fruits are difficult to replace.

KEYWORDS

- **Fruits**
- **Postharvest Quality**
- **Ripening**
- **Biochemical attributes**
- **Shelf life**
- **Composition**
- **Resistance**

REFERENCES

Amoros, A.; Serrano, M.; and Riquelme, F.; Importancia del etileno en el desarrollo y maduracion del albaricoque (*Prunus armeniaca* L. cv. Bulida). *Fruits.* **1989**, *44*, 171–175.

Baldwin, E. A.; Fruit flavor, volatile metabolism and consumer perceptions. In: *Fruit Quality and Its Biological Basis;* Knee, M., Ed.; Sheffield Academic Press: Sheffield, UK, **2002**; pp 89–106.

Bankapur, V. M.; and Habib, A. F.; Mutation studies in papaya (*Carica papaya* L.)-Radiation sensitivity. *Mysore. Agric. Sci.* **1979**, *13*, 18–21.

Barritt, B. H.; Apple quality for consumers. *Int. Dwarf. Fruit. Tree. Assoc.* **2001**, *34*, 54–46.

Bhat, K. M.; Farooqui, K. D.; and Sharma, M. K.; Lal Ambri – an ideal apple for export promotion. *Indian Hortic.* **2005**, *50*(2), 111.

Bourne, M. C.; Rupture tests vs. small-strain tests in predicting consumer response to texture. *Food Technol.* **1979a**, *10*, 67–70.

Bourne, M. C.; Fruit texture-an overview of trends and problems. *J. Text. Stud.* **1979b**, 10, 83–94, **1979b**.

Bourne, M. C.; Texture of temperate fruits. *J. Text. Stud.* **1979c**, *10*, 25–44.

Brieder, H.; On the exploitation of breeding and the practical utilization of X-ray induced somatic mutations in long lived and vegetatively propagated cultivated plants (*Vitis vinifera*). *Mitt. Kohforsch Inst. Prag.* **1964**, *14*, 165–171.

Brooks, A. J.; Artificial cross fertilization of the mango. *West Indies Bull.* **1912**, *12*, 567–569.

Brown, A. G.; and Harvey, D. M.; The nature and inheritance of sweetness and acidity in the cultivated apple. *Euphytica.* **1971**, *20*, 68–80.

Brown, S. K.; and Maloney, K. E.; Genetic improvement of apple: breeding, markers, mapping and biotechnology. In: *Apples: Botany, Production and Uses;* Ferree, D., Warrington, I., Eds.; CAB International: Wallingford, UK, **2003**; pp 31–59.

Brown, S. K.; and Maloney, K. E.; *Malus x domestica* apple. In: *Biotechnology of Fruit and Nut Crops;* Litz, R. E., Ed.; CABI Publishing: Cambridge, MA, **2005**; pp 475–511.

Bruhn, C. M.; Consumer and retailer satisfaction with the quality and size of California peaches and nectarines. *J. Food Qual.* **1995**, *18*, 241–256.

Brummel, D. A.; and Harpster, M. H.; Cell wall metabolism in fruit softening and quality and its manipulation in transgenic plants. *Plant. Mol. Biol.* **2001**. *47*, 311–340.

Burge, G.; Morgan, E.; and Scelye, J.; Opportunities for synthetic plant chimeral breeding. Past and Future. *Plant Cell, Tiss. Org. Cult.* **2002**, *70*(1), 13–21

Burns, W.; and Prayag, S. H.; *The Book of Mango.* Bombay Department of Agriculture Bulletin 103; **1921**.

Byrne, D. H.; Nikolic, A. N.; and Burns, E. E.; Variability in sugars, acids, firmness, and color characteristics of 12 peach genotypes. *J. Am. Soc. Hortic. Sci.* **1991**, *116*, 1004–1006.

Callahan, A. M.; Dardic, K. C.; and Scorza, R.; Characterization of 'Stoneless': a naturally occurring partially stoneless plum cultivar. *J. Am. Soc. Hortic. Sci.* **2009**, *134*, 120–125.

Cameron, J. W.; and Frost, H. B.; Genetics, breeding and nucellar embryony. In: *The Citrus Industry,* 2nd ed.; Reuther, W., Batchelor, L. D., Webber, H. J., Eds.; **1968**, Vol. 2, pp 325–370.

Chadha, T. R.; Breeding of new apple varieties by hybridization. *Himachal. Hortic.* **1962**, *2–3*(1), 116–121

Chadha, T. R.; Temperate fruit in India: a retrospect and prospect. *Indian Hortic.* **1987**, *32*, 2–5.

Chadha, T. R.; and Mukherjee, S. K.; Performance of some induced tetraploids in grapes. Abst. 3rd International symposium on subtropical and Tropical Horticulture, HIS, Bangalore, **1972**.

Chadha, T. R.; and Sharma, R. P.; Phenotypic variability in some intercultivar crosses of apple (*Malus pumila* Mill). *Himachal J. Agric. Res.* **1975**, *2*(2), 17–22.

Chadha, T. R.; and Sharma, Y. D.; Breeding of apple varieties in Himachal Pradesh. *Indian J. Hortic.* **1978**, *35*(3), 178–183.

Chaikiattiyos, S.; Kurubunjerdjit, R.; Akkaravessapong, P.; Rattananukul, S.; Chueychum, P.; and Anuput, P.; Improvement and evaluation of the selected 'Kaew Sisaket' mango in Thailand. *Acta Hortic.* **2000**, *509*, 185–192.

Coombe, B. G.; The development of fleshy fruits. *Ann. Rev. Plant. Physiol.* **1976**, *27*, 207–228.

Couranjou, J.; Genetic studies of 11 quantitative characters in apricot. *Sci. Hortic. 61*, 61–75.

Crassweller, R.; and Green, G.; Production and handling techniques for processing apples. In: *Apples: botany, production and uses;* Ferree, D., Warrington, I., Eds.; CAB International: Wallingford, UK, **2003**; pp 615–633.

Daniells, J. W.; The Cavendish subgroup: Distinct and less distinct cultivars. In: *entification of Genetic Diversity in the Genus Musa;* Jarret, R. J., Ed.; INIBAP: Montpellier (FRA), **1990**; pp 29–35.

Das, P. K.; and Mukherjee, S. K.; Induction of autotetraploidy in grapes. *Indian J. Genet. Plant Breed.* **1967**, *27*(1), 107–116.

Deng, X. X.; *Citrus Varieties in China.* China Agriculture Press: Beijing; **2008.**

Deng, X. X.; Guo, W. W.; and Sun, X. H.; Progress of seedless citrus breeding in China. *Acta Hortic Sin.* **1996a,** *23*(3), 235–240.

Deng, Z. N.; La Malfa, S.; and Tribulato, E.; Prove di diradamento dei frutti di mandarino 'Simeto' Atti VI Giornate scientifiche Societa Orticola Italiana. **2002,** *22,* 195–196.

Dong, Y. H.; Mitra, D.; Kootstra, A.; Lister, C.; and Lancaster, J.; Postharvest stimulation of skin color in royal gala apple. *J. Am. Soc. Hortic. Sci.* **1995,** *120,* 95–100.

Dou, H. T.; and Gmitter, F. G.; Post harvest quality and acceptance of LB 8-9 mandarin as a fresh fruit cultivar. *Horttechnology.* **2007,** *17*(1), 72–77.

Downing, D. L.; Apple cider. In: *Processed Apple Products;* Downing, D. L., Ed.; Avi Book: New York, **1989;** pp 169–187.

Elia, L.; Producing a profitable peach. *Fruit Grower.* **2001,** (June), 6.

Falconer, D. S.; and Mackay, T. F. C.; *Introduction to Quantitative Genetics,* 4th ed.; Longman: London; **1996.**

Farooqui, K. D.; and Dalal, M. A.; Breeding scab resistance in apple: an ecofriendly strategy. *Acta Hortic.* **2005,** *696,* 33–37.

Farooqui, K. D.; Fazili, M. A.; and Gaffar, S. A.; Resistance apple hybrids show promise for adoption by growers. *SKUAST. J. Res.* **2006,** *8,* 1–8.

Finn, C. F.; and Clark, J.; Emergence of blackberry as a world crop. *Chronica Hortic.* **2011,** *3,* 13–18.

Francis, F. J.; Quality as influenced by color. *Food Qual. Pref.* **1995,** *6,* 149–155.

Frost, H. B.; and Soost, R.; Seed reproduction: development of gametes and embryos. In: *The Citrus Industry,* 2nd ed.; Reuther, W.; Batchelor, L. D.; and Webber, H. J. Eds.; **1968,** Vol. 2, pp 290–324.

Gazit, S.; and Kadman, A.; '13-1' mango rootstock selection. *Hortic. Sci.* **1980,** *15,* 669.

Gil-Izquierdo, A.; Riquelme, M.; and Porras, I.; Effect of the rootstock and interstock grafted in lemon tree (*Citrus limon* (L.) Burm.) on the flavonoid content of lemon juice. *J. Agric. Food Chem.* **2004,** *52*(2), 324–331.

Gopalakrishna, N.; and Phadnis, N. A.; Selection-94-A new variety of grape (*Vitis vinifera* L.) *Poona. Agric. Coll. Magaz.* **1960,** *48*(1), 22–24.

Gorinstein, S.; Drzewiecki, J.; Park, Y.; Characterization of blond and Star Ruby (red) Jaffa grapefruits using antioxidant and electrophoretic methods. *Int. J. Food. Sci. Technol.* **2006,** *41*(3), 311–319.

Gross, R. A.; and Simmonds, N. W.; Mutations in the Cavendish banana group. *Trop. Agric.* (Trinadad). **1954,** *31,* 131–132.

Gunjate, R. T.; and Burondkar, M. M.; Parthenocarpic mango developed through hybridization. *Acta Hortic.* **1993,** *341,* 107–111.

Gupta, O. P.; Seedlings selection in grapewine (*Vitis vinifera* L.). M.Sc. Thesis. Mahatma Phule Krishi Vidyapeeth, Rahuri; **1971.**

Hampson, C. R.; Quamme, H. A.; Hall, J. W.; MacDonald, R. A., King, M. C., and Cliff, M. A. Sensory evaluation as a selection tool in apple breeding. *Euphytica.* **2000,** *111,* 79–90.

Hayes, W. B.; *Fruit Growing in India,* 3rd ed.; Kitabistan: Allahabad; **1957,** pp 299.

Hillebrand, S.; Schwarz, M.; and Winterhalter, P.; Characterization of anthocyanins and pyrano-anthocyanins from blood orange (*Citrus sinensis* L.) Osbeck juice. *J. Agric. Food Chem.* **2004**, *52*(24), 7331–7338.

Hoda, M. N.; and Ram Kumar; Improvement of mango. In: *Proceedings of the National Seminar on Irregular Bearing in Mango-Problems and Strategies* (12–13 July 1991). Kamal Printing Press, Muzafarpur, Bihar, India, **1993**, pp 34–35.

Hodgson, R. W.; Horticultural varieties of citrus. In: *The Citrus Industry*, 2nd ed.; Reuther, W.; Batchelor, L. D.; and Webber, H. J. Eds.; **1967**, Vol. 2, pp 431–591.

Hofmeyr, J. D. J.; Further studies of tetraploidy in *C. papaya* L. *South African J. Hortic.* **1945**, *41*, 225–230.

Hofmeyr, J. D. J.; Inheritance of dwarfness in *Carica papaya* L. *South African J. Hortic.* **1942**, *45*, 96–99.

Horvat, R. J.; and Chapman, G. W.; Comparison of volatile compounds from peach fruit and leaves (cv. Monroe) during maturation. *J. Agric. Food. Chem.* **1990**, *38*, 1442–1444.

Hulamani, N. C.; Mokashi, A. N.; Sulikeri, G. S.; and Rao, M. M.; National Symposium on Tropical Subtropical Fruit Crop. Bangalore, **1981**, p 18.

Iglesias, I.; Carbo, J.; Bonany, J.; Dalmau, R.; Guanter, G.; Montserrat, R.; Moreno, A.; and Pages, J.; Manzano, las variedades de ma´s intere´s. Institut de Recerca i Tecnologia Agroalimenta`ries, Barcelona, **2000**, 240 pp.

Infante, R.; Farcuh, M.; and Meneses, C.; Monitoring the sensorial quality and aroma through an electronic nose in peaches during cold store. *J. Sci. Food Agric.* **2008a**, *88*, 2073–2078.

Infante, R.; Martinez-Gomez, P.; and Predieri, S.; Quality oriented fruit breeding: Peach. *J. Food Agric. Environ.* **2008b**, *6*(2), 342–355.

Iyer, C. P. A.; and Randhawa, G. S.; Increasing colchicines effectiveness in woody plants with special reference to fruit crops. *Euphytica.* **1965d**, *14*, 293–295.

Iyer, C. P. A.; and Subramanyam, M. D.; IIHR Selection-8 and improved guava cultivar. *South. Indian Hortic.* **1988**, *36*(5), 258–259.

Iyer, C. P. A.; and Subramanyan, M. D.; Improvement of mango. In: *Advances in Horticulture*, Chadha, K. L.; and Pareek, O. P. Eds.; Malhotra Publishing: New Delhi; **1993**, Vol. 1, pp 267–278.

Janick, J.; History of fruit breeding. In: Temperate Fruit Breeding. *Fruit, Vegetable Cereal Sci Biotechnol.* **2011**, *5*(Special Issue 1), pp 1–7.

Janick, J.; The origins of fruits, fruit growing and fruit breeding. *Plant Breed. Rev.* **2005**, *25*, 255–326.

Janick, J.; and Moore, J. N.; Eds. *Advances in Fruit Breeding.* Purdue University Press: West Lafayette, Indiana; **1975**.

Janick, J.; and Moore, J. N.; *Fruit Breeding.* Wiley: New York; **1996**, Vol. 3.

Janick, J.; Cummins, J. N.; Brown, S. K.; and Hemmat, M.; Apples. In: *Fruit Breeding, Tree and Tropical Fruits*, Janick, J.; and Moore, J. N. Eds.; John Wiley & Sons, Inc., **1996**, Vol. 1, pp 1–77.

Jimenez-Bermudez, S.; Redondo-Nevado, J.; Munoz-blanco, J.; Caballero, J. L.; LopezAranda, J. M.; Valpuesta, V.; Pliego-Alfaro, F.; Quesada, J. A.; and Mercado, J. A.; Manipulation of a strawberry fruit softening by antisense expression of a pectate lyase gene. *Plant. Physiol.* **2002**, *128*, 751–759.

Jindal, P. C.; Grape In: *Fruits of India: Tropical & Subtropical,* Bose, T. K. Ed.; Naya Prakash: Calcutta; **1985**, pp 219–276.

Jindal, P. C.; Singh, K.; and Pandey, S. N.; *Hary. J. Hortic. Sci.* **1983**, *12*, 32–39.

Jintanawongse, S.; Chunwongse, J.; Chumpong, S.; and Hiranpradit, H.; Improvement of existing commercial mango cultivars in Thailand. (Abstract). 6th International Mango Symposium, 6–9 April, Pattaya, Thailand; **1999**.

Kader, A. A.; Fruits in the global market. In: *Fruit Quality and its Biological Basis*, Knee, M. Ed.; Sheffield Academic Press: Sheffield, UK; **2002**, pp 1–16.

Kato, M.; Ikoma, Y.; and Matsumoto, H.; Accumulation of carotenoids and expression of carotenoid biosynthetic genes during maturation in citrus fruit. *Plant Physiol.* **2004**, *134*(2), 824–837.

Kays, S. J.; Preharvest factors affecting appearance. *Postharvest Biol. Technol.* **1999**, *15*, 233–247.

Kostermans, A. J. G. H.; and Bompard, J. M.; *The Mangoes: Their Botany, Nomenclature, Horticulture and Utilization.* Academic Press: London; **1993**.

Kumar, A.; and Ellis, B. E.; The phenylalanine ammonia-lyase gene family in raspberry. Structure, expression, and evolution. *Plant Physiol.* **2001**, *127*, 230–239.

Kurian , R. M.; and Iyer, C. P. A.; Stem anatomical characters in relation to tree vigour in mango (Mangifera indica L.). *Sci. Hortic.* **1992**, *50*, 245–253.

La Rosa, G.; and Tribulato, E.; Etna, un nuovo ibrido tra Satsuma e Clementine. *Frutticoltura.* **2002**, 12, 27–32.

Laurens, F.; Review of the current apple breeding programs in the world: objectives for scion cultivar development. *Acta Hortic.* **1999**, *484*, 163–170.

Lavi, U.; Sharon, D.; Tomer, E.; Adato, A.; and Gazit, S.; Conventional and modern breeding of mango cultivars and rootstocks. *Acta Hortic.* **1993**, *341*, 145–151.

Le Lezec, M.; Babin, J.; and Belouin, A.; Varietes inscrites an catalogue. *Arboricult. Fruitiere.* **1996**, *496*, 25–32.

Lea, A.; Amarganess and astringency: the procyanidins of fermented apple ciders. In: *Amarganess in Foods and Beverages*, Rouseff, R., Ed.; Elsevier: Amsterdam; **1990**, pp123–143.

Lee, H. S. Characterization of major anthocyanins and the colour of red fleshed budd blood orange (*Citrus sinesis*). *J. Agric. Food Chem.* **2002**, *50*(5), 1243–1246.

Lespinasse, Y.; and Frédéric, B.; Breeding for fruits. http:/wchr.agrsci.unibo.it/wc2/ lespinas.html; **1998**.

Lewinsohn, E.; Schalechet, F.; Wilkinson, J.; Matsui, K.; Tadmor, Y.; Nam, K. H.; Amar, O.; Lastochkin, E.; Larkov, O.; Ravid, U.; Hiatt, W.; Gepstein, S.; and Pichersky, E.; Enhanced levels of the aroma and flavor compound S-linalool by metabolic engineering of the terpenoid pathway in tomato fruits. *Plant Physiol.* **2001**, *127*, 1256–1265.

Lo Piero, A.; Puglisi, I.; and Rapisarda, P.; Anthocyanins accumulation and related gene expression in red orange fruit induced by low temperature storage. *J. Agric. Food Chem.* *53*(23), 9083–9088.

Lui, Q.; Xu. J.; and Liu, Y.; A novel bud mutation that confers abnormal patterns of lycopene accumulation in sweet orange fruit (*Citrus sinensis* L. Osbeck). *J. Exp. Bot.* **2007**, *58*(15–16), 4161–4171.

Majumder, P. K.; and Mukherjee, S. K.; Aneuploidy in guava I. Mechanism of variation in number of chromosome. *Cytologia.* **1972a**, *37*, 541–548.

Majumder, P. K.; and Mukherjee, S. K.; Aneuploidy in guava II. The occurrence of trisomics, tetrasomics and higher aneuploids in the progeny of triploid. *Nucleus.* **1972b**, *13*, 42–47.

Marais, Z.; Mango evaluation for breeding. Citrus and Subtropical Fruit Research Institute (CSFRI) Information Bulletin, **1992**, 234, 7.

Matsumoto, R.; Okudai, N.; and Yamamoto, M.; A new citrus cultivar 'Youkou'. *Bull. Fruit Tree Res. Sci.* **1999**, *33*, 67–76.

Matsumoto, R.; 'Shiranuhi': a late-maturing Citrus cultivar. *Bull. Fruit Tree Res. Sci.* **2001,** *35,* 115–120.

Matsumoto, R.; Yamamoto, M.; and Okudai, N.; A new citrus cultivar 'Amaka'. *Bull. Fruit Tree Res. Sci.* **2001a,** 35, 47–56.

Matsumoto, R.; Yamamoto, M.; and Kuniga, T.; New citrus cultivar 'Setoka'. *Bull. Fruit Tree Res. Sci.* **2001b,** *2,* 25–31.

Matsumoto, R.; Yamamoto, M.; and Kuniga, T.; New citrus cultivar 'Nishinokaori'. *Bull. Fruit Tree Res. Sci.* **2003,** *2,* 17–23.

Moore, J. N.; and Janick, J.; *Methods in Fruit Breeding.* Purdue University Press: West Lafayette, Indiana; **1983.**

Moore, J. N.; Brown, G. R.; and Brown, E. D.; Relationships between fruit size and seed number and size in blackberries. *Fruit Var. J.* **1974a,** *28,* 40–45.

Morpurgo, R.; Brunner, H.; Grasso, G.; Duren, M.; Roux, N.; and Afeza, R.; Enigma of banana breeding: a challenges for biotechnology. *Agro. Food. Indian Hi-Tech.* **1997,** *8*(4), 16–21.

Mukherjee, S. K.; Improvement of mango, grapes and guava. In: *Fruit Breeding in India,* Nijjar, G. S. Ed.; Oxford & IBH: New Delhi; **1977,** pp 15–20.

Mukherjee, S. K.; Chakraborty, S.; Sadhukhan, S. K.; and Saha, P.; Survey of mangoes of West Bengal. *Indian J. Hortic.* **1983,** *40,* 7–13.

Mukherjee, S. K.; Singh, R. N.; Majumdar, P. K.; and Sharma, D. K.; An improved technique of mango hybridization. *Indian J. Hortic.* **1961,** *18,* 302–304.

Mukherjee, S. K.; Singh, R. N.; Majumdar, P. K.; and Sharma, D. K.; Present Position regarding breeding of Mango (*Mangifera indica* L.) in India. *Euphytic.* **1968,** *17,* 462–467.

Mukunda, G. K.; Studies on the performance of certain clones of mango cv. Alphonso. PhD Thesis, University of Agricultural Sciences, Bangalore, India; **2003.**

Naik, K. C.; Improvement of the mango (*Mangifera indica* L.) by selection and hybridization. *Indian J. Agric. Sci. 18,* 35–41.

Nakano, M.; Nesumi, H.; and Yoshioka, T.; Segregation of plants with undeveloped anthers among hybrids derived from seed parent 'Kiyomi' (*Citrus unshiu* x *C. sinensis*). *J. Jpn. Soc. Hortic. Sci.* **2001,** *70*(5), 539–545.

Negi, S.S.; Rajan, S.; Ram Kumar, Sinha, G. C.; Yadav, I. S.; and Agarawal, P. K.; Development of new varieties of mango through hybridization. 5th International Mango Symposium, Tel Aviv, **1996,** 43 pp.

Noiton, D. A. M.; and Alspach, P. A.; Founding clones, inbreeding, coancestry and status number of modern apple cultivars. *J. Am. Soc. Hortic. Sci.* **1996,** *121,* 773–782.

Nybom, N.; On the inheritance of acidity in cultivated apples. *Heriditae.* **1959,** *45,* 332–350.

Olmo, H. P.; Breeding tetraploid grapes. *Proc. Am. Soc. Hortic. Sci.* **1952,** *59,* 285–290.

Oppenheimer, C.; Study tour report on subtropical fruit growing and research in India and Ceylon. Special Bulletin No. 3 State of Israel Ministry of Agriculture. Agricultural Research Station, Rehovot, Israel; **1956.**

Ourecky, D. K.; Pratt, C.; and Einset, J.; Fruiting behavior of large-berried and large-clustered sports of grapes. *Proc. Am. Soc. Hortic. Sci.* **1967,** *91,* 217–223.

Pandey, S. N.; Mango cultivars. In: *Mango Cultivation,* Srivastava, R. P. Ed.; International Book Distributing Co.: Lucknow, India; **1998,** pp 39–99.

Parker, D. D.; Retail price response to quality characteristics of fresh peaches by store type. *Agribusiness.* **1993,** *9,* 205–215.

Parker, D. D.; Zilberman, D.; and Moulton, K.; How quality relates to price in California fresh peaches. *Calif. Agric.* **1991,** *45,* 14–16.

Pandey, R. M.; and Pandey, S. N.; *The Grape in India.* ICAR: New Delhi; **1990,** pp 115.

Pathak, R. A.; and Dwivedi, R.; Report of Fruit Research work. Subtropical and Temperate Fruits. RAU, Pusa; **1988,** pp 76–77.

Pereira-Lorenzo, S.; Ramos-Cabrer, A. M.; Ascasıbar-Errasti, J.; and Pineiro-Andion, J.; Analysis of apple germplasm in Northwestern Spain. *J. Am. Soc. Hortic. Sci.* **2003,** *128*(1), 67–84.

Pereira-Lorenzo, S.; Ramos-Cabrer, A. M.; and Dıaz-Hernandez, M. B.; Evaluation of genetic identity and variation of local apple cultivars (*Malus x domestica*) from Spain using microsatellite markers. *Genet. Resour. Crop. Evol.* **2007,** *54,* 405–420.

Phadnis, N. A.; Improvement of guava (*P. guajava* L.) by selection in Maharashtra. *Indian J. Hortic.* **1970,** *27,* 99–105.

Phadnis, N. A.; and Gupta, O. P.; Seedling selection in the grape vine (*Vitis vinifera* L.) 3rd International symposium on subtropical Horticulture, HIS, Bangalore; **1972,** p 14.

Phadnis, N. A.; Kunte, Y. N.; Chougare, I. B.; and Bagde, T. R.; History of development of Selection-7 a promising variety of grapes for the Deccan. *Punjab. Hortic. J.* **1968,** *8*(2), 65–69.

Pinto, A. C. Q.; and Byrne, D. H.; Mango hybridization studies in tropical savannah ('Cerrados') of Brazil Central region. *Acta Hortic.* **1993,** *341,* 98–106.

Pinto, A. C. Q.; Andrade, S. R. M., Ramos, V. H. V. and Cordeiro, M. C. R.; Intervarietal hybridization in mango: techniques, main results and their limitations. *Acta Hortic.* **2004,** *645,* 327–331.

Pope, W. T.; Mango Culture in Hawaii. Hawaii Agricultural Experiment Station Bulletin 58; **1929.**

Prasad, A.; Grape in Uttar Pradesh. In: *Grape India.* Andhra Pradesh Grape Growers Association: Hyderabad; **1968,** pp 31–32.

Pratt, C.; Reproductive anatomy in cultivated grapes- a review. *Am. J. Enol. Vatic.* **1971,** *22,* 92–109.

Radha Devi, D. S.; and Nayar, N. K.; Gamma ray-influenced variation in polygenic character expression in banana. *Indian Bot. Contactor.* **1992,** *9*(4), 165–171.

Rajan, S.; Negi, S. S.; and Ram Kumar.; Improvement of guava by selection and hybridization. Annual Report (1996–1997) of Central Institute of Sub-tropical Horticulture, Lucknow, **1997,** p 10.

Rajput, M. S.; Chadha, K. L.; and Negi, S. S.; Dashehari – 51, a regular bearing and high yielding clone of mango cv. Dashehari. (Abstract). 5th International Mango Symposium, 1–6 September, Tel Aviv, Israel; **1996,** p 42.

Ram, M.; and Majumder, P. K.; Dwarf mutant of papaya (*Carica papaya* L.) Induced by gamma rays. *J. Nuclear. Agric. Biol.* **1981,** *10,* 72–74.

Rama Rao, M.; and Dayanand, T.; A note on the promising guava hybrids of Anantharajupet. *Andhra Agric. J.* **1977,** *24,* 1–2, 53–54.

Ramaswamy, N.; Survey and isolation of 'plus trees' of mango. *Acta Hortic.* **1989,** *231,* 93–96.

Randhawa, G. S.; and Chadha, K. L.; Grapes can grow in a big way in northern India. *Indian Hortic.* **1964,** *8*(2), 9–13.

Randhawa, G. S.; Nath, N.; and Yadav, I. S.; Improvement of grapes by selection and hybridization. *Indian J. Hortic.* **1965,** *22,* 219–230.

Rapisarda, P.; Bellon, S.; and Fabroni, S.; Juice quality of two new mandarin-like hybrids (*Citrus clementina* Hort. ex Tan. x *Citrus sinensis* L. Osbeck) containing Anthocyanins. *J. Agric. Food Chem.* **2008,** *56*(6), 2074–2078.

Rapisarda, P.; Fabroni, S.; and Peterek, S.; Juice of new citrus hybrids (*Citrus clementina* Hort. ex Tan. x *C. sinensis* L. Osbeck) as a source of natural antioxidants. *Food Chem.* **2009**, *117*(2), 212–218.

Rapisarda, P.; Pannuzzo, P.; and Romano, G.; Juice components of a new pigmented citrus hybrid *Citrus sinensis* (L.) Osbeck x *Citrus clementina* Hort ex Tan. *J. Agric. Food Chem.* **2003**, *51*(6), 1611–1616.

Redgwell, R. J.; and Fischer, M.; Fruit texture, cell wall metabolism and consumer perceptions. In: *Fruit Quality and its Biological Basis*, Knee, M. Ed.; Sheffield Academic Press: Sheffield, UK; **2002**, pp 46–88.

Redgwell, R. J.; MacRae, E.; Hallett, I.; Fischer, M.; Perry, J.; and Harker, R.; *In-vivo* and *in-vitro* swelling of cell walls during fruit ripening. *Planta.* **1997**, *203*, 162–173.

Reece, P.; Hearn, C.; and Gardner, F.; Nova tangelo: an early ripening hybrid. *Fla State Hortic Soc.* **1964**, *7*, 109–110.

Rodrigo, M.; Marcos, J.; and Alferez, F.; Characterization of Pinalate, a novel Citrus sinensis mutant with a fruit-specific alteration that results in yellow pigmentation and decreased ABA content. *J. Exp. Bot.* **2003**, *54*(383), 727–738.

Rowe, P. R.; Breeding banana and plantains. *Plant Breed. Rev.* **1984**, *2*, 135–155.

Rowe, P. R.; and Rosales, F.; Banana and plantains. In: *Fruit Breeding-Tree and Tropical Fruits*, Janick, J.; Moore, J. N. Eds.; John Wiley & Sons: New York; **1996**; pp 167–211.

Roy, B.; A mango chimera. *Curr. Sci.* **1950**, *19*, 93.

Roy, B.; and Visweswariya, S. S.; Cytogenetics of mango and banana. Report of Maharashtra Association for Cultivation of Science, Pune, India; **1951**.

Roy, R. S.; Mallik, P. C.; and Sinha, R. P.; Mango breeding in Bihar, India. *Proc. Am. Soc. Hortic. Sci.* **1956**, *68*, 259–264.

Ruiz-Altisent, M.; Lleo, L.; and Riquelme, F.; Instrumental quality assessment of peaches: fusion of optical and mechanical parameters. *J. Food Eng.* **2006**, *74*, 490–499.

Sachan, S. C. P.; Katrodia, J. S.; Chundawat, B. S.; and Patel, M. N.; New mango hybrids from Gujarat. *Acta Hortic.* **1988**, *231*, 103–105.

Salvi, M. J.; and Gunjate, R. T.; Mango breeding work in Konkan region of Maharashtra state. *Acta Hortic.* **1988**, *231*, 100–102.

Satyanarayana, M.; Tatachari, Y.; and Dasarathi, T. B.; Effect of chemical mutagens on sprouting and survival of banana corms. In: *Proceedings of National Seminar on Banana*, Muthukrishnana, C. R.; and Abdul-Khader, J. B. M. M. Eds.; TNAU: Coimbatore; **1980**, pp 65–66.

Saunt, J.; *Citrus Varieties of the World*, 2nd ed.; Sinclair International Limited: Jarrold Way, Bowthorpe, Norwich, England; **2000**.

Scorza, R.; May, L. G.; Purnell, B.; and Upchurch, B.; Differences in number and area of mesocarp cells between small and large fruited peach cultivars. *J. Am. Soc. Hortic. Sci.* **1991**. *116*, 861–864.

Sen, P. K.; Mallik, P. C.; and Ganguly, B. D.; Hybridization of the mango. *Indian J. Hortic.* **1946**, *4*, 4.

Seth, J. N.; and Ghidyal, P. C.; Some promising apple hybrids evolved and developed at Chaubattia. *Prog. Hortic.* **1984**, *15*, 239–243.

Seymour, G. B.; and Manning, K.; Genetic control of fruit ripening. In: *Fruit Quality and its Biological Basis*, Knee, M., Ed.; Sheffield Academic Press: Sheffield, UK; **2002**.

Shanmugavelu, K. G.; Selvaraj, M.; and Thamburaj, S.; Review of research on fruit crops in Tamil Nadu. *South Indian Hortic.* **1987**, *35*, 1–3.

Sharma, D. K.; and Majumder, P. K.; Induction of variability in mango through physical and chemical mutagens. *Acta Hortic.* **1988b,** *231,* 112–116.

Sharma, S. D.; and Uppal, D. K.; Transmission of seedlessness in inter-varietal crosses of grapes. *Indian Hortic.* **1978,** *26*(4), 184–185.

Sharma, Y. D.; Chadha, T. R.; and Gupta, G. K; Breeding of apple varieties with better keeping quality and disease resistance. In: Advances in Research on Temperate Fruits. Symposium held at Dr Y. S. Parmar University of Horticulture and Forestry, Solan, **1986,** pp 65–68.

Shun, C. J. and Chen, Q. Y.; Shuitu seedless red pomelo. *Fruit Tree South China.* **1998,** *27,* 3–8.

Siddiqui, S. H.; Induced somatic mutation in mango (*Mangifera indica* L.) cv. Langra. *Pakistan J. Bot.* **1985,** *17,* 75–79.

Siddiqui, S. H.; Mujeeb, K. A.; and Vati, S. M.; Evolution of new varieties of mango (*Mangifera indica* L.) through induced somatic mutations by ionizing radiations. In: Proceedings of the 1st Agricultural Symposium Atomic Energy Commission, Dhakka, Bangladesh, **1966,** pp 34–37.

Simmonds. N. W.; *Bananas,* 2nd ed.; Longman: London, **1966,** pp 512.

Simmonds, N. W.; Megasporogenesis and female fertility in three edible triploid bananas. *Genetics.* **1960,** *57,* 269–278.

Singh, L. B.; S$_1$: a new promising selection of guava (*P. guajava* L.). Annual Report, Fruit Research Station, Saharanpur, **1959,** pp 58–60.

Singh, L. B.; *Mango Botany: Cultivation and Utilization.* Leonard Hill: London; **1960.**

Singh, R. N.; Hybridization and mango improvement. *Indian J. Hortic.* **1954,** *11,* 69–88.

Singh, H.; and Chadha, K. L.; Improvement of Dashehari by clonal selection. (Abstract). National Symposium on Tropical and Subtropical Fruit Crops, HIS, Bangalore; **1981,** p 5.

Singh, R.; and Murthy, B. N. S.; Improvement of grape. In: *Advances in Horticulture,* Chadha, K. L.; and Pareek, O. P. Eds.; Malhotra Publishing House: New Delhi; **1993,** pp 349–381.

Singh, J. P.; Daulta, B. S.; and Godara, N. R.; An early clone of perlette grapes. *Haryana J. Hortic. Sci.* **1974a,** *31* (1/2), 92–93.

Singh, R.; Murthy, B. N. S.; and Srinivas, T. R.; Arka Neelamani, Arka Sweta, Arka Majestic and Arka Chitra: New grape hybrids. *Indian Hortic.* **1998,** *43*(2), 28–29.

Singh, R. N.; Gorakh, S.; Rao, O. P.; and Mishra, J. S.; Improvement of Banarsi Langra through clonal selection. *Prog. Hortic.* **1985,** *17,* 273–277.

Singh, R. N.; Majumder, P. K.; Sharma, D. K.; and Mukherjee, S. K.; Some promising mango hybrids. *Acta Hortic.* **1972,** *24,* 117–119.

Smith, D. L.; Abbott, J. A.; and Gross, K. C.; Down-regulation of tomato β-galactosidase 4 results in decreased fruit softening. *Plant Physiol.* **2002,** *129,* 1–8.

Sparta, G.; Pasciuta, G.; and Maugeri, V.; Scelte varietali in Agrunicoltura. ed. Regione Siciliana Assessorato Agricoltura e Foreste IX Servizio Regionale Servizi allo Sviluppo, Edi BO, Sicily, Italy; **2006.**

Stockwin, W.; Boost stone fruit quality. *Fruit Grower,* **1996,** (June), 13–16.

Sturrock, D.; Final report on some mango hybrids-1969. *Proc. Fla. State Hortic. Soc.* **1969,** *82,* 318–321.

Subramanyam, M. D.; and Iyer, C. P. A.; Improvement of guava In: *Advances in Horticulture;* Chadha, K. L., Pareek, O. P. Eds.; Malhotra Pub.: New Delhi; **1993,** Vol. 1, pp 337–347.

Swamy, G. S.; Ramarao, B. V.; and Reddy, P. J.; Hybrid mangoes of Andhra Pradesh. *Acta Hortic.* **1972,** *24,* 139–142.

Thonte, G. T.; and Chakrawar, V. R.; The variability and correlation studies of guava strain. National symposium on subtropical fruit crops, Bangalore, **1981**; p 17.Tribulato, E.; and La Rosa, G.; 'Desiderio' e 'Bellezza': Due nuovi ibridi di mandarino. *Rivista di Frutticoltura.* **1996a**, *2*, 45–48.

Tribulato, E.; and La Rosa, G.; 'Ippolito': Un nuovo clone di Tarocco. *Frutticoltura* **1994**, *1*, 34–35.

Tribulato, E.; and La Rosa, G.; L'arancio Tarocco ed I suoi cloni. *Frutticoltura.* **1994**, *11*, 9–14.

Tribulato, E.; and La Rosa, G.; 'Primosole' e 'Simeto': Due nuovi ibridi di mandarino. *Italus Hortus,* **1993**, *1*, 21–25.

Tribulato, E.; Inglese, G.; and Reforgiato, R. G.; Liste Varietali Agrumi, Mandarino-simili. Suppl. *L'informatore. Agrario.* **2002**, *46*, 9–14.

Tulmann Neto, A.; *In vitro* methods for the induction of mutations in 'Maca' banana improvement programmes. *Revista Brasileira de Genetica.* **1989**, *12*(4), 871–879.

Ufuk, K.; Mark, A.; and Febres, K.; Decreasing unpalatable flavonoid components in *Citrus*: The effect of transformation construct. *Physiologia Plantarum.* **2009**, *137*(2), 101–114.

Vardi, A.; Conventional and novel approaches to Citrus breeding. *Proc. Int. Soc. Citric.* **1992**, *1*, 34–43.

Vijaya Kumar, M.; Ramaswamy, N.; and Rajagopalan, R.; Exploiting natural variability in mango. In: Proceedings of the National Seminar on Irregular Bearing in Mango – Problems and Strategy, RAU, Sabour, Bihar, India, **1991**, pp 55–56.

Visser, T.; Schaap, A. A.; and De Vries, D. P.; Acidity and sweetness in apple and pear. *Euphytica.* **1968**, *17*, 153–167.

Vrebalov, J.; Ruezinsky, D.; Padmanabhan, V.; White, R.; Medrant, D.; Drake, R.; Schuch, W.; and Giovannoni, J.; A MADS-box gene necessary for fruit ripening at the tomato Ripening-Inhibitor (Rin) locus. *Science.* **2002**, *296*, 343–346.

Wagner, E.; On spontaneous tetraploid mutants of *Vitis vinifera* L. *Vitis.* **1958**, *1*, 197–217.

Wang, C. T.; and Xiang, Z. X.; Selection of high sugar and early ripe mutants of 'Miyagawa' wase by irradiation. *Zhejiang Citrus.* **1998**, *15*(1), 16–18.

Watada, A. E.; Abbott, J. A.; Hardenburg, R. E.; and Lusby, W.; Relationships of apple sensory attributes to headspace volatiles, soluble solids and titratable acids. *J. Am. Soc. Hortic. Sci.* **1981**, *106*, 130–132.

Weinberger, J. H.; and Harmon, F. N.; Seedlessness in vinifera grapes. *Proc. Am. Soc. Hortic. Sci.* **1964**, *85*, 270–274.

Whiley, A. W.; Mayers, P. E.; Saranah, J.; and Bartley, J. P.; Breeding mangoes for Australian conditions. *Acta Hortic.* **1993**, *341*, 136–145.

Williams, T.; Recently released and promising new Citrus varieties from the University of California Breeding Programme available at www.citrusvariety.ucr.edu/documents/recently_ released. pdf; **2010**.

Wilson, G. B.; Cytological studies in the Musae I. Meiosis in some triploid clones. II. Meiosis in some diploid clones. *Genetics.* **1946**, *31*, 241–258/475–482.

Winkel-Shirley, B.; Flavonoid biosynthesis. A colorful model for genetics, biochemistry, cell biology and biotechnology. *Plant. Physiol.* **2001**, *126*, 485–493.

Xu, J.; and Deng, X. X.; Red juice sacs of citrus and its main pigments. *J. Fruit. Sci.* **2002**. *19*, 307–313.

Xu, J.; Tao, N.; and Liu, Q.; Presence of diverse ratios of lycopene beta-carotene in five pink or red fleshed citrus cultivars. *Sci. Hortic.***2006**. *108*(2), 181–184.

Yadav, I. S.; Pandey, S. K.; and Nigam, K. S.; Performance of new introduction. IV. Italian grapes (*V. vinifera*). *Prog. Hortic.* **1977,** *8*(4), 5–13.

Yadav, I. S.; Pandey, S. N.; and Singh, R. N.; Performance of new introduction. III. Australian and Japanese grapes. *Prog. Hortic.* **1971,** *3*(3), 71–83.

Yamamoto, M.; Matsamoto, R.; and Yamado, Y.; Relationship between sterility and seedlessness in Citrus. *J. Jpn. Soc. Hortic. Sci.* *64*(1), 23–29.

Yoshida, T.; and Yamada, Y.; New citrus cultivar 'Akemi'. *Bull. Fruit Tree Res. Sci.* **2001,** *32*, 53–62.

Yoshida, T.; Matsumoto, R.; Okudai, N.; New citrus cultivar 'Reikou'. *Bull. Fruit Tree Res. Sci.* **2009,** *8*, 5–23.

Yoshida, T.; Yamada, Y.; and Nesumi, H.; New citrus cultivar 'Harumi'. *Bull. Fruit Tree Res. Sci.* **2000,** *32*, 43–52.

Young, T. W.; and Ledin, R. B.; Mango breeding. *Proc. Fla. State. Hortic. Soc.* **1954,** *67*, 241–244.

Yukinori, S.; and Keiji, K.; New citrus cultivar 'Himekoharu'. *Bull. Bhune Fruit Tree Sci.* **2008,** *22*, 5.

Zerpa, D. M.; Triploides De *Carica papaya* L. *Agronomy Tropical.* **1957,** *7*, 83–86.

Zhang, J.; Zhang, M.; and Deng, X.; Obtaining autotetraploids in vitro at a high frequency in *Citrus sinensis*. *Plant. Cell. Tiss. Organ. Cult.* **2007.** *89*, 211–216.

CHAPTER 4

ADVANCES IN CONVENTIONAL BREEDING APPROACHES FOR POSTHARVEST QUALITY IMPROVEMENT IN VEGETABLES

SHIRIN AKHTAR[1]

[1]Department of Horticulture (Vegetable and Floriculture),
Bihar Agricultural University, Sabour, Bhagalpur, Bihar (813210) India;
Email: shirin.0410@gmail.com

CONTENTS

The word "quality" is derived from the Latin word "*qualitas*" meaning attribute, property, or basic nature of an object. Quality may otherwise be defined as "degree of excellence or superiority" and is a combination of attributes, properties, or characteristics that make the commodity valuable as food (Kader, et al., 1999). Hence, a product is can be rated as of better quality when it is superior in one or several attributes that are valued objectively or subjectively. When it comes to vegetables, producers would want higher yield, better appearance, ease of harvest and withstanding transportation. For wholesalers and retailers vegetables should be attractive in appearance, firm and the shelf life should be high. However, the final judgment of the quality of the product comes from the satisfaction of the consumers. From this point of view, quality may be defined as degree of fulfillment of a number of conditions that determine its acceptance by the consumer. Consumers are concerned about appearence, freshness, firmness, flavor as well as nutritive values. It can be said that the ultimate aim of vegetable producers is to satisfy consumers and that depends on the quality of the produce that reaches him. (Shewfelt et al., 1997). This can also be referred to as postharvest quality of the vegetables. Hence it may be said that quality of vegetables includes the sensory properties, *viz*, appearance, texture, odor, aroma, taste, and flavor as well as safety, chemical constituents, nutritive values, mechanical properties, functional property and defects (Abbott, 1999).

Vegetable breeding programmes were initiated with commercial production components such as yield, uniformity, and resistance to diseases and pests. Later on consumer preference components such as appearance, postharvest durability, flavor, color, etc. came to the attention of vegetable breeders. Breeding for improved quality initially requires a definition of the major parameters that contribute to quality. Moreover, the criteria of good quality vary among consumer groups. The subjectiveness of quality introduces a great deal of complications in its genetic improvement. Therefore, quality has often taken a backseat in the breeding programmes. Recently, the concern over providing nutritional security has given consumption of vegetables an immense significance and has made the content of vitamins, minerals, and other phytonutrients in the vegetables a major breeding objective.

4.1 COMPONENTS OF QUALITY

Quality of vegetables is a broad term. It encompasses the following:
  Cosmetic quality
  Taste quality
  Nutritional factors
  Antinutritional factors

4.2   COSMETIC QUALITY

Cosmetic quality of vegetables is includes the appearance and the texture or feel. Appearance includes shape, size (dimension, weight and volume), form, smoothness, compactness, uniformity, color, freedom from any defects. Texture, on the other hand, includes firmness, hardiness, softness, crispiness, succulence, juiciness, mealiness, grittiness, toughness, fibrousness of the vegetable.

Appearance is the first impression that the consumer receives and the most important component of the acceptance and eventually of the purchase decision. Different studies indicate that almost 40 percent of the consumers decide what to buy inside the supermarket (Camelo, 2004). Shape is one of the subcomponents more easily perceived, although in general, it is not a decisive aspect of quality, except in case of deformations or morphological defects. Compactness is the most relevant feature in vegetables where the inflorescence is the marketable organ, *viz.*, broccoli or cauliflower or where leaves form "heads" like lettuce, cabbage, endive, etc. Compactness, though not associated with organoleptic characteristics, is an indicator of the degree of development at harvest.Open inflorescences indicate that they were picked too late while noncompact heads result due to premature harvest. Besides, compactness also indicates freshness of the produce since compactness decreases with dehydration.

Uniformity of the produce is a concept conforming to all the components of quality (size, form, color, ripeness, compactness, etc.). The total quantity of the produce belonging to one grade should be similar. It is a relevant feature for the consumer depicting that the product has already been categorized according to the official standards of quality for the concerned product. The main activity in preparation for the market is making products uniform.

One of the main components of appearance is the absence of defects and often it becomes the deciding factor for purchase of vegetables. In many cases, internal or external defects do not affect product excellence, but the consumer rejects them. Different causes during growth (climate, irrigation, soil, variety, fertilization, etc.) can lead to morphological or physiological defects. Cat face in tomatoes, knobby tubers and hollow heart in potatoes, root ramifications in carrots, etc. are some examples of physical defects. Tip burn on leafy vegetables and black heart in celery, blossom end rot in tomato due to calcium deficiencies as well as the internal rots in various vegetables due to boron deficiencies are examples of physiological defects. More serious are those physical or physiological defects that originate during or after preparation for the fresh market and that show up at retail or consumer's level such as the mechanical damages, bruises, or wounds that take place during the handling of the product and that are the entrance doors to most pathogens causing postharvest rots. Chilling injury, ethylene effects as well as sprouting and rooting, are physiological responses to inadequate storage conditions.

Freshness is another important component of cosmetic quality and also contributes to appearance. They are also indicative of the expected flavor and aroma when products are consumed. Freshness, the condition of being fresh indicates how much time has elapsed since harvest. In vegetables harvest is the point of maximum organoleptic quality characterized by the greatest turgidity, color, flavor, and crispness.

In determining the freshness of the vegetable, color (both intensity and uniformity) is the external aspect most easily evaluated by the consumer. It is the decision making factor in case of leafy vegetables or immature fruit vegetables such as cucumber, snapbeans, etc. where an intense green is associated with freshness and pale green or yellowing to senescence. Colour is also an indicator of maturity and very important in case of nonclimacteric fruit vegetables, such as pepper, eggplant, and cucurbits those where no substantial changes take place after harvest in general. In climacteric ones where changes occur after harvest, color is less decisive and basically indicates the degree of maturity, as in tomato. In tomato color also indicates the stage of harvest for different purposes, that is, mature green and breaker stage for long distance transportation, pink stage for local market and red ripe stage for processing.

Size of the vegetable has its own importance for vegetable cosmetic quality and in general, intermediate sizes are preferred. In vegetables that are naturally large such as pumpkins, watermelons, melons, etc., there is a much defined trend toward sizes like 1–2kg mass that can be consumed by a family in a relatively short period of time (1 week). Size is one of the main indicators of the moment of harvest and in many cases it is directly associated to other aspects of quality such as flavor or texture. This is especially observed in case of zucchinis, peas, haricot beans and miniature vegetables where consumers particularly value small sizes.

Gloss enhances the color of most products, and it is particularly valued in vegetables like pepper, eggplant, tomato, etc., to such an extent that they may be waxed and polished to improve their shine. In vegetables, gloss is associated in a certain way to turgidity: a brilliant shiny green is often an indicator of freshness. Glossiness can be used as a harvest index in eggplants, cucumbers, squash, etc. that are harvested unripe where the decrease in shine indicates that they are overmature and have lost part of their characteristics of flavor and texture. On the contrary, in melons it indicates that the crop has reached proper maturity stage for harvest.

Texture is defined by different sensations as perceived by humans. Thus, firmness as well as the nature of the surface (hairy, waxy, smooth, rough, etc.) is perceived with the hands, together with the lips, while teeth determine the rigidity of the structure that has been chewed. The tongue and the rest of the mouth cavity detect the type of particles that are crushed by teeth (soft, creamy, dry, juicy, etc.). The ears also contribute to the sensation of texture, for example, the noises generated when chewing in those species where crispness is an important aspect (Wills, et al., 1981).

Texture in combination with flavor and aroma, constitutes the eating quality. An overripe tomato, for example, is mainly rejected by its softening and not because important changes in the flavor or aroma have taken place. Although it is decisive for the quality of some vegetables, in others it has a relative importance. In terms of texture, each product is valued differently; for its firmness (tomato, pepper), the absence of fibers (asparagus, globe artichoke), crispness (lettuce, celery, carrot), etc.

4.3   TASTE QUALITIES

4.3.1   FLAVOR

Flavor is the combination of the sensations perceived by the tongue (i.e., taste) and by the nose (i.e., aromas) (Wills, et al., 1981). Although these sensations can be perfectly separated from each other, the sensitive receptors being at close proximity, with the act of bringing near the mouth, biting, chewing, and tasting, the aromas are distinctly perceived, particularly those that are liberated with the crushing of tissues. People also have anticipation of flavor that vegetables might have from a preconceived notion about it.

In vegetables, taste is usually expressed in terms of the combination of sweetness and sourness which is estimated by the total soluble solids (TSS) and the acidity content. Organic acids, particularly, citric, malic, oxalic, and tartaric, are the very important components of taste, particularly in relation with TSS. Organic acid content is quantified as titratable acidity. With maturity the quantity of organic acids tend to decrease and the TSS:titratable acidity ratio increases.

Astringency (sensation of loss of lubrication in the mouth) and bitter tastes are due to different compounds. They are rare and specific to vegetables. When these traits exist, they usually diminish with maturity, excepting in cases as bitter gourd-where it is the characteristic feature and appear naturally due to presence of cheratin.

There are also certain specific compounds that characterize different vegetables, for example, pungency in hot peppers is due to capsaicin content and other four-structurally similar compounds. In some vegetables the compounds responsible for the taste are compartmentalized in healthy tissues and they only get in contact by cutting, chewing, or crushing. This is found in case of pungency in garlic and onion where the compound alliin is present, which on cutting and crushing is converted to allicins by the enzyme allinase causing the typical pungency. Cooking these vegetables as a whole without cutting or crushing prevents these reactions and the resulting taste is different.

The aroma of vegetables is due to the human perception of numerous volatile substances which are generally aromatic compounds present in them. On refrigeration the liberation volatile aromatic compounds reduce at low temperature resulting in lesser aroma. It is often seen that these aromas are frequent when the tissues degenerate.

4.4 NUTRITIONAL FACTORS

Vegetables, being low in dry matter content are insufficient to satisfy daily nutritional requirements. They have high water content and are generally low in carbohydrates (except potatoes, sweet potatoes and cassava), proteins (except for legumes and some crucifers) and fat. However, they are good sources of minerals vitamins and dietary fibers. The actual content of nutrients varies according to conditions of cultivation, varieties, climate, etc.

The discovery that certain foods have biologically active compounds, beneficial to health beyond basic nutrition opened a new horizon for food and nutrition science. These compounds or their metabolites that have been designated as "functional", help in prevention of diseases like cancer, cardiovascular problems, neutralize free radicals, reduce cholesterol and hypertension, and prevent thrombosis, along with other manifold beneficial effects. As most of these compounds are of plant origin, they are referred to as phytochemicals. Vegetables are rich sources of these phytochemicals, such as, terpenes (carotenoids in yellow, orange, and red vegetables like tomatoes and carrots), phenols (red and purple colors as in eggplant and beet), lignans (broccoli), thiols (sulfur compounds present in garlic, onion, leek and other alliums, cabbages, and other crucifers) and others.

4.5 ANTINUTRITIONAL FACTORS

However, vegetables also contain certain compounds which may have adverse effect on health, may contribute to off-flavor, acridity, astringency, bitterness that render them unacceptable to consumers and unfit for consumption. Some of these are present naturally and sometimes under adverse growing conditions or at improper stage of harvest are responsible for presence of these substances. The aims of all breeding programmes have been to remove these antinutritional factors.

A list of quality attributes specific to different vegetables has been given in the Table 4.1.

TABLE 4.1   Quality attributes specific to different vegetables

Crop	Cosmetic and Taste Qualities	Nutritional Factors	Antinutritional Factors
Tomato	Colour, shape, firmness index, number of locules, pericarp thickness, alcohol insoluble solids, dry matter, total soluble solids, titratable acidity, TSS: acidity ratio, pH, etc.	Carotenoids especially lycopene, beta-carotene, Vitamin C, Vitamin K, potassium, etc.	Oxalates, caffeic acid derivatives, etc.

TABLE 4.1 *(Continued)*

Crop	Cosmetic and Taste Qualities	Nutritional Factors	Antinutritional Factors
Eggplant	Texture, shape, size, color, less seeded, pulp softness, flavor, etc.	Fiber, carbohydrates, potassium, manganese, calcium, Vitamin C, anthocyanin.	Steroidal alkaloids (solanine, solasonine, solamargine, solasodine), trypsin inhibitors in fruit peel, phenols, amide proteins, etc.
Hot pepper	Color, fruit smoothness, capsaicin content, capsarubin content, oleoresin content, etc.	Carotenoids especially capsanthin, beta-carotene, Vitamin C, Vitamin B6, fiber content, etc.	None major
Sweet pepper	Fruit color, fruit weight, firmness, sugar, organic acids, etc.	Carotenoids especially capsanthin, beta-carotene, Vitamin C, Vitamin B6, fiber content, etc.	Phenolic compounds like chlorogenic acid.
Potato	Tuber shape, size, color, flavor, starch quality, sweetening under cold storage	Carbohydrates, proteins, minerals, Vitamin A, Vitamin C.	Alkaloids like alpha-solanine, alpha-chaconine, lectins, trypsin inhibitors
Cauliflower	Color, shape, texture, compactness, flavor of curd	Vitamin C, beta-carotene, Vitamin B6, calcium, magnesium, Vitamin K, folate, fiber.	Glucosinolates (sulfur containing glucosides)
Cabbage	Compactness, crispness, flavor, shape, and appearance of head	Vitamin C, beta-carotene, Vitamin B6, calcium, magnesium, protein, Vitamin K, folate, fiber.	Phytic acid, tannic acid, trace amounts of oxalic acid
Chinese cabbage	Crispness, flavor, shape, and appearance of head	Vitamin C, beta-carotene, Vitamin B6, calcium, magnesium, protein, Vitamin K, folate, fiber.	Trace amounts of phytate and oxalate
Onion	Flavor, color, pungency, and shape of bulb, TSS	Allicin, sulfur compounds, flavonoids, chromium, fiber	Tannins, saponins, oxalates, cynogenic glucosides

TABLE 4.1   *(Continued)*

Crop	Cosmetic and Taste Qualities	Nutritional Factors	Antinutritional Factors
Garlic	Flavor, color, pungency, shape of cloves and bulb, TSS	Allicin, sulfur compounds, flavonoids, chromium, fiber	Tannins, saponins, oxalates, cynogenic glucosides
Okra	Fibrousness, mucilage content, dry matter content (high for dehydration purpose, low for freezing and canning purposes)	Protein, minerals, carotenoids, fiber, Vitamin C, Vitamin K, iodine content	Phytate, saponin contents
Pumpkin	Appearance, thickness of rind and flesh, color, TSS	Carotenoids especially beta-carotene, sugar, Vitamin C, Vitamin B6.	Stachyose, raffinose, verbascose, trypsin inhibitor, phytic acid, tannin, saponin
Melons	Appearance, color, flavor, odor, sweetness, texture, flesh color, TSS, TSS: acidity ratio, flesh firmness	Beta-carotene, Vitamin C, niacin, folic acid, sodium, magnesium, protein, carbohydrate, dietary fiber, potassium	Tannin, saponin
Bitter gourd	Shape, presence of tubercles on surface, unique taste, flavor	Vitamin C, iron, protein, phosphorus, medicinal properties	Phytic acid, phytin, polyphenol, saponin
Bottle gourd	Appearance, firmness, TSS, shape, color	Carbohydrates, dietary fiber, proteins, Vitamin C, iron, sodium	Phytic acid, phytin, slight traces of lead in seeds
Cucumber	Flesh color, flesh firmness, smell of fruit, juiciness of fruit, sweet, bitter, or off taste	Carbohydrates, sugars, dietary fiber, thiamine, riboflavin, niacin, pantothenic acid, Vitamin C, calcium, iron, magnesium, phosphorus, potassium, zinc	Trypsin inhibitor, phytate, lectin, tannin in seed
Carrot	Shape, size, color, sweetness, harshness, bitterness	Carotenoids especially beta-carotene, Vitamin C, Vitamin K, fiber, potassium.	None major
Cowpea	Color, tenderness, glossiness of pods	Protein, methionine, Vitamin B, minerals	Trypsin inibitors, tannins, phytic acids, hemagglutinin

TABLE 4.1 *(Continued)*

Crop	Cosmetic and Taste Qualities	Nutritional Factors	Antinutritional Factors
Hyacinth bean	Stringiness, fleshy pod pericarp, fibrousness, sugar: polysaccharide ratio	Protein, flavonoids, sugar, Vitamin C, Vitamin K, manganese, fiber, potassium, folate content	Proteolytic inhibitors, phytohaemagglutinin, lathyrogens, cyanogenetic compounds affecting digestibility, saponins
Garden pea	Texture, color, sweetness	Protein, Vitamin C, manganese, fiber, Vitamin B1, folate	Chymotrypsin
Lettuce	Greenness of leaf, turgidity, crispiness	Carbohydrates, dietary fiber, protein, beta-carotene, folate, vitamin C, Vitamin K, Vitamin E, iron	Nitrate, oxalate, saponin
Sweet potato	Colour, shape, size, flavor, starch, cooking attributes	Carbohydrates, carotenoids especially beta-carotene, Vitamin C, manganese	Saponin, oxalate, tannin, phytate, trypsin inhibitors
Colocasia	Corm and cormel shape, starch chemistry, cooking attributes	Ascorbic acid, reducing and nonreducing sugar, starch, Vitamin B6, niacin, potassium, copper, manganese	Oxalic acid, trypsin inhibitor
Leafy amaranth	Colour of leaf, texture, tenderness, flavor	Beta-carotene, vitamin C, fiber, minerals like iron, calcium, phosphorus, sodium, potassium	Oxalates and nitrates

Source: Hazra and Dutta, (2011).

In order to improve vegetable quality a breeder must have proper knowledge about the specific traits to target for different vegetables so as to provide the best quality product to the consumer. The genetics of the character, the environmental influences, the best methodology of crop production and postharvest activities must be recognized and adopted.

4.6 SOLANACEOUS VEGETABLES

4.6.1 *TOMATO*

Major fruit quality traits of interest to both fresh market and processing tomato include fruit size, shape, texture, color, firmness, locule number, total solids, nutritional quality and flavor, along with pH, titratable acidity, and vitamin contents (Garg et al., 2008; Foolad, 2007). High genetic variation for all the traits have been reported (Siddiqui et al. 2013) which provides ample scope for improvement of these traits.

Tomato fruit shape varies greatly, and may be round, flatty, rectangular, ellipsoid, ovoid, obovoid,oblong, heart-shaped and pyriform. Semidominance and gene interactions were operative for the control of fruit shape (Zdravkovic et al., 2003; Akhtar and Pandit, 2009). Fruit shape is very often influenced by fruit size and it has been found that bigger sized fruits are flattened in shape, while in varieties with smaller sized fruits round shape is more common (Rodriguez-Burruezo et al., 2005).

Texture is one of the critical components for the consumer's perception of tomato fruit quality (Causse et al., 2003; Serrano-Megias and Lopez-Nicolas, 2006). Many traits are involved in fruit texture, mainly sensory attributes such as flesh firmness, mealiness, meltiness, juiciness, and crispness (Redgwell and Fischer, 2002; Szczesniak, 2002). Major changes in texture occur during fruit ripening are mainly associated with softening which considerably influences postharvest performance i.e., transportation, storage, shelf life, and pathogen resistance, (Brummell and Harpster, 2001). Firmness (mechanical properties) of fruit or vegetables has been found to be cultivar dependent and round fruits generally are lesser firm than the oval and elongated fruits. Varieties having higher fruit weight have lower firmness (Rodriguez–Burruezo et al., 2005). Varieties, maturity stage, cultivation practice, and environmental conditions influence tomato texture.

Pericarp thickness also plays a significant role in governing fruit firmness. Thicker pericarp generally enhances the firmness and ultimately the shelf life of the tomato as well as it can withstand long distance transport. Oval shaped varieties have a higher pericarp thickness than others (Thakur and Kaushal, 1995; Chakraborty et al., 2007). The relative thickness of the pericarp, showed somewhat less variation among genotypes, however, a large variation existed in the same genotype at different maturity stages (Chaib et al., 2007).

Locule number has negative association with fruit firmness. Pear shaped cultivars are known to have two locules whereas round fruited cultivars have a higher locules number (Chakraborty et al., 2007). However, it has been found that the genotypes with large locular portion and with high concentration of acid and sugar possess better sugar flavor than those with a smaller locular portion (Stevens et al., 1979). The lower number of locules is preferred as ascorbic acid and sugar content are markedly higher in locules than pericarp tissue (Dhaliwal et al., 1999).

Total soluble solid content in ripe tomato fruit is the chief determinant of commercial value.Average TSS contents in the range of 3.9–5.2°Brix among different cultivars have been found, while in high pigment cultivars the range varied between 5.87 and 6.20° Brix (Akhtar and Hazra, 2013b; Mane et al., 2010; Ilahy et al., 2011a). Both additive and dominance gene action govern this character, dominance component being more important, while role of epistasis has also been found (Akhtar and Hazra, 2013). Narrow sense heritability was reported to be moderate suggesting that simple selection would be insufficient to improve the character. Single seed descent method with progeny row testing and deferred selection will be the best breeding method to improve the character. Several wild relatives having high content of TSS may be utilized in the breeding programme such as *S. cheesmanii, S. chmeilwskii, S. pennellii, S. pimpinellifolium.*

Malic and citric acids are the major organic acids present in tomato and they account for about 15 percent of the dry content of fresh tomatoes, measured as titratable acidity (expressed as percent anhydrous citric acid) and the malic to citric acid ratio is a varietal attribute and is known to be responsible for significant variations in the acidity as found in different cultivars (Davies and Hobson, 1981; Gonzalez-Cebrino et al., 2011). Different values of acidity from 0.25 to 0.84 percent for different tomato genotypes have been reported by different scientists (George et al. 2004; Raffo et al., 2002; Ilahy et al., 2011a), in which maximum value has been recorded for Cherry tomatoes (0.51 to 0.70%).

The acidity of tomatoes is felt more due to the pH rather than titratable acidity. pH is important for tomato processing and pH below 4.5 is a desirable trait, because it halts the proliferation of microorganisms in the final product during industrial processing. (Giordano et al., 2000; Garcia and Barret, 2006a). pH does not differ significantly among cultivars and the value ranges between 4.1 and 4.6 in general except in case of high pigment cultivars where the pH may go up to 4.9 (Frusciante et al., 2007; Ilahy et al., 2011b).

TSS: titratable acidity (TA) ratio an useful indicator of tomato taste since acids influence the perception of sweetness (Hernandes-Suarez et al., 2008; Beckles, 2012), as well as it gives a good indication of tomato ripeness (Gonzalez-Cebrino et al., 2011). A minimum TSS of 5 and a minimum TA of 0.4, respectively (TSS: TA of 12.5), is considered desirable to produce a good-tasting table tomato (Beckles, 2012). The interaction between total sugar (TS) content and acidity are the most important characteristics influencing tomato taste (Rodica et al., 2008; Tadesse et al., 2012).

The total sugar contents of fruit ranged from 1.67 to 4.70 percent (Turhan and Seniz, 2009; Melkamu et al., 2008; Jongen, 2002) and it is mainly (about 75%) glucose and fructose (reducing sugar) and some sucrose (nonreducing sugar). The concentration of total sugar, reducing sugar and nonreducing sugar varies significantly in different genotypes (Gonzalez-Cebrino et al., 2011; Gupta et al., 2011). Nonadditive genetic variance was predominant in governing this trait (Kaul and Nandpuri, 1972). For reducing sugar, generation mean analysis demonstrated that additive effects were equal to dominance effects. A single major gene, dominant for a high percentage of reducing sugar, regulates the percentage of reducing sugar in tomatoes, designated *sucr* (Stommel and Haynes, 1993).

Tomato fruit is also an important source of the water-soluble antioxidants, particularly ascorbic acid (AsA). Tomato contains moderate amounts of ascorbic acid (on an average, 20 mg/100 g) (Gould, 1992), however, significant variation is found. The variation in ascorbic acid (AsA) content in tomatoes depends mainly on the agronomic conditions and the varietal differences (Singh et al., 2010; Chakraborty et al., 2007; Gonzalez-Cebrino et al., 2011). Profound influence of temperature and light intensity have also been observed and the as a content is more at lower temperature (Chakraborty et al., 2007).

Tomatoes are rich sources of carotenoids, comprising of lycopene (70–90%), β-carotene (5–26%), and lutein (1–5%) with trace amounts (<1%) of other carotenoids (Ronen et al., 1999; Kotíková et al. 2009). Other carotenoids present in ripe tomato fruits include phytoene, phytofluene, δ-carotene, ζ-carotene, neorosporene, and lutein (Khachik et al., 2002). Tomato is the richest source of lycopene, an important antioxidant with several health benefits and is responsible for the red color of tomatoes. The lycopene content of tomatoes exhibit wide variation among genotypes, stages of maturity and growing conditions (Sahlin et al., 2004; Odriozola-Serrano et al., 2008; Akhtar and Hazra, 2013b). Lycopene content in red fruited genotypes is much higher than yellow or orange fruited genotypes. Wild relatives such as *Solanum pimpinellifolium*, which contains very high levels of lycopene may be used in breeding programmes for enhancing lycopene content. β- carotene is of special interest since it is the main provitamin-A. Significant genotypic variation for this trait in tomato has been observed (Raffo et al., 2006; Gupta et al., 2011). Orange fruited genotypes possess high beta-carotene. Cherry tomato (*Solanum lycopersicum* var. *cerasiforme*) is a source of very high quantity of β-carotene and hence can be used as a potential source to improve the trait.Some of color mutants such as high pigment (*hp*), dark green (*dg*) and old gold crimson (*og^c*), beta-carotene (*B*) have been discovered that may be sources to improve the color of normal or local genotypes. (Ilahy et al., 2011b; Rubio-Diaz et al., 2010). These quality parameters are under the control of both fixable and nonfixable gene effects, the nonfixable gene effects combined with gene interactions being more important. Duplicate epistasis as well as significant additive x additive type nonallelic interaction with negative sign for the characters has been reported for the traits, which will hinder the pace of progress through simple selection (Akhtar and Hazra, 2013a). Single seed descent method with progeny row testing and deferred selection will be the best breeding method to develop line bred varieties with good fruit quality character.

Tomato fruit flavor is the sum of the interaction between sugars, acids, and multiple volatile compounds. While several hundred volatiles have been identified in tomato, only about 15–20 actually impact our perception of the fruit. Most of these important volatiles are derived from the oxidative cleavage of carotenoids and are referred to as apocarotenoids (Vogel et al., 2008).

For tomato extended shelf life is a very important quality. Tomato is a climacteric fruit and the process of ripening occurs very quickly. A few pleiotropic, single gene ripening mutants such as slow ripening alcobaca (*alc*) (Almeida, 1961), ripening inhibitor (*rin*) (Robinson and Tomes, 1968), and nonripening (*nor*) (Tigchelaar et al., 1973) inhibit or greatly slow down a wide range of processes related to ripening of normal tomato fruit. These mutants in heterozygous form are known to extend fruit availability period and postharvest fruit shelf life (Garg et al., 2008) and may be used in breeding programmes to develop hybrids with longer shelf life.

4.6.2 EGGPLANT

The major traits concerning eggplant appearance are size, shape, external color, smoothness, uniformity, softness of fruits, absence of prickles, less seeds, freedom from defects and little or no browning of flesh after cutting. The biochemical constituents such as anthocyanins, total phenols, polyphenol oxidase activity and glycoalkaloid content are important for eggplant taste related and nutritional quality.

Nine different colors have been reported in brinjal, namely, deep green, light green, purple, dark purple, light purple, pink, purple with green stripes or streaks, green with violet streaks or patches and white (Singh et al., 1999, Hazra et al., 2003). Anthocyanin that imparts the violet and purple coloration is the main pigment along with chlorophyl which imparts green color. Combination of these two pigments has darkening effect (Nothman et al., 1976). Color of initial ovary color does not have any relation to the final fruit color (Hazra et al., 2003, Nothman et al., 1976). Different reports by different workers are available about fruit color in eggplant. The fruit color has been reported to have monogenic inheritance (Nolla 1932; Jannaki and Ammal 1933; Swamay Rao 1970; More and Patil 1982; Patil and More 1983); complex inheritance (Fukusawa 1964; Gopinath et al.1986); complimentary digenic inheritance (Thakur et al. 1969); nonallelic trigenic inheritance (Khapre et al 1988). The purple color has been reported to be dominant over green (Tatebe 1944; Khan and Ramzan 1954), which is dominant over white color (Swamay rao 1970; Choudhari 1972, 1977; More and patil 1982; Gopinath et al., 1986; Joshi 1989).

Green flesh is dominant over white flesh and controlled monogenically (Wanjari and Khapre, 1977). Consumers and the industry prefer varieties with a luminous white flesh color (Prohens et al., 2005). A wide variation in the distance of the fruit flesh color to the pure white among the materials studied, and surprisingly there are many landraces that have a flesh color that is closer to the pure white than commercial varieties. The flesh color is highly heritable and, in consequence, the exploitation of the variation present in landraces could lead to new commercial varieties with a more luminous white flesh color (Haynes et al., 1996).

Elongated fruit shape is dominant over round (Swamay Rao 1970; Choudhari 1977) which in turn is dominant over oval shape (Patil and More 1983). Involvement of four genes has been reported of which one may be of complementary type and three may be of duplicate complementary type (Nimbalkar and More, 1981).

Presence of prickles on stem leaves or petiole has been reported to be a dominant character having monogenic control (Rangaswamy and Kadam Bavanasundarm, 1973).

Fruit length, number of fruits per plant, little leaf incidence, total phenol content and fruit yield per plant recorded high amount of genetic variability along with heritability and genetic advance and hence there is ample scope for improving these characters by simple phenotypic selection (Kumar et al., 2013).

High genetic variability has been observed among of different chemical constituents of the eggplant cultivars (Hazra et al., 2005; Patil et al., 1994). The stage of fruit development as well as environmental conditions also plays an important role (Sidhu et al., 1982). The elongated genotypes have been found to be rich in dry matter, phenolics, crude protein and glycoalkaloid contents, while the oblong types are rich in total sugar content. Small fruited genotypes (50–60g weight) have been reported to contain high phenol content, low sugar, low crude protein as compared to genotypes with big sized fruits (more than 100g weight) that contain low total phenol, high sugar and low crude protein (Hazra et al., 2004). This scenario indicates that both high palatability and nutrient contents were associated with increased fruit size. The inheritance of total phenol content was controlled by additive gene effect whereas for protein it was dominance gene effect and for sugar both additive and nonadditive, the additive gene effect being greater in magnitude than the nonadditive counterpart (Hazra et al., 2005).

Processing of fruits also affects nutrient composition. Vitamin C content is reduced by canning. Toasting of brinjal fruits increases the content of nicotinic acid since the trigonellin is converted to nicotinic acid in course of toasting (Hazra and Dutta, 2011).

Both discoloration and bitterness in eggplant increase with increasing percentage of total phenols (Chadha et al., 1990). Some phenolics have a bitter taste (Macheix et al., 1990). However, the bitterness and off flavor of some eggplant varieties seems to be incited by saponins and glycoalkaloids and not by the phenolic compounds characteristic of eggplant (Aubert et al., 1989).

The high antioxidant capacity of eggplant is attributed to its high content of phenolic compounds (Cao et al., 1996; Sato et al., 2011). The main class of phenolics in eggplant includes hydroxycinnamic acid conjugates and, of these, chlorogenic acid (5-O-caffeoylquinic acid and its isomers) (CGA) typically accounts for 80 percent and 95 percent of the total hydroxycinnamic acids present in the fruit flesh (Prohens et al., 2013; Stommel and Whitaker, 2003; Whitaker and Stommel, 2003) and 30 percent and 75 percentof the total phenolics of the fruit when harvested at the commercially mature stage (Luthria, 2012; Mennella et al., 2012). CGA content in eggplant flesh is highly correlated with total phenolics and antioxidant activity (Luthria et al., 2010, 2012). The beneficial effects on health of chlorogenic acid are manifold, viz., free radical scavengers and antitumoral activities (Sawa et al.,

1998; Triantis et al., 2005), anticarcinogenic (Yang et al., 2012; Burgos-Morón et al., 2012), anti-obesity effect (Cho et al., 2010), antidiabetic effect (Ong et al., 2012; Coman et al., 2012).

Ascorbic acid also attributes to antioxidant activities of eggplant. Although there is variation among varieties for the concentration in ascorbic acid, the content of ascorbic acid is 27 times lower than that of phenolics. Such lower concentration, together with the fact that chlorogenic acid and ascorbic acid have similar antioxidant activities (Kim et al., 2002; Triantis et al., 2005), demonstrates that phenolics and, in particular, chlorogenic acid account for most of the antioxidant capacity of eggplant. Therefore, a breeding program directed at enhancing the antioxidant capacity of eggplant by increasing the ascorbic acid concentration would produce limited results. Furthermore, the narrow-sense heritability of ascorbic acid concentration is lower than that of phenolics.

High variability among different genotypes for phenolics have revealed that selection of eggplant accessions with an increased concentration of phenolics as a way to develop new varieties with improved nutritional quality (Stommel and Whitaker, 2003). The high intraspecific variation for CGA content and total phenolic content can be used in several ways in conventional breeding programmes (Hanson et al., 2006; Mennella et al., 2012; Okmen et al., 2009; Prohens et al., 2007; Raigón et al., 2008; Stommel and Whitaker, 2003; Whitaker and Stommel, 2003). Selection among the accessions or varieties with highest CGA content can result in the identification of materials with higher content in CGA. However, very likely, landraces with high content in CGA will not present agronomic and commercial characteristics competitive with present modern varieties, and its practical utility as commercial varieties may be limited. An alternative is the development of hybrids between accessions or lines with high content in CGA and complementary for agronomic traits. Eggplant hybrids are known to be heterotic for yield (Rodriguez-Burrruezo et al., 2008) and competitive with commercial hybrids in open field conditions (Muñoz-Falcón et al., 2008). Also, these hybrids can be used, through several breeding methods (Acquaah, 2012), to select and develop inbred lines with higher content in CGA and improved agronomic and commercial characteristics or for introgression of this trait in elite lines. Cultivated eggplant can be hybridized, although with different degrees of success, with a group of related species, including wild species, for example, cultivated scarlet eggplant (*S. aethiopicum* L.), gboma eggplant (*S. macrocarpon* L.), *S. incanum*(), resulting in introgression of traits like high CGA, resistance to disease-pests into eggplant (Daunay, 2008; Ma et al., 2011; Prohens et al., 2013; Stommel and Whitaker, 2003; Gisbert et al., 2006). However, the fact that the hybrids between *S. melongena* and *S. macrocarpon* have a reduced fertility (Bletsos et al., 2004) restricts its use to a long-term breeding program.

However, a drawback of increasing the concentration of these antioxidants in eggplant is that the oxidation of phenolics causes the browning of the fruit flesh after its exposure to the air, and this may lead to a reduction in the apparent quality

(Macheix et al., 1990). The activity of polyphenol oxidases varies among eggplant varieties (Dogan et al., 2002), which suggests that selection for a combined high concentration of phenolics and low browning should be feasible. Other factors, like intracellular pH, which affects the activity of the polyphenol oxidase (PPO) enzymes (Concellon et al., 2004; Dogan et al., 2002), or the presence of ascorbic acid in the fruit flesh tissues, which prevents the oxidation of orthodiphenols (Macheix et al., 1990), might also have a role in the modification of the browning process in eggplant. Given the preference of consumers and the industry for varieties with white flesh and a low degree of browning (Prohens et al., 2005), new varieties with a high content of phenolics should also have a moderate flesh browning.

Degree of browning and color difference have high heritability, which means that to obtain offspring with a low degree of browning or color difference, parents should also have low values for these traits. As in many other crops, the degree of browning and color difference is positively correlated with the phenolic concentration (Amiot et al., 1992; Hansche and Boynton, 1986, Prohens et al., 2007). Ascorbic acid, which has antibrowning properties because it reduces enzymatically formed o-quinones to their precursor diphenols (Macheix et al., 1990), does not show any correlation with either degree of browning or color difference in eggplant. This is probably because of the very low concentrations of ascorbic acid in the eggplant fruit compared with the concentration of phenolics. Regarding the pH, which also has an influence on the activity of polyphenol oxidases (Yoruk and Marshall, 2003), it does not show a correlation with either degree of browning or color difference.

Substantial genetic variability and heritability in the minerals studied to warrant selection in the eggplant accessions for improvement depict that the level of genetic variability observed for different minerals would be useful for breeding program for developing mineral rich varieties of eggplant. The high genetic variance components and heritability estimates couple with significantly positive correlation could be used as selection criteria for identification of mineral rich eggplant germplasm (Arivalagan et al., 2013).

4.6.3   PEPPERS (HOT AND SWEET)

The most commonly cultivated pepper species encompassing both hot and sweet pepper is *Capsicumannuum*. The other cultivated species being *Capsicum frutescens, Capsicum pendulum* and *Capsicum frutescens*. The fruits have a typical taste, pungency, and color. It is used worldwide both as vegetable and spices for flavor, aroma, and addition of color to foods (Zhuang, et al., 2012). There are many properties that differ in chilli from other fruit vegetables, such as in their shape, size, color, flavor, and pungency. Peppers may have hot, sweet, fruity, earthy, smoky, and floral flavor. The most important quality attribute for peppers is pungency, and the factor responsible for this is a group of alkaloids named capsicinoids, synthesized exclusively in the placenta tissue of the peppers and accumulated in the epidermal cells in the placenta (Iwai et al., 1979). The synthesis of the capsicinoids occurs around 20 days after anthesis (Fujiwake et al., 1982). Pepper that are fresh is known as the very good source of vitamin C and E and as well as provitamin A and carotenoids and total soluble phenolics, these are responsible for its antioxidant properties (Krinsky, 2001; Marinova et al., 2005; Perucka and Masterska, 2007; Navarro, et al., 2006; Chatterjee, et al., 2007; Conforti, et al., 2007; Deepa, et al., 2007; Serrano-Martinez et al., 2008; Kumar et al., 2009). Vitamin C, including ascorbic acid and its oxidation product (dehydroascorbic acid), has many biological activities in the human body due to its antioxidant properties (Davy et al., 2000; Yahia et al., 2001). It has been reported that consumption of 100g FW of peppers provides 100–200 percent of the RDA (recommended daily administration) of Vitamin C (Lee & Kader, 2000). The ascorbic acid content showed high variability (ranged 55–200 mg/ 100g FW) and an increasing trend was observed from green to red fruits, while a decreasing trend in red partially dried and red fully dried fruits (Kumar et al., 2009; Tilahun et al, 2013).

Varieties and stages of maturity have great influence on chillies pungency (Kanner et al., 2006; Sanatombi and Sharma, 2008). *Capsicum frutescens* species possessed the higher capsaicin content and was found to be hotter than all the *Capsicum annuum* accession/varieties Antonious and Jarret, 2006; Gnayfeed et al., 2001; Juliana et al., 1997; Mathur et al. 2000; Tilahun et al, 2013) and capsaicin content was also highly variable (0.3 – 4.5 mg/ g) between varieties and species (Tilahun et al., 2013). Table 4.2 depicts the ranking of different pepper genotypes of the world according to their pungency. These genotypes can be used in breeding programmes.

TABLE 4.2 Ranking of world *Capsicum* based on the pungency level

Scoville Heat Units	Pod Type	Species	Name
0	Bell, bydagi	*C.annum*	Bell , bydagi kaddi
1000	ancho mulato	*C.annum*	100 ancho mulato
4000	Seranno	*C.annum*	400Seranno
5000	New maxican	*C.annum*	Sandi
8000	Cayenne	*C.annum*	Cayenne
17000	Aji	*C.baccatum*	Aji Escabeeche
21000	Hungarian	*C.annum*	Fe Grande
22000	Jalapeno	*C.annum*	Mitla
23000	Cayenne	*C.annum*	Long-slim Cayenne
25000	Jalapeno	*C.annum*	Jalapeno M

TABLE 4.1 *(Continued)*

Scoville Heat Units	Pod Type	Species	Name
30000	Ellachipur	*C.annum*	Ellachipur Sannam
36000	Hindpur	*C.annum*	Hindpur 57
60000	Asian	*C.annum*	Thai Hot jwala
7000	Tepin	*C.annum*	Chiltepin
75600	kanthari	*C.annum*	Kanthari white
88350	Tabasco	*C. fruitescens*	Bird eye chilli
120000	Tabasco	*C.annum*	Tabasco
150000	Habanero	*C chinense*	Red Habanero
210000	Habanero	*C chinense*	Yellow Habanero
455000	--	*C chinense*	Naga jolokia
855000	--	*C. annum*	Tezpur

Source: Mathur et al. (2000)

Heterosis for the pungency trait, assessed by the capsaicin and dihydrocapsaicin contents in fruits, was found, indicating the existence of epistasis, overdominance, or dominance complementation. Nonpungent parent alleles contributed to the capsaicin and dihydrocapsaicin contents since transgressive segregation did occur. Furthermore, the type of gene action varied between capsaicin and dihydrocapsaicin, and a seasonal effect during fruit development could affect gene action (Garceä S-Claver, 2007).

The variability in the pepper germplasm for capsaicin and ascorbic acid content can be exploited for breeding of cultivars with improved nutritional qualities. All the chilli types can be used as a potential source of capsaicin, especially the *Capsicum frutescens*1 accession. Likewise, Bayadagi Kaddi can be used as a source of vitamin C for enhancing the nutritive value of human diets (Tilahun et al., 2013).

Colour is another important attribute of pepper. The attractive red color of peppers is attributed to capsanthin and capsarubin collectively called oleoresin (Bosland and Votava, 2000), which is used as a natural color additive in food, drugs, and cosmetics. These pigments are also rich in bioflavonoids, which are powerful antioxidants and inhibit the progression of chronic diseases such as muscular degeneration, cardiovascular diseases and cancer. Oleoresin extracted from dried and ground chillies is the total flavor extract which has gained industrial importance through its utilization in processed products and pharmaceutical formulations. Besides it offers uniform quality, longer shelf life and freedom from micro-organisms.

Bell pepper of different colors has a unique array of nutritional benefits. Green peppers feature an abundance of chlorophyl. Yellow peppers have more of the lutein and zeaxanthin carotenoids. The most abundant carotenoids in ripe orange pepper fruits are β-carotene, capsanthin, and capsarubin (Guzman et al., 2010). Yellow-or-ange colors of pepper fruits are mainly due to the accumulation of α- and β-carotene, zeaxanthin, lutein and β-cryptoxanthin. Carotenoids such as capsanthin, capsorubin, and capsanthin-5,6-epoxide confer the red colors. Three loci, c1, c2, and y control the synthesis of carotenoids (Gómez-García 1 and Ochoa-Alejo, 2013).

Dry fruit yield/plant showed significant and positive association with plant height, plant spread, number of fruits/plant, fruit girth, seeds/fruit and capsanthin content (Kumari et al, 2011). Based on magnitude of correlation coefficient values, number of fruits/plant, plant height, plant spread, fruit girth, number of seeds/fruit and capsanthin content may be regarded as very closely related characters with dry fruit yield/plant in paprika types (Kumari et al, 2011), in hot pepper (Gogoi and Gowtham, 2003; Hari et al., 2005; Karad et al., 2006 and Chatterjee et al., 2007) and also in sweet pepper (Islam and Singh, 2009). Higher yield could be obtained by operating selection pressure over any of these traits. Similar results were reported and in sweet pepper by Islam andSingh (2009). Path coefficient studies revealed that number of fruits/plant had the highest positive direct effect on dry fruit yield/plant. Number of fruits/plant, plant height, plant spread, number of seeds/fruit, days to maturity and capsanthin content (in the given order) were important yield and qual-ity components having direct bearing on dry fruit yield and hence can be considered while breeding for improved yield in paprika.

4.6.4   POTATO

Color, size and shape of potato tubers are crucial quality aspects for consumers that are determining factors for their purchase. Color is of utmost importance when it comes to cosmetic quality of potato that determines consumer preference. Next to the color traits, other obvious quality traits are the morphological traits like tuber size, shape, regularity of shape and eye depth.

The skin color of potato tubers can range from whitish yellow, orange, brown andred to deep purple. The accumulation of anthocyanin pigments in the phelum and epidermis layers of the tuber is primarily responsible for the skin coloration. The earliest report mentions three loci affecting skin color in tetraploid potato (Sala-man, 1910), D, R, and P localized on the potato chromosomes 10, 2, and 11 (Geb-hardt et al., 1989; Van Eck et al., 1994b; Van Eck et al., 1993). Red pigments have been shown to be derivatives of pelargonidin and the purple anthocynanins are de-rived from delphinidin (Lewis et al., 1998; Naito et al., 1998; Rodriguez-Saona et al., 1998). Molecular studies have revealed the enzymatic factors related to expres-sion of the color at the loci (Zhang et al., 2009; Jung et al., 2005, 2009).

In cultivated potato, the flesh color is predominantly white or yellow and con-sumer preference is also for either white or yellow, making breeding for potato flesh color very significant. The yellow to orange coloring of the potato tuber flesh is caused by the presence of certain carotenoids. Lutein, violaxanthin, zeaxanthin, and antheraxanthin are the forms predominantly present in cultivated potato (Breithaupt and Bamedi, 2002; Brown et al., 1993; Iwanzik et al., 1983; Nesterenko and Sink, 2003). Yellow flesh color is caused by a dominant allele at the *Y* (yellow) locus, which was mapped on chromosome 3 of potato (Bonierbale et al., 1988). On the same location as the *Y* locus a beta-carotene hydroxylase gene (*bch* or *Chy2*) was mapped, thereby indicating that this is the most important candidate gene for yellow flesh (Thorup et al., 2000; Kloosterman et al., 2010; Wolters et al., 2010). Although *Chy2* has a major allelic effect, there are apparently more genes contributing to the variation in tuber flesh color in potato (Brown et al., 2006). Next to the yellow flesh color, a darker coloring toward orange is also found. Allelic analysis showed that in a set of different genotypes, all the accessions with orange flesh color were homozygous for one specific *Zep* allele (*zeaxanthin epoxidase*) (Wolters et al., 2010). Transformation of yellow fleshedpotatoes with *Zep* constructs showed an increase in the level of the carotenoid zeaxanthin (Römer et al., 2002). The group referred to as "Papa Amarilla", composed of the diploid groups *S. phureja*, *S. stenotomum* and *S. goniocalyx*, contains high levels of carotenoids, exceeding those in modern tetraploid cultivars (Brown et al., 2007). These may be used in breeding programmes to enhance carotenoids in flesh.

Tuber shape varies from long to compressed/round and is measured by the ratio between length and width. The genetics of tuber morphology still remains unclear despite several years of research. For tuber shape a single locus on chromosome 10 was mapped with a dominant allele Ro conferring round tuber shape (Van Eck et al., 1994a). Other reports using populations with different genetic backgrounds mention QTLs on chromosomes 2, 5, and 11(Bradshaw et al., 2008), 2 and 11 (Śliwka et al., 2008) and 7, 12 and an unassigned linkage group (Sørensen, 2006). Thus, there are clearly more factors controlling this trait, probably depending on the genetic background. One of the other morphological traits, eye depth, appears to be controlled by a single locus, which is closely linked with the Ro locus at chromosome 10 at a distance of 4 cm (Li et al., 2005b; Śliwka et al., 2008). The correlation between tuber shape and eye depth is shown by the linkage of deep eyes with round tubers (Li et al., 2005b). Little is known on regularity of shape. It depends on several components such as depths of indentations, various tuber defects and uniformity of shape and therefore will probably behave as a complex trait. Two QTLs on chromosomes 3 and 5 were reported for regularity of shape (Śliwka et al., 2008).

Dormancy and shelf life are additional important quality traits for fresh market potatoes. Apart from the use of chemicals like CIPC and DMN (Campbell et al., 2010) to delay sprouting and hence extend shelf life and dormancy, breeding programmes can be effectively used to improve the trait. The trait is governed by quantitative inheritance (Freyre et al., 1994; Van den Berg et al., 1996). Dormancy is a very complex trait influenced by many different factors (Senning et al., 2010; Campbell et al., 2010). High variation in related species is also found for dormancy. In *S. jamessii* a dormancy period of eight years is observed (Bamberg, 2010) and it can be used as a donor source for the trait.

In potato tubers the main dry matter component are carbohydrates, comprising of around 75% of the total dry matter. The largest portion of these carbohydrates is made up of starch (Camire et al., 2009). There is a large variation in starch content present in cultivated potato, 11.0–34.4 percent starch of fresh weight, and an even larger variation in the wild related species, 3.8–39.6 percent starch of fresh weight (Jansen et al., 2001). Although starch content has a strong genetic basis, a significant portion of the variation is due to the life cycle length of the potato plant. Late maturing cultivars tend to have higher starch yield than early ones (Van Eck, 2007).

Potato is not especially high in vitamin levels. Some of the B vitamins are observed in potato, although in low concentrations. Vitamin B1, B2, B3, and B6 were investigated in a series of cultivars. The observed variation ranged from 0.08 to 0.10, 0.022–0.36, 0.62–2.07 and 0.15–0.3 mg/100g fresh weight, respectively (Burlingame et al., 2009). However, the predominant vitamin in potatoes is vitamin C. The amount of vitamin C ranges in content between 84 and 145 mg per 100 g fresh weight depending on cultivar, agronomic practice and storage conditions (Augustin, 1975). High variation within genotypes, with a four-fold difference in vitamin C content among different varieties has been reported (Love et al., 2004).

Potatoes are on the other hand an excellent source of minerals, particularly, K, P, Fe (18–65 µg/g dry weight), Zn (12.5–20 µg/g dry weight), Mg and Mn (White et al., 2009; Brown, 2008). Potatoes are a particularly good source of dietary iron. However, absorption of $Fe2+$ is inhibited by the presence of polyphenols and phytic acid. Potato is naturally low in these antinutrients, allowing a good uptake of iron from potato tubers (Brown, 2008). Moderate to high heritability for iron content has been reported suggesting environmental influence and genetic background specificity (Brown et al., 2010). The relatively high amounts of minerals and observed heritabilities combined with a broad observed range of amounts per mineral in different genetic material indicates that with breeding for mineral content there is a huge possibility for improvement.

Processing quality of potato tuber also has great importance. While producing French fries or potato chips, reducing sugars (mainly glucose and fructose) react with the free amino acids in a nonenzymatic Maillard reaction leading to the production of brown- to black-pigmented products rendering the chips or fries unacceptable for consumers. These reducing sugars accumulate in the cold storage of the potatoes prior to processing. The process of cold sweetening is a large problem for the potato processing industry. Most of the genes involved in the processes underlying cold sweetening have been identified and molecular characterization has been done (Kumar et al., 2004; Sowokinos, 2001). Selection for lower reducing sugar content is the way for controlling this problem through conventional breeding (Shepherd et al., 2010).

4.7 BULB CROPS

4.7.1 ONION

Onion is an important bulb crops. The important traits for onion bulb quality are its size, shape, color, pungency, firmness, dormancy, and the total soluble solids. Onion bulbs may be deep red, light red, white, yellow, or brown in color and its shape may vary from round to pyriform to flat and its texture may be soft or firm. High TSS is preferred for dehydration products of onion.

Onion bulb color has been found to be controlled by five major loci I, C, G, L, and R (Clarke et al., 1944; El-Shafie and Davis, 1967; Rieman, 1931). The inheritance pattern is complex involving epistatic interaction among these five loci. Onion bulb color pigments are known as flavonoid compounds and the red color is attributable to anthocyanin, a type of flavonoid (Fossen et al., 1996; Rhodes and Price, 1996). Flavonoids are major plant secondary metabolites and play such a role as UV protection and pigmentation in plants (Shirley, 1996). Epistatic interaction among loci, in addition to their sequential involvement in red color development, suggests that the loci might involve genes encoding enzymes involved in the anthocyanin synthesis pathway (El-Shafie and Davis, 1967; Koops et al., 1991). Recent studies on the characterization of onion bulb color inheritance at the molecular level revealed that some major color differences are caused by mutations of the genes involved in the anthocyanin synthesis pathway. Kim et al. (2004b, 2005b) reported that inactivation of the DFR (dihydroflavonol 4-reductase) gene in the US-type yellow onion resulted in a lack of anthocyanin production. A new allele of the anthocyanidin synthase gene that controls a pink bulb color was identified, and an underlying mechanism of pink color development was revealed by a significant reduction of ANS gene transcription (Kim et al., 2004a, 2005a). Another nonfunctional allele of the ANS gene was identified in Brazilian yellow cultivars (Kim et al., 2005c). In addition, a gold bulb color is attributable to a mutant allele that contains a premature stop codon on the onion CHI (chalcone isomerase) gene (Kim et al., 2004c). The yellow color in onion is due to the flavonoid quercetin. Quercetin levels tend to be highest in red and yellow onions and lowest in white onions (Patil et al. 1995, Lombard et al. 2002). Amounts of quercetin in onions vary with bulb color, type, and variety (Leighton et al. 1992, Patil and Pike 1995, Lombard 2000).

The consumer perception of onion pungency depends on the presence of organosulphur compounds that are synthesized from enzymatic reactions involving S-alk (en) yl-L-cysteine S-oxide decomposition (Lancaster et al., 2000). The consumer demand for pungency varies and in western countries it has been oriented toward onion cultivars with low pungency (mild cultivars), while in Asian countries higher pungency is preferred. The accumulation of organosulphur compounds in onions depends upon many factors, especially sulfur-based fertilization, the environment, and the genotype of the cultivars (Yoo et al., 2006; Chope et al., 2007). However, sulfur-based fertilization and onion pungency are not always positively correlated, and contrary results have been reported (Randle 1992; Randle and Bussard, 1993). However, the pyruvic acid content is influenced by environmental growing conditions, genotypes and storage (Yoo et al., 2006), and the genetic background seems to be the most important factor because different cultivars have different abilities to control sulfur uptake and assimilation in the biosynthesis pathway that results in the flavor (Randle, 1992; Randle et al., 1994). The pyruvic acid content in the bulb can be used as an objective parameter for screening purposes in breeding programmes. The pyruvic acid level in bulbs is affected by both the tunic color and precocity. The early genotypes have been found to be less pungent and accumulated less pyruvate in their bulbs (Gallina et al., 2012). Long storage onions for temperate regions are typically more pungent than short storage short-day type onions. Incomplete dominance prevails for low pungency (Peterson et al. 1986). Pal and Singh (1987) reported pyruvic acid content to be positively associated with storage and drying quality and it was controlled by both additive and dominance gene effects, the former being more important. The quality of dehydrated onion mainly depends on uniform white color, globe shape for convenient handling, high solid contents (both soluble and insoluble), high nonreducing sugars and high pungency and its better retention after drying (Sethi et al., 1973). Studies have shown that total yield, mean yield, average bulb weight and equatorial diameter are positively correlated, polar diameter and T.S.S. are nonsignificantly correlated however percent marketable bulbs, neck thickness, plant height, collar thickness, number of leaves are strongly negatively correlated (Hosamani et al. 2010; Singh et al., 2013). These factors are to be kept in mind for breeding of onion for yield and quality.

4.7.2   GARLIC

For garlic, color and size of the bulb, number and size of the cloves, nonsprouting of bulbs and its flavor are the important quality traits. Garlic bulb is generally creamish in color and this color is due to quercetin. Some bulbs may be pinkish in color attributed to anthocyanin. The pinkish colored garlic is said to have better resistance to diseases in storage. The typical flavor of garlic originate in damaged tissues from enzymatic cleavage of preformed stable flavor precursors by alliinase. The flavor precursors are secondary metabolites, the alk(en)yl cysteine sulfoxides (CSOs), primarily allyl cysteine sulfoxide (ACSO, alliin, 2-propenyl cysteine sulfoxide) and methyl cysteine sulfoxide (MCSO, methiin). The synthesis and storage of these compounds, and related g-glutamyl cysteine sulfoxide peptides (g-GP), results in garlic containing the highest amount of sulfur recorded in any plant species (Nielsen et al., 1991). Garlic sulfur compounds have shown potential in combating cardiovascular disease and some cancers (Kik et al., 2001), and there is therefore interest in improving the content of these compounds in garlic.

Outside its centers of origin, garlic reproduction occurs by vegetative propagation. This type of reproduction does not allow meiotic recombination, making it impossible to obtain genetic recombinants and hindering the development of cultivars that are adapted to different growing conditions (Vieira and Nodari, 2007). Thus, the available cultivars of garlic are a result of the accumulation of somatic mutations in the basic crop material (Morales et al, 2013). Clonal selection is the main breeding method for modern garlic (Jo et al., 2012), since plant sterility usually precludes crop improvement through cross-hybridization (Lampasona et al., 2003). In absence of the common sexual reproduction system available in garlic, random and induced mutations, polyploids (Volk et al., 2004; Koul et al., 1997) and the production of true seeds (Kamenetsky et al., 2004) could be the only sources of genetic variability available for the breeding programs. The production of true seeds was achieved in Japan (Etoh, 1997), Germany (Simon, 1993) and the U.S.A. (Pooler and Simon, 1993a; Jenderek and Hannan, 2004). This situation opens a new and a greater possibility for the future genetic improvement of garlic (Simon, 1993; Koul et al., 1997).

4.8   ROOT CROPS

4.8.1   RADISH

The important quality traits for radish are root length, root diameter, root color, root shape, pungency, and freedom from pithiness.

Root color in radish was controlled by two genes, one conditioning the presence or absence of color (A) and the other gene conditioning the type of color (B) (Makarova and Ignatova, 1983). According to Makarova and Ivanova (1983), anthocyanin coloration of the roots was controlled by recessive epistatic genes, with purple: red: white roots segregating in the ratio 9:3:4. The studies further suggested that one gene, A, was responsible for the presence/absence of coloration and another gene, B, was responsible for the type of coloration, recessive a being epistatic to B, i.e. blocking the expression of color irrespective of whether it was purple (BB, Bb) or red (bb). Evidence was examined for the existence of a series of A genes, one responsible for coloration of the whole plant (Apl) and one for coloration of the root (Ar). From studies of Ahmed and Tanki (1994) it was found that purple root color was dominant over pink, partially purple, and white root color, and was under the control of two genes P and C. These two genes in double dominant state produce purple roots. The genotype of purple parent was tentatively designated as PPCC. The pink and partially purple root colors were dominant over white and they were under the control of single dominant genes P and C, respectively. Postulated genotypes were PPCC (purple), PPcc (pink), ppCC (partially purple), and ppcc (white). Root shape was studied in two crosses involving parents with either round or flat round roots. Round root shape was dominant to flat round root shape in the F1 generation and was under the control of a single gene, designated as R. Root length, root width, root weight were governed by overdominance (Pandey et al., 1981). The content of thiocyanate in radish roots was found to be positively correlated with their dry weight, and the content of total protein, crude fiber, and soluble sugar. A strong relationship was found between the content of thiocyanate and dry weight of radish leaves. The negative correlation between the thiocyanate content in the leaves and the firmness of storage roots and the positive correlation with their pithiness might indicate the translocation of this compound into green plant parts during the aging of root tissue. The root thiocyanate content and the percentage bolting correlated significantly only in the case of Tokinashi. The closeness of relations between the ratio of leaves:storage root and thiocyanate content, though in general small, was affected also by cultivar. A similar effect was observed for the correlations between the thiocyanate contents in leaves and storage roots (Capecka, 1998).

4.8.2   CARROT

The most obvious visual characteristics associated with quality of carrots are the root length, root diameter, root color, root shape, presence of carotenoids and dry matter are also important from quality point of view in carrots (Hussain et al., 2008).

Moderate heritability values for average root length were reported by Brar and Sukhija (1981), Silva et al. (2009) which may be attributed to additive genetic variance (Vieira et al., 2012). Root mass has been found to be highly positively correlated with root diameter (Vieira et al., 2012).

Carrot roots may exhibit a range of colors including white, yellow, orange, red, and purple (Banga 1964). Purple pigmentation is due to the presence of anthocyanins, whereas orange, red, and yellow pigmentation is due to carotenoids. The primary carotenoids in orange carrot tissue are alpha and beta-carotene (Laferriere and Gabelman, 1968). The primary carotenoids in yellow carrot tissue are xanthophylls (Imam and Gabelman, 1968). Carrot pigmentation is of particular interest for nutritional reasons since animals convert beta-carotene into provitamin A. Carrot roots are an important source of vitamin A in the human diet. HPLC (high-performance liquid chromatography) analyses of the carotenoids of carrot leaf tissue demonstrated xanthophylls and β-carotene as the carotenoids present in all carrots, but significantly higher levels of α-carotene was identified in carrot germplasm with orange and dark orange storage roots (Bowman, 2012). Bowman (2012) also reported that chlorophyl fluorescence analysis identified genotype specific variation in photosystem II efficiency, significantly impacted by the time of harvest and expression of three candidate genes involved in photosynthesis was significantly higher in the orange rooted carrot germplasm. Kust (1970) and Laferriere and Gabelman (1968) determined that lack of pigmentation is controlled by dominant alleles. Kust (1970) described an epistatic relationship between the alleles Y Y_1, and Y_2 with two pigment enhancing alleles, IO, and O. He hypothesized that the number of IO and O alleles had to be greater than the number of Y, Y_1, and Y_2 alleles for the presence of orange root color. He further suggested that the recessive genotype $yyy_1y_1y_2y_2ioiooo$ should be white since it did not have the dominant pigment enhancing alleles. QTL analyses have shown that two major interacting loci, Y and Y_2 on linkage groups

2 and 5, respectively, control much variation for carotenoid accumulation in carrot roots (Just eta la., 2009). Heritability and number of genetic factors have also been estimated for total carotenes, β-carotene, α-carotene, ξ-carotene, lycopene, and phytoene and each trait showed high heritability of around 90% (Santos and Simon 2006). The red rooted cultivars have higher lycopene than orange rooted ones.

4.8.3   BEET ROOT

Beet root is a vegetable that can be easily cultivated and its good storability ensures availability of fresh product year around without the need of applying expensive storage equipment (Szopińska, 2013). The consumer preference is for dark red, uniformly colored roots. Uniform root shape is also important. Internal zoning, that is, presence of alternate white rings on cutting the roots lowers the root quality, so that should be absent.

The red color of beet root is due to the pigments betalains that comprise of two groups: red-violet betacyanins and yellow betaxanthins. The betacyanins, a natural group of plant antioxidant, include betanine and isobetanine (Cai et al., 2003). Among betaxanthins, the main position is occupied by vulgaxanthines (I and II). Different redness shades of the root flesh are determined by the ratio of betacyanins to betaxantins, and their relative content depends on root size, variety, weather conditions during cultivation and fertilization (Felczyński and Elkner, 2008).

Dominant alleles at two tightly linked loci (R and Y) condition production of betalain pigment in the beet plant (Keller, 1936). However, several alleles at the R and Y locus influence pigment amount and distribution Wolyn and Gabelman (1989). The following table shows the phenotypic expression of root according to different genotypic constitution.

Genotype	Phenotype
R-Y-	Red roots, hypocotyls, petioles
rrY-	Yellow roots, hypocotyls, petioles
R-yy	White roots and red hypocotyls
R^hR^hY-	Red hypocotyls
rryy	White roots and yellow hypocotyls
R^tR^tY-	Striped petioles
R^tR^tyy	Striped petioles
R-Y^r	Red roots and green leaves
rrY^r	Yellow roots and green leaves
R^t-Y^r-	Red roots and striped petioles
R^hR^t	Pink roots, hypocotyls, and petioles

Wolyn and Gabelman, (1989).

Small negative genotypic and phenotypic correlations between pigment concentration and sucrose content in red beet, which suggested that selection for high pigment and low sucrose was a feasible breeding goal (Watson and Gabelman, 1984). In addition, recurrent selection for pigment concentration has been effective at increasing pigment concentration, suggesting other modifying genes play an important role in betalain synthesis (Wolyn and Gabelman, 1989; Goldman et al., 1996). Alleles at the Y locus affect not only root color but also the color of petioles and hypocotyls.

Internal zoning, characterized by irregular sectors of blotchy red and white root color has been observed in many beetroot cultivars. While temperature fluctuation has influence on this type of coloration, it is also conditioned by a single recessive gene *bl* (Watson and Goldman, 1997).

4.9 CUCURBITS

4.9.1 CUCUMBER

Cucumber fruit quality encompasses many variables, viz., skin color, skin texture, spine color, fruit shape, presence, or absence of warts, etc. are. Skin characteristics include thickness, stomata number, and other factors that may influence palatability and fruit processing (Breene et al., 1972; Kingston and Pike, 1975). Skin tenderness is a character that creates dilemma in breeders since tender skin is desirable for edible purposes, but offers little resistance to rough handling and is prone to dehydration leading to loss in weight. On the other hand, tough and thick skin is unpalatable, but can tolerate nuances of rough handling and transportation. Fruit firmness is a function of both flesh firmness and seed cavity (endocarp) or locule size. As seed cavity diameter becomes more compared to diameter of the fruit, the fruit becomes less firm. Besides slender fruits with higher ratio of length:diameter has greater fraction of diameter in mesocarp and exocarp than in endocarp, whereas those with lesser length:diameter ratio has proportionally large endocarp areas. Outer wall thickness also gives an indication of fruit firmness. Seed cavity acceptability depends on ratio of the cavity diameter to the fruit diameter, carpel separation, placental hollowness, occurrence of two and four carpel off types and rate of seed maturation.

Selection may be carried out for simply inherited traits like, skin color, spine color, fruit shape, presence or absence of warts, fruit confrontation (length:diameter) under the control of single gene. However, it is difficult to manipulate the traits like skin characteristics, outer wall, seed cavity (exo, meso and endocarp). The primary genetic effects controlling fruit firmness are additive, in combination with epistatic effects (Cook, 1995).

Bitterness is a trait that destroys the taste of cucumbers. Bitterness is due to the formation of two cucurbitacins (terpenoid compounds) that impart a bitter flavor to seedlings, roots, stems, leaves, and fruit. Usually the bitter principal does not accumulate very heavily in the fruit. When it does, it accumulates nonuniformly among fruits and within the fruit. The cucurbitacins are likely to concentrate at the stem end and in and just under the skin of the fruit. Two genes are involved in controlling bitterness in cucumber; a dominant one produces extremely bitter fruit and a recessive one inhibits the formation of curcurbitacin in foliage and fruit. An enzyme, elaterase, will hydrolyze cucurbitacins to nonbitter compounds. Elaterase activity is, however, believed to be controlled independent of the genes controlling bitterness.

The new trend for enhancing cucumber nutritional quality is by developing cucumber enriched with beta-carotene. Xishuangbanna gourd (XIS; Cucumis sativus var. xishuangbannanesis Qi et Yuan; $2n = 2x = 14$) is cultivated in the mountain regions (800–1,200 m elevation) of southern Yunnan province, P.R. China, produces fruit that possess an orange-colored endocarp/mesocarp (i.e., beta-carotene content), and is cross-compatible with cultivated cucumber (Yang et al. 1991). XIS is being used in breeding programmes to develop orange colored cucumbers rich in beta-carotene content. Additive genetic effects condition carotenoid accumulation in immature fruits, and additive and nonadditive factors are important for orange color expression in mature fruits (Navazio and Simon, 2001; Cuevas et al., 2010). However, growing environment has a dramatic influence on the expression of carotenoid pigmentation in immature fruits, while it does not appreciably influence the internal color of mature fruit.

4.9.2   PUMPKIN AND SQUASHES

Shape, color and skin texture, fruit weight and fruit diameter, carotenoids, vitamin C and proteins along with carbohydrates are important traits to be considered for quality of pumpkin. Prudent use of the right pumpkin will help growers meet yield and quality demand for market and thus maximize returns (Searle et al., 2003). Fruit shape, color, and skin texture are highly variable (Bisognin, 2002, Oynishi et al., 2013; Zinash et al., 2013). The genotypic coefficient of variability (GCV) was considerably high for maximum traits which offer adequate scope for effective selection criteria and improvement especially with the combination of high genotypic and phenotypic coefficients (Oynishi et al., 2013).

Dry matter content and TSS along with sucrose content are the major traits affecting the palatable quality of pumpkin (Cordenunsi et al., 2003; Sturm et al., 2003 Monforte et al., 2004; Sinclair et al., 2006). TSS content in different the pumpkin accessions vary a greatly from 4.1 to 10.3 °Brix as reported by various workers (Murakami et al., 1992; Sudhakar et al., 2003; Burger and Schaffer, 2007; Zinash et al., 2013). Sugar is the major component of TSS and it determines the flavor and sensory quality of pumpkin fruit. Reducing sugar is highly related to the sweetness of the fruit (Murakami et al., 1992; Zinash et al., 2013). Ascorbic acid content in pumpkins and squashes was also found to have high genetic variation. A gradual rise in the amount of vitamin C from 4.3 mg/100 g FW in young fruit to 15 mg/100g

FW in ripe fruit has been reported by Sharma and Rao (2013). Ascorbic acid was significantly positively correlated with dry matter, TSS, total sugar, reducing sugar, titratable acidity and sugar to acid ratio. Acidity of the fruits also had wide variation and it has been found that bigger sized fruits with higher dry matter had greater acidity (Tittonell et al., 2001).

Wide variability in carotenoid content has also been found. Total carotenoids ranged from 2.34 mg to 14.85 mg per 100g of fresh weight as reported by Pandey et al. (2003). Content of carotenoids in Spanish pumpkin was higher than that of other pumpkin varieties and even higher than that of beta-carotene in carrots (Wu et al. 1998). Carotenoid content also increase with ripening.

Polyphenols are the most prominent determining factors for the quality traits like color, aroma, bitterness, and astringency of fresh fruits (Sharma and Rao, 2013). The total phenols in pumpkin fruit were highest at its young stage and have important role in several protection mechanisms in order to prevent their early senescence (Sharma and Rao, 2013). Similar findings had reported in the fruit of *C. moschata* by Pandya and Rao (2010).

It has been observed that the dry matter content, TSS, total sugar, reducing sugar, ascorbic acid and titratable acidity are positively correlated with each other (Hazzard, 2006; Zinash et al., 2013). Hence, for improvement of quality, selection could be made based on one of these traits for ease of screening pumpkin and squash genotypes. Due to governance of nonadditive gene action, carotenoids, and ascorbic acid can be improved through hybridization followed by recurrent selection.

4.9.3   MELONS

Melons refer to a group of fruit vegetables that are used as dessert including *Cucumis melo* (muskmelon) and *Citrullus lanatus* (watermelon). The quality of melons is described by its shape, size, skin color, flesh color, texture, andflavor including sweetness and aroma. In watermelons crispness is another quality important for consumer preference.

Great variation has been observed in flesh color (orange, orange light or pink, green, white, or even mixture of these colors), rind color (green, yellow, white, orange, red, gray, or blend of these colors), rind texture (smooth, warty, striped, netted, rough, or combination of these textures), form (round, flattened, or elongated), and size (from 4 up to 200 cm) (Kirkbride, 1993; Goldman, 2002) in muskmelon. Due to this immense variation, *Cucumis melo* has been classified into different botanical varieties as follows (Guis et al., 1998):

1. *C. melo* var. *cantaloupensis* Naud. Medium-size fruits, rounded in shape, smooth surface or warty, and often have prominent ribs and sutures, if there is netting, it is sparse. Orange-fleshed, aromatic flavor and high in sugars.

2. *C. melo* var. *reticulatus* Ser. Medium-size fruits, and netted surface. If ribs are present, they are not well marked, flesh color from green, white to red-orange. Most are sweet and have a musky odor.

3. *C. melo* var. *saccharinus* Naud. Medium-size fruits, round or oblong shape, smooth with grey tone sometimes with green spots, very sweet flesh.

4. *C. melo* var. *inodorus* Naud. Smooth or netted surface, flesh commonly white, green or orange, lacking the typical musky flavor. These fruits are usually later in maturity and longerkeeping due to reduced or no ethylene production compared to cantaloupensis or reticulatus.

5. *C. melo* var. *flexuosus* Naud. Long and slender fruit, green rind, and finely wrinkled or ribbed. Green-fleshed and usually eaten as an alternative to cucumber. Low level of sugars.

6. *C. melo* var. *conomon* Mak. Small fruits, smooth surface, crisp white-fleshed. These melons ripen very rapidly and develop high sugar content but little aroma.

7. *C. melo* var. *dudaim* Naud. Small fruits, yellow rind with red streak, white to pink-fleshed.

However, such wide variation is not found in watermelon.

The flesh color of muskmelons is another important quality attribute for consumer appeal (Yamaguchi, 1977). In general, four basic and distinctive flesh colors can be observed in melon fruits: orange, light orange or pink, green, and white. In orange-fleshed melons theprincipal pigment is beta-carotene along with traces of delta-carotene, alpha-carotene, phytofluene, phytoene, lutein, violaxanthin, and traces of other carotenoids (Seymour and McGlasson, 1993). In green fleshed and white-fleshed melons low levels of beta-carotene and xanthophylls and chlorophylls are present (Watanabe et al., 1991; Flugel and Gross, 1982). All nongreen melon cultivars, with flesh color ranging from cream to intense orange, vary only in their beta-carotene content, and have only low levels of other carotenoids (Burger et al., 2009). The flesh color has polygenic inheritance (Harel-Beja et al., 2010; Cuevas et al., 2009).

The flesh color of watermelon may be red, orange, salmon yellow/orange, canary yellow and diverse carotenoids that are attributable to the different flesh colors. The major carotenoid in red-fleshed watermelon is lycopene containing an average of 48.7 μg·g-1 fresh weight (Holden et al., 1999) which is approximately 60 percent more than a tomato fruit. In watermelon, the amount of lycopene is variable across maturity, genotype, and ploidy level. Perkins-Veazie et al. (2002b) reported that about 20 percent more lycopene was accumulated in fully mature watermelon than immature ones. In several studies, triploid watermelon has been shown to contain more lycopene than diploid watermelons (Perkins-Veazie et al., 2001; Leskovar et al., 2004). The major pigments of orange-fleshed watermelon appear to be prolycopene and/or ζ-carotene (Tadmor et al., 2004). Natural genetic variation exists for watermelon flesh color and this can be utilized in improvement of this trait.

Sugar concentration is the major factor influencing melon fruit quality (Yama-guchi et al., 1977) and sucrose is the main sugar accumulated during fruit ripening in both muskmelon and watermelon (Burger and Schaffer, 2007; Hubbard at al., 1989). TSS is a common method of predicting fruit sugar content. A minimum of 10 percent total soluble solids is required for considering a dessert muskmelon as good, but high-quality melons can reach a soluble solid content as high as 17 per-cent (Sykes, 1991). Additive and polygenic control of sucrose levels has been found (Harel-Beja et al., 2009).

The aroma or fragrance of muskmelon fruits, brought about by the release of volatiles, are essential quality factors strongly linked to the ripening process and genetically controlled (Yahyaoui et al., 2002). The aroma and taste of most melon fruits are influenced considerably by ester compounds (ethyl esters and acetate es-ters) as well as to a certain extent by sulfur compounds (Hector et al., 2008).

Commercial melon varieties, with sweet, nonbitter, and low-acidic fruits, carry three genes (*suc/suc, so/so, bif/bif*), which control high-quality-fruit traits, in reces-sive form (Burger et al., 2003). Therefore, any intraspecific crosses, using tradi-tional breeding methods, between melon land races (Seshadri and More, 2002) and commercial melon cultivars will produce hybrid fruit with low-quality characteris-tics, because of the effect of dominant genes controlling low-sweetness, high acidity and high-bitterness levels in the melon land race fruit. Consequently, back-crossing may be adopted to develop desired melon genotype.

4.9.4 COLE CROPS

Cole crops such as cauliflower, cabbage, broccoli, Brussels sprouts, kale, collards, and kohlrabi are important vegetables crops grown for their own significance due to their high nutritive properties and wider adaptability. Wide range of vitamins, minerals, fiber, and carbohydrate can be obtained from the cole crops as well as high amounts of vitamin C, soluble fiber and multiple potential anticancer nutrients (Chun et al. 2004). Cole crops are rich in cyanidin, indole-3-carbinol, that boost DNA repair in cells and appears to block the growth of the cancer cells. These sub-stances along with the phenolic compounds, ascorbic acid, β-carotene and α- and β-tocopherols and glucosinolates in the crop have the beneficial properties (Singh et al. 2007).

Cauliflower curds of high quality should be white to cream in color, firm, and compact. The curds should be free of mechanical damage, decay, browning, or yel-lowing, which can be caused by sun exposure. Curds should be surrounded by a whorl of trimmed green turgid leaves. Curd size is also an important trait. It is a rich source of phytochemicals and contains Vitamin A, Thiamine, Riboflavin, Niacin, Vitamin C, Calcium, Iron, and Phosphorous and helps in fighting diseases (Fennema and Owen, 1996). The main bioactive phytochemicals in cauliflower include poly-phenols, terpenoids, glucosinolates, and other sulfur-containing compounds (Scal-bert et al., 2002). Dominance (h) and dominance x dominance (l) component of genetic variation along with duplicate type of epistasis conditioned curd weight, curd diameter and curd (Saha and Kalia, 2012). Curd color has monogenic control, with white being recessive.

Purple and orange cauliflowers are recent discoveries, arising from mutation. An interesting and unique Purple (Pr) gene mutation in cauliflower (Brassica oleracea var botrytis) confers an abnormal pattern of anthocyanin accumulation, giving the striking mutant phenotype of intense purple color in curds and a few other tissues (Chiu et al., 2010). Or gene, from a high beta-carotene orange cauliflower mutant has been identified that controls carotenoid accumulation by inducing the formation of chromoplasts, which provide a metabolic sink to sequester and deposit carotenoids in cauliflower(Zhou et al., 2008). Purple cauliflower can be a potential source for anthocyanin, while orange cauliflower for beta-carotene, both being excellent antioxidant sources.

In cabbage, head shape, head size, core width, core length, core solidity, head compactness and the amount of ascorbic acid and beta-carotene determine the quality. Flat, round, or conical heads are available and the shape is determined by the polar diameter:equatorial diameter ratio. If the ratio is 0.8–1, the head is termed round, 0.6 or less it is drum head and more than 1 it is conical head. Generally round head cabbage of medium size, narrow core width, short core length (less than 25 percent of the head diameter) with soft core is preferred by consumers. Compact head is a very important trait of cabbage. The head compactness can be judged by the position of the wrapper leaves; if it covers two-thirds of the head, it is said to be compact. Besides, a compact head should have a small core. Compactness is also positively correlated with net weight of head, i.e., the more the net weight of head, the more the compactness of head. High variability and high heritability have been reported for head compactness, core length, net head weight, equatorial diameter, suggesting that these characters can be improved easily through effective selection (Singh et al., 2011).

Cabbage is a rich source of vitamins, minerals, and fiber, along with secondary metabolites called glucosinolates, which are known to possess anticarcinogenic properties (Sarikamiş et al. 2009). It is also a source of antioxidant phytochemicals (ascorbic acid, lutein, β-carotene, DL-α- tocopherol and phenolics). High variability in these substances between cultivars has been reported (Faltusová et al., 2011). Carotenoid content varied greatly by 15-fold, and ascorbic acid content ranged from 25.47 to 57.60 mg/100 g fresh weight (2.3-fold) (Singh et al., 2011). High heritability has been reported for carotenoid content suggesting selection as an effective tool in improving the trait. Plants with higher contents can be utilized as a potential genetic source for breeding.

Cole crops in general are good sources of minerals like Ca, Mg, K, Fe, and Zn. Significant variability in mineral content has been reported in each of cabbage, broccoli, knolkhol, kale, collards. Mineral content has been found to be negatively correlated with weight of curd/head (Davis et al. 2004, Farnham et al. 2000; Broadley et al. 2008) and selection for medium size and average mineral composition can be done.

4.10   LEGUMINOUS VEGETABLES

Peas and beans are leguminous vegetables and seed protein content is the most important quality parameter for this group of crops.

For garden peas, optimal green color, size and shape of seeds, high content of resistance starch and high content of proteins is considered important for quality.

The major component of pea is starch, which accounts for up to 50 percent of the seed dry matter (Borowska et al. 1996; Wang et al. 1998). Starch is made up of amylose and amylopectin, the former having a linear molecular structure while the latter having a branched structure. Unlike most plants where amylose content is 20–25 percent, garden pea starch has 70–80percent amylose. Content of resistant starch is related to content of amylose (Dostálová et al. 2009). Protein and total dietary fiber account for about 24percent and 20 percent of dry matter, respectively, whereas lipids are present in lower amounts (2.5% of dry matter) (Black et al. 1998). High variation in starch and protein contents are often observed, whereas the variations in the contents of other components are usually lower (Borowska et al. 1996). Peas have large genetic variation with respect to their seed reserves, starch, and proteins, which makes it possible to select new materials with appropriate properties for particular applications.

In the other leguminous vegetables such as cowpea, French bean, faba bean, the protein content is highly variable (Hazra et al., 1996; Mutschler et al., 1980; Robertson et al., 1985). However, in peas and beans the seed protein content is negatively correlated with the yield (Evans and Gridley, 1979). Thus, selection should be directed at high yield with average protein content. Nonadditive gene action has also been found to condition protein content, hence enhancing this trait by simple selection is difficult.

4.11   OTHER VEGETABLES

4.11.1   OKRA

Okra is cultivated for its green nonfibrous fruits or pods containing round seeds. Okra is an important source of vitamins, calcium, potassium, and other mineral matters which are often lacking in the diet in developing countries (IBPGR, 1990). The ideal okra should have be deep green in color with less fiber, less mucilaginous substance along with high dry matter for dehydration purpose, while for canning and freezing purposes the dry matter should be low. However, in some countries of West Africa and especially in Nigeria, high mucilage is preferred. Deep green colored fruits are popularly believed to also have high mucilaginous character and fruit skin color in okra has been reported to be monogenically controlled and the recessive genes involved form a multiple allelic series by Udengwu, (2008). Kalia and Padda (1962) reported that in inter- varietal crosses between three varieties of okra having purple, green, and creamy fruits, a multiple allelic series appeared to control fruit color, green being dominant over cream while purple was dominant over both green and cream. Kolhe and D'Cruz (1966) indicated that pigmentation on the fruit of okra was monogenically controlled with reddish pigmentation being completely dominant over nonpigmentation of fruits. White skin was reported by Jassim (1967) to be dominant over green with a pair of genes involved. Nath and Dutta (1970) on the other hand stated that fruit skin color was controlled digenically, with light cream color and pink shading both being dominant over green color.

4.11.2   SWEET POTATO

Starch and dry matter content, sweetness (taste), fiber content, protein content, level of flesh oxidation, root shape, skin color and flesh color are important traits pertaining to quality of sweet potato. Smooth shapes are preferred to cracked, rough, and wrinkled shape. Dry matter content generally varies between 28 and 35 percent, but orange-fleshed have relatively low dry matter content (less than 18%) (Rees et al., 2003). White flesh color is more acceptable to consumers than orange flesh color (Rees et al., 2003). But, orange-fleshed varieties are rich in beta-carotene which is important from nutritional point of view. Protein content is positively correlated to dry matter content. Sugar content is negatively correlated with starch content. Most of the traits are quantitatively inherited (Jones, 1986). Heritability for tuber shape has been found to be low, while that for heritability of growth habit, skin color, β-carotene content, starch content and protein content were found to be moderate to high (Wilson et al., 1989). Hence, evaluation and characterization of populations with high frequency of favorable alleles for desirable traits can support breeding programs through selection and using them in crossing programme is likely to have a positive effect on the desired traits. Such an evaluation allows breeders to concentrate efforts in those populations with higher potential of producing superior progenies.

Self-incompatibility and high level of cross-incompatibility, polyploidy level (hexaploid) makes improvement of this crop by conventional means somewhat difficult. For further improvement of sweet potato, other Ipomoea species may play an important role in providing new genes, such as those for flesh color and protein content of the storage roots (Li, 1982). A better knowledge of genetic diversity and relationships between sweet potato and its wild relatives will aid in the development of breeding programs that efficiently utilize wild Ipomoea germplasm (Lin et al., 2007).

4.11.3 FUTURE STRATEGIES

Vegetable breeding has come a long way. The process is theoretically simple, and it does create novelty. The art of breeding lies in identifying and combining the specific traits for each purpose. The goals of public and private breeding programmes vary widely depending on location, need, and resources. In general, breeding goals in vegetables have gone through various phases: breeding for yield, for resistance to biotic stresses, for resistance to abiotic stresses, for shelf life, for taste and currently for nutritional quality (Bai and Lindhout 2007). Recently, recovery of these quality phenotypes in food crops has benefited from renewed research activity (Goff and Klee 2006), which is driven both by the efforts of public health agencies and health professionals to add more nutritious, "functional" foods to our diets and by the willingness of consumers to pay for them. The key challenge is to subsequently introduce this specific genetic material (DNA) into elite production varieties through breeding while retaining their performance attributes (Giovannoni, 2006). The advent of genomics has brought a real boost to the generation of data, knowledge, and tools that can be applied to breeding and this transformed breeding from individual based activity to a multidisciplinary teamwork. Hence, it is possible exploit genes from vegetable germplasm in an efficient way. As a result, it is expected that the improvement of vegetable cultivars will continue in the future. With advances in genome mapping and quantitative genetic analyses, the genetic basis is being dissected for traits that are related to domestication in many crops (Poncet et al., 2004). The advent of molecular markers and linkage maps has made it possible to find associations between markers and phenotypes. Breeders can use a known molecular markers associated with a specific trait or a chromosome segment to select the presence of that particular molecular marker rather than the phenotype. These markers can be used to a great extent with the basic target of increasing the efficiency of breeding programmes. Due to marker-assisted selection, the paradigm of plant breeding has changed from selection of phenotypes toward selection of genes, either directly or indirectly. Nowadays, genomes of different vegetables have been mapped and have been made available online. With the advance of genome sequencing and genomics, the genetic basis of plant growth and development are expected to be better understood. Knowing the candidate genes for important traits and having the knowledge of exact functional nucleotide polymorphism within the gene, breeders can easily identify useful alleles in the wild germplasm and create novel genotypes by introgressing and pyramiding favorable unused natural alleles and/or even by shuffling and reorganization of genomic sequences. Learning from domestication and with more and more available knowledge on genomics, plant breeders might consider manipulating transcription and regulation factors in the genome to generate a gene pool of new trait variation. Manipulation of a plant regulatory gene can influence the production of several phytonutrients generated from independent biosynthetic pathways and lead to a novel genotype that cannot be achieved by a conventional breeding approach.

KEYWORDS

- **Vegetables**
- **Postharvest quality**
- **Genetics**
- **Ripening**
- **Quality improvement**
- **Antinutritional attributes**
- **Shelf life**
- **Composition**
- **Breeding**

REFERENCES

Abbott, J. A.; Quality measurement of fruits and vegetables. *Postharvest. Biol. Technol.* 15(3), 207–225.

Ahmed, N.; and Tanki, M. I.; Inheritance of root characters in turnip (Brassica rapa L.). *Indian J. Genet. Plant. Breed.* **1994**, *54*, 247–252.

Akhtar, S.; and Hazra, P.; Nature of gene action for fruit quality characters of tomato (Solanum lycopersicum). *Afr. J. Biotechnol.* **2013**, *12*(20), 2869–2875. doi: 10.5897/AJB12.1231

Akhtar, S.; and Pandit, M. K.; F$_2$ segregation pattern for fruit shape and pedicel character in tomato. *J. Crop. Weed.* **2009**, *5*(1), 154–57.

Almeida, J. L. F.; Um novo aspecto de melhoramento do tomato. *Agricultura.* **1961**, *10*, 43–44.

Amiot, M. J.; Tacchini, M.; Aubert, S.; and Nicolas, J.; Phenolic composition and browning susceptibility of various apple cultivars at maturity. *J. Food. Sci.* **1992**, *57*, 958–962.

Ano, G.; Hebert, Y.; Prior, P.; and Messiaen, C. M.; A new source of resistance to bacterial wilt of eggplants obtained from a cross: *Solanum aethiopicum* L. *Solanummelongena* L. *Agronomie.* **1991**, *11*, 555–560.

Antonious, G. F.; and Jarret, R. T.; Screening Capsicum accessions for capsaicinoids content. *J. Environ. Sci. Health.* **2006**, *41*(5), 717–729.

Arivalagan, M.; Bhardwaj, R.; Gangopadhyay, K. K.; Prasad, T. V.; and Sarkar, S. K.; Mineral composition and their genetic variability analysis in eggplant (*Solanum melongena* L.) germplasm. *J. Appl. Bot. Food Qual.* **2013**, *86*, 99–103, doi:10.5073/JABFQ.2013.086.014.

Aubert, S.; Daunay, M. C.; and Pochard, E.; Saponsides st´eroidiques de l'aubergine (Solanum melongena L.) I. Int´ere´t alimentaire, m´ethodologie d'analyse, localisation dans le fruit. *Agronomie.* **1989**, *9*, 641–651.

Augustin, J.; Variations in nutritional composition of fresh potatoes. *J. Food. Sci.* **1975**, *40*(6), 1295–1299.

Bai, Y.; and Lindhout, P.; Domestication and breeding of tomatoes: what have we gained and what can we gain in the future? *Ann. Bot.* **2007**, *100*, 1085–1094.

Bamberg, J. B.; Tuber dormancy lasting eight years in the wild potato *Solanum jamesii* . *Am. J. Potato. Res.* **2010**, *87*(2), 226–228.

Banga, O.; Origin and distribution of the western cultivated carrot. *Genet. Agraria. 17*, 357–370.

Beckles, D. M.; Factors affecting the postharvest soluble solids and sugar content of tomato (*Solanum lycopersicum*L .) fruit. *Postharvest Biol. Technol.* **2012**, *63*(1), 129–140.

Benjamin, L. R.; and Sulherland, R. A.; Storage-root weight, diameter and length relationships in carrot (*Daucus carota*) and red beet (*Beta vulgaris*). *J. Agric. Sci.* **1989**, *113*, 73–80.

Bisognin, D. A.; Origin and evolution of cultivated cucurbits. *Ciecia Rural.* **2002**, *32*, 715–723.

Black, R. G.; Brouwer, J. B.; Meares, C.; and Iyer, L.; Variation in physico-chemical properties of field peas (*Pisum sativum*). *Food Res Int.* **1998**, *31*, 81–86.

Bletsos, F.; Roupakias, D.; Tsaktsira, M.; and Scaltsoyjannes, A.; Production and characterization of interspecific hybrids between three eggplant (*Solanum melongena* L.) cultivars and *Solanum macrocarpon* L. *Sci. Hortic.* **2004**, *101*, 11–21.

Bonierbale, M. W.; Plaisted, R. L.; and Tanksley, S. D.; Rflp Maps based on a common set of clones reveal modes of chromosomal evolution in potato and tomato. *Genetics.* **1988**, *120*(4), 1095–1103.

Borowska, J.; Zadernowski, R.; and Konopka, I.; Composition of some physical properties of different pea cultivars. *Nahrung.* **1996**, *40*, 74–78.

Bosland, P. W.; and Votava, E. J.; *Peppers: Vegetable and Spice Capsicums.* CABI Publishing: Wallinford, U.K; **2000**.

Bowman, M. J.; Gene Expression and Genetic Analysis of Carotenoid Pigment Accumulation in Carrot (Daucus carota L.) Ph.D. Thesis. University of Wisconsin-Madison; **2012**.

Bradshaw, J. E.; Hackett, C. A.; Pande, B.; Waugh, R.; and Bryan, G. J.; QTL mapping of yield, agronomic and quality traits in tetraploid potato (*Solanum tuberosum* subsp *tuberosum*). *Theor. Appl. Genet.* **2008**, *116*(2), 193–211.

Brar, J. S.; and Sukhija, B. S.; Studies on genetic parameters in carrot (*Daucus carota* L.). *J. Res Punjab Agric. Univ.* **1981**, *18*, 287–291.

Breene, W. M.; Davis, D. W.; and Chou, H.; Texture profile analysis of cucumbers. *J. Food Sci.* **1972**, *37*, 113–117.

Breithaupt, D. E.; and Bamedi, A.; Carotenoids and carotenoid esters in potatoes (*Solanum tuberosum* L.): New insights into an ancient vegetable. *J. Agric. Food. Chem.* **2002**, *50*(24), 7175–7181.

Broadley, M. R.; Hammond, J. P.; King, G. J.; Astley, D.; Bowen, H. C.; Meacham, M. C., Mead, A., Pink, D. A. C.; Teakle, G. R.; Hayden, R. M.; Spracklen, W. P.; and White P. J.; Shoot calcium and magnesium concentrations differ between subtaxa, are highly heritable, and associate with potentially pleiotropic loci in *Brassica oleracea*. *Plant. Physiol.* **2008**, *146*, 1707–1720.

Brown, C. R.; Breeding for phytonutrient enhancement of potato. *Am. J. Potato. Res.* **2008**, *85*(4), 298–307.

Brown, C. R.; Culley, D.; Bonierbale, M.; and Amoros, W.; Anthocyanin, carotenoid content, and antioxidant values in native south American potato cultivars. *Hortic. Scie.* **2007**, *42*(7), 1733–1736.

Brown, C. R.; Edwards, C. G.; Yang, C. P.; and Dean, B. B.; Orange flesh trait in potato – inheritance and carotenoid content. *J. Am. Soc. Hortic. Sci.* **1993**, *118*(1), 145–150.

Brown, C. R.; Haynes, K. G.; Moore, M.; Pavek, M. J.; Hane, D. C.; Love, S. L.; Novy, R. G.; and Miller, J. C.; Stability and broad-sense heritability of mineral content in potato: iron. *Am. J. Potato Res*. **2010**, *87*(4), 390–396.

Brown, C. R.; Kim, T. S.; Ganga, Z.; Haynes, K.; De Jong, D.; Jahn, M.; Paran, I.; and De Jong, W.; Segregation of total carotenoid in high level potato germplasm and its relationship to beta-carotene hydroxylase polymorphism. *Am. J. Potato Res*. **2006**, *83*(5), 365–372.

Brummell, D. A.; and Harpster, M. H.; Cell wall metabolism in fruit softening and quality and its manipulation in transgenic plants. *Plant. Mol. Biol*. **2001**, *47*, 311–340.

Burger, Y. and Schaffer, A. A.; The contribution of sucrose metabolism enzymes to sucrose accumulation in *Cucumis melo. J. Am. Soc. Hortic. Sci.* **2007**, *132*, 704–712.

Burger, Y.; Paris, H.; Cohen, R.; Katzir, N.; Tadmor, Y.; Lewinsohn, E.; and SchaVer, A. A.; Genetic diversity of Cucumis melo. *Hortic. Rev.* **2009**, *36*, 165–198.

Burger, Y.; Sa'ar, U.; Distelfeld, A.; Katzir, N.; Yeselson, Y.; Shen, S.; and Schaffer, A. A.; Development of sweet melon (*Cucumis melo*) genotypes combining high sucrose and organic acid content. *J. Am. Soc. Hortic. Sci*. **2003**, *128*, 537–540.

Burlingame, B.; Mouille, B.; and Charrondiere, R.; Nutrients, bioactive non-nutrients and anti-nutrients in potatoes. *J. Food Compos. Anal*. **2009***, 22*(6), 494–502.

Cai, Y.; Sun, M.; and Corke, H.; Antioxidant activity of betalains from plants of the Amaranthaceae. *J. Agri. Food Chem*. **2003**, *51*, 2288–2294.

Camelo, A. F. L.; Manual for the preparation and sale of fruits and vegetables from field to market. FAO Agricultural Services Bulletin 151; **2004**.

Camire, M. E.; Kubow, S.; and Donnelly, D. J.; Potatoes and human health. *Crit. Rev. Food. Sci. Nut*. **2009**, *49*(10), 823–840.

Campbell, M. A.; Gleichsner, A.; Alsbury, R.; Horvath, D.; and Suttle, J.; The sprout inhibitors chlorpropham and 1,4-dimethylnaphthalene elicit different transcriptional profiles and do not suppress growth through a prolongation of the dormant state. *Plant Mol. Biol.* **2010**, *73*(1–2), 181–189.

Cao, G.; Sofic, E.; and Prior, R. L.; Antioxidant capacity of tea and common vegetables. *J. Agric. Food Chem*. **1996**, *44*, 3426–3431.

Capecka, E.; Thiocyanate content in relation to the quality features of radish *Raphanus sativus* L. *Acta-Physiologiae-Plantarum*. **1998**, *20*(2), 143–147.

Causse, M.; Buret, M.; Robini, K.; and Verschave, P.; Inheritance of nutritional and sensory quality traits in fresh market tomato and relation to consumer preferences. *J. Food Sci.* **2003***, 68*, 2342–2350.

Chaib, J.; Devaux, M. F.; Grotte, M. G.; Robini, K.; Causse, M.; Lahaye, M.; and Marty, I.; Physiological relationships among physical, sensory, and morphological attributes of texture in tomato fruits. *J. Exp. Bot*. **2007**, *58*, 1915–1925.

Chakraborty, I.; Vanlalliani, Chattopadhyay, A.; and Hazra, P.; Studies on processing and nutritional qualities of tomato as influenced by genotypes and environment. *Veg. Sci*. **2007**, *34*, 26–31.

Chatterjee, B.; Chenga Reddy, V.; Ramana, J. V.; Ravi Sankar, C.; and Panduranga Rao, C.; Correlation and path analysis in chilli (*Capsicum annuum* L). *Andhra Agric. J*. **2007**, *54*, 36–39.

Chatterjee, S.; Niaz, Z.; Gautham, S.; Adhikari, S.; Pasad, S. V and Sharma, A.; Antioxidant activity of some phenolic constituents from green pepper (Piper nigrum L.) and fresh nutmeg mace (Myristica fragrans). *Food Chem*. **2007**, *102*, 515–523.

Chiu, L. W.; Zhou, X.; Burke, S.; Wu, X.; Prior, R. L.; and Li, L.; The purple cauliflower arises from activation of a MYB transcription factor. *Plant. Physiol*. **2010**, *154*, 1470–1480.

Chope, G. A.; Terry, L. A.; White, P. J.; Preharvest application of exogenous abscisic acid (ABA) or an ABA analogue does not affect endogenous ABA concentration of onion bulbs. *Plant Growth Regul.* **2007**, *52*, 117–129.

Chun, O. K.; Smith, N.; Sakagawa, A.; and Lee, C. Y.; Antioxidant properties of raw and processed cabbages. *Int. J. Food Sci. Nutr.* **2004**, *55*(3), 191–199.

Clarke, A. E.; Jones, H. A.; and Little, T. M.; Inheritance of bulb color in the onion. *Genetics.* **1944**, *29*, 569–575.

Concell'on, A.; An~ 'on, M. C.; and Chaves, A. R.; Characterization and changes in polyphenol oxidase from eggplant fruit (Solanum melongena L.) during storage at low temperature. *Food Chem.* **2004**, *88*, 17–24.

Conforti, F.; Stati, G. A.; and Menichini, F.; *Food Chem.* **2007**, *102*, 1094–1104.

Cook, K. L.; Influence of genetic factors and pollination on fruit firmness in parthenocarpic and non-parthenocarpic pickling cucumbers. Masters dissertation. Oregon State University; **1995**.

Cuevas, H. E.; Song, H.; Staub, J. E.; and Simon, P. W.; Inheritance of beta-carotene-associated flesh color in cucumber (Cucumis sativus L.) fruit. *Euphytica.* **2010**, *171*, 301–311. doi 10.1007/s10681-009-0017-2.

Cuevas, H. E.; Staub, J. E.; Simon, P. W.; and Zalapa, J. E.; A consensus linkage map identiWes enomic regions controlling fruit maturity and beta-carotene-associated Xesh color in melon (Cucumis melo L.). *Theor. Appl. Genet.* **2009**, *119*, 741–756.

Daunay, M. C.; Lester, R. N.; and Ano, G.; Les aubergines, In: L'am'elioration des plantes tropicales; Charrier, A., Jacquot, M., Hamon, S., and Nicolas, D., Eds.; Cirad et Orstom, Montpellier, France, **1997**, pp 83– 107.

Davey M. W.; Montagu, M. V.; Inze, D.; Sanmartin, M.; Kanellis, A.; Smirnoff, N.; Benzie, I. J. J.; Strain, J. J.; Favell, D.; Fletcher, J.; Plant l-ascorbic acid: chemistry, function, metabolism, bioavailability and effects of processing. *J. Sci. Food Agric.,* **2000**, *80*, 825–860.

Davies, J.; and Hobson, G. E.; The constituents of tomato fruit-the influence of environment, nutrition and genotype. *Crit. Rev. Food Sci. Nutr.* **1981**, *15*(3), 205–280.

Davis, D. R.; Epp, M. D.; and Riordan, H. D.; Changes in USDA food composition data for 43 garden crops 1950 to 1999. *J. Am. Coll. Nutr.* **2004**, *23*, 669–682.

Deepa, N.; Kaur, C.; George, B.; Singh, B.; and Kapoor, H. C.; Antioxidant constituent in some sweet pepper (Capsicum annuum L.) genotypes during maturity. *LWT-Food. Sci. Technol.* **2007**, *40*, 212–219.

del Rocío Gómez-García, M.; and Ochoa-Alejo, N.; Biochemistry and molecular biology of carotenoid biosynthesis in chili peppers (*Capsicum* spp.). *Int. J. Mol. Sci.* **2013**, *14*, 19025–19053; doi:10.3390/ijms140919025

Dhaliwal, M. S.; Singh, S.; Badhan, B. S.; Cheema, D. S.; and Singh, S.; Diallel analysis for total soluble solids contents, pericarp thickness and locule number in tomato. *Veg. Sci.* **1999**, *26*(2), 120–122.

Dogan, M.; Arslan, O.; and Dogan, S.; Substrate specificity, heat inactivation and inhibition of polyphenol oxidase from different aubergine cultivars. *Int. J. Food Sci. Technol.* **2002**, *37*, 415–423.

Dostálová, R.; Horáček, J.; Hasalová, I.; and Trojan, R.; Study of resistant starch (RS) content in peas during maturation. *Czech J. Food Sci.* **2009**, *27*, 120–S124.

El-Shafie, M. W.; and Davis, G. N.; Inheritance of bulb color in the onion (Allium cepa L.). *Hilgardia.* **1967**, *38*, 607–622.

Etoh, T.; True seeds in garlic. *Acta Hortic.* (ISHS). **1997**, *433*, 247–255.

Faltusová, Z.; Kučera, L.; and Ovesná, J.; Genetic diversity of *Brassica oleracea* var. *capitata* gene bank accessions assessed by AFLP. *Electr. J. Biotechnol.* **2011**, ISSN: 0717-3458 http://www. ejbiotechnology.info,doi: 10.2225/vol14-issue3-fulltext-4.

Farnham, M. W.; Grusak, M. A.; and Wang, M.; Calcium and magnesium concentration of inbred and hybrid broccoli heads. *J. Am. Soc. Hortic. Sci.* **2000**, *125*(3), 344–349.

Felczyński, K.; and Elkner, K.; Effect of long-term or-ganic and mineral fertilization on the yield and quality of red beet (*Beta vulgaris* L.). *Veget. Crops. Res. Bull.* **2008**, *68*, 111–125. doi: 10.2478/v10032-008-0010-7

Fennema and Owen, R.; *Food Chemistry,* 3rd ed.; Marcel Dekkar, Inc.: New York; **1996**.

Flugel, M.; and Gross, J.; Pigment and plastid changes in mesocarp and exocarp of ripening muskmelon, *Cucumis melo* var. Galia. *Angew. Botanik.* **1982**, *56*, 393–406.

Foolad, M. R.; Genome mapping and molecular breeding of tomato. *Int. J. Plant. Genomics.* **2007**, pp. 1–52.

Fossen, T.; Andersen, O. M.; Ovstedal, D. O.; Pedersen, A. T.; and Raknes, A.; Characteristic anthocyanin pattern from onions and other *Allium* spp. *J. Food Sci.* **1996**, *61*, 703–706.

Freyre, R.; Warnke, S.; Sosinski, B.; and Douches, D. S.; Quantitative trait locus analysis of tuber dormancy in diploid potato *Solanum* spp). *Theor. Appl. Genet.* **1994**, *89*(4), 474–480.

Frusciante, L.; Carli, P.; Ercolano, M. R.; Pernice, R.; Matteo, A. D.; Fogliano, V.; and Pellegrini, N.; Antioxidant nutritional quality of tomato. *Mol. Nutr. Food Res.* **2007**, *51*, 609–617.

Fuijwake, H.; Suzuki, T.; and Lwai, K.; Intracellular distributions of enzyems and intermediates involved in biosynthesis of capsicum and its analogues in capsicum fruits. *Agric. Bio. Chem.* **1982**, *46*, 2685–2689.

Gallina, P. M.; Cabassi, G.; Maggioni, A.; Natalini, A.; and Ferrante, A.; Changes in the pyruvic acid content correlates with phenotype traits in onion clones. *Austr. J. Crop. Sci.* **2012**, *6*(1), 36–40.

Garceäs-Claver, A.; Gil-Ortega, R.; Aälvarez-Fernaändez, A.; and Mariäa Soledadarnedo-Andreäs, M.; Inheritance of Capsaicin and Dihydrocapsaicin, Determined by HPLC-ESI/MS, in an Intraspecific Cross of *Capsicum annuum* L. *J. Agric. Food Chem.* **2007**, *55*, 6951–6957.

Garcia, E.; and Barrett, D. M.; Evaluation of processing tomatoes from two consecutive growing seasons: quality attributes peelability and yield. *J. Food Process. Preservest.* **2006a**, *30*, 20–36.

Garg, N.; Cheema, D. S.; and Dhatt, A. S.; Utilization of rin, nor, and alc alleles to extend tomato fruit availability. *Int. J. Veg. Sci.* **2008**, *14*, 41–54.

Gebhardt, C.; Ritter, E.; Debener, T.; Schachtschabel, U.; Walkemeier, B.; Uhrig, H.; and Salamini, F.; RFLP Analysis and Linkage Mapping in *Solanum tuberosum. Theor. Appl. Genet,* **1989**, *78*(1), 65–75.

George, B.; Kaur, C.; Khurdiya, D. S.; and Kapoor, H. C.; Antioxidants in tomato (Lycopersium esculentum) as a function of genotype. *Food. Chem.* **2004**, *84*(1), 45–51.

Giordano, L. B.; Silva, J. B. C.; and Barbosa, V.; Escolha de cultivars e plantio. In: Silva JBC and Guarding LB (org) Tomatoe para processamento industrial. *Brasilia: Emrapa, CNPH*; **2000**, pp 36–59.

Giovannoni, J. J.; Breeding new life into plant metabolism. *Nat. Biotechnol.* **2006**, *24*, 418–419.

Gisbert, C.; Prohens, J.; and Nuez, F.; Efficient regeneration in two potential new crops for subtropical climates, the scarlet (*Solanum aethiopicum*) and gboma (*S. macrocarpon*) eggplants. *New Zealand J. Crop. Hortic. Sci.* **2006**, *34*, 55–62.

Goff, S.; and Klee, H.; Plant volatile compounds: sensory cues for health and nutritional value? *Science.* **2006**, *311*, 815–819.

Gogoi, D.; and Gautam, B. P.; Correlation and path coefficient analysis in chilli (Capsicum annuum L). *Agric. Sci. Digest.* **2003**, *23*, 162–166.

Goldman, A.; *Melons, for the Passionate Grower.* Artisan: New York; **2002**.

Goldman, I. L.; Eagen, K. A.; Breltbach, D. N.; and Gabelman, W. R.; Simultaneous selection is effective In increasing betalaln pigment concentration but not total dissolved solids in red beet (*Beta vulgaris* L). *J. Am. Soc. Hortic. Sci.* **1996**, *121*, 23–26.

Gonzalez-Cebrino, F.; Lozano, M.; Ayuso, M. C.; Bernalte, M. J.; Vidal-Aragon, M. C.; and Gonzalez-Gomez, D.; Characterization of traditional tomato varieties grown in organic conditions. *Spanish J. Agric. Res.* **2011**, *9*(2), 444–452.

Gould, W. A.; *Tomato production, processing and technology.* CTI Publ: Baltimore; **1992**.

Gupta, A.; Kawatra, A.; and Sehgal, S.; Physical-chemical properties and nutritional evaluation of newly developed tomato genotypes. *Afr. J. Food Sci. Technol.* **2011**, *2*(7), 167–172.

Guzman, I.; Hamby, S.; Romero, J.; Bosland, P. W.; and O'Connell, M. A.; Variability of Carotenoid Biosynthesis in Orange Colored Capsicum spp. *Plant Sci.* **2010**, *179*(1–2), 49–59. doi:10.1016/j.plantsci.2010.04.014\

Hallard, J.; L'aubergine au Japon. *PHM Revue Horticole.* **1996**, *374*, 55–56.

Hansche, P. E.; and Boynton, B.; Heritability of enzymatic browning in peaches. *Hortic. Sci.* **1986**, *21*, 1194–1197.

Harel-Beja, R.; Tzuri, G.; Portnoy, V.; Lotan-Pompan, M.; Lev, S.; Cohen, S.; Dai, N.; Yeselson, L.; Meir, A.; Libhaber, S. E.; Avisar, E.; Melame, T.; van Koert, P., Verbakel, H., Hofstede, R., Volpin, H., Oliver, M., Fougedoire, A., Stalh, C., Fauve, J., Copes, B., Fei, Z., Giovannoni, J., Ori, N., Lewinsohn, E., Sherman, A., Burger, J., Tadmor, Y., SchaVer, A. A. and Katzir, N.; A genetic map of melon highly enriched with fruit quality QTLs and EST markers, including sugar and carotenoid metabolism genes. *Theor Appl Genet.* **2010**, *121*, 511–533. doi: 10.1007/s00122-010-1327-4.

Hari, G. S.; Rao, P. V.; and Reddy, Y. N.; Correlation studies in paprika (Capsicum annuum L). *Crop Res.* **2005**, *29*, 495–498.

Hayman, M.; and Kam, P. C. A.; Capsaicin: a review of its pharmacology and clinical applications. *Curr. Anaes. Crit. Care*, **2008**, *19*, 338–343.

Haynes, K. G.; Sieczka, J. B.; Henninger, M. R.; and Fleck, D. L.; Clone experiment interactions for yellow-flesh intensity in tetraploid potatoes. *J. Am. Soc. Hortic. Sci.* **1996**, *121*, 175–177.

Hazra, P.; and Dutta, A. K.; Vegetable breeding for quality traits. In: *Breeding and Protection of Vegetables*, Rana, M. K., Ed.; New India Publishing Agency: New Delhi, **2011**; pp 274–314.

Hazra, P.; Das, P. K.; and Som, M. G.; Combining ability for yield and seed protein in cowpea. *Indian J. Genet. Plant Breed.* **1996**, *56*, 516–518.

Hazzard, R.; Pumpkin crop. Vegetable Notes for Vegetable Farmers in Massachusetts, **2006**, *17*(20), 20–32.

Hernandez Suarez, M.; Rodriguez Rodriguez, E.; and Diaz Romero, C.; Analysis of organic acid content in cultivars of tomato harvested in Tenerife. *Eur. Food Res. Technol.* **2008**, *226*(3), 423–435.

Holden, J. M.; Eldridge, A. L.; Beecher, G. R.; Buzzard, I. M.; Bhagwat, S. A.; Davis, C. S.; Douglass, L. W.; Gebhardt, S. E.; Haytowitz, D. B.; and Schakel, S.; Carotenoid content of U.S. foods: an update of the database. *J. Food Comp. Anal.* **1999**, *12*, 169–196.

Hosamani, R. M.; Patil, B. C.; and Ajjappalavara, P. S.; Genetic variability and character association studies in onion (*Allium cepa* L). *Karnataka J. Agric. Sci.* **2010**, *23*(2), 302–305.

Hubbard, N. L.; Huber, S. C.; and Pharr, D. M.; Sucrose phosphate synthase and acid invertase as determinants of sucrose concentration in developing muskmelon (*Cucumis melo* L.) fruits. *Plant. Physiol.* **1989**, *91*, 1527–1534.

Ilahy, R.; Hdider, C.; Lenucci, M. S.; Tlili, I.; and Dalessandro, G.; Phytochemical composition and antioxidant activity of high-lycopene tomato (*Solanum lycopersicum* L.) cultivars grown in Southern Italy. *Sci. Hortic.* **2011b**, *127*(3), 255–261.

Ilahy, R.; Hdider, C.; Lenucci, M. S.; Tlili, I.; and Dalessandro, G.; Antioxidant activity and bioactive compound changes during fruit ripening of high-lycopene tomato cultivars. *J. Food. Compos. Anal.* **2011a**, *24*(4–5), 588–595.

Imam, M. K.; and Gabelman, W. H.; Inheritance of a number of phenotypes in *Daucus carota* L. (PhD dissertation). University of Wisconsin, Madison, Wisconsin; **1968**.

Islam, S.; and Singh, R. V.; Correlation and path analysis in sweet pepper (*Capsicum annuum* L). *Veg. Sci.* **2009**, *36*, 128–130.

Iwai, K. T.; and Fhjiwake, H.; Formation and metabolism of pungent principles of capsicum fruits. IV. Formation and accumulation of pungent principles in *Capsicum annum* var. *annum* cv. Karayatsubsa at different growth stages after flowering. *Agric. Biol. Chem.* **1979**, *43*, 2493–2498.

Iwanzik, W.; Tevini, M.; Stute, R.; and Hilbert, R.; Carotenoid content and composition of different potato varieties and their importance for the color of the flesh of the tuber. *Potato Res.* **1983**, *26*(2), 149–162.

Jansen, G.; Flamme, W.; Schuler, K.; and Vandrey, M.; Tuber and starch quality of wild and cultivated potato species and cultivars. *Potato. Res.* **2001**, *44*(2), 137–146.

Jasim, A. J.; Inheritance of certain characters in okra (*Hibiscus esculentus* L.) Unpublished Ph.D thesis, Louisiana State University, U.S.A.; **1967**, pp 65.

Jo, M. H.; Ham, I. K.; Moe, K. T.; Kwon, S. W.; Lu, F. H.; Park, Y. J.; Kim, W. S.; Won, M. K.; Kim, T. I.; Lee, E. M.; Classification of genetic variation in garlic (*Allium sativum* L.) using SSR markers. *Aust. J. Crop. Sci.* **2012**, *6*(4), 625–631.

Jongen, W.; Fruit and vegetables processing. Wood Head Publishing in Food Science and Technology. Wagenningen University: Netherlands, **2002**, pp 350.

Juliana. D.; Oen, S.; Azizahwati, L. H.; and Winarno, F. G.; Capsaicin content of various varieties of Indonesian chillies, *Asia. Pac. J. Clin. Nutr.* **1997**, *6*(2), 99–101.

Jung, C. S.; Griffiths, H. M.; De Jong, D. M.; Cheng, S.; Bodis, M.; Kim, T. S.; and De Jong, W. S.; The potato developer (D) locus encodes an R2R3 MYB transcription factor that regulates expression of multiple anthocyanin structural genes in tuber skin. *Theor. Appl. Genet.* **2009**, *120*(1), 45–57.

Jung, C. S.; Griffiths, H. M.; De Jong, D. M.; Cheng, S. P.; Bodis, M.; and De Jong, W. S.; The potato P locus codes for flavonoid 3 ',5 '-hydroxylase. *Theor. Appl. Genet.* **2005**, *110*(2), 269–275.

Kader, A.; Fruit maturity, ripening and quality relationships. *Acta. Hortic.* **1999**, *485*, 203–208.

Kalia, H. R.; and Padda, D. S.; Inheritance of some fruit characters in okra. *Indian J. Genet. Plant. Breed.* **1962a**, *18*(1), 57–68.

Kamenetsky, R.; Shafir, I. L.; Baizerman, M.; Kik, C.; Khassanov, F.; and Rabinowitch, H. D.; Garlic (*Allium sativum* L.) and its wild relatives from Central Asia: evaluation for fertility potential. *Acta Hortic.* (ISHS). **2004**, *637*, 83–91.

Kanner, J.; Mendel, H.; and Budowski, P.; Carotene oxidizing factors in red pepper fruits (*Capsicum annuum* L.): Peroxidase activity. *J. Food. Sci.* **2006**, *42*(6), 1549–1551.

Karad, S. R.; Navale, P. A.; and Kadam, D. E.; Variability and path-coefficient analysis in chilli (*Capsicum annuum* L.). *Int. J. Agric. Sci.* **2006**, *2*(1), 90–92.

Kaul, D. L.; and Nandpuri, K. S.; Combining ability studies in tomato. *Punjab. Agric. Univ. J. Res.* **2006,** *9,* 15–18.

Keller, W.; Inheritance of some major color types In beets. *J. Agric. Res.* **1936,** *52,* 27–38.

Khachik F.; Carvalho L.; Bernstein, P. S.; Muir, G. J.; Zhao, D. Y.; and Katz, N. B.; Chemistry, distribution, and metabolism of tomato carotenoids and their impact on human health. *Exp. Biol. Med. (Maywood).* **2002,** *227,* 845–851.

Kik, C.; Kahane, R.; and Gebhardt, R.; Garlic and Health. *Nutr. Metab. Cardiovasc Dis.* **2001,** *11,* 57–65.

Kim, D. O.; Lee, K. W.; Lee, H. J.; and Lee, C. Y.; Vitamin C equivalent capacity (VCEAC) of phenolic phytochemicals. *J. Agric. Food Chem.* **2002,** *50,* 3713–3717.

Kim, S.; Binzel, M.; Yoo, K.; Park, S.; and Pike, L. M.; *Pink (P),* a new locus responsible for a pink trait in onions (*Allium cepa*) resulting from natural mutations of anthocyanidin synthase. *Mol. Gen. Genom.* **2004a,** *272,* 18–27.

Kim, S.; Binzel, M.; Yoo, K.; Park, S.; and Pike, L. M.; Inactivation of DFR (Dihydroflavonol 4-reductase) gene transcription results in blockage of anthocyanin production in yellow onions (*Allium cepa*). *Mol. Breed.* **2004b,** *14,* 253–263.

Kim, S.; Jones, R.; Yoo, K.; and Pike, L. M.; Gold color in onions (*Allium cepa*), a natural mutation of the chalcone isomerase gene harboring a pre-mature stop codon. *Mol. Gen. Genom.* **2004c,** *272,* 411–419.

Kim, S.; Yoo, K.; and Pike, L. M.; Development of a codominant PCR-based marker for an allelic selection of the pink trait in onions (*Allium cepa*) based on the insertion mutation in the promoter of the anthocyanidin synthase gene. *Theor. Appl. Genet.* **2005a,** *110,* 573–578.

Kim, S.; Yoo, K.; and Pike, L. M.; Development of a PCR-based marker utilizing a deletion mutation in the DFR (dihydroflavonol 4-reductase) gene responsible for the lack of anthocyanin production in yellow onions (*Allium cepa*). *Theor. Appl. Genet.* **2005b,** *110,* 588–595.

Kim, S.; Jones, R.; Yoo, K.; and Pike, L. M.; The L locus, one of complementary genes required for anthocyanin production in onions (*Alliumcepa*), encodes anthocyanidin synthase. *Theor. Appl. Genet.* **2005c,** 111, 120–127.

Kingston, B. D.; and Pike, L. M.; Internal fruit structure of warty and non-warty cucumbers and their progeny. *Hortic. Sci.* **1975,** *10,* 319.

Kirkbride, J. H.; *Biosystematic Monograph of the Genus Cucumis (Cucurbitaceae),* Parkway Publishers: Boone, NC; **1993,** p 159.

Kloosterman, B.; Oortwijn, M.; Uitdewilligen, J.; America, T.; de Vos, R.; Visser, R. G. F.; and Bachem, C. W. B.; From QTL to candidate gene: Genetical genomics of simple and complex traits in potato using a pooling strategy. BMC Genomics; **2010,***11.*

Kolhe, A. K.; and D' Cruz, R.; Inheritance of pigmentation in okra. *Ind. J. Genet. Plant. Breed.* **1966,** *26,* 112–117.

Koops, A. J.; Hall, R. D.; and Kik, C.; Bolkleurvorming bij de ui: een biochemisch genetisch model. *Prophyta.* **1991,** *6,* 20–22.

Kotikova, Z.; Hejtmankova, A.; and Lachman, J.; Determination of the influence of variety and level of maturity on the content and development of carotenoids in tomatoes. *Czech. J. Food. Sci.* **2009,** *27,* S200–S203.

Koul, A. K.; Gohil, R. N.; and Langer, A.; Prospects of breeding improved garlic in the light of its genetic and breeding systems. *Euphytica.* **1997,** *28,* 457–464.

Krinsky, N. I.; Carotenoids as antioxidants. *Nutrition.* **2001***, 17,* 815–817.

Kumar, D.; Singh, B. P.; and Kumar, P.; An overview of the factors affecting sugar content of potatoes. *Ann. Appl. Biol.* **2004,** *145*(3), 247–256.

Kumar, O. A.; and Tata, S. S.; Ascorbic acid contents in chili peppers (*Capsicum* L.). *Notulae Scientia Biologicae.* **2009,** *1,* 50–52.

Kumar, S. R.; Arumugam, T.; Anandakumar, C. R. and Premalakshmi, V.; Genetic variability for quantitative and qualitative characters in Brinjal (*Solanum melongena* L.). *Afr. J. Agric. Res.* **2013,** *8*(39), pp. 4956–4959.

Kumari, S. S.; Jyothi, K. U.; Reddy, V. C.; Srihari, D.; Siva Sankar, A.; and Ravi Sankar, C.; Character association in paprika (Capsicum annuum L.). *J. Spices. Aromatic. Crops.* **2011,** *20*(1), 43–47.

Kust, A. F.; Inheritance and differential formation of color and associated pigments in xylem and phloem of carrot, *Daucus carota* L. (PhD dissertation). University of Wisconsin, Madison, Wisconsin; **1970.**

Laferriere, L.; and Gabelman, W. H.; Inheritance of color, total carotenoids, alphaorotene, and beta-carotene in carrots, *Daucus carota* L. *Proc. Am. Soc. Hortic. Sci.* **1968,** *93,* 408–418.

Lampasona, G. S.; Martı́nez, L.; Burba, J. L.; Genetic diversity among selected Argentinean garlic clones (*Allium sativum* L.) using AFLP (Amplified Fragment Length Polymorphism). *Euphytica.* **2003,** *132,* 115–119.

Lancaster, J. E.; Shaw, M. L.; Joyce, M. D. P.; McCallum, J. A.; and McManus, M. T.; A novel alliinase from onion roots. Biochemical characterization and cDNA cloning. *Plant. Phyiol.* **2000,** *122,* 1269–1280.

Lee, S. K.; and Kader, A. A.; Preharvest and postharvest factors influencing vitamin C content of horticultural crops. *Postharvest Biol. Technol.* **2000,** *20,* 207–220.

Leighton, T.; Ginther, C.; Fluss, L.; Harter, W. K.; Cansado, J.; and Notario, V.; Molecular characterization of quercetin and quercetin glycosides in *Allium* vegetables phenolic compounds in food and their effects on health II. *ACS. Symp. Ser.* **1992,** *507,* 220–238.

Leskovar, D. I.; Bang, H.; Crosby, K. M.; Maness, N.; Franco, J. A.; and Perkins-Veazie, P.; Lycopene, carbohydrates, ascorbic acid, and yield components of diploid and triploid watermelon cultivars are affected by deficit irrigation. *J. Hortic. Sci. Biotechnol.* **2004,** *79,* 75–81.

Lester, R. N.; and Hasan, S. M. Z.; Origin and domestication of the brinjal eggplant, Solanum melongena, from S. incanum, in Africa and Asia. In: *Solanaceae III: Taxonomy, chemistry, evolution,* Hawkes, J. G., Lester, R. N., Nee, M., and Estrada, N., Eds.; Linnean Society of London: London; **1991,** pp 369–387.

Lewis, C. E.; Walker, J. R. L.; Lancaster, J. E.; and Sutton, K. H.; Determination of anthocyanins, flavonoids and phenolic acids in potatoes. I: Coloured cultivars of *Solanum tuberosum* L. *J. Sci. Food. Agric.* **1998,** *77*(1), 45–57.

Li, L.; Breeding for increased protein content in sweet potatoes. In: *Sweet Potato,* Villareal, R. L., and Griggs, T. D., Eds.; Proceedings of the First International Symposium AVRDC, Tainan, Taiwan: AVRDC, **1982,** pp 345–354.

Li, X. Q.; De Jong, H.; De Jong, D. M.; and De Jong, W. S.; Inheritance and genetic mapping of tuber eye depth in cultivated diploid potatoes. *Theor. Appl. Genet.* **2005b,** *110*(6), 1068–1073.

Lin, K. H.; Lai, Y. C.; Chang, K. Y.; Chen, Y. F.; Hwang, S. H.; and Lo, H. F.; Improving breeding efficiency for quality and yield of sweet potato. *Bot. Stud.* **2007,** *48,* 283–292.

Lombard, K. A.; Investigation of the flavonol quercetin in onion (Allium cepa L.) by high-performance liquid chromatography (HPLC) and spectrophotometric methodology. M.S. Thesis. Texas Tech University; **2000.**

Lombard, K. A.; Geoffriau, E.; and Peffley, E.; Flavonoid quantification in onion (Allium cepa L.) by spectrophotometric and HPLC analyses. *Hortic. Sci.* **2002**, *37*(4), 682–685.

Love, S. L.; Salaiz, T.; Shafii, B.; Price, W. J.; Mosley, A. R.; and Thornton, R. E.; Stability of expression and concentration of ascorbic acid in North American potato germplasm. *Hortic. Sci.* **2004**, *39*(1), 156–160.

Macheix, J. J.; Fleuriet, A.; and Billot, J.; *Fruit phenolics*. CRC Press: Boca Raton, FL; **1990**.

Makani, J. M.; Monpara, B. A.; and Dhameliya, H. R.; Inheritance of Certain Traits in Brinjal (*Solanum melnogena* L.). *Natnl. J. Pl. Improv.* **2007**, *9*(1), 32–35.

Makarova, G. A.; and Ivanova, T. I.; Inheritance of characters of the root and leaf in radish. *Genetika, 19*, 304–311. In Russian with English summary; **1983**.

Mane, R.; Sridevi, O.; Salimath, P. M.; Deshpande, S. K.; and Khot, A. B.; Performance and stability of different tomato (Solanum lycopersicum) genotypes. *Ind. J. Agric. Sci.* **2010**, *80*(10), 898–901.

Marinova, D.; Ribarova, F.; and Atanassova, M.; Total phenolics and total flavonoids in Bulgarian fruits and vegetables. *J. Univ. Chem. Tech. Metall.* **2005**, *40*(3), 255–260.

Mathur, R.; Dangi, R. S.; Dass, S. C.; and Malhotra, R. C.; The hottest chilli variety in India. *Curr. Sci.* **2000**, *79*(3), 287–288.

Mathur, R.; Dangi, R. S.; Dass, S. C.; and Malhotra, R. C.; The hottest chilli variety in India. *Curr. Sci.* **2000**, *79*(3), 287–288.

Melkamu, M.; Seyoum, T.; and Woldetsadik, K.; Effects of pre- and post harvest treatments on changes in sugar content of tomato. *Afr. J. Biotechnol.* **2008**, *7*(8), 1139–1144.

Morales, R. G. F.; Resende, J. T. V.; Resende, F. V.; Delatorre, C. A.; Figueiredo, A. S. T.; and Da-Silva, P. R.; Genetic divergence among Brazilian garlic cultivars based on morphological characters and AFLP markers. *Genet. Mol. Res.* **2013**, *12*(1), 270–281.

Mutschler, M. A.; Bliss, F. A.; and Hall, T. C.; Variation in accumulation of seed storage protein among genotypes of *Phaseolus vulgaris* L. *Plant. Physiol.* **1980**, *65*, 627–630.

Naito, K.; Umemura, Y.; Mori, M.; Sumida, T.; Okada, T.; Takamatsu, N.; Okawa, Y.; Hayashi, K.; Saito, N.; and Honda, T.; Acylated pelargonidin glycosides from a red potato. *Phytochemistry.* **1998**, *47*(1), 109–112.

Nath, P.; and Dutta, O. P.; Inheritance of fruit hairiness, fruit stem colour and leaf lobing in okra, *Abelmoschus esculentus. Can. J. Genet. Cytol.* **1970**, *12*, 589–593.

Navarro, J. M.; Flores, P.; Garrido, C.; and Martinez, V.; Changes in the contents of antioxidant compounds in pepper fruits at different ripening stages, as affected by salinity. *Food Chem.* **2006**, *96*, 66–73.

Navazio, J. P.; and Simon, P. W.; Diallel analysis of high carotenoid content in cucumbers. *J. Am. Soc. Hortic. Sci.* **2001**, *126*(1), 100–104.

Nesterenko, S.; and Sink, K. C.; Carotenoid profiles of potato breeding lines and selected cultivars. *Hortic. Sci.* **2003**, *38*(6), 1173–1177.

Nielsen, K. K.; Mahoney, A. W.; Williams, L. S.; and Rogers, V. C.; X-ray fluorescence measurements of Mg, P, S, Cl, K, Ca, Mn, Fe, Cu and Zn in fruits, vegetables and grain products. *J. Food. Comp. Anal.* **1991**, *4*, 39–51.

Nimbalkar, V. S.; and More, D. C.; Genetic stuidies in brinjal (*Solanum melogena* L.) cross White Green x Manjari Gota. *J. Maharashtra Agric. Univ.* **1981**, *6*, 202.

Nunez-Palenius, H. G.; Gomez-Lim, M.; Ochoa-Alejo, N.; Grumet, R.; Lester, G.; and Cantliffe, D. G.; Fruits: genetic diversity, physiology, and biotechnology features. *Crit. Rev. Biotechnol.* **2008**, *28*, 13–55. doi: 10.1080/07388550801891111.

Odriozola-serrano, I.; Soliva-fortuny, R.; Gimeno-ano, V.; and Martin-belloso, O.; Modeling changes in health-related compounds of tomato juice treated by high-intensity pulsed electric fields. *J. Food. Eng.* **2008,** *89,* 210–216.

Onyishi, G. C.; Ngwuta, A. A.; Onwuteaka, C.; and Okporie, E. O.; Assessment of Genetic Variation in Twelve Accessions of Tropical Pumpkin (*Cucurbita maxima*) of South Eastern Nigeria. *World. Appl. Sci. J.* **24**(2), 252–255. doi: 10.5829/idosi.wasj.2013.24.02.13011.

Pal, N.; and Singh, N.; Analysis of genetic architecture for pungency (pyruvic acid) in onion. *Curr. Sci.* **1987,** *56,* 719–720.

Pandey, S.; Singh, J.; Upadhyay, A. K.; Ram, D.; and Rai, M.; Ascorbate and Carotenoid Content in an Indian Collection of Pumpkin (*Cucurbita moschata* Duch. ex Poir.). *Cucurbit Genetics Cooperative Report.* **2003,** *26,* 51–53.

Pandey, S. C.; Pandita, M. L.; and Dixit, J.; Genetics of yield and yield contributing characters in radish (*Raphanus sativus* L.). *Haryana. Agric. Univ. J. Res.* **1981,** *11,* 384–388.

Pandya, J. B.; and Rao, T. V. R.; Analysis of certain biochemical changes associated with growth and ripening of Cucurbita moschata Duch. fruit in relation to its seed development. *Prajna.-J. Pure. Appl. Sci.* **2010,** *18,* 34–39.

Patil, B. S.; and Pike, L. M.; Distribution of quercetin content in different rings of various coloured onion (Allium cepa L.) cultivars. *J. Hortic. Sci.* **1995,** *70*(4), 643–650.

Patil, S. K.; and More, D. C.; Inheritance studies in brinjal. *J. Maharashtra. Agric. Univ.* **1983a,** *8,* 43.

Perkins-Veazie, P.; Collins, J. K.; Pair, S. D.; and Roberts, W.; Lycopene content differs among red-fleshed watermelon cultivars. *J. Sci. Food. Agric.* **2001,** *81,* 983–987.

Perkins-Veazie, P.; Collins, J. K.; Pair, S. D.; and Roberts, W.; Watermelon; Lycopene content changes with ripeness stage, germplasm, and storage. In: Maynard, D, N., Ed.; Cucurbitaceae 2002, ASHS Press: Naples, Florida; **2002b,** pp427–430.

Perucka, I.; and Matercska, M.; *Acta Sci. Pol., Technol. Aliment.* **2007,** *6*(4), 67–74.

Peterson, C. E.; Simon, P. W.; and Ellerbrock, L. A.; 'Sweet Sandwich' onion. *Hortic. Sci.* **1986,** *21,* 1466–1468.

Polignano, G. B.; Laghetti, G.; Margiotta, B.; and Perrino, P.; Agricultural sustainability and underutilized crop species in southern Italy. *Plant. Genet. Resour. Characteriz. Utiliz.* **2004,** *2,* 29–35.

Poncet V.; Robert T.; Sarr, A.; and Gepts, P.; Quantitative trait loci analyses of the domestication syndrome and domestication process. In: *Encyclopedia of plant and crop science,* Goodman, R., Ed.; Marcel Dekker: New York, NY, **2004,** pp 1069–1073.

Pooler, M. R.; and Simon, P. W.; Characterization and classification of isozyme and morphological variation in a diverse collection of garlic clones. *Euphytica.* **1993,** *68,* 121–130.

Prohens, J.; Mun͂oz, J. E.; Rodriguez-Burruezo, A.; and Nuez, F.; Mejora gen´etica de la berenjena. *Vida. Rural.* **2005,** pp 52–56.

Prohens, J.; Rodrı´guez-Burruezo, A.; Raig´on, M. D.; and Nuez, F.; Total Phenolic Concentration and Browning Susceptibility in a Collection of Different Varietal Types and Hybrids of Eggplant: Implications for Breeding for Higher Nutritional Quality and Reduced Browning. *J. Am. Soc. Hortic. Sci.* **2007,** *132*(5), 638–646.

Raffo, A.; La Malfa, G.; Fogliano, V.; Maiani, G.; and Quaglia, G.; Seasonal variations in antioxidant components of cherry tomatoes (*Lycopersicon esculentum* cv. *Naomi* F$_1$). *J. Food. Compos. Anal.* **2006,** *19,* 11–19.

Raffo, A.; Leonardo, C.; Fogliano, V.; Ambrosino, P.; Salucci, M.; Gennaro, L.; Bugianesi, R.; Giuffrida, F.; and Quaglia, G.; Nutritional value of cherry tomatoes (*Lycopersicon esculentum* cv. *Naomi* F$_1$) harvested at different ripening stages. *J. Agric. Food Chem.* **2002**, *50*, 6550–6556.

Ramirez, E. C.; Whitaker, J. R.; and Virador, V. M.; Polyphenol oxidase. In: *Handbook of food enzymology*, Whitaker, J. R., Voragen, A. G. J., and Wong, D. W. S., Eds.; Marcel Dekker: New York; **2002**, pp 509–523.

Randle, W. M.; Onion germplasm interacts with sulfur fertility for plant sulfur utilization and bulb pungency. *Euphytica.* **1992**, *59*, 151–156.

Randle, W. M.; and Bussard, M. L.; Pungency and sugars of short-day onions as affected by S nutrition. *J. Am. Soc. Hortic. Sci.* **1993**, *118*, 766–770.

Randle, W. M.; Block, E.; Littlejohn, M. H.; Putnam, D.; and Bussard, M. L.; Onion (*Allium cepa* L.) thiosulfinates respond to increasing sulfur fertility. *J Agric. Food Chem.* **1994**, *42*(10), 2085–2088.

Redgwell, R. J.; and Fischer, M.; Fruit texture, cell wall metabolism and consumer perceptions. In: *Fruit Quality and Its Biological Basis,* Knee, M. Ed.; Blackwell: Oxford; **2002**, pp 46–88.

Rees, D.; Van Oirschot, Q.; and Kapinga, R.; Sweet potato Post–Harvest Assessment: Experiences from East Africa. Chatham, UK; **2003**.

Rhodes, M. J. C.; and Price, K. R.; Analytical problems in the study of flavonoid compounds in onions. *Food Chem.* **1996**, *57*, 113–117.

Rieman, G. H.; Genetic factors for pigmentation in the onion and their relation to disease resistance. *J. Agric. Res.* **1931**, *42*, 251–278.

Robertson, L. D.; Nakkoul, H.; and Williams, P. C.; The possibility of selection for higher protein content in faba bean (*Vicia faba* L.). *FABIS Newslett.* **1985**, *11*, 11–14.

Robinson, R. W.; and Tomes, M. L.; Ripening inhibitor: a gene with multiple effects on ripening. *Rep. Tomato Genet. Coop.* **1968**, *18*, 36–37.

Rodıca, S.; Apahıdean, S. A.; Apahıdean, M.; and Manıtıu, P. L.; Yield, physical and chemical characteristics of greenhouse tomato grown on soil and organic substratum. 43rd Croatian and 3rd Int. Symposium on Agric. Opatija. Croatia; **2008**, pp 439–443.

Rodriguez-Burruezo, A.; Prohens, J.; Rosello, S.; and Nuez, F.; Heirloom varieties as sources of variation for the improvement of fruit quality in greenhouse-grown tomatoes. *J. Hortic. Sci. Biotechnol.* **2005**, *80*, 453–460.

Rodriguez-Saona, L. E.; Giusti, M. M.; and Wrolstad, R. E.; Anthocyanin pigment composition of red-fleshed potatoes. *J. Food. Sci.* **1998**, *63*(3), 458–465.

Römer, S.; Lübeck, J.; Kauder, F.; Steiger, S.; Adomat, C.; and Sandmann, G.; Genetic Engineering of a Zeaxanthin-rich Potato by Antisense Inactivation and Co-suppression of Carotenoid Epoxidation. *Metab. Eng.* **2002**, *4*(4), 263–272.

Ronen, G.; Cohen M.; Zamir D.; and Hirschberg, J.; Regulation of carotenoid biosynthesis during tomato fruit development: expression of the gene for lycopene epsilon-cyclase is downregulated during ripening and is elevated in the mutant Delta. *Plant J.* **1999**, 17, 341–351.

Ru, W. Z.; Jin, T. M.; Wu, J. R.; and. Jin, T. M.; Determination of beta-carotene in different pumpkin varieties by HPLC. *Acta. Agriic. Boreali Sinica.* **1998**, *13*, 141–144.

Rubio-Diaz, D. E.; De Nardo, T.; Santos, A.; de Jesus, S.; Francis, D.; and Rodriguez-Saona, L. E.; Profiling of nutritionally important carotenoids from genetically-diverse tomatoes by infrared spectroscopy. *Food. Chem.* **2010**, *120*(1), 282–289.

Saha, P.; and Kalia, P.; Genetics of yield and quality traits in cauliflower (Brassica oleracea var. botrytis L.). *Agrotechnol.* **2012**, *1*, 2. http://dx.doi.org/10.4172/2168-9881.S1.004.

Sahlin, E.; Savage, G. P.; and Lister, C. E.; Investigation of the antioxidant properties of tomatoes after processing. *J. Food. Compos. Anal.* **2004**, *17*, 635–647.

Sakata, Y.; and Lester, R. N.; Chloroplast DNA diversity in brinjal eggplant (Solanum melongena) and related species. *Euphytica.* **1997**, *97*, 295–301.

Salaman, R.; The inheritance of colour and other characters in the potato. *J. Genet.* **1910**. *1*(1), 7–46.

Sanatombi, K.; and Sharma, G. J.; Capsaicin content and pungency of different capsicum spp. cultivars. *Not. Bot. Hort. Agrobot. Cluj.* **2008**, *36*(2), 89–90.

Santos, C. A. F.; and Simon, P. W.; Heritabilities and minimum gene number estimates of carrot carotenoids. *Euphytica.* **2006**, *151*, 79–86.

Sarikamiş, G., Balkaya, A.; and Yanmaz, R.; Glucosinolates within a collection of white head cabbages (*Brassica oleracea* var. *capitata* sub. var. *alba*) from Turkey. *Afr J Biotechnol.* **2009**, 8(19), 5046–5052.

Sawa, T.; Nakao, M.; Akaike, T.; Ono, K.; and Maeda, H.; Alkylperoxyl radical-scavenging activity of various flavonoids and other phenolic compounds: Implications for the anti-tumor-promoter effect of vegetables. *J. Agric. Food Chem.* **1998**, *47*, 397–402.

Scalbert, A.; Morand, C.; Manach, C.; and Rémésy, C.; Absorption and metabolism of polyphenols in the gut and impact on health. *Biomed. Pharmacother.* **2002**, *56*, 276–282.

Searle, B.; Renquist, R.; and Bycroft, B.; Agronomic factors affecting the variability of squash fruit weight. *Agronomy N.Z.* **2003**, p 32.

Senning, M.; Sonnewald, U.; and Sonnewald, S.; Deoxyuridine triphosphatase expression defines the transition from dormant to sprouting potato tuber buds. *Mol. Breed.* **2010**, *26*(3), 525–531.

Serrano-Martinez, A.; Fortea, F. M.; Del Amor, F. M.; and Nufiez-Delicado. E.; *Food Chem.* **2008**, *107*, 193–199.

Serrano-Megias, M.; and Lopez-Nicolas, J. M.; Application of agglomerative hierarchical clustering to identify consumer tomato preferences: influence of physicochemical and sensory characteristics on consumer response. *J. Sci. Food Agric.* **2006**, *86*, 493–499.

Seshadri, V. S.; and More, T. A.; Indian land races in Cucumis melo. In: *Second International Symposium on Cucurbits, Nishimura*, S., Ezura, H., Matsuda, T., and Tazuke, A., Eds., ISHS.: Tsukuba, Japan, **2002**; pp 187–192.

Sethi, V.; Anand, J. C.; and Bhagchandani, P. M.; Quality screening of white onion cultivars for use in dehydration. *Indian Food Packer.* **1973**, *27*, 5–8.

Seymour, G. B.; and McGlasson, W. B.; Melons. In: *Biochemistry of Fruit Ripening, Seymour*, G., Taylor, J., and Tucker, G., Eds.; Chapman & Hall: London; **1993**, pp 273–290.

Shepherd, L. V. T.; Bradshaw, J. E.; Dale, M. F. B.; McNicol, J. W.; Pont, S. D. A.; Mottram, D. S.; and Davies, H. V.; Variation in acrylamide producing potential in potato: Segregation of the trait in a breeding population. *Food. Chem.* **2010**, *123*(3), 568–573.

Shewfelt, R. L.; Erickson, M. E.; Hung, Y. C.; and Malundo, T. M. M.; Applying quality concepts in frozen food development. *Food. Technol.* **1997**, *51*(2), 56–59.

Shirley, B. W.; Flavonoid biosynthesis: 'new' functions for an 'old' pathway. *Trends. Plant. Sci.* **1996**, *1*, 377–382.

Siddiqui, M. W.; Ayala-Zavala, J. F.; and Dhua, R. S.; Genotypic variation in tomatoes affecting processing and antioxidant attributes. *Crit. Rev. Food. Sci. Nutr.* **2013**, doi: 10.1080/10408398.2012.710278.

Silva, G. O.; Vieira, J. V.; and Vilela, M. S.; Seleção de caracteres de cenoura cultivada em dois sistemas de produção agroecológicos no Distrito Federal. *Revista. Ceres.* **2009**, *56*, 595–601.

Simon, P. W.; Breeding carrot, cucumber, onion and garlic for improved quality and nutritional value. *Hort. Bras.* **1993**, *11*, 171–173.

Singh, B. K.; Sharma, S. R.; Kalia, P.; and Singh, B.; Genetic variability for antioxidants and horticultural traits in cabbage. *Indian. J. Hort.* **2011**, *68*(1), 51–55.

Singh, J.; Upadhyay, A. K.; Prasad, K.; Bahadur, A.; and Rai, M.; Variability of carotenes, vitamin C, E and phenolics in Brassica vegetables. *J. Food Compos. Anal.* **2007**, *20*(2), 106–112.

Singh, M.; Walia, S.; Kaur, C.; Kumar, R.; and Joshi, S.; Processing characteristics of tomato (Solanum lycopersicum) cultivars. *Indian J. Agric. Sci.* **2010**, *80*, 174–176.

Singh, S. R.; Ahmed, N.; Lal, S.; Ganie, S. A.; Amin, M.; Jan, N.; and Amin, A.; Determination of genetic diversity in onion (*Allium cepa* L.) by multivariate analysis under long day conditions. *Afr. J. Agric. Res.* **2013**, *8*(45), 5599–5606.

Śliwka, J.; Wasilewicz-Flis, I.; Jakuczun, H.; and Gebhardt, C.; Tagging quantitative trait loci for dormancy, tuber shape, regularity of tuber shape, eye depth and flesh colour in diploid potato originated from six *Solanum* species. *Plant. Breed.* **2008**, *127*(1), 49–55.

Sørensen, K. K.; Mapping of morphological traits and associations with late blight resistance in Solanum tuberosum and *S. vernei*, Danish Institute of Agricultural Sciences, Royal Veterinary and Agricultural University, Danmark, Frederiksberg; **2006**.

Sowokinos, J. R.; Biochemical and molecular control of cold-induced sweetening in potatoes. *Am. J. Potato. Res.* **2001**, *78*(3), 221–236.

Stevens, M. A.; Kader, A. A.; and Albright, M.; Potential for increasing tomato flavour via sugar and acid contents. *J. Am. Soc. Hortic. Sci.* **1979**, *104*, 40–42.

Stommel, J. R.; and Haynes, K. G.; Genetic control of fruit sugar accumulation in a *Lycopersicon esculentum* x *L. hirsutum cross. J. Am. Soc. Hortic. Sci.* **1993**, *118*, 859–863.

Stommel, J. R.; and Whitaker, B. D.; Phenolic acid composition of eggplant fruit in a germplasm core subset. *J. Am. Soc. Hortic. Sci.* **2003**, *128*, 704–710.

Sykes, S.; Melons: New varieties for new and existing markets. *Agric. Sci.* **1990**, *3*, 32–35.

Szczesniak A. S. Texture is a sensory property. *Food Qual. Pref* **2002**, *13*, 215–225.

Szopińska, A. A.; and Gawęda, M.; Comparison of yield and quality of red beet roots cultivated using conventional, integrated and organic method. *J. Hortic. Res.* **2013**, *21*(1), 107–114. doi: 10.2478/johr-2013-0015.

Tadesse, T.; Workneh, T. S.; and Woldetsadik, K.; Effect of varieties on changes in sugar content and marketability of tomato stored under ambient conditions. *Afr. J. Agric. Res.* **2012**, *7*(14), 2124–2130.

Tadmor, Y.; Katzir, N.; King, S.; Levi, A.; Davis, A.; and Hirschberg, J.; Fruit coloration in watermelon: lesson from the tomato. Proc. Cucurbitaceae 2004, the 8th EUCARPIA Meeting on Cucurbit Genetics and Breeding, Olomouc, Czch Republic; **2004**, pp 181–185.

Thakur, N. S.; and Kaushal, B. B.; Study of quality characteristics of some commercial varieties and F[1] hybrids of tomato (*Lycnpersicon esculentum* Mill.) grown in Himachal Pradesh in relation to processing. *Ind. Food Pack.* **1995**, *49*(3), 25–31.

Thorup, T. A.; Tanyolac, B.; Livingstone, K. D.; Popovsky, S.; Paran, I.; and Jahn, M.; Candidate gene analysis of organ pigmentation loci in the Solanaceae. *Proc. Natl. Acad. Sci. USA.* **2000**, *97*(21), 11192–11197.

Tigchelaar, E. C.; Tomes, M. L.; Kerr, E. A.; and Barman, R. J.; A new fruit ripening mutant, nonripening (nor). *Rep. Tomato. Genet. Coop.* **1973**, *23*, 33.

Tilahun, S.; Paramaguru, P.; and Rajamani, K.; Capsaicin and ascorbic acid variability in chilli and paprika cultivars as revealed by HPLC analysis. *J. Plant Breed. Genet.* **2013**, *01*(02), 85–89.

Tittonell, P.; DeGrazia, J.; and Chiesa, A.; Effect of nitrogen fertilization and plant population during growth on lettuce (lactuca sativa L.) postharvest quality. Proceedings of the Fourth International Conference on Postharvest Science. *Acta Hortic.* **2001**, *553*(1), 67–68.

Triantis, T.; Stelakis, A.; Dimotikali; and Papadopoulos, K.; Investigations on the antioxidant activity of fruit and vegetable aqueous extracts on superoxide radical anion using chemiluminiscence techniques. *Anal. Chim. Acta.* **2005**, *536*, 101–105.

Turhan, A.; and Seniz, V.; Estimation of certain chemical constituents of fruits of selected tomato genotypes grown in Turkey. *Afr. J. Agric. Res.* **2009**, *4*(10), 1086–1092.

Udengwu, O. S.; Inheritance of fruit colour in Nigerian local okra, *Abelmoschusesculentus* (L.) Moench, cultivars. *J. Trop. Agric. Food. Environ. Extension.* **2008**, *7*(3), 216–222.

Van den Berg, J. H.; Ewing, E. E.; Plaisted, R. L.; McMurry, S.; and Bonierbale, M. W.; QTL analysis of potato tuber dormancy. *Theor. Appl. Genet.* **1996**, *93*(3), 317–324.

Van Eck, H. J.; Genetics of morphological and tuber traits. In: *Potato Biology and Biotechnology: Advances and Perspectives*, Elsevier: London; **2007**, pp 91–115.

Van Eck, H. J.; Jacobs, J. M. E.; Stam, P.; Ton, J.; Stiekema, W. J.; and Jacobsen, E.; Multiple aleles for tuber Sshape in diploid potato detected by qualitative and quantitative genetic-analysis using RFLPs. *Genetics.* **1994a**, *137*(1), 303–309.

Van Eck, H. J., Jacobs, J. M. E., Vandenberg, P. M. M. M., Stiekema, W. J. and Jacobsen, E.: The Inheritance of Anthocyanin Pigmentation in Potato (*Solanum tuberosum* L.) and mapping of tuber skin color loci using RFLPs, *Heredity.* **1994b**, *73*, 410–421.

Van Eck, H. J.; Jacobs, J. M. E.; Vandijk, J.; Stiekema, W. J.; and Jacobsen, E.; Identification and mapping of 3 flower color loci of potato (*Solanum tuberosum* L) by RFLP Analysis. *Theor. Appl. Genet.* **1993**, *86*(2–3), 295–300.

Vieira, J. V.; Silva, G. O.; and Boiteux, L. S.; Genetic parameter and correlation estimates of processing traits in half-sib progenies of tropical-adapted carrot germplasm. *Horticultura Brasileira.* **2012**, *30*, 7–11.

Vieira, R. L.; and Nodari, R. O.; Diversidade genética de cultivares de alho avaliada por marcadores RAPD. *Ciênc. Rural.* **2007**, *37*, 51–57.

Vogel J. T.; Tan, B.; McCarty, D. R.; and Klee, H. J.; The Carotenoid Cleavage Dioxygenase 1 enzyme has broad substrate specificity, cleaving multiple carotenoids at two different bond positions. *J. Biol. Chem.* **2008**, *283*, 11364–11373.

Volk G. M.; Henk, A. D.; and Richards, C. M.; Genetic diversity among U.S. garlic clones as detected using AFLP methods. *Am. Soc. Hortic. Sci.* **2004**, *129*, 559–569.

Wang, T. L.; Bogracheva, T. Y.; and Hedley, C. L.; Starch: as simple as A, B, C? *J. Exp. Bot.* **1998**, *49*, 481–502.

Watanabe, K.; Saito, T.; Hirota, S.; Takahashi, B.; and Fujishita, N.; Carotenoid- pigments in orange, light orange, green and white flesh colored fruits of melon (Cucumis melo L). *J. Jpn. Soc. Food Sci. Tech-Nippon Shokuhin Kagaku Kogaku Kaishi.* **1991**, *38*, 153–159.

Watson, J. F.; and Gabelman, W. H.; Genetic analysis of betacyanln, betaxanthin, and sucrose concentrations in roots of table beet. *J. Am. Soc. Hortic. Sci.* **1984**, *109*, 386–391.

Watson, J. F.; and Goldman, I. L.; Inheritance of a Gene Conditioning Blotchy Root Color in Table Beet (Beta vulgaris L.). *J. Heredity.* **1997**, *88*(6), 540–543.

Whitaker, B. D.; and Stommel, J. R.; Distribution of hydroxycinnamic acid conjugates in fruit of commercial eggplant (Solanum melongena L.) cultivars. *J. Agric. Food Chem.* **2003**, *51*, 3448–3454.

White, P. J.; Bradshaw, J. E.; Dale, M. F. B.; Ramsay, G.; Hammond, J. P.; and Broadley, M. R.; Relationships between yield and mineral concentrations in potato tubers. *Hortic. Sci.* **2009,** *44*(1), 6–11.

Wills, R. B.; H., Lee, T. H.; Grahan, D.; McGlasson, W. B.; and Hall, E. G.; Postharvest. An introduction to the physiology and handling of fruits and vegetables. New South Wales University Press Limited, Kensington, Australia. pp 150; **1981.**

Wilson, J. E.; Pole, F. S.; Smit, N. E. J.; and Taufatofua, P.; *Sweet Potato Breeding.* IRETA publications, Apia, Western Samoa; **1989.**

Wolters, A. M. A.; Uitdewilligen, J.; Kloosterman, B. A.; Hutten, R. C. B.; Visser, R. G. F.; and van Eck, H. J.; Identification of alleles of carotenoid pathway genes important for zeaxanthin accumulation in potato tubers. *Plant. Mol. Biol.* **2010,** *73*(6), 659–671.

Wolyn, D. J.; and Gabelman, W. H.; Inheritance of root and petiole pigmentation In red table beet. *J. Hered.* **1989,** *80,* 33–38.

Yahia E. M.; Contreras-Padilla, M.; Gonzalez-Aguilar, G.; Ascorbic acid content in relation to ascorbic acid oxidase activity and polyamine content in tomato and bell pepper. *LWT-Food Sci. Tech.* **2001,** *34,* 452–457.

Yahyaoui, F. E. L.; Wongs-Aree, C.; Latche, A.; Hackett, R.; Grierson, D.; and Pech, J. C.; Molecular and biochemical characteristics of a gene encoding an alcohol acyl-transferase involved in the generation of aroma volatile esters during melon ripening. *Eur. J. Biochem.* **2002,** *269,* 2359–2366.

Yamaguchi, M.; Hughes, D. L.; Yabumoto, K.; and Jennings, W. G.; Quality of cantaloupe muskmelons—variability and attributes. *Sci Hortic.* **1977,** *6,* 59–70.

Yoo, K. S.; Pike, L.; Crosby, K.; Jones, R.; and Leskovar, D.; Differences in onion pungency due to cultivars, growth environment, and bulb sizes. *Sci. Hortic.* **2006,** *110,* 144–149.

Yoruk, R.; and Marshall, M. R.; Physicochemical properties and function of plant polyphenol oxidase: A review. *J. Food. Biochem.* **2003,** *27,* 361–422.

Zdravkovic, J.; Markovic, Z.; Zecevic, B.; Zdravkovic, M.; and Setenovic-Rajicic, T.; Epistatic gene effects on the fruit shape of the parents of F_1, F_2, BC_1 and BC_2 progeny. *Acta. Horticulturae.* **2003,** *613,* 321–325.

Zhang, Y. F.; Cheng, S. P.; De Jong, D.; Griffiths, H.; Halitschke, R.; and De Jong, W.; The potato R locus codes for dihydroflavonol 4-reductase. *Theor. Appl. Genet.* **2009,** *119* (5), 931–937.

Zhou, X.; Van Eck, J.; and Li, L.; Use of the cauliflower Or gene for improving crop nutritional quality. *Biotechnol. Annu. Rev.* **2008,** *14,* 171–190. doi: 10.1016/S1387-2656(08)00006-9.

Zhuang, Y. L.; Chen, L.; Cao, J. X.; *J. Funct. Foods.* **2012,** *4,* 331–338.

Zinash, A.; Workneh, T. S.; and Woldetsadik, K.; Effect of accessions on the chemical quality of fresh pumpkin. *Afr. J. Biotech.* **2013,** *12*(51), 7092–7098. doi: 10.5897/AJB10.1751.

CHAPTER 5

ADVANCES IN STORAGE OF FRUITS AND VEGETABLES FOR QUALITY MAINTENANCE

G. R. VELDERRAIN RODRÍGUEZ[1], M. L. SALMERÓN-RUIZ,[1], G. A. GONZÁLEZ AGUILAR[1], MD. WASIM SIDDIQUI[2], and J. F. AYALA ZAVALA[1]

[1]Centro de Investigaciyn en Alimentaciyn y Desarrollo, AC (CIAD, AC), Carretera a la Victoria Km 0.6, La Victoria, Hermosillo, Sonora, 83000, México

[2]Department of Food Science and Technology, Bihar Agricultural University, BAC, Sabour, Bhagalpur, Bihar, 813210, India

CONTENTS

5.1   INTRODUCTION

Nowadays, the increased awareness of diet and health, is leading to a greater consumption of fruits and vegetables around the world (Zheng et al., 2013). In this context, consumer tends to include in their diet foods considered rich sources of essential nutrients, such as vitamins and minerals. In order that, several studies have asserted that fruits and vegetables, mainly fruits and vegetables, are rich source of a broad variety of essential nutrients and bioactive compounds related to human health and well-being (Haminiuk et al., 2012; Dembitsky et al., 2011; Joshipura et al., 1999; Sommerburg et al., 1998; Kaur and Kapoor, 2001). Hence, following this global trend in healthy diets, food industry promotes the intake of fresh fruits and vegetables to deliver or provide to human diet a broad variety of bioactive compounds, especially antioxidants like phenolic compounds (PC) or carotenoids.

According to this, the global production of fruits and vegetables has been increasing by 94 percent from 1980 to 2004, increasing on average 4.5 percent yearly between 1990 and 2004 (Olaimat and Holley, 2012). Meanwhile, the importation of fruits and vegetables in the US doubled to $12.7 billion from 1994 to 2004, reaching daily sales in 2005 of sixmillion packages in North America (Jongen, 2005). However, fruits and vegetables are highly perishable items due to their highly susceptibility to microbiological spoilage or even due to their own ripening processes. Fruit ripening and softening are the major attributes that leads to fruit spoilage in fleshy or climacteric fruits, reducing its shelf life for a few days being considered as inedible due to overripening (Bapat et al., 2010). The spoilage changes in overripening fruits includes excessive softening and changes in taste, aroma and skin color (Rana 2006). However, fruit ripening is an irreversible process that can only be delayed once it starts (Srivastava and Dwivedi, 2000; Jeong et al., 2003; Hofman et al., 2001). Thus, the spoilage prevention results in economic saving of millions of dollars to the fruits and vegetable industry.

In order to ensure the consumer choice, the fruits and vegetable industry have to offer products of high quality and value. The quality of fresh-produce, as it is shown in Figure 5.1, is measured as a combination of attributes, properties or characteristics that determine their value to the consumer. Within the most valuable parameters for the consumer are included the appearance, texture, flavor, and nutritive value (Lamikanra, 2002). These parameters depend upon the commodity or the product and whether it is eaten fresh or cooked (Kader, 2002). In this context, the interpretation or acceptance of the different quality parameter may vary from one consumer to other (Tijskens, 2000). Hence, fruit ripening or maturity at harvest is the most important factor that determines storage life and final fruit quality. Immature fruits are more subject to shrivel and mechanical damage and are of inferior quality when ripe (Kader, 1995). Overripe fruits are likely to become soft and mealy with insipid flavor soon after harvest. Any fruit picked too early or too late in its season is more susceptible to physiological disorders and has a shorter storage life than fruit picked at the proper maturity (Ferguson et al., 1999).

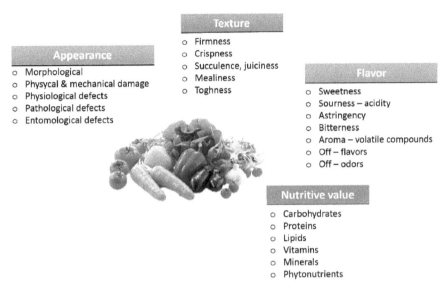

Appearance
- Morphological
- Physycal & mechanical damage
- Physiological defects
- Pathological defects
- Entomological defects

Texture
- Firmness
- Crispness
- Succulence, juiciness
- Mealiness
- Toghness

Flavor
- Sweetness
- Sourness – acidity
- Astringency
- Bitterness
- Aroma – volatile compounds
- Off – flavors
- Off – odors

Nutritive value
- Carbohydrates
- Proteins
- Lipids
- Vitamins
- Minerals
- Phytonutrients

FIGURE 5.1   The quality attributes of fruits and vegetables products and its symptoms of the disorder.

In general, fruit ripening is characterized by a variety of evidentiary changes in morphology, biochemistry, and physiology of tissues, however, one of the most dramatic changes associated with fruit ripening is the accumulation and loss of pigments resulting in distinct changes in color (Rana 2006). Moreover, the ripening process requires large energy and prolonged membrane integrity to perform catabolic and anabolic changes. However, the changes associated with fruit ripening involves modifications in structure and cell wall composition affecting the fruit firmness, metabolism of sugars and acids involved in flavor determination, biosynthesis and deposition of carotenoids determining fruit color (Wakabayashi, 2000; Merzlyak et al., 1999). Hence, to preserve climacteric fruits at their highest quality levels, it's important to consider all the parameters or changes during its ripening process.

Firmness, which is a determinant parameter of fruit quality, depends on the strength and stability of cell wall and it is affected during fruits ripening. Moreover, the fruit tissues softening typically accompanies the ripening of many fruits. This process is based on the modification in the structure of cell walls, particularly by the degradation of cell wall polymers. Furthermore, the ripening process contains a large amount of hydrolases that are involved in the degradation of cell wall polymers, resulting in a fruit texture rejected for the consumers (Fischer and Bennett, 1991). For example, as the fruit cell walls generally contain large amounts of pectins, the pectin degradation may cause the collapse of the cell adhesion decreasing the tissue strength (Toivonen and Brummell, 2008). Nevertheless, fruit cell walls also contain a significant amount of xyloglucans and their depolymerization has been observed in early stages of softening in some fruits, such as tomato, avocado, melon, and kiwi fruit (Yakushiji et al., 2001). However, as the polymer degradation is a product of their ripening, firmness, loss, or tissue softening can be diminished if ripening is delayed.

Moreover, the above mentioned quality parameters as well as microbial spoilage can diminish if are applying the proper produce sanitization, transport and storage conditions being this last one the most important. Thus, to extend the shelf life of fruits and vegetables, the storage conditions as well as novel emerging technologies should be applied before and under their storage process.

5.2   PERSISTENT PROBLEMS IN PRESERVATION OF FRUITS AND VEGETABLES

Nowadays, due to its importance in the human diet, the fruits and vegetable industry continues to be one of the most important and ever-growing sectors of the global market. However, as soon as the fruits and vegetables are cut off from their natural nutrient supply, their quality begins to diminish leading to spoilage, which results in the consumer rejection of the product and several economic losses (James, 2003). At the present time, many procedures such as freezing, smoking, and heating are used alone or in combination to preserve food products, but there are still faced with heavy losses over the 40 percent in the case of fruits and vegetables (Lacroix and Ouattara, 2000). The adequate manage and storage of fruits and vegetables has acquired such an importance in both health and economic sectors, being the cost of foodborne illness in the US over $50 billion per year involving more than 48 million people (Bermúdez-Aguirre and Barbosa-Cánovas, 2013).

The postharvest storage of fruits and vegetables is a process that changes the quality attributes of fruit after harvest in terms of skin color, pulp firmness, and shelf life (Kader, 2002; Burton, 1982). Once the fruits or vegetables are removed from their plants tissue they are exposed to losses of water, minerals, and sustenance (Kader, 2002; Mitra, 1997). Continuation of respiration in harvested fruit tissues, using available sugars and organic acids, results in a rapid senescence. Kienzle et al. (2011) suggest that in order to keep the quality attributes of fruits and vegetables, optimum postharvest handling and storage temperatures between 0 and 10°C results in the delay of fruit discoloration and fruit ripeness (Kienzle et al., 2011). Likewise, in recent years several improvements have been made in storage technologies and the postharvest manipulations in order to reduce losses, maintain taste and nutritional properties, and enhance market value.

As it has been mentioned, the fruit and vegetables properties including color, firmness, and taste change over time. Hence, the shelf life of any fruits and vegetables refers to the time before the product attributes drops below the acceptable limit under standardized storage conditions (Tijskens, 2000). In order that, new technologies and storage procedures are rising and being applied by the fruits and vegetable industry, from guidelines for packing fruits and vegetables to new sanitizing methods and emergent technologies to extend shelf life of any product (Gorny, 2001). Usually, food spoilage includes physical damage and chemical changes, such as oxidation and color changes, or the appearance of off-flavors and off-odors. The microbial growth during storage is the most common cause of spoilage and it is responsible of textural changes due to polymer degradation within the food matrix (Gram et al., 2002). Therefore, the temperature effects, which will be addressed further this chapter, are essential to reduce either microbial and metabolism spoilage.

However, besides the fruits texture and do that consumers take product appearance into consideration as a primary criterion, color has been considered to have a key role in food choice, food preference and acceptability, and may even influence taste thresholds, sweetness perception and pleasantness (Clydesdale, 1993). As it has been mentioned, along with texture, color is one of the main attributes that characterizes the freshness of most vegetables (Rico et al., 2007). For example, some vegetables like lettuce and carrots can undergo changes in color due to different biochemical processes, mainly chlorophyll degradation and browning appearance in the case of the lettuce, and carotene degradation whiteness and browning in carrots (Koca et al., 2007; Desobry, 1998; Degl'Innocenti, 2005; Castaner et al., 1996). These browning problems are common in a number of fruits such as apricots, apples, pears, peaches, bananas, and grapes or vegetables such as potatoes, mushrooms and lettuce. This browning or discoloration in fruits and vegetables, limits their shelf life, especially in dehydrated and frozen fruits and vegetables (Jiang et al., 2004). Therefore, to retain the color of minimally processed fruits a pretreatment is usually required to inhibit either enzymatic or nonenzymatic browning.

According to this, to inhibit the nonenzymatic browning in fruit and vegetable products, these can be refrigerated, controlled the water activity in dehydrated food, reduction by reducing sugar contents in potatoes by storage or glucose oxidase treatment, reduction of amino nitrogen content in juices by ion exchange, packaging with oxygen scavengers and use of sulfites. However, the use of sulfites additives has been listed as an important risk factor for the initiation and progression of liver diseases due to oxidative damage (Wang et al., 2014; Bai et al., 2013) Likewise, to prevent the excessive use of sulfites as food additives and extend the shelf life, other alternatives have been applied to prevent enzymatic and nonenzymatic browning such as ascorbic acid formulations, Polyphenol Oxidase (PPO) inhibitors, complexing agents (e.g., EDTA, sodium acid pyrophosphate), sulfyhydryl-containing amino acids, organic halides and edible coatings.

Moreover, the exploitation of the synergy effects of different conditions/treatments has been studied for years ago (Tonutti, 2012). The most common example of these studies is the application of controlled atmospheres (CA) in addition to optimal refrigeration (Thompson, 2010). The combination of both low temperature and low O_2 and high CO_2 concentrations is highly effective in slowing general metabolism, ethylene production and respiration rate, thereby delaying the loss of fruit quality parameters (Mahajan and Goswami, 2004). Likewise, the use of 1-methyl-cyclopropene (1-MCP) in combination and interplay of different factors such as low temperatures, oxygen has been applied to fruits like apples and pears for maintaining firmness and other fruit quality parameters (Tonutti, 2012).

5.3 TEMPERATURE EFFECTS ON OVERALL QUALITY AND PHYTOCHEMICAL

The consumer's preference for fresh fruits is challenging because they have a very short shelf life, due to their sensitivity to fungal attack and excessive texture softening caused by the natural ripening process (Cordenunsi et al., 2005). Even when the microbiological food spoilage has been previously described, the postharvest life of fruit and vegetables is also essential for a successful storage of fruits and vegetables. Therefore, as it has been mentioned, the postharvest life of fruit and vegetables has been traditionally defined in terms of visual appearance (freshness, color, and absence of decay or physiological disorders) and texture (firmness, juiciness, and crispness). Diverse studies reported that a number of important postharvest fruit disorders are caused or increased by temperature conditions being these disorders often visible at the time of harvest, or soon afterwards (Snowdon, 2010; Hopkirk et al., 1994; Wade, 1979; Holland et al., 2005; Streif et al., 1993). In fruits and vegetables, maturity is referred to a combination of properties such as firmness, skin color and sugar contents and they are affected in response to metabolic pathways of fruits and vegetable physiology.

Ethylene (C_2H_4) is naturally produced by plants, but only ripening climacteric fruits and diseased or wounded tissue produce it in sufficient amounts to affect adjacent tissues (Yang and Oetiker, 1994). However, since C_2H_4 exerts its effects through metabolic reactions, the exposure of fruit tissue at their lowest recommended storage temperature will reduce the response. Therefore, the reduction of C_2H_4 affects the quality properties involve in maturity of fruit and vegetables (Saltveit, 1999). Depend on the fruit or vegetables metabolism, the effect of temperature over its quality properties. For example, O'Hare in 1994 reports that the postharvest life of a pineapple is dependent on temperature, from a few hours at 20°C to several weeks at 1°C (O'Hare, 1994). Likewise, Watada et al. in (1996) reports that when temperature increases from 0 to 10°C, respiration rate increases substantially, with the Q_{10} ranging from 3.4 to 8.3 among various fresh-cut products (Watada et al., 1996). For example, the Q_{10} of zucchini, tomato and kiwi y over 3.5 while bell pepper, muskmelon and Crenshaw had values of 8.3.

The Q_{10} ratio is commonly used to describe the temperature dependence of biological processes. The Q_{10} function assumes an exponential relationship with temperature in which Q_{10} is the ratio of the rate at one temperature to that at 10°C lower (Tjoelker et al., 2001). High Q_{10} values, particularly in the 10–20°C range indicates the importance of handling and storing both intact and fresh-cut products at near 0°C, if the product is not sensitive to chilling injury (CI) (Watada et al., 1996). The symptoms associated with CI (uneven ripening, surface pitting, increased fungal infection and water-soaked areas) usually appear only upon returning fruit to ambient temperature (Hong and Gross, 2000). As ethylene is a plant hormone produced in response of various types of stress, including mechanical wounding (cutting and peeling), chemicals (herbicides), and temperature (freezing, chilling), it has been proved that ethylene is involved in CI of a number of fruits. For example, Chaplin et al. (1983) found that ethylene treatment increased development of CI symptoms and decreased shelf life of an avocado (*Persea Americana* Mill) (Chaplin et al., 1983). In contrast, Zhou et al. (2001) reports that in nectarine fruits ethylene treatments may be beneficial against symptoms of CI, like woolliness, due that respiration is associated with chilling injury (Zhou et al., 2001). Moreover, it has been suggested that ethylene is essential to promote a proper sequence of cell wall hydrolysis necessary for normal fruit softening.

Jasenka and Dunja (2011) studied the effect of temperature over shelf life of small fruits at 25 and 4°C. They found that storage of at 25°C as opposed to 4°C, facilitated faster spoilage of small fruits, being the fruits stored at 4°C the ones retaining marketable qualities, on average, 9.2 longer than the fruits stored at 25°C. This study, supports the fact that low temperature is essential for maintaining a good quality due to the metabolic decrease of respiration rate, which increase is comparable to deterioration rates (Watada and Qi, 1999). Moreover, fresh-cut fruits and vegetables, which have higher respiration rates than the intact product, generally are more perishable than intact products because they have been subjected to severe physical stress, such as peeling, cutting, slicing, shredding, trimming, and/or coring, and removal or protective epidermal cells (Watada et al., 1996). Several studies reported that the effect of slicing in some fresh-cut products, as tomato slices, includes a rapid rise in CO_2 and C_2H_2 production which reduced shelf life (Gil et al., 2002). However, about the 40 percent of fruits and vegetables in the market is chilling sensitive, thus chilling injury is a concern with fresh-cut products held at chilling temperatures due to their loss of quality properties (Watada and Qi, 1999).

Furthermore, Javanmardi and Kubota (2006) have shown that tomatoes storage at 5°C in comparing to 12°C inhibited weight loss and enhancement of lycopene. Their study showed that the rate of weight loss in tomatoes storage at 12°C was 49 percent per day and for 5°C was 0.15 percent per day, whereas their water loss rate at room temperature was of 0.68 percent per day (Javanmardi and Kubota, 2006). This higher weight loss in room temperature could be attributed to higher respiration rates compared to those tomatoes stored at lower temperatures. Likewise, Meng et al. (2008) reported that low temperatures of storage decrease weight loss due to the decrease in the respiratory metabolism of grapes (Meng et al., 2008).

On the other hand, there is a low critical temperature below which an increase of phenylpropanoid metabolism is stimulated. According to diverse studies, the phenolic metabolism is enhanced under chill stress and that behavior of the same metabolism is further dependent on the storage temperature (Lattanzio, 2003). However, diverse studies asserts that hot-water treatments protect fruits against chilling injury and delay ripening due the inhibition of ethylene synthesis at temperatures near or above 35°C (Lu et al., 2010). These hot-water treatments have been used by the fresh-cut industry to extend the shelf life of diverse products without compromising their phytochemical quality. McDonald et al. (1999) reports that all treatments temperatures greater than 27°C reduced decay following chilling and ripening of tomato fruits, achieving decays of 60 percent with the 42°C temperature treatments (McDonald et al., 1999). Moreover, these technologies have proven to be efficient for extending the postharvest life of fruits like mangoes without changing the nutritional profile of fruits, either their phytochemical profile. According to this, Kim et al. (2007) found that the PC as well as the antioxidant capacity of mango fruit were unaffected by hot-water treatments (Kim et al., 2007).

Furthermore, the stability of PC during storage of fruits and vegetables has been recently addressed by several studies. Anthocyanins are well-known natural colorants, which provide bright red color in foods, and therefore, due to their high reactivity they readily degrade and form colorless or undesirable brown-colored compounds (Kirca et al., 2007). However, the rates of anthocyanin degradation during storages significantly varied among sources and likely occur due to factors such as varying molar ratios between reactants (anthocyanins and or PC with peroxide), nonanthocyaninpolyphenolic concentration, secondary free radical formation, or other oxidation reactions such as o-quinone formation involving phenolics and anthocyanins (Özkan et al., 2002). For example, .Kirca et al. asserts that storage temperature had a strong influence on the degradation of anthocyanins, showing results of their study where storage at 37°C promotes a faster anthocyanin degradation as compared to refrigerated storage at 4°C (Kirca et al., 2007). Similar results were reported by Del Pozo-Insfran et al., showing that increasing the reaction temperature from 10 to 20 °C significantly increased color degradation (Q_{10}~1. 6) in food source of açai, hibiscus, purple potato, black carrot and red cabbage, whereas in red grape was observed a 1.9-fold increase (Del Pozo-Insfran et al., 2004).

In general, fruits and vegetables, visually spoil before any significant antioxidant capacity loss occurs (Kevers et al., 2007). According to a study perform by Klimczak et al.about the storage effects in orange juices, the results showed that among all compounds analyzed in orange juice (favanones, hydroxycinnamic acids, and free phenolics), the vitamin C was the one most affected by temperature and time of storage. They found that increase of temperature by each 10°C caused a distinct decrease in the concentration of vitamin C, and after 6 months of storage at 18, 28 and 38 °C the content of vitamin C decreased by 21, 31 and 81 percent, respectively. In addition, Patthamakanokporn et al. perform a study about the changes in antioxidant activity during storage, establishing as temperature 5°C as representative of the common storage temperature for fruits used by consumers. Likewise, the period of time of storage of guava, makiang, and maloud fruits was 10 days, due that is the regular period of time consumers keeps these fruits under refrigeration. In this study, they report that storage of guava fruit has no effect over the moisture content during storage at 5°C, however, the antioxidant activity by ORAC and FRAP methods in guava tended to increase with 10 days of storage, whereas makiang and maloud shown a decrease in both methods.

5.4 RELATIVE HUMIDITY EFFECTS ON OVERALL QUALITY CHANGES OF FRUITS AND VEGETABLES

In order to maintain the quality of stored fruits and vegetables, they are kept under humid conditions, being in some cases the higher the humidity the better. In order that, under refrigerated conditions, the removal of moisture from the store air reduces its relative humidity and increases its vapor pressure deficit. This means that the stored crop will be losing water, reducing its quality attributes (Thompson, 2008). The water loss is a sign of deterioration identified by the consumer as a visual and sensory impairment in fruits and vegetables products, especially in fresh-cut products. Fresh-cut are products full of juices rich in nutrients which enhance the microbial growth, leading to fruit spoilage (Brecht 2006). Additionally, when a whole fruit is cut or sliced, the protection afforded by the fruit skin is lost, leaving the fruit tissue susceptible to pathogens attack and water loss (Ayala-Zavala et al., 2008). Water loss is critical on the fresh-cut vegetables, and this transpiration results in a series of biochemical and physiological changes, including substrate-enzyme contact. Moreover, some studies report that peeled fruits or fresh-cut products may present browning highly correlated with weight loss (Jiang and Fu, 1999).

The relative humidity (RH) expresses the degree of saturation of the air as a ratio of the actual (e_a) to the saturation ($e°(T)$) vapor pressure at the same temperature (T) and it is calculated with the following equation: $RH = 100[^{e_a}/_{e°}(T)]_{(10)}$ (Allen et al., 1998). Moreover, it has been suggested that the impact of RH on quality, such as appearance and texture, was no doubt ascribed to water loss (Paull, 1999). Along with the temperature, the RH is considered as a major criterion to define critical limits in monitoring programs associated with hazard analysis critical control point (HACCP) system (Paull, 1999). However, even when the storage temperature can be controlled, maintaining RH in large storage rooms within a narrow range at high relative humidities presents some practical difficulties. Nevertheless, the RH is involved in the shelf life of fresh-cut produce, thus it's essential to keep it in the optimal conditions. For example, Frazier and Whesthoff (1993) showed that most of the microorganisms affecting the shelf life of fresh-cuts need and optimal environment with an RH higher than 80 percent of their growth (Westhoff and Frazier 1993). Therefore, the reduction of humidity levels in the critical equilibrium could result into an effective limitation of microbial growth (Ayala-Zavala et al., 2008).

In addition, the water activity (a_w) is directly related to equilibrium RH of a food, and describes the degree to which water is "bound" in the system, controlling its availability to act as a solvent and participate in chemical/biochemical reactions and growth of microorganisms. The complete concept of a_w is the ratio of the vapor pressure of water in equilibrium with a food to the saturation vapor pressure of water at the same temperature. The a_w is equal to the equilibrium RH, expressed as a fraction. This important property can be used to predict the stability and safety of food with respect to microbial growth, rates of deteriorative reactions, and chemical/physical properties. The a_w principle has been incorporated by various regulatory agencies (FDA CFR Title 21) in definitions of safety regulations regarding the growth and proliferation of undesirable microorganisms, standards of several preserved foods, and packaging requirements.

5.5   USE OF MODIFIED OR CONTROL ATMOSPHERE STORAGE OF FRUITS AND VEGETABLES

The demand for fruits and vegetables has led to an increase in the quantity and variety of products available for the consumer. However, fruits and vegetables are very perishable and usually have a shelf life of 5–7 days at 1–7°C (Zhang et al., 2013). Due to the complexities involved with produce, that is, varying respiration rates, which are product and temperature dependent, different optimal storage temperatures for each commodity, water absorption, and so on, many considerations are involved in choosing an acceptable packaging technology (Farber et al., 2003). Modified atmosphere (MA) and controlled atmosphere (CA) are preservations techniques that can be used to extend shelf life and marketability of fruits and vegetables. MA packing relies on modification of the atmosphere inside the package, achieved by the respiration of the product and the transfer of gases through the packaging, which leads to an atmosphere rich in CO_2 (7–15%) and poorer in O_2 (2–6%) (Zhang et al., 2013; Pandey and Goswami, 2012). CA, is a technique preservation used for bulk storages, in this the composition of gases is maintained in the package, so it requires continuous monitoring of gases (Sandhya 2010). In CA storage facilities, sophisticated mechanisms of O_2 removal, CO_2 production and CO_2 removal have been applied to control the room atmosphere within a narrow target range at optimal storage temperature (Thompson 2010). Potentially, MA packaging and CA can reduce respiration rate, ethylene sensitivity and production, physiological changes, delaying enzymatic browning and retaining visual appearance (Kader et al., 1989; Waghmare et al., 2013).

MA packaging success depends on different factors such as: good quality of the initial product, good manufacturing practices, correct gas mixture, adequate refrigeration of the product and appropriate packaging material (Brody 1995; Martínez-Ferrer et al., 2002). In fact, there are a wide variety of polymers and gas mixtures available for packaging fresh-cut produce that should be optimized for each commodity. In this sense Soliva-Fortuny and Martín-Belloso (2003) evaluated fresh-cut conference pears packaged under different MA conditions and stored in refrigeration and found that the use of plastic bags of a permeability of 15 cm^3 O$_2$/m$_2$/bar/24 h and initial atmospheres of 0 kPa O$_2$ extended their microbiological shelf life for at least 3 weeks of storage. In 2004, González-Aguilar et al., (2004) reported 14 days at 10 °C for the same cultivar under 12–15 percent O$_2$ and 2–5 percent CO$_2$ while that Marrero and Kader (2006) in fresh-cut pineapple from 4 days at 10 °C to over two weeks at 0 °C under 10 percent CO$_2$ combined with a maximum of 8 percent O$_2$ The fresh-cut cantaloupe cubes placed in film sealed containers, when flushed with 4 percent O$_2$ + 10 percent CO$_2$, maintained better their saleable quality at 5°C (Bai et al., 2001). Corbo et al. (2004) investigated the changes in sensory quality and proliferation of spoilage microorganisms on lightly processed and packaged cactus pear as a function of storage temperature and MA packaging (65% N$_2$, 30% CO$_2$, 5 % O$_2$). It was found that cactus pear fruit packed under modified atmosphere had a longer shelf life at 4°C.

In addition, the effect of different MA packaging in combination whit chemical treatments have been successfully studied to increase the postharvest shelf life of fresh-cut apple, strawberry, jackfruit bulbs and papaya [90–93]. "Golden Delicious" apple slices sealed were sealed in polypropylene boxes with nonconventional gas mixtures (90% N$_2$ + 5% O$_2$ + 5% CO$_2$ 90% N$_2$O+ 5% O$_2$+ 5% CO$_2$ or 65% N$_2$O + 25% Ar + 5% O$_2$ + 5% CO$_2$) in combination with dipping in an aqueous solution of ascorbic acid (0.5 %), citric acid (0.5%) and of CaCl$_2$ (0.5%) was tested during refrigerated storage at 4°C for 12 days by Rocculi, Romani [91]. Atmospheres with high argon and nitrous oxide levels showed beneficial effect on the product quality during the 10 day period of storage.Aguayo, Jansasithorn [92] evaluated the shelf life of fresh-cut strawberries to determine the effects of 1-MCP (1 μ LL^{-1} for 24 h at 5°C), CaCl$_2$ to 1 percent during 2 min and/or CA to 3 KPa O$_2$ + 10 Kpa CO$_2$ applied to whole product and/or wedges. They found that the combined treatment of 1-MCP + CaCl$_2$ + CA slowed down softening, deterioration rate and microbial growth and the shelf life of fresh-cut strawberries subjected to the combination treatment was extended to 9 days at 5°C.

In 2008 [90] investigated the effect of different MA packaging for extending the shelf life of fresh-cut jackfruit kept under low temperature conditions. A MA packaging consisting in 3 kPa O + 5 kPa CO$_2$ was flushed into polyethylene bags. Prior to MA packaging, fresh-cut jackfruits were given a postcutting phytosanitation wash followed by a dip pretreated with calcium chloride, ascorbic acid and sodium benzoate under mild acidified conditions. The data highlighted the efficacy of pretreatments in interaction with the MA packaging in restricting the microbial load. In particular, the samples pretreated with a combination of CaCl$_2$ (1%), ascorbic acid (0.02%), citric acid (1% w/v) and sodium benzoate (0.045% w/v) showed higher shelf life compared to the samples treated only with CaCl$_2$ (1% w/v) and ascorbic acid (0.02% w/v). As expected, the MA packaging conditions were helpful in reducing microbial counts. In 2011, [94] also evaluated quality changes of fresh-cut jackfruit bulbs as effect of an additive pretreatment with $CaCl_2$, ascorbic acid, citric acid, and sodium benzoate followed by chitosan coating. Different types of samples were subjected to CA storage (3 kPa O2 + 6 or 3 kPa CO2; N2 balance) or normal air at 6°C. CA conditions, pretreatment, as well as a Chitosan coating in synergy with each other, minimized the loss in total phenolics and ascorbic acid content to levels of around 5 percent and 17 percent, respectively, during extended storage up to 50 days. In addition, the CA condition of 3 kPa O2 + 6 kPa CO2 had higher efficacy in retaining quality attributes of the samples.

Finally, Waghmare and Annapure [93] evaluated fresh-cut papaya dipped in a solution of calcium chloride (1% w/v) and citric acid (2% w/v), packed in an atmosphere of 5% O_2, 10% CO_2, 85% N_2 during storage at 5°C for 25 days. Chemical treatment followed by MA packaging showed the best results among the treatment in terms of retaining sensory quality characteristics as improve the surface color, maintain firmness and reduces microbial counts and extending shelf life of fresh-cut papaya up to 25 day at 5°C. It is clear that atmosphere concentration adequate for preservation of fruits and vegetables depend on the characteristics of the fruit, but fortunately good overviews about the use of this approach to extend shelf life of different fruits and vegetables are provided in the literature. This preservation technique has been commercially implemented and additionally, could contribute to, encourage greater consumption of fruits and vegetables (Brody, 1995; Martínez-Ferrer et al., 2002).

5.6   APPLICATION OF ANTIMICROBIAL COMPOUNDS DURING THE STORAGE OF FRUITS AND VEGETABLES

Fruits and vegetables are considered vehicles for foodborne illness due to that during the minimal processing (sorting, cutting, blending and other handling procedures), the natural protection of fruit are removed and hence, they become highly susceptible to microbial contamination (Gombas et al., 2003; Doyle and Erickson, 2008; Oms-Oliu et al., 2010; Niemira and Boyd, 2013). In addition, leakage of juices and sugars from damaged tissues allows the growth and fermentation of some species of yeasts such as *Saccharomyces cerevisiae* and *Saccharomyces exiguous* (Heard, 2002). Similarly, a variety of pathogenic bacteria such as *Listeria monocytogenes*, *Salmonella* spp. and *Shigella* spp., *Aeromonas hydrophila*, *Yersinia enterocolitica*, and *Staphylococcusaureus* as well as some pathogenic *Escherichia coli* strains may be present on fruits and vegetables (Oms-Oliu et al., 2010; Breidt and Fleming, 1997).

Once micro-organisms grow into a firm biofilm, cleaning and disinfection become much more difficult (Wirtanen et al., 2001). The microbial biofilms are complex structures in which bacterial populations are enclosed in a matrix due to cells forming aggregates by adhering to each other or to the food matrix (Costerton et al., 1995). The ability to stick to surfaces and to engage in a multistep process (attach-

ment, growth and dispersal) leading to the formation of a biofilm is almost ubiqui-
tous among bacteria, as it is shown in Figure 5.2 (Van Houdt and Michiels, 2010).
The cells associated with the biofilms have advantages in growth and survival over
planktonic cells and these advantages are due to the formation of exopolysaccharide
(EPS) matrix which surrounds the biofilm. These EPS matrix is a protective mechanism of bacteria against their environment, therefore, it protects the biofilm from
sanitizers and supplies nutrients (James et al., 1995). Nevertheless, these proliferation of biofilm formation by microorganisms on fruits and vegetables is currently
retarded or inhibited by using antimicrobial substances as natural organic acids and
plant essential oils (EOs) (Rojas-Grau and Martin-Belloso, 2008; Ayala-Zavala et
al., 2009; Senhaji et al., 2007; Viuda-Martos et al., 2008; Ayala-Zavala et al., 2008).
Some treatments used for incorporating these substances consist in surface treatments or dipping fruit pieces and their incorporation into edible coatings (ECs).

Organic acids as citric acid, lactic acid, and ascorbic acid are usually applied as
a dip and have been described as strong antimicrobial agents against psychrophilic
and mesophilic microorganisms in fruits and vegetables (Uyttendaele et al., 2004;
Bari et al., 2005). Their antimicrobial effect is due to a pH reduction in the environment, anion accumulation and disruption of membrane transport and/or permeability, (Rico et al., 2007). The growth of a microorganism is inhibited when the pH falls
below the range of pH values that allows its development (Oms-Oliu et al., 2010).
Fruits and vegetables with high pH (>4.6) and aw (>0.85) are considered susceptible
to microbial contamination and are highly perishable, therefore, the acidification of
the product surface with organic acids is recommended (Soliva-Fortuny and Martín-
Belloso, 2003).

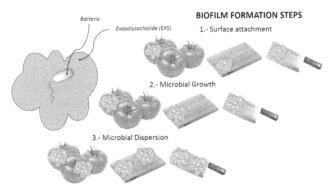

FIGURE 5.2   Bacterial biofilm formation steps on fruits and vegetable surface.

Infusion of fruits with citric acid solution (0.1, 0.25, 0.5, and 1.0% w/v) during
the peeling process reduced the surface pH of peeled oranges (from 6.0 to <4.6)
and citric acid at 0·5 percent w/v was effective to preserve fresh-cut oranges dur-

ing 8 days at 21 °C. (Pao and Petracek 1997). The extension of shelf life resulted primarily from the inhibition of spoilage bacteria. Also, citric acid (6 g/L) was used on fresh-cut cilantro to inactivate E. *coli* O157:H7 and was achieved reduction than 1 log (Allende et al., 2009). On grated carrots, Uyttendaele et al. (2004) found that lactic acid (2% w/v) at pH 3.5 reduced 2.5 log of *Aeromonas (ca. 10^4 CFU/mL)*. As well, dipping of iceberg lettuce in 0·5% lactic acid reduced 1·9 $\log_{10}$ CFU $g^{-1}$and 1·5 $\log_{10}$ CFU g^{-1} the populations of *E. coli* (6·3 $\log_{10}$ CFU g^{-1} initial) and L. *monocytogenes* (5.2 $\log_{10}$ CFU g^{-1} initial) (Akbas and Ölmez, 2007). Similarly, Ascorbic acid (3.40% and pH 2.48) was effective to inactivation of *Salmonella* inoculated at levels of 7.1 to 7.4 log CFU/g on Roma tomato halves, bacterial populations remained below detectable levels (<1.3 log CFU/g) prior to storage and until the 7th day (Yoon et al., 2004).

EOs also have been studied for their antimicrobial activity against many microorganisms, including several pathogens (Dorman and Deans, 2000; Delaquis et al., 2002; Gutierrez et al., 2009). The mechanisms of action of EOs are associated with degradation of the cell wall, damage to cytoplasmic membrane and membrane proteins, leakage of cell contents, coagulation of the cytoplasm and depletion of the proton motive force (Burt, 2004). The inactivation of *S. typhurmurium* on tomatoes using oregano oil (100 ppm) was investigated by Gündüz et al. (2010), the maximum log reduction for *S. typhurmurium* was 2.78 CFU/g. Myrtle oil at 1000 ppm was used on fruits and vegetables also against *S. typhimurium*and log reductions seen were 1.66 CFU/g to 1.89 CFU/g for tomatoes and iceberg lettuce (Gündüz et al., 2009). Ngarmsak et al. (2006) and Delaquis et al. (2002) applied a vanillin dip (0.12%, w/v) to delay the development of total aerobic bacteria and yeast and mold populations of fresh-cut mangoes stored at 5 and 10 °C for up to 14 and 7 day, respectively. Vainillin also was incorporated with a commercial antibrowning dipping solution (calcium ascorbate, NatureSeal™) on fresh-cut apples, 12 mM vanillin inhibited the total aerobic microbial growth by 37 and 66percent in fresh-cut "Empire" and "Crispin" apples, respectively, during storage at 4°C for 19 days (Rupasinghe et al., 2006).

Addition to the above, Tzortzakis (2009) reported antifungal activity of cinnamon oil, preexposing tomato fruit to 500 ppm cinnamon vapors for 3 days, and then inoculated with fungi, reduced B. *cinerea* and C. *coccodes* lesion development. Dipping of fresh-cut kiwifruit in carvacrol solutions at 5–15 mM reduced total viable counts from 6.6 to <2 log CFU/g for 21 d at 4°C (Roller and Seedhar, 2002). Also, it has been reported that hat the addition of 0.02% (v/v) citrus, mandarin, cider, lemon and lime EOs to a minimally processed fruit mix inhibited the proliferation of the naturally occurring microbiota and reduced the growth rate of inoculated *S. cerevisiae* populations, thus extending the shelf life of the fruit salad (Lanciotti et al., 2004). In addition, citrus essential oils on mixture of fruit slices with apple, pear, grape, peach and kiwi fruit inhibited the proliferation of naturally occurring microbiota population and to reduce the growth rates of a *S. cerevisiae* strain inoculated at levels of 10^2 CFU/g at 13 °C during 18 day of storage.

The application of ECs to deliver active substances as EOs is one of the recent major advances made in order to inhibit the growth of microorganisms present on the surface and increase the shelf life of fruits and vegetables. Several types of ECs have been used for extending shelf life of fresh commodities alginate, Chitosan, pectin and methylcellulose (Rojas-Graü et al., 2008; Alvarez et al., 2013; Chen et al., 2012). ECs an alginate-based with cinnamon, clove, and lemongrass EOs at 0.7% (w/v) applied on apple pieces reduced the *E. coli* O157:H7 population by more than 4 log CFU/g, and extended the microbiological shelf life by more than 30 day (Raybaudi-Massilia et al., 2008). The same authors also evaluated an alginate coating as a carrier of malic acid and EOs (cinnamon, Palmarosa and lemongrass) to improve the shelf life and safety of fresh-cut melon. They found that the incorporation of 0.3% v/v Palmarosa oil in the coating looks promising, since it inhibited growth of the native microbiota and reduced the population of inoculated *Salmonella enteritidis* (Raybaudi-Massilia et al., 2008). Respect to Chitosan has been well known for its excellent film-forming properties, broad antimicrobial activity, and a high compatibility with other incorporated compounds (Shahidi et al., 1999; Dong et al., 2004). EC quitosan-based on broccoli reduced in total mesophilic and psychrotrophic bacteria counts during 20 days storage, similarly, decreased in total *E. coli* counts (endogenous and O157:H7) (Moreira et al., 2011).

In addition, incorporating EOs into chitosan ECs can improve its antimicrobial efficiency, as the diffusion of the oil compounds would compensate the nonmigrated antimicrobial power of chitosan (Aider, 2010; Friedman and Juneja, 2010). ECs Chitosan-based with EOs plus bioactive compounds significantly inhibited the growth of mesophilic and psychrotrophic bacteria, and also controlled *E. coli* and *L. monocytogenes* survival (Alvarez et al., 2013). Adding lemon essential oil enhanced the Chitosan antifungal activity both during cold storage in strawberries inoculated with a spore suspension of *Botrytis cinerea* (Perdones et al., 2012). On other EC pectin-based (0 and 36.1 g L−1) was incorporated cinnamon leaf oil to concentration 36.1 g L^{-1} and was applied to grapes stored at 10 °C, during 15 days no fungal decay appeared (Melgarejo-Flores et al., 2013). Sangsuwan et al. (2008) evaluated the effect of chitosan/methyl cellulose films on microbial characteristics of fresh-cut cantaloupe and pineapple. Chitosan/methyl cellulose film provided an inhibitory effect against *Escherichia coli* (TISTR 780) on fresh-cut cantaloupe, the microbial counts decreased by 5.18 at 0 log CFU/piece during 8 days of storage at 10 °C. While that on pineapple slices, *S. cerevisiae* (TISTR 5240) decreased 2.56 log CFU/piece at 12 days of storage.

Finally, it is well established that pathogenic microorganisms are associated with fruits and vegetables. The potential of acid organics, EOs and ECs as antimicrobial agents on fruits and vegetables has been shown and future research will continue to focus on the search for new antimicrobial compounds especially from natural sources that appear to be healthier for consumers. As well as, in new treatments that allow a better preservation of fruits and vegetables (Table 5.1). However, many antimicrobials must be used at high concentrations to achieve activity against target microorganisms. Therefore, those compounds in higher concentrations that negatively affect the flavor and odor or contribute inappropriate flavors and odors would be unacceptable (Davidson et al., 2010). Likewise, in addition to adverse effects on organoleptic parameters of fruits and vegetables, it would be unacceptable for food antimicrobial to mask spoilage leading to the consumers to intake of spoilage products without knowing.

TABLE 5.1　Antimicrobial substances applied to fresh-cut fruits

Substance or Compound	Concentration	Fruits and Vegetables	Treatment	Affected Microorganism	Reference
Acidified sodium chlorite	0.2 g L^{-1}	Fresh-cut cilantro	Washed	Escherichia coliO157:H7	Allende et al. (2009)
Sodium chlorite	0.3 g L^{-1}	Fresh-cut apples	Dip	Escherichia coliHB101	Allende et al. (2009)
trans-cinnamaldehyde	2 g/100 g	Fresh-cut papaya	Microencapsulated into beta-cyclodextrin	Total aerobic plates, psychrotrophics and yeast	Brasil et al. (2012)
Ethanol	5 g/Kg	Fresh-cut mango	Vapor	Yeasts and mold	Plotto et al. (2006)
Sodium chlorite and chitosan edible coatings	Sodium chlorite: 1,000 mg/L Chitosan: 1%, w/v	Fresh-cut pears	Dip followed by packing bags	Escherichia coli O157:H7	Xiao et al. (2011)
Hot water and paracetic acid	80 mg/L	Fresh-cut melon	Dip	Psychrotrophic and mesophilic bacteria	Silveira et al. (2011)
Chorine	0.1 g/mL	Lettuce pieces	Dip	L. monocytogenes	Beuchat et al. (2004)
Combination Methyl jasmonate-ethanol	300 µL/L-333 µL/L	Slices tomatoes	Dip	Total aerobic mesophilic microorganisms, yeast and mold	Ayala-Zavala et al. (2008)
Peroxyacetic acid	40 ppm	Fresh-cut carrots	Dip	Salmonella spp.	Ruiz-Cruz et al. (2007)
Ozonated water	0.18 ppm	Fresh-cut celery	Dip	Total bacterial	Zhang et al. (2005)

CONCLUDING REMARKS

Storage of fruits and vegetables has had an exponential grow and enhancement over the last years, with the development of novel sanitization procedures and the use of emergent technologies to assure the quality of several fresh-produce fruits and vegetables. The use of combined technologies as it has been formerly explained, and the positive effect they have in the fresh-produce industry may lead future researches to the elucidation of mechanisms of action against diverse microbiological organisms which lead to food spoilage. As well as the use of emergent technologies like application of natural antimicrobial agents on fresh-produce products without compromising its quality or its health benefits. Moreover, the use of natural antimicrobial agents like phytochemicals, in plenty of cases results in an addition of beneficial characteristics provided by bioactive properties of phytochemicals like PC or carotenoids. Hence, the application of these types of phytochemicals has been broadly used in several fresh-cut fruits and vegetables extending its shelf life and beneficial properties to the consumers. Nevertheless, the storage temperature of fruits and vegetables in general must be taken into count due that it has a significant relevance over the stability of the bioactive compounds within the fresh-produce fruits.

However, future studies must be focused on the extraction of natural antimicrobial agents, as well as novel disinfectants or effective techniques like controlled atmospheres, edible coatings or UV irradiation for its application in foods or surface of storage to promote the larger shelf –life of the products keeping their freshness and overall quality intact.

KEYWORDS

- **Postharvest Quality**
- **Storage**
- **Antimicrobial Compounds**
- **Temperature**
- **Browning**
- **Biofilm**
- **Modified Atmosphere Storage**

REFERENCES

Aguayo, E.; Jansasithorn, R.; and Kader, A.; Combined effects of 1-methylcyclopropene, calcium chloride dip, and/or atmospheric modification on quality changes in fresh-cut strawberries. *Postharvest Biol. Technol.* **2006**, *40*(3), 269–278.

Aider, M.; Chitosan application for active bio-based films production and potential in the food industry: Review. LWT. *Food Sci. Technol.* **2010**, *43*(6), 837–842.

Akbas, M. Y.; and Ölmez, H.; Inactivation of Escherichia coli and Listeria monocytogenes on iceberg lettuce by dip wash treatments with organic acids. *Lett. Appl. Microbiol.* **2007**, *44*(6), 619–624.

Allen, R. G.; et al.; Crop evapotranspiration-Guidelines for computing crop water requirements-FAO Irrigation and drainage paper *56*. FAO, Rome; **1998**, 300, 6541.

Allende, A.; et al.; Antimicrobial effect of acidified sodium chlorite, sodium chlorite, sodium hypochlorite, and citric acid on Escherichia coli O157:H7 and natural microflora of fresh-cut cilantro. *Food Cont.* **2009**, *20*(3), 230–234.

Alvarez, M. V.; Ponce, A. G.; and Moreira, M. d. R.; Antimicrobial efficiency of chitosan coating enriched with bioactive compounds to improve the safety of fresh-cut broccoli. *LWT - Food Sci. Technol.* **2013**, *50*(1), 78–87.

Ayala-Zavala, J.; et al.; High relative humidity in-package of fresh-cut fruits and vegetables: advantage or disadvantage considering microbiological problems and antimicrobial delivering systems? *J. Food Sci.* **2008**, *73*(4), R41–R47.

Ayala-Zavala, J. F.; González-Aguilar, G. A.; and Del-Toro-Sánchez, L.; Enhancing safety and aroma appealing of fresh-cut fruits and vegetables using the antimicrobial and aromatic power of essential oils. *J. Food Sci.* **2009**, *74*(7), R84–R91.

Ayala-Zavala, J. F.; et al.; Bio-preservation of fresh-cut tomatoes using natural antimicrobials. *Eur. Food Res. Technol.* **2008**, *226*(5), 1047–1055.

Bai, J.; et al.; Sulfite exposure-induced hepatocyte death is not associated with alterations in p53 protein expression. *Toxicology.* **2013**, *312*(0), 142–148.

Bai, J. H.; et al.; Modified atmosphere maintains quality of fresh-cut Cantaloupe (Cucumis melo L.). *J. Food Sci.* **2001**, *66*(8), 1207–1211.

Bapat, V. A.; et al.; Ripening of fleshy fruit: Molecular insight and the role of ethylene. *Biotechnol. Adv.* **2010**, *28*(1), 94–107.

Bari, M.; et al.; Combined efficacy of nisin and pediocin with sodium lactate, citric acid, phytic acid, and potassium sorbate and EDTA in reducing the Listeria monocytogenes population of inoculated fresh-cut produce. *J. Food Protect.* **2005**, *68*(7), 1381–1387.

Bermúdez-Aguirre, D.; and Barbosa-Cánovas, G. V.;Disinfection of selected vegetables under nonthermal treatments: Chlorine, acid citric, ultraviolet light and ozone. *Food Cont.* **2013**, *29*(1), 82–90.

Beuchat, L. R.;. Adler. B. B.;and Lang, M. M.; Efficacy of chlorine and a peroxyacetic acid sanitizer in killing Listeria monocytogenes on iceberg and romaine lettuce using simulated commercial processing conditions. *J. Food Protect.* **2004**, *67*(6), 1238–1242.

Brasil, I. M.; et al.; Polysaccharide-based multilayered antimicrobial edible coating enhances quality of fresh-cut papaya. LWT *–Food Sci. Technol.* **2012**, *47*(1), 39–45.

Brecht, J.; Shelf-life limiting quality factors in fresh-cut (sliced) tomatoes: anti-ethylene treatment and maturity & variety selection to ensure quality retention. In: Oral Presentation on the 2006 Tomato Breeders Round Table & Tomato Quality Workshop Tampa, Fla.:University of Florida. **2006.**

Breidt, F.; and Fleming, H. P.; Using lactic acid bacteria to improve the safety of minimally processed fruits and vegetables. *Food Technol.* **1997**, *51*(9), 44–51.

Brody, A. L.; A perspective on MAP products in North America and Western Europe. Principles of modified atmosphere and sous-vide packaging. Lancaster (PA): *Technomic.* **1995**, pp 13–36.

Burt, S.; Essential oils: Their antibacterial properties and potential applications in foods-a review. *Int. J. Food Microbiol.* **2004**, *94*(3), 223–253.

Burton, W. G.; *Post-harvest physiology of food crops,* Longman Group Ltd., **1982.**

Castaner, M.; et al.; Inhibition of browning of harvested head lettuce. *J. Food Sci.* **1996**, *61*(2), 314–316.

Chaplin, G. R.; Wills, R.; and Graham, D.; Induction of chilling injury in stored avocados with exogenous ethylene. *Hortic. Sci.* **1983**, *18*(6), 952–953.

Chen, W.; et al.; Inactivation of Salmonella on whole cantaloupe by application of an antimicrobial coating containing chitosan and allyl isothiocyanate. *Int. J. Food Microbiol.* **2012**, *155*(3), 165–170.

Clydesdale, F. M.; Color as a factor in food choice. *Crit. Rev. Food Sci. Nutr.* **1993**, *33*(1), 83–101.

Corbo, M.; et al.; Effect of temperature on shelf life and microbial population of lightly processed cactus pear fruit. *Postharvest. Biol. Technol.* **2004**, *31*(1), 93–104.

Cordenunsi, B. R.; et al.; Effects of temperature on the chemical composition and antioxidant activity of three strawberry cultivars. *Food Chem.* **2005**, *91*(1), 113–121.

Costerton, J. W.; et al.; Microbial biofilms. *Ann. Rev. Microbiol.* **1995**, *49*(1), 711–745.

Davidson, P. M.; Sofos, J. N.; and Branen, A. L.; *Antimicrobials in Food.* CRC Press, **2010.**

Degl'Innocenti, E.; et al.; Biochemical study of leaf browning in minimally processed leaves of lettuce (Lactuca sativa L. var. Acephala). *J. Agric. Food Chem.* **2005**, *53*(26), 9980–9984.

Del Pozo-Insfran, D.; Brenes, C. H.; and Talcott, S. T.; Phytochemical composition and pigment stability of Açai (Euterpe oleracea Mart.). *J. Agric. Food Chem.* **2004**, *52*(6), 1539–1545.

Delaquis, P. J.; et al.; Antimicrobial activity of individual and mixed fractions of dill, cilantro, coriander and eucalyptus essential oils. *Int. J. Food Microbiol.* **2002**, *74*(1), 101–109.

Dembitsky, V. M. et al.; The multiple nutrition properties of some exotic fruits: Biological activity and active metabolites. *Food Res. Int.* **2011.**

Desobry, S. A.; Netto, F. M.; and Labuza, T. P.; Preservation of β-carotene from carrots. *Crit. Rev. Food Sci. Nutr.* **1998**, *38*(5), 381–396.

Dong, H.; et al.; Effects of chitosan coating on quality and shelf life of peeled litchi fruit. *J. Food Eng.* **2004**, *64*(3), 355–358.xl

Dorman, H.; and Deans, S.; Antimicrobial agents from plants: antibacterial activity of plant volatile oils. *J. Appl. Microbiol.* **2000**, *88*(2), 308–316.

Doyle, M.; and Erickson, M.; Summer meeting 2007–the problems with fruits and vegetables: an overview. *J. Appl. Microbiol.* **2008**, *105*(2), 317–330.

Farber, J.; et al.; Microbiological safety of controlled and modified atmosphere packaging of fresh and fresh-cut produce. *Comprehen. Rev. Food Sci. Food Saf.* **2003**, *2*(s1), 142–160.

Ferguson, I.; Volz, R.; and Woolf, A.; Preharvest factors affecting physiological disorders of fruit. *Postharvest Biol. Technol.* **1999**, *15*(3), 255–262.

Fischer, R. L.; and Bennett, A.; Role of cell wall hydrolases in fruit ripening. *Annu. Rev. Plant. Biol.* **1991**, *42*(1), 675–703.

Friedman, M.; and Juneja, V. K.; Review of antimicrobial and antioxidative activities of chitosans in food. *J. Food Protect.* **2010**, *73*(9), 1737–1761.

Gil, M. I.; Conesa, M. A.; and Artés, F.; Quality changes in fresh cut tomato as affected by modified atmosphere packaging. *Postharvest Biol. Technol.* **2002**, *25*(2), 199–207.

Gombas, D. E.; et al.; Survey of Listeria monocytogenes in ready-to-eat foods. *J. Food Protect.* **2003**, *66*(4), 559–569.

González-Aguilar, G. A.; et al.; Physiological and quality changes of fresh-cut pineapple treated with antibrowning agents. *LWT. Food Sci. Technol.* **2004**, *37*(3), 369–376.

Gorny, J. R.; A summary of CA and MA requirements and recommendations for fresh-cut (minimally processed) fruits and vegetables. VIII International Controlled Atmosphere Research Conference 600. **2001**.

Gram, L.; et al.; Food spoilage—interactions between food spoilage bacteria. *Int. J. Food Microbiol.* **2002**, *78*(1–2), 79–97.

Gündüz, G. T.; Gönül, Ş. A.; and Karapinar, M.; Efficacy of myrtle oil against Salmonella Typhimurium on fruits and vegetables. *Int. J. Food Microbiol.* **2009**, *130*(2), 147–150.

Gündüz, G. T.; Gönül, Ş. A.;and Karapinar, M.; Efficacy of sumac and oregano in the inactivation of Salmonella Typhimurium on tomatoes. *Int. J. Food Microbiol.* **2010**, *141*(1–2), 39–44.

Gutierrez, J.; Barry-Ryan, C.; and Bourke, P.; Antimicrobial activity of plant essential oils using food model media: Efficacy, synergistic potential and interactions with food components. *Food Microbiol.* **2009**, *26*(2), 142–150.

Haminiuk, C. W.; et al.; Phenolic compounds in fruits–an overview. *Int. J. Food Sci. Technol.* **2012**, *47*(10), 2023–2044.

Heard, G. M.; Microbiology of fresh-cut produce. Fresh-cut fruits and vegetables: Science, technology, and market, **2002**; pp 187–248.

Hofman, P.; et al.; Ripening and quality responses of avocado, custard apple, mango and papaya fruit to 1-methylcyclopropene. *Anim. Prod. Sci.* **2001**, *41*(4), 567–572.

Holland, N.; Menezes, H. C.; and Lafuente, M. T.; Carbohydrate metabolism as related to high-temperature conditioning and peel disorders occurring during storage of citrus fruit. *J. Agric. Food Chem.* **2005**, *53*(22), 8790–8796.

Hong, J. H.; and Gross, K. C.; Involvement of ethylene in development of chilling injury in fresh-cut tomato slices during cold storage. *J. Am. Soc. Hortic. Sci.* **2000**, *125*(6), 736–741.

Hopkirk, G.;et al.; Influence of postharvest temperatures and the rate of fruit ripening on internal postharvest rots and disorders of New Zealand 'Hass' avocado fruit. *New Zealand J. Crop. Hortic. Sci.* **1994**, *22*(3), 305–311.

James, G.; Beaudette, L.; and Costerton, J.; Interspecies bacterial interactions in biofilms. *J. Indian Microbiol.* **1995**, *15*(4), 257–262.

James, I. F.; AD03E Preservation of fruit and vegetables, Agromisa Foundation; **2003**.

Javanmardi, J.; and Kubota, C.; Variation of lycopene, antioxidant activity, total soluble solids and weight loss of tomato during postharvest storage. *Postharvest Biol. Technol.* **2006**, *41*(2), 151–155.

Jeong, J., Huber, D. J.; and Sargent, S. A.; Delay of avocado (Persea americana) fruit ripening by 1-methylcyclopropene and wax treatments. *Postharvest Biol. Technol.* **2003**, *28*(2), 247–257.

Jiang, Y.; Li, Y.; and Li, J.; Browning control, shelf life extension and quality maintenance of frozen litchi fruit by hydrochloric acid. *J. Food Eng.* **2004**, *63*(2), 147–151.

Jiang, Y. M.; and Fu, J. R.; Postharvest browning of Litchi fruit by water loss and its prevention by controlled atmosphere storage at high relative humidity. *LWT – Food Sci. Technol.* **1999**, *32*(5), 278–283.

Jongen, W.; *Improving the Safety of Fresh Fruit and Vegetables*; CRC Press, **2005**.

Joshipura, K. J.; et al.; Fruit and vegetable intake in relation to risk of ischemic stroke. *JAMA.* **1999**, *282*(13), 1233–1239.

Kader, A., Maturity, ripening, and quality relationships of fruit-vegetables. Strategies for Market Oriented Greenhouse Production 434, **1995**, pp 249–256.

Kader, A. A.; et al.; Modified atmosphere packaging of fruits and vegetables. *Crit. Rev. Food Sci. Nutrit.* **1989**, *28*(1), 1–30.

Kader, A. A.; Quality parameters of fresh-cut fruit and vegetable products. Fresh-cut fruits and vegetables, **2002**, 11–20.

Kaur, C.; and Kapoor,H. C.;Antioxidants in fruits and vegetables–the millennium's health. *Int. J. Food Sci. Technol.* **2001**, *36*(7), 703–725.

Kevers, C.; et al.; Evolution of antioxidant capacity during storage of selected fruits and vegetables. *J. Agric. Food Chem.* **2007**, *55*(21), 8596–8603.

Kienzle, S.; et al.; Harvest maturity specification for mango fruit (*Mangifera indica*L.'Chok Anan') in regard to long supply chains. *Postharvest Biol. Technol.* **2011**, *61*(1), 41–55.

Kim, Y.; Brecht, J. K.;and Talcott, S. T.; Antioxidant phytochemical and fruit quality changes in mango (Mangifera indica L.) following hot water immersion and controlled atmosphere storage. *Food Chem.* **2007**, *105*(4), 1327–1334.

Kirca, A., Özkan, M.; and Cemeroğlu, B.; Effects of temperature, solid content and pH on the stability of black carrot anthocyanins. *Food Chem.* **2007**, *101*(1), 212–218.

Koca, N.; Burdurlu, H. S.;and Karadeniz, F.; Kinetics of colour changes in dehydrated carrots. *J. Food Eng.* **2007**, *78*(2), 449–455.

Lacroix, M.; and Ouattara, B.;Combined industrial processes with irradiation to assure innocuity and preservation of food products—a review. *Food Res. Int.* **2000**, *33*(9), 719–724.

Lamikanra, O.; *Fresh-Cut Fruits and Vegetables: Science, Technology, and Market,* CRC Press, **2002**.

Lanciotti, R.; et al.; Use of natural aroma compounds to improve shelf-life and safety of minimally processed fruits. *Trend. Food Sci. Technol.* **2004**, *15*(3–4), 201–208.

Lattanzio, V.; Bioactive polyphenols: their role in quality and storability of fruit and vegetables. *J. Appl. Bot.* **2003**, *77*(5/6), 128–146.

Lu, J.; et al.; Effect of heat treatment uniformity on tomato ripening and chilling injury. *Postharvest Biol. Technol.* **2010**, *56*(2), 155–162.

Luo, Y., et al.; Dual effectiveness of sodium chlorite for enzymatic browning inhibition and microbial inactivation on fresh-cut apples. LWT -. *Food Sci. Technol.* **2011**, *44*(7) 1621–1625.

Mahajan, P.; and Goswami, T.; Extended storage life of litchi fruit using controlled atmosphere and low temperature. *J. Food Process. Preserv.* **2004**, *28*(5), 388–403.

Marrero, A.; and Kader, A. A.;Optimal temperature and modified atmosphere for keeping quality of fresh-cut pineapples. *Postharvest Biol. Technol.* **2006**, *39*(2), 163–168.

Martínez-Ferrer, M.; et al.; Modified atmosphere packaging of minimally processed mango and pineapple fruits. *J. Food Sci.* **2002**, *67*(9), 3365–3371.

Martínez-Ferrer, M. et al.; Modified atmosphere packaging of minimally processed mango and pineapple fruits. *J. Food Sci.* **2002**, *67*(9), 3365–3371.

McDonald, R. E.; McCollum, T. G.; and Baldwin, E. A.; Temperature of water heat treatments influences tomato fruit quality following low-temperature storage. *Postharvest Biol. Technol.* **1999**, *16*(2), 147–155.

Melgarejo-Flores, B. G.; et al.; Antifungal protection and antioxidant enhancement of table grapes treated with emulsions, vapors, and coatings of cinnamon leaf oil. *Postharvest Biol. Technol.* **2013**, *86*(0), 321–328.

Meng, X.; et al.; Physiological responses and quality attributes of table grape fruit to chitosan preharvest spray and postharvest coating during storage. *Food Chem.* **2008**, *106*(2), 501–508.

Merzlyak, M. N.; et al.; Non-destructive optical detection of pigment changes during leaf senescence and fruit ripening. *Physiologia. Plantarum.* **1999**, *106*(1), 135–141.

Mitra, S. K.; *Postharvest Physiology and Storage of Tropical and Subtropical Fruits*, Cab International, **1997**.

Moreira, M. d. R.; Roura, S. I.; and Ponce, A.; Effectiveness of chitosan edible coatings to improve microbiological and sensory quality of fresh-cut broccoli. LWT – Food. *Sci. Technol.* **2011**, *44*(10), 2335–2341.

Ngarmsak, M.; et al.; Antimicrobial activity of vanillin against spoilage microorganisms in stored fresh-cut mangoes. *J. Food Protect.* **2006**, *69*(7), 1724–1727.

Niemira, B. A.; and Boyd, G.; Influence of modified atmosphere and varying time in storage on the irradiation sensitivity of Salmonella on sliced roma tomatoes. *Radiat. Phys. Chem.* **2013**, *90*, 120–124.

O'Hare, T. J.; *Respiratory Characteristics of Cut Pineapple Tissue*. Postharvest Group, DPI: Queensland, Australia, **1994**.

Olaimat, A. N.; and R. A.; Holley, Factors influencing the microbial safety of fruits and vegetables: a review. *Food Microbiol.* **2012**, *32*(1), 1–19.

Oms-Oliu, G.; et al.; Recent approaches using chemical treatments to preserve quality of fresh-cut fruit: a review. *Postharvest Biol. Technol.* **2010**, *57*(3), 139–148.

Özkan, M.; et al.; Degradation Kinetics of Anthocyanins from Sour Cherry, Pomegranate, and Strawberry Juices by Hydrogen Peroxide. *J. Food Sci.* **2002**, *67*(2), 525–529.

Pandey, S. K.; and Goswami, T. K.; Modelling perforated mediated modified atmospheric packaging of capsicum. *Int. J. Food Sci. Technol.* **2012**, *47*(3), 556–563.

Pao, S.; and Petracek, P.; Shelf life extension of peeled oranges by citric acid treatment. *Food Microbiol.* **1997**, *14*(5), 485–491.

Paull, R.; Effect of temperature and relative humidity on fresh commodity quality. *Postharvest Biol. Technol.* **1999**, *15*(3), 263–277.

Perdones, A.; et al.; Effect of chitosan–lemon essential oil coatings on storage-keeping quality of strawberry. *Postharvest Biol. Technol.* **2012**, *70*(0), 32–41.

Plotto, A.; et al.; Ethanol vapor prior to processing extends fresh-cut mango storage by decreasing spoilage, but does not always delay ripening. *Postharvest. Biol. Technol.* **2006**, *39*(2), 134–145.

Rana, M. K.; Ripening changes in fruits and vegetables –a review. *Haryana J. Hortic. Sci.* **2006**, *35*(3/4), 271–279.

Raybaudi-Massilia, R. M.; et al.; Comparative study on essential oils incorporated into an alginate-based edible coating to assure the safety and quality of fresh-cut fuji apples. *J. Food Protect.* **2008**, *71*(6), 1150–1161.

Rico, D.; et al.; Extending and measuring the quality of fresh-cut fruit and vegetables: a review. *Trend. Food Sci. Technol.* **2007**, *18*(7), 373–386.

Rocculi, P.; Romani, S.; and Rosa,M. D.; Evaluation of physico-chemical parameters of minimally processed apples packed in non-conventional modified atmosphere. *Food Res. Int.* **2004**, *37*(4), 329–335.

Rojas-Grau, M. A.; and Martin-Belloso, O.; Current advances in quality maintenance of fresh-cut fruits. Stewart. *Postharvest Rev.* **2008**, *4*(2), 1–8.

Rojas-Graü, M. A.; Tapia, M. S.; and Martín-Belloso, O.; Using polysaccharide-based edible coatings to maintain quality of fresh-cut fuji apples. LWT-Food. *Sci. Technol.* **2008**, *41*(1), 139–147.

Roller, S.; and Seedhar, P.; Carvacrol and cinnamic acid inhibit microbial growth in fresh-cut melon and kiwifruit at 4° and 8°C. *Lett. Appl. Microbiol.* **2002**, *35*(5), 390–394.

Ruiz-Cruz, S.; et al.; Efficacy of sanitizers in reducing Escherichia coli O157:H7, Salmonella spp. and Listeria monocytogenes populations on fresh-cut carrots. *Food Contr.* **2007**, *18*(11), 1383–1390.

Rupasinghe, H. P. V.; et al.; Vanillin inhibits pathogenic and spoilage microorganisms in vitro and aerobic microbial growth in fresh-cut apples. *Food Res. Int.* **2006**, *39*(5), 575–580.

Saltveit, M. E.; Effect of ethylene on quality of fresh fruits and vegetables. *Postharvest Biol. Technol.* **1999**, *15*(3), 279–292.

Sandhya; Modified atmosphere packaging of fruits and vegetables: current status and future needs. *LWT-Food Sci. Technol.* **2010**, *43*(3), 381–392.

Sangsuwan, J.; Rattanapanone, N.; and Rachtanapun, P.; Effect of chitosan/methyl cellulose films on microbial and quality characteristics of fresh-cut cantaloupe and pineapple. *Postharvest Biol. Technol.* **2008**, *49*(3), 403–410.

Saxena, A.; et al.; Effect of controlled atmosphere storage and chitosan coating on quality of fresh-cut jackfruit bulbs. *Food Bioproc. Technol.* **2011**, 1–8.

Saxena, A.; Bawa, A. S.;and Srinivas Raju, P.; Use of modified atmosphere packaging to extend shelf-life of minimally processed jackfruit (Artocarpus heterophyllus L.) bulbs. *J. Food Eng.* **2008**, *87*(4), 455–466.

Senhaji, O.; Faid, M.; and Kalalou, I.; Inactivation of Escherichia coli O157: H7 by essential oil from Cinnamomum zeylanicum. *Braz. J. Infect. Dis.* **2007**, *11*(2), 234–236.

Shahidi, F.; Arachchi, J. K. V.; and Jeon, Y. -J.; Food applications of chitin and chitosans. *Trend Food Sci. Technol.* **1999**, *10*(2), 37–51.

Silveira, A. C.; et al.; Hot water treatment and peracetic acid to maintain fresh-cut Galia melon quality. *Innov. Food Sci. Emer. Technol.* **2011**, *12*(4), 569–576.

Snowdon, A. L.; Post-harvest diseases and disorders of fruits and vegetables: Volume 2: Vegetables. Vol. 2, Manson Publishing; **2010.**

Soliva-Fortuny, R. C.; and Martín-Belloso, O.; New advances in extending the shelf-life of fresh-cut fruits: a review. *Trend. Food Sci. Technol.* **2003**, *14*(9), 341–353.

Sommerburg, O.; et al.; Fruits and vegetables that are sources for lutein and zeaxanthin: the macular pigment in human eyes. *Br. J. Ophthalmol.* **1998**, *82*(8), 907–910.

Srivastava, M. K.; and Dwivedi, U. N.; Delayed ripening of banana fruit by salicylic acid. *Plant. Sci.* **2000**, *158*(1), 87–96.

Streif, J.; Retamales, J.; and Cooper T.; Preventing cold storage disorders in nectarines. In: International Symposium on Postharvest Treatment of Horticultural Crops, *368*, **1993**.

Thompson, A. K.; Controlled atmosphere storage of fruits and vegetables, **2010**, CABI.

Thompson, K.; Fruit and vegetables: harvesting, handling and storage, John Wiley & Sons; **2008.**

Tijskens, L.; *Acceptability.* Fruit and vegetable quality: An integrated view. Technomic Publishing INC: Lancaster Pennsylvania, USA, **2000**; pp 144–157.

Tjoelker, M. G.; Oleksyn, J.; and Reich, P. B.;Modelling respiration of vegetation: evidence for a general temperature-dependent Q10.*Glob. Change. Biol.* **2001**, *7*(2), 223–230.

Toivonen, P.; and Brummell, D. A.;Biochemical bases of appearance and texture changes in fresh-cut fruit and vegetables. *Postharvest Biol. Technol.* **2008**, *48*(1), 1–14.

Tonutti, P.; Innovations in Storage Technology and Postharvest Science. In VII International Postharvest Symposium, *1012*, **2012.**

Tzortzakis, N. G.; Impact of cinnamon oil enrichment on microbial spoilage of fruits and vegetables. *Innov. Food Sci. Emerg. Technol.* **2009**, *10*(1), 97–102.

Uyttendaele, M.; et al.; Control of Aeromonas on minimally processed vegetables by decontamination with lactic acid, chlorinated water, or thyme essential oil solution. *Int. J. Food Microbiol.* **2004**, *90*(3), 263–271.

Van Houdt, R.; and Michiels, C. W.; Biofilm formation and the food industry, a focus on the bacterial outer surface. *J. Appl. Microbiol.* **2010**, *109*(4), 1117–1131.

Viuda-Martos, M.; et al.; Antibacterial activity of different essential oils obtained from spices widely used in Mediterranean diet. *Int. J. Food Sci. Technol.* **2008**, *43*(3), 526–531.

Wade, N.; Physiology of cool-storage disorders of fruit and vegetables. In: *Low Temperature Stress in Crop Plants*. Press Academic: New York; **1979**, pp 81–96.

Waghmare, R.; Mahajan, P.; and Annapure, U.; Modelling the effect of time and temperature on respiration rate of selected fresh-cut produce. *Postharvest Biol. Technol.* **2013**, *80*, 25–30.

Waghmare, R. B.; and Annapure, U. S.; Combined effect of chemical treatment and/or modified atmosphere packaging (MAP) on quality of fresh-cut papaya. *Postharvest Biol. Technol.* **2013**, *85*(0) pp. 147–153.

Wakabayashi, K.; Changes in cell wall polysaccharides during fruit ripening. *J. Plant. Res.* **2000**, *113*(3), 231–237.

Wang, C.; et al.; A new fluorescent turn-on probe for highly sensitive and selective detection of sulfite and bisulfite. *Sens. Actuat. B. Chem.* **2014**, *190*(0), 792–799.

Watada, A. E.; and Qi, L.; Quality of fresh-cut produce. *Postharvest Biol. Technol.* **1999**, *15*(3), 201–205.

Watada, A. E.; Ko, N. P.;and Minott, D. A.; Factors affecting quality of fresh-cut horticultural products. *Postharvest Biol. Technol.* **1996**, *9*(2), 115–125.

Westhoff, D. C.; and Frazier, W. C.; *Microbiología de los alimentos*. Editora Acribia, Zaragoza, **1993**.

Wirtanen, G.; et al.; Microbiological methods for testing disinfectant efficiency on Pseudomonas biofilm. *Colloid. Surf. B. Biointerf.* **2001**, *20*(1), 37–50.

Xiao, Z.; et al.; Combined effects of sodium chlorite dip treatment and chitosan coatings on the quality of fresh-cut d'anjou pears. *Postharvest Biol. Technol.* **2011**, *62*(3), 319–326.

Yakushiji, H.; Sakurai, N.; and Morinaga, K.; Changes in cell-wall polysaccharides from the mesocarp of grape berries during veraison. *Physiologia. Plantarum.* **2001**, *111*(2), 188–195.

Yang, S.; and Oetiker, J.; The role of ethylene in fruit ripening. *Postharvest Physiol. Fruits.* **1994**, *398*, 167–178.

Yoon, Y., et al.; Inactivation of Salmonella during drying and storage of roma tomatoes exposed to predrying treatments including peeling, blanching, and dipping in organic acid solutions. *J. Food Protect.* **2004**, *67*(7), 1344–1352.

Zhang, B.-Y.; et al.; Effect of high oxygen and high carbon dioxide atmosphere packaging on the microbial spoilage and shelf-life of fresh-cut honeydew melon. *Int. J. Food Microbiol.* **2013**, *166*(3), 378–390.

Zhang, L.; et al.; Preservation of fresh-cut celery by treatment of ozonated water. *Food Contr.* **2005**, *16*(3), pp 279–283.

Zheng, L.; et al.; Antimicrobial activity of natural antimicrobial substances against spoilage bacteria isolated from fruits and vegetables. *Food Cont.* **2013**.

Zhou, H.-W.; et al.; The role of ethylene in the prevention of chilling injury in nectarines. *J. Plant. Physiol.* **2001**, *158*(1), 55–61.

CHAPTER 6

ACTIVE AND SMART PACKAGING FILM FOR FOOD AND POSTHARVEST TREATMENT

IDA IDAYU MUHAMAD[1]; ERARICAR SALLEH[1];
NOZIEANA KHAIRUDIN[1]; MOHD HARFIZ SALEHUDIN[1]; and
NORSUHADA ABDUL KARIM[1]

[1]Food and Biomaterial Eng. Research Group, Bioprocess Engineering
Department, Faculty of Chemical Engineering, Universiti Teknologi Malaysia,
81300 Johor Bahru, Johor, Malaysia

CONTENTS

6.1 INTRODUCTION TO ACTIVE, SMART, AND INTELLIGENT PACKAGING

Packaging has been defined as a socio-scientific discipline, which operates in society to ensure delivery of goods to the ultimate consumer of those goods in the best condition intended for their use. The International Institute of Packaging defines packaging as the enclosure of products, item, or packages in a wrapped pouch, bag, box, cup, tray, can, tube, bottle, or other container form to perform one or more of the following functions such as containment, protection, preservation, communication utility and performance (Roberstson, 2006). Packaging has a significant role in the food processes and the whole food supply chain. Food packaging has to perform several tasks as well as fulfilling many demands and requirements. It has to protect food from environmental conditions such as light, oxygen, moisture, microbes, mechanical stresses and dust. Other basic tasks have been to ensure adequate labeling for providing information to the customer; packaging should be easy to open, has reclosable lids and suitable dosing mechanism. Basic requirements for packaging are good marketing properties, reasonable price, technical feasibility (e.g., suitability for automatic packaging machines, seal ability), suitability for food contact, low environmental stress and suitability for recycling or refilling. A package has to satisfy all these various requirements effectively and economically (Ahvenainen, 2003).

6.1.1 HISTORICAL DEVELOPMENT OF PACKAGING

Food packaging has evolved from simply a container to hold food to something today that can play an active role in food quality. Many packages are still simply containers, but they have properties that have been developed to protect the food. These include barriers to oxygen, moisture, and flavors (Risch, 2009).

Paper is one of the oldest forms of "flexible packaging" today. Sheets of treated mulberry bark were used by the Chinese to wrap foods as early as the First or Second century B.C. During the next 1,500 years, the paper-making technique was refined and transported to the Middle East, then Europe and finally into the United Kingdom in 1310. Eventually, the technique arrived in America in Germantown, Pennsylvania, in 1690. But these first papers were somewhat different from those used today. Early paper was made from flax fibers and later old linen rags. It was not until 1867 that paper originating from wood pulp was developed. Although commercial paper bags were first manufactured in Bristol, England, in 1844, Francis Wolle invented the bag- making machine in 1852 in the United States. Further advancements during the 1870s included glued paper sacks and the gusset design. After the turn of the century (1905), the machinery was invented to automatically produce in- line printed paper bags. With the development of the glued paper sack, the more expensive cotton flour sacks could be replaced. But a sturdier multiwalled paper sack for larger quantities could not replace cloth until 1925 when a means of

sewing the ends was finally invented. The first commercial cardboard box was produced in England in 1817, more than two hundred years after the Chinese invented
cardboard. Corrugated paper appeared in the 1850s; about 1900, shipping cartons of
faced corrugated paperboard began to replace self-made wooden crates and boxes
used for trade.

As with many innovations, the development of the carton was accidental. Robert
Gair was a Brooklyn printer and paper bag maker during the 1870s. While he was
printing an order of seed bags, a metal rule normally used to crease bags shifted in
position and cut the bag. Gair concluded that cutting and creasing paperboard in one
operation would have advantages; the first automatically made carton, now referred
to as "semi-flexible packaging," was created. The development of flaked cereals
advanced the use of paperboard cartons. The Kellogg brothers were first to use cereal cartons at their Battle Creek, Michigan, Sanatorium. When this "health food" of
the past was later marketed to the masses, a waxed, heat sealed bag of Waxtite was
wrapped around the outside of a plain box. The outer wrapper was printed with the
brand name and advertising copy. Today, the plastic liner protects cereals and other
products within the printed carton. Paper and paperboard packaging increased in
popularity well into the twentieth century. With the advent of plastic as a significant
player in packaging (late 1970s and early 1980s), paper and its related products
tended to fade in use. Then the trend has halted as designers try to respond to environmental concerns (Hook and Heimlich; Risch, 2009).

Although glass-making began in 7000 B.C. as an offshoot of pottery, it was first
industrialized in Egypt in 1500 B.C. Made from base materials (limestone, soda,
sand, and silica), which were in plentiful supply, all ingredients were simply melted
together and molded while hot. Since that early discovery, the mixing process and
the ingredients have changed very little, but the molding techniques have progressed
dramatically. At first, ropes of molten glass were coiled into shapes and fused together. By 1200 B.C., glass was pressed into molds to make cups and bowls. When
the blowpipe was invented by the Phoenicians in 300 B.C., it not only speeded
production but allowed for round containers. Colors were available from the beginning, but clear, transparent glass was not discovered until the start of the Christian
Era. During the next 1,000 years, the process spread steadily, but slowly, across
Europe. The split mold developed in the seventeenth and eighteenth centuries further provided for irregular shapes and raised decorations. The identification of the
maker and the product name could then be molded into the glass container as it was
manufactured. As techniques were further refined in the eighteenth and nineteenth
centuries, prices of glass containers continued to decrease. One development that
enhanced the process was the first automatic rotary bottle- making machine, patented in 1889. Current equipment automatically produces 20,000 bottles per day. While
other packaging products, such as metals and plastics, were gaining popularity in the
1970s, packaging in glass tended to be reserved for high- value products. As a type
of "rigid packaging" glass has many uses today (Hook and Heimlich; Risch, 2009).

Ancient boxes and cups, made from silver and gold, were much too valuable for common use. Other metals, stronger alloys, thinner gauges and coatings were eventually developed. The process of tin plating was discovered in Bohemia in 1200 A.D. and cans of iron, coated with tin, was known in Bavaria as early as the fourteenth century. However, the plating process was a closely guarded secret until the 1600s. Thanks to the Duke of Saxony, who stole the technique, it progressed across Europe to France and the United Kingdom by the early nineteenth century. After William Underwood transferred the process to the United States via Boston, steel replaced iron, which improved both output and quality. In 1764, London tobacconists began selling snuff in metal canisters, another type of today's "rigid packaging". But no one was willing to use metal for food since it was considered poisonous. The safe preservation of foods in metal containers was finally realized in France in the early 1800s. In 1809, General Napoleon Bonaparte offered 12,000 francs to anyone who could preserve food for his army. Nicholas Appert, a Parisian chef and confectioner, found that food sealed in tin containers and sterilized by boiling could be preserved for long periods. A year later (1810), Peter Durand of Britain received a patent for tinplate after devising the sealed cylindrical can. Since food was now safe within metal packaging, other products were made available in metal boxes. In the 1830s, cookies and matches were sold in tins and by 1866 the first printed metal boxes were made in the United States for cakes of Dr. Lyon's tooth powder (Hook and Heimlich; Risch, 2009).

The first cans produced were soldered by hand, leaving a 1 1/2-inch hole in the top to force in the food. A patch was then soldered in place but a small air hole remained during the cooking process. Another small drop of solder then closed the air hole. At this rate, only 60 cans per day could be manufactured. In 1868, interior enamels for cans were developed, but double seam closures using a sealing compound were not available until 1888. Aluminum particles were first extracted from bauxite ore in 1825 at the high price of $545 per pound. When the development of better processes began in 1852, the prices steadily declined until the low price of 14 per pound in 1942. Although commercial foils entered the market in 1910, the first aluminum foil containers were designed in the early 1950s while the aluminum can appeared in 1959. After cans were invented and progressively improved, it was necessary to find a way to open them. Until 1866, a hammer and chisel was the only method. It was then that the key wind metal tear-strip was developed. Nine years later (1875), the can opener was invented. Further developments modernized the mechanism and added electricity, but the can opener has remained, for more than 100 years, the most efficient method of retrieving the contents. In the 1950s, the pop top/tear tab can lid appeared and now tear tapes that open and reseal are popular. Collapsible, soft metal tubes, today known as "flexible packaging," were first used for artists' paints in 1841. Toothpaste was invented in the 1890s and started to appear in collapsible metal tubes. But food products really did not make use of this packaging form until the 1960s. Later, aluminum was changed to plastic for such food items as sandwich pastes, cake icings and pudding toppings (Hook and Heimlich; Risch, 2009).

Plastic is the youngest in comparison with other packaging materials. Although discovered in the 19th century, most plastics were reserved for military and wartime use. Styrene was first distilled from a balsam tree in 1831. But the early products were brittle and shattered easily. Germany refined the process in 1933 and by the 1950s foam was available worldwide. Insulation and cushioning materials as well as foam boxes, cups, and meat trays for the food industry became popular. Vinyl chloride, discovered in 1835, provided for the further development of rubber chemistry. For packaging, molded deodorant squeeze bottles were introduced in 1947 and in 1958, heat shrinkable films were developed from blending styrene with synthetic rubber. Today some water and vegetable oil containers are made from vinyl chloride. Another plastic was invented during the American Civil War. Due to a shortage of ivory, a United States manufacturer of billiard balls offered a $ 10,000 reward for an ivory substitute. A New York engineer, John Wesley Hyatt, with his brother Isaiah Smith Hyatt, experimented several years before creating the new material. Patented in 1870, "celluloid" could not be molded, but rather carved and shaped, just like ivory (Hook and Heimlich; Risch, 2009).

Cellulose acetate was first derived from wood pulp in 1900 and developed for photographic uses in 1909. Although DuPont manufactured cellophane in New York in 1924, it wasn't commercially used for packaging until the late 1950s and early 1960s. In the interim, polyethylene film wraps were reserved for the military. In 1933, films protected submarine telephone cables and later were important for World War II radar cables and drug tablet packaging. Other cellophanes and transparent films have been refined as outer wrappings that maintain their shape when folded. Originally clear, such films can now be made opaque, colored, or embossed with patterns. The polyethylene terephthalate (PETE) container only became available during the last two decades with its use for beverages entering the market in 1977. By 1980, foods and other hot-fill products such as jams could also be packaged in PETE. Current packaging designs are beginning to incorporate recyclable and recycled plastics but the search for reuse functions continues (Risch, 2009).

There are different types of active packaging. One type, referred to as a susceptor, is used for microwave foods, including popcorn. The first bag of microwave popcorn was sold in 1971. The package was a simple paper bag. It was not until the package including a microwave susceptor was introduced in the mid-1980s that the product became a large success. The package consists of two layers of paper with a metalized PET film (susceptor) laminated between the layers of paper in a position so that it lies on the floor of the microwave oven. The metalized film is produced in the same way described earlier but with a thinner layer of metal that interacts with the microwave energy and heat to temperatures of 200°C or higher (a thicker layer such as that used for packaging for overwrap would reflect microwave energy instead of absorbing it). The heat generated gives the energy needed to get the kernels to pop. Without the susceptor, the product will have a large number of unpopped kernels. One initial patent for the popcorn bag was issued in 1988. One other type of active packaging material is one that can absorb oxygen. As was mentioned earlier in relation to the liners of crowns for beer bottles, oxygen absorbers can be built into packaging to remove residual oxygen from around product or a sachet with material (typically iron oxide) can be placed inside the package. Some companies are exploring means of incorporating flavors into packaging to maintain the quality of the flavor and have it release at the time of consumption. One package has been developed by Lee Reedy. Flavors and nutritional supplements are sealed into the cap for a bottle. When the cap is twisted to open the bottle, a small plastic blade cuts the seal and releases the nutrients and flavor into the beverage. This preserves the quality and freshness of the flavors and supplements until the time of consumption (Risch, 2009).

6.1.2 CLASSIFICATION OF PACKAGING

Packaging can be broadly categorized into several types such as passive, active, intelligent, and smart (Table 6.1).

TABLE 6.1   Classification of food packaging

Types of Packaging	Definition
Passive packaging	The traditional packaging that involves the use of a covering material, characterized by some inherent insulting, protective, or ease-of-handling qualities (Teixeira, 2010).
Active packaging	Entails the concept of the package reacting to various stimuli—to keep the internal environment favorable for the products. A typical example would be a packaging with oxygen scavenger (an oxygen scavenger can absorb oxygen inside a package to increase the shelf life of the item) (Teixeira, 2010).
	It changes the condition of the packed food to extend shelf life or to improve safety or sensory properties while maintaining the quality of the packaged food (Ahvenainen, 2003).
	Food packaging which has an extra function in addition to that of providing a protective barrier against external influence. It can control and even react to, phenomena taking place inside the package (Fabech et al., 2000).

TABLE 6.1   *(Continued)*

Types of Packaging	Definition
Intelligent packaging	Increase the functionality of the package by simply changing the structure of the package, without the addition of any technology (Teixeira, 2010).
	Monitors and provides the consumer with information about the quality of the packed food (Fabech et al., 2000).
	Gives information on product quality directly (freshness indicators), the package and its headspace gases (leakage/gas indicators) and the storage conditions of the package (time-temperature indicators).
	Monitor the condition of packaged food to give information about the quality of the packaged food during transport and storage (Ahvenainen, 2003).
	Provides information on the product and its origin as such and identify, for example, rough physical handling of the package and protect the product from tampering and pilferage. As far as the safety and quality of perishable food products is concerned, microbiological quality must play a remarkable role (Kerry, 2006; Kerry, 2012)..
Smart packaging	Packaging that is made much more functional and useful; it involves the use of technology that adds features such that packaging becomes an irreplaceable part of the whole product. It performs additional functions, responds to stimuli generated by the environment or from the product being packaged, and reflects the change in a manner that makes the product more convenient and useful for the consumer or firms in the supply chain. It relies on the use of chemical, electrical, electronic, or mechanical technology, or any combination of them (Teixeira, 2010).
	A packaging which has an inherent ability to gather information on its operating environment or history, to process that information on its operating environment or history, to process that information in order to draw intelligent inferences from it and to act on those inferences by changing its characteristics in an advantageous manner (Fabech et al., 2000).

6.2   GLOBAL MARKET OF ACTIVE, SMART, AND INTELLIGENT PACKAGING

The increasing demand for fresh and quality packaged food, consumer convenience and manufacturers' concern for longer shelf life of the food products are driving the market for global active and smart packaging technology for food and beverage market. The global market for active and smart packaging technology for food and beverage is expected to grow from $15.798 million in 2010 to $23,474 million in 2015, at a CAGR of 8.2 percent from 2010 to 2015. Modified atmosphere packaging commands the largest share of the overall active and smart packaging technology in terms of value, while smart and intelligent packaging technology is witnessing the fastest growth at a CAGR of 12.1 percent. In 2010, advanced packaging accounted for just 5 percent of the overall packaging market. The advanced packaging market is dominated by MAP (Modified Atmosphere Packaging), which accounted for over 50 percent of the total advanced packaging market in 2010 (Figure 6.1). Active and smart packaging technology offers tremendous potential to fulfill the growing demand of food safety in various applications which include dairy products, meat, and poultry, ready-to-eat meal segment. In active packaging, oxygen scavengers and moisture absorbers form the two largest product segments. Both are estimated to grow at a CAGR of 8 percent and 11.9 percent respectively. North America is the major market for active and smart packaging technology due to increasing health awareness among the consumers. Therefore, it holds the largest share (35.1%) of the global active and smart packaging market. Europe forms the second largest market for active and smart packaging technology due to increased demand for sustainable packaging and stringent regulations. Currently, the market players are focusing on development of new products. Due to this reason, new product development accounted for the highest share of the total competitive developments in the global advanced packaging technology market for food and beverage from June 2008 to September 2010. Maximum developments are seen in oxygen scavenger product segment (Source: Anonymous, Markets and Markets, 2011).

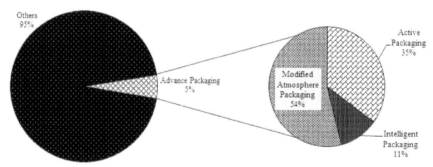

FIGURE 6.1   Global Active, Smart, and Intelligent Packaging Market (Source: Anonymous, Markets and Markets, 2011).

Over the past decade, active and intelligent packaging have experienced significant growth and change as new products and technologies have challenged the status quo of the traditional forms of food and beverage packaging (Kotler and Keller, 2006). Firstly introduced in the market of Japan in the mid-1970s, active and intelligent packaging materials and articles, only in the mid-1990s raised the attention of the industry in Europe and in the USA. The global market for food and beverages of active and intelligent coupled with controlled/modified atmosphere packaging (CAP/MAP) increased from $15.5 billion in 2005 to $16.9 billion by the end of 2008 and it should reach $23.6 billion by 2013 with a compound annual growth rate of 6.9 percent. The global market is broken down into different technology applications of active, controlled, and intelligent packaging; of these, CAP/MAP has the largest share of the market estimated to comprise 45.4 percent in 2008, probably decreasing slightly to approximately 40.5 percent in 2013 in Figure 6.2 (Restuccia et al., 2010).

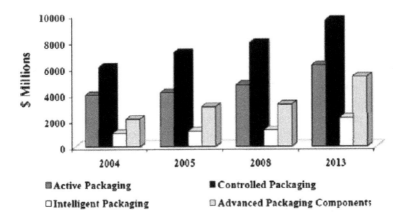

FIGURE 6.2 Growth of active, controlled, and intelligent packaging for the food and beverage industry 2004-2014 ($ million) (Restuccia et al. 2010).

6.3 ACTIVE PACKAGING TECHNOLOGY

The development of a whole range of active packaging technologies and systems, some of which may have applications in both new and existing food products, is fairly new (Kerry et al., 2006). The active packaging technologies will be able to aid in the preservation and quality retention of commercially processed and packaged (Brody et al., 2001). Active packaging has been variously defined in the literature and it packaging can be classified as following (Day, 2008; Brody et al., 2001):-

   i.   Oxygen (O_2) scavengers
   ii. Carbon dioxide (CO_2) scavengers/emitters
   iii. Ethylene scavengers
   iv. Preservative releaser

   v.   Ethanol absorber/emitter
   vi. Moisture absorbers
   vii. Flavor/odor absorbers and releasers
   viii. Temperature control
   ix. Antimicrobial packaging

6.4   SMART AND INTELLIGENT PACKAGING TECHNOLOGY

Intelligent packaging (also more loosely described as smart packaging) is packaging that in some way senses some properties of the food it encloses or the environment in which it is kept and which is able to inform the manufacturer, retailer, and consumer of the state of these properties. Although distinctly different from the concept of active packaging, features of intelligent packaging can be used to check the effectiveness and integrity of active packaging systems (Hutton, 2003). Smart and intelligent packaging systems can monitor and provide information about the quality of packed food. These emerging systems, attached as labels or incorporated into food packaging material, have been seen as potential breakthrough technologies offering enhanced possibilities to monitor the product quality trace the critical points and provide more complex information throughout the supply chain. A more effective package/product quality control system can eventually result in more efficient production and higher quality. Fewer complaints and returns from retailers and consumers mean cost-savings and better brand and image for the manufacturer and also more product information can be delivered to consumers. Major influencing technologies in smart and intelligent packaging include (Kerry et al., 2006):-

   i.   Radio Frequency Identification (RFID)
   ii. Time-temperature Indicators
   iii. Gas Indicators
   iv. Visual Oxygen Indicators
   v.   Invisible Oxygen Indicators
   vi. Freshness Indicators
   vii. Moisture Indicators
   viii. Physical shock indicator

6.5   ACTIVE AND SMART PACKAGING USING A BIO-SWITCH CONCEPT

The term "active packaging" changes the condition of the packed food to extend shelf life or to improve safety or sensory properties, while a "smart" packaging coordination monitors the state of packaged foods to give information about the quality of the packaged food during transfer and storage (Han, 2003; Gontard, 2005). Without tasting directly, however, there have been few methods for the consumer to assess the readiness of individually packaged fermented food or to ensure if a food product has been spoiled by contaminating bacteria. Hence a combined technology is developed to perceive the degree of continuous natural fermentation in food products or the spoilage of nonfermented food by microorganisms. Based on pH and titratable acidity of the packaged food, a color indicator has been developed and evaluated on its applicability to food packaging. While the bioactive substances protect the food from external microbial contamination, the color changes of the developed indicators represented properly the quality of the internal packaged food and sustainability of the bioactive substances. With a proper understanding of the quantitative correlation, commercially practical use of the indicator for nonfermented and fermented food packaging could be accomplished. Active packaging can be prepared based on two main concepts that is active releasing (e.g., antimicrobial film, antioxidant film) and active scavenging (e.g., oxygen scavenging, ethylene scavenging) (Han, 2003). The development of a whole range of active packaging technologies and systems, some of which may have applications in both new and existing food products, is fairly new (Kerry et al., 2006). The active packaging technologies will be able to aid in the preservation and quality retention of commercially processed and packaged (Brody et al., 2001).

The invented system consists of an integral part of the package (bioactive compound) and another form of component placed inside the package (smart bio-switch indicator). Bioactive substances will control the internal conditions of the package while the color indicator gives the information regarding the chemical and microbiological quality of the packed food. The objective of this invention is to provide further shelf life extension and to maintain the quality, safety, and integrity of packaged food, with an application of an active component based on modified biopolymer using a newly evolved bio-switch concept. Antimicrobial film is a form of active packaging that applies active releasing concept which consists of antimicrobial and antioxidant film. The general concept of bio-switch (Figure 6.3) describes a system that is able to detect and automatically give responses if there is any change or external stimulus in the related environment. The bio-switch will convert the stimulus into a particular functionality (Boumans, 2003.

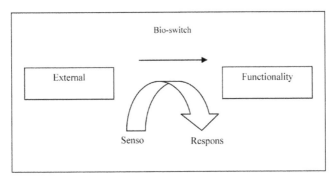

FIGURE 6.3 General concept of bio-switch.

The general concept of bio-switch describes a system that is able to detect and respond automatically to changes (external stimuli) in the environment. For instance, the external stimulus may be a change in pH, or the presence of certain metabolites from biological activity. The bio-switch converts this stimulus into a particular functionality. Materials are created that are able to entrap compounds with a specific function which are released on an external stimulus from the environment. For use in active packaging, biopolymer-based particles are prepared that contain an antimicrobial that will only be released in the case of initial microbial contamination. The bio-switch particles monitor the releasing system by the stimulus of a microbial contamination that actively add or emit compounds (i.e., antimicrobials, antioxidants, and preservatives to the packaged food or onto the surface of the package). The invention is a combination of active packaging technology and smart packaging concept where the stimulus of a microbial contamination is further incorporated with an indicator to signal the conditions in the packaged food (Boumans, 2003).

6.5.1 ANTIMICROBIAL PACKAGING AND MECHANISM OF ACTION

In recent years, the demand for minimally processed, easily prepared and ready-to-eat "fresh" food products, globalization of food trade, and distribution from centralized processing pose major challenges for food safety and quality (Vermeiran et al., 1999; Suppakul et al., 2003). Currently, food-borne microbial outbreaks are driving a search for innovative ways to inhibit microbial growth in the foods while maintaining quality, freshness, and safety. Active packaging interacts with the product or the headspace between the package and the food system, to obtain a desired outcome. Likewise, antimicrobial (AM) food packaging acts to reduce, inhibit, or retard the growth of microorganisms that may be present in the packed food or packaging materials itself (Appendini and Hotchkiss, 2002). Food packaging materials traditionally used to provide only barrier and protective functions. However, various kinds of active substance can now be incorporated into the packaging material to improve its functionality and give it new or extra functions such as the incorporation AM agents into a polymer which can limit or prevents the microbial growth. Such active packaging technologies are designed to extend the shelf life of foods, while maintaining their nutritional quality and safety. This application could be used for foods effectively not only in the form of films but also as suitable containers and utensil. AM food packaging is one of the special applications of active food packaging that controls inside food and atmospheric conditions actively and responsively. Active packaging technologies are designed to extend the shelf life, while maintaining the nutritional quality and safety of food which involve interactions between the food, the packaging material and the internal atmosphere. Figure 6.4 represents the mechanism of how the AM Active Packaging inhibit the microbial growth. Microorganism will first try to hydrolyze the starch-based particles causing the release of the AM compound which finally resulting inhibition of the microbial growth (Boumans, 2003).

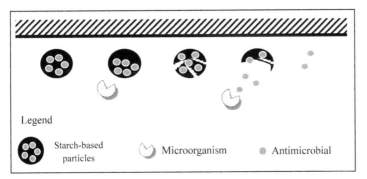

FIGURE 6.4   The antimicrobial active packaging action applying bio-switch concept.

6.5.2   ANTIMICROBIAL STARCH-BASED FILM/FOOD PACKAGING

Recently, the demands for Malaysian traditional food product pose their major problem which, their less marketing over worldwide and this is due to the shelf life of the food itself (e.g., keropok lekor, tempoyak, etc.). The introduction of antimicrobial agents into food packaging material could help to prolong the shelf life of food products by inhibiting the growth of microorganism. AM polymers can be used in several food related applications including packaging (Hotchkiss, 1997). It can extend the shelf life and promote safety by reducing the rate of growth of specific microorganisms from direct contact of the package with the surface of solid foods (e.g., meats, cheese, etc.) or in the bulk of liquids (e.g., milk or meat exudates).

Besides, AM packaging materials could also be self-sterilizing. If the packaging materials have self-sterilizing ability because of their own antimicrobial activity, they may eliminate chemical sterilization of packages using peroxide and simplify the aseptic packaging process (Hotchkiss, 1997). The self-sterilizing materials could be widely applied for clinical uses in hospitals, biological lab ware, biotechnology equipment and biomedical devices, as well as food packaging. Such AM packaging materials greatly reduce the potential for decontamination of processed products and simplify the treatment of materials in order to eliminate product contamination. AM polymers might also be used to cover surfaces of food processing equipment so that they self-sanitize during use (e.g., filter gaskets, conveyers, gloves, garments, and other personal hygiene equipment) (Appendini and Hotchkiss, 2002). The effectiveness of AM film to inhibit the microbial growth had been reported by many researchers. The application of AM agent into food packaging could be effectively used not only in the form of film but also as containers and utensil (Han, 2000).

6.5.3 MICROBIOLOGICAL STUDY OF ANTIMICROBIAL STARCH-BASED FILM

The present subtopic discusses the effectiveness of AM starch-based film in various tests. The present test will focus on the efficacy of film to inhibit the growth of selected Gram positive food pathogenic bacteria (i.e., *Bacillus subtilis*) and Gram negative food pathogenic bacteria (i.e. *Escherichia coli*). Agar plate test also known as zone inhibition assay was performed as a preliminary step to screen the antibacterial activity of all films formulations, in an effort to select film formulations with high antibacterial activity against test bacteria. The inhibitory zone in agar diffusion test can be affected by the solubility and diffusion rate of the test compounds in agar medium, thus agar diffusion test does not accurately reflect the antimicrobial effectiveness of the test compounds (Kim et al., 2003; Han, 2003). Therefore, the liquid culture test had been done to support the result of agar plate test also known as agar diffusion test. The liquid culture test determines the antimicrobial activity of the test compounds by viable count and provides information on microbial growth kinetics, thus being more sensitive than the agar diffusion method (Han, 2003; Mann and Markham 1998).

6.5.4 INHIBITION OF ESCHERICHIA COLI AND BACILLUS SUBTILIS ON AGAR PLATE TEST

Antibacterial activity of AM film against two pathogenic bacteria was expressed in terms of zone inhibition. The agar diffusion test simulates wrapping of foods, and therefore can be used to estimate how much the antimicrobial agent migrates from the film to the food when the film contacts contaminated surfaces (Appendini and Hotchkiss, 2002; Jaejoon, 2006). All samples were examined for possible inhibition zones after incubation at 37°C for 48 hrs. Table 6.2 lists calculated inhibition area for each plate test. The control films showed no inhibition area and colonies were formed all over the plate. A clear zone formed on the plate indicates the inhibitory effect as shown in Figure 6.5.

TABLE 6.2   Analysis of the zone of inhibition data in agar plate test for *E.coli* and *B. subtilis* at 37°C in the presence of HEC-wheat starch-based film incorporated with thymol

	AMI zone of inhibition				
	0.5 %	1.0 %	1.5 %	2.0 %	2.5 %
Escherichia coli	0.772 ± 0.01	0.769 ± 0.01	0.763 ± 0.03	0.719 ± 0.02	0.772 ± 0.02
Bacillus subtilis	0.771 ± 0.03	0.762 ± 0.02	0.801 ± 0.01	0.751 ± 0.01	0.740 ± 0.03

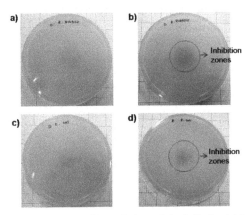

FIGURE 6.5   Inhibition of (a,b) *Bacillus subtilis* and (c,d) *Escherichia coli* on solid media by HEC-wheat starch-based film incorporated with thymol after incubation for 24 hours at 37°C with (a,c) no AM agent and (b,d) with AM agent.

Active packaging technologies involve interactions between the food, the packaging material and the internal gaseous atmosphere (Labuza and Breene, 1988). Food packaging materials used to provide only barrier and protective functions. However, various kinds of active substances can now be incorporated into the packaging material to improve its functionality and give it new or extra functions. Such active packaging technologies are designed to extend the shelf life of foods, while maintaining their nutritional quality and safety (Han, 2000). The most promising active packaging systems are oxygen scavenging system (Rooney, 1981) and antimicrobial system (Rooney, 1996). AM packaging materials have to extend the lag period and reduce the growth rate of microorganisms to prolong the shelf life and maintain food safety. They have to reduce microbial growth of nonsterile foods or maintain the stability of pasteurized foods without post-contamination. The AM packaging developed in the form of coating managed to prolong shelf life of selected foods which comparison as shown in Figure 6.6. In Figure 6.6a, AM coated tomatoes on the right managed to sustain their shape and color. Similarly in Figure 6.6b, tofu coated with AM film edible solution (on the left) also shows preserved properties in comparison to noncoated sample.

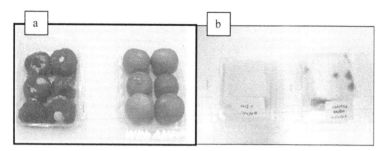

FIGURE 6.6   Shelf life extension test on (**a**) tomatoes after storing at room temperature (28°C) for 42 days and (**b**) tofu after storing at 8°C in refrigerator for 30 days

6.5.5   APPLICATION OF AM STARCH-BASED FILM TO EXTEND SHELF LIFE OF TOMATO

The food that not been consumed immediately after production such as vegetables need to be contained in package to retain its shelf life and quality. The roles of food packaging are to protect the food from dust, light, oxygen, pathogenic microorganism, moisture, and other harmful substance from contaminating it hence extent the food shelf life. In response to the dynamic changes in current consumer demand and market trends, the concept of using edible film and coatings to extend food shelf has been emphasized. AM starch-based packaging and coating is edible film that provide additional protective coating for fresh products and can also give the same effect as modified atmosphere storage in modifying internal gas composition (Park et al., 1994). It also demonstrates ability to improving and prolonging shelf life of food product (Castillo and Serrano, 2005). Recently, antimicrobial film is having an impact on several aspects of the food industry comprises food safety, and the health benefits food delivers. The application of smart and active bio-packaging is not new. Numerous plastic films from different sources natural and synthetic are being developed that will allow the food to stay fresh longer. These films incorporated with additive to allow it to release and work as pathogen killer.

6.5.5.1   WEIGHT LOSS OF TOMATO

Tomato is a climacteric fruit and continues to ripen after harvest (Athmaselvi et al., 2013). During ripening, the green pigment chlorophyll degrades and carotenoids are synthesized (Liu et al., 2009). As ripening continues, tomatoes can become overripe and lose its nutrient very rapidly due to bacteria or fungi spoilage. There is also increasing consumer concern about the eating quality of tomatoes. Hence, antimicrobial (AM) edible films and coatings are used for improving the shelf life of food products without impairing consumer acceptability. It designed to prevent surface contamination while providing a gradual release of the active substance (Buonocore et al., 2003). Food storage and shelf life were studied between tomato that uncoated and coated with starch-based film coating. Quality parameters of the foods were determined such as weight loss, color, and mold observation in 28 days of period. During the storage period, moisture and the quality of the food will reduce. It is necessary to keep the moisture of the food at certain level in order to maintain its quality before reach to consumer's hand. In order to do so, the food must be packed. Figure 6.7 showed the effect of packaging types on tomato weight loss in room condition. In overall, it showed tomato wrapped with starch/chitosan film gives lower weight loss compared to control starch film.

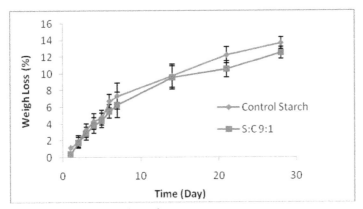

FIGURE 6.7   Percentage weight loss of tomato.

On the very last day of the experiment, the starch/chitosan composite film showed lower weight loss which at 11.76 percent compared to control starch film 11.83. It also demonstrates that weight loss of tomato wrapped with starch/chitosan film was reduced at a slow pace means that the films can retain the food's moisture better than control starch film. Addition of organic or inorganic fillers to the starch matrix as well as the addition of functional compounds to form composite film is one of the approaches for reinforcement of barrier and mechanical properties (Jimenez, 2012). Physicochemical properties of the starch films showed a great variability depending on the compounds added to the matrix and the processing method. Previous study shows chitosan-based films have good mechanical properties and selective gas permeability (Campos et al., 2010). Chitosan can form semipermeable coatings, which can modify the internal atmosphere, thereby delaying ripening and decreasing transpiration rates in fruits and vegetables (Bourtoom, 2008). It also showed that films from chitosan were rather stable and their mechanical and barrier properties changed only slightly during storage (Butler et al., 1996).

6.5.5.2   COLOUR CHANGE OF TOMATO

The ripening process of tomato is due to biochemical changes subsequently lead-ing to remarkable changes in color (Pinheiro et al., 2013). After the ripening pro-cess end, the tomato will start to decay, the surface of the tomato become injured and darkened its color. Different packaging systems control starch film and starch/chitosan composite film was applied on tomato and its color change was observed for 4 weeks. Figure 6.8 showed changes of the tomato color (a*) with storage time for control starch and starch/chitosan composite. At final day, control starch film has higher red color intensity than starch/chitosan composite film. Colour transi-tion in starch film is also higher due to the tomato that decays more rapidly. It was because there is no antimicrobial agent presence in the control starch film to inhibit microbes from spoiling the tomato. However, in starch/chitosan composite (S:C, 9:1 film), the a* color coordinate was changed from 5.2 on day 0 to 7.2 in day 6 then the value increased slowly until the end of the experiment. It showed that the color intensity of the tomato wrapped with and starch/chitosan composite (S:C, 9:1) film was less than control starch film. Lightness, L* of tomato that wrapped with control starch and starch/chitosan (S: C 9:1) was also observed. It showed that the tomato that wrapped with control starch film was darkened more rapidly compared to other films indicates rapid ripening and spoilage of the fruit. It caused by the presence of chitosan that inhibit the bacterial activities thus slow down the ripening subse-quently reduce rotting process of the tomato. Several studies analyzed the effect of chitosan on starch-based films (Bourtoom and Chinnan 2008; Shen et al., 2010; Zhong et al., 2011) and underlined the known antimicrobial property of chitosan. The findings also shown that addition of chitosan into starch film not only increase its mechanical resistance, but also decrease the oxygen permeability, water vapor permeability, and water solubility. Hence, slow ripening process in starch-based film incorporated with chitosan result may also due to lower water loss and a more controlled rate of respiration (Sanchez-Garcia et al., 2010).

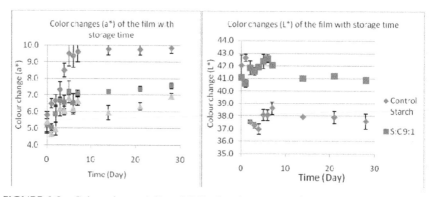

FIGURE 6.8   Colour change (a*) and (L*) of package tomato in 4 weeks duration.

6.5.5.3   VISUAL OBSERVATION OF MICROFLORA APPEARANCE

Storage study was conducted in 4 weeks to measure the effectiveness of the film against the food spoilage. Tomato spoilage for an example is due to decaying process that caused by microbes ie; mold. Figure 6.9 showed visual observation of tomato wrapped with control starch and starch/chitosan (S: C, 9:1) film for 4 weeks duration. In first 2 weeks, there are no obvious changes in all three groups of wrapped tomatoes. In the tomato packed with control starch film, observable color changed was detected in tomato wrapped with control starch as the color have changed from orange-red to red-orange color where starch/chitosan composite relatively shows no difference. Tomato surface was deformed at the third weeks for starch/chitosan (S: C 9:1) but more severely in control film. At fourth weeks (28th days), the tomatoes started to molds in both package but more severely in starch-based package which shows the tomato deformed and burst. Shelf life of fresh produced goods is greatly related to transpiration rate, a physiological process that subsequently influences the relative humidity and also condensation inside the food package (Gallagher et al., 2013). The right gas and humidity combination can slow respiratory metabolism, and delay compositional changes in color, flavor, and texture. It can also inhibit or delay microbial growth where it could be achieve because chitosan have selective gas permeability.

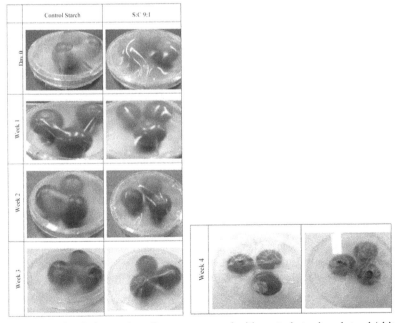

FIGURE 6.9   Visual observation of tomato wrapped with control starch and starch/chitosan composite film.

6.6   KEY ISSUES IN FOOD SAFETY INDICATORS

A current trend in the food industry is the manufacture of mildly preserved, healthy, and easy-to-prepare products driven by consumer demands for fresh 'natural' convenience food. It is crucial to maintaining food safety and quality while reducing costs and centralizing activities especially in longer distribution distances. These factors put high demands on the shelf life extending capacity of food packaging systems. In extending shelf life and maintaining safety toward, for example bacteria and fungi, traditional packaging systems are reaching their limits. New packaging technologies have been continually developed to prolong and indicate the quality of fresh, refrigerated, and processed foods. Especially, packaging may be termed active when it performs some roles in the preservation of foods other than providing an inert barrier to external conditions. Thus the role of an active smart packaging could be defined as "doing more than just protection, interacting with the products, and actually responding to changes".

Food condition includes various aspects that may play a role in determining the shelf life of packaged foods, including microbiological aspects such as degree of fermentation or spoilage by microorganisms. During fermentation, microorganisms such as bacteria initially grew under micro aerobic conditions to produce organic acids, CO_2, and ethanol as the by-products. Consequently, the changes in CO_2 concentration had a good correlation with both pH and titratable acidity of the fermented food (Ouattara et al., 2000). Based on these results, the color indicator has been developed and evaluated on its applicability to food packaging. Figure 6.10 illustrates the mechanism of AM with color indicator as active smart packaging. From Figure 6.10, it shows that the developed system consists of two main mechanisms. The first mechanism is the bioactive substances could help to protect the food from external microbial contamination. The second mechanism is the color changes of the developed indicator represented properly the quality of the internal packaged food and sustainability of the bioactive substances. Therefore, the shelf life of the food will be extended and the quality can be monitored when the bio-switch technology is incorporated to the plastic.

"A smart packaging for monitoring safety using a bio-switch concept" is a kind of food safety monitoring kit. It is developed in plastic film form with pH-color indicator resulted from the combination technology of modified polymer (i.e., starch-LDPE), substances with antimicrobial characteristics (i.e., Lauric Acid extracted from palm oil and potassium chloride), and encapsulated colorant substances extracted from selective plants. The suggested applications are as shown in Figure 6.11. This smart packaging is developed in thermoform and blow molded films can be used widely especially in perishable foods like vegetables and fruits, Malaysian traditional foods such as keropok lekor, tempoyak, or cencaluk as a monitoring film to indicate the ripeness of the food, or to show the most suitable time for consumption especially for exported fermented produce. It can be used to monitor the spoilage of perishable produce such as fish and meat from deterioration. It can also be used in dipping solution form which is edible, convenient, and cost-effective.

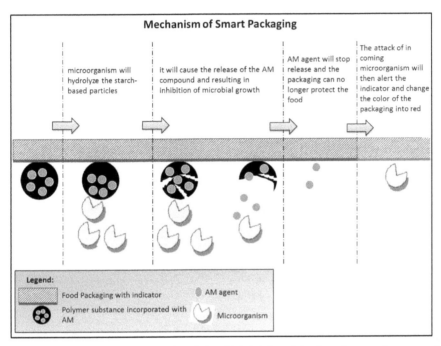

FIGURE 6.10 The mechanism of antimicrobial (AM) activity in polymer substances of an active and smart packaging incorporated with a newly developed color indicator for pH change resulted from the activity of microorganism that contaminating the food.

It could be developed for two purposes: to detect the pH-based degree of continuous natural fermentation in food products and also the spoilage of nonfermented food by microorganisms. The color changes of the developed indicators represented properly the quality of the internal packaged food and sustainability of the bioactive substances while the modified polymer accommodated with antimicrobial characteristics helps to protect the food from external microbial contamination. Hence, the role of a smart packaging could be defined as "doing more than just protection, interacting with the products, and actually responding to changes".

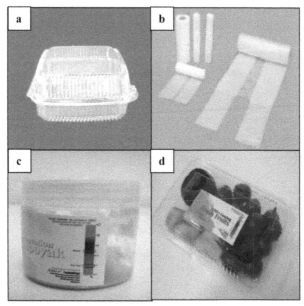

FIGURE 6.11 The smart packaging using bio-switch concept in (**a**) thermoform container and indicator film; and (**b**) blow molded bags and wrappers; (**c**) On-cap label film; (**d**) As container layer or use for dipping or coating.

Active and smart packaging methods have been studied widely and innovations have been developed. However, very few of them have been developed into commercially available products and yet there is no implementation of this technology in Malaysia. But this technology could be implemented widely in order to extend shelf life, increase the safety assurance for better preservation and stability of food products. Since Malaysia has a wide food variety, food producers and manufacturers highly using plastic as packaging tools, hence smart packaging is a suitable technology to be implemented and has big potential in commercialized use for food and beverages, cosmeceuticals, nutraceuticals, and pharmaceutical products.

6.7 CONCLUSIONS

The objective of this invention is to monitor the quality, safety, and integrity of packaged food, with an application of an active natural component based on modified biopolymer while providing further shelf life extension using a bio-switch concept. The purpose to develop and characterize an AM starch-based film of this research is successfully accomplished using wheat as starch-based and lysozyme as the antimicrobial agent. Antimicrobial packaging materials are required to extend the lag period and reduce the growth rate of microorganisms to prolong shelf life and maintain food safety. Food manufacturers may be able to maintain the minimum inhibitory concentration of an antimicrobial to prevent growth of pathogenic and spoilage microorganisms by using controlled release packaging. Lysozyme as one of the AM in developed active packaging film can help prolong the shelf life of food product while maintaining the quality, safety, and integrity of packaged food. The adopted approach has a potential relevance on practical aspects of packaging because the modeling of the lysozyme release from food packaging material can be considered as a useful tool needed to simulate and consequently predict the behavior of the release device in the actual working situation. In conclusion, AM starch-based packaging is a promising form of active food packaging. Although most packaged perishable food products had been protected by various way such as heat sterilized or by having a self-protecting immune system, microbial contamination could occur on the surface or damaged area of the food through package defect or restorage after opening. Hence an indicator of such changes is useful to be incorporated into packaging materials to indicate, control inhibit the microbial contamination and inform the consumer of the safety and edibility status of the packaged food. In this chapter, the challenges in meeting some of these requirements have been discussed.

KEYWORDS

- **Postharvest quality**
- **Passive packaging**
- **Active packaging**
- **Intelligent packaging**
- **Scavengers/emitters**
- **Antimicrobial packaging**

REFERENCES

Ahvenainen, R.; Active and intelligent packaging: an introduction. In: *Novel Food Packaging Techniques;* Ahvenainen, R., Ed.; Woodhead Publishing Ltd; Cambridge, UK, **2003**; pp 5–21.

Anonymous; Global Active, Smart and Intelligent Packaging Market by Products, Applications, Trends and Forecasts (2010–2015). marketsandmarkets.com, 2011.

Appendini, P.; and Hotchkiss, J. H.; Immobilization of lysozyme on synthetic polymers for the application to food packages. Presented at Annual Meeting of Institute of Food Technologists, New Orleans, La., June 22–26, **1996**.

Appendini, P.; and Hotchkiss, J. H.; Immobilization of lysozyme on food contact polymers as potential antimicrobial films. *Packag. Technol. Sci.* **1997**, *10*, 271–279.

Appendini, P.; and Hotchkiss, J. H.; Review of antimicrobial packaging food packaging. *J. Innov. Food. Sci. Emerg. Technol.* **2002**, *3*, 113–126.

Appendini, P.; and Hotchkiss, J. H.; Surface Modification of poly(styrene) by the Attachment of an Antimicrobial Peptide. *J. Appl. Polym. Sci.* **2001**, *81*, 609–616.

Athmaselvi, K. A.; Sumitha, P.; and Revathy, B.; Development of Aloe vera based edible coating for tomato. *Int. Agrophys.* **2013**, *27*, 369–375.

Boumans, H.; Release-on-Command: Bioswitch. *Lead. Life. Sci.* **2003**, *22*, 4–5.

Bourtoom, T.; Edible films and coatings: characteristics and properties. *Int. Food. Res. J.* **2008**, *15*, 1–12.

Bourtoom, T. and Chinnan, M. S.; Preparation and properties of rice starch-chitosan blend biodegradable film. LWT- *Food. Sci. Technol.* **2008**, *41*(9), 1633–1641.

Brody, A. L.; Strupinsky, E. R.; and Kline, L. R.; *Active Packaging for Food Applications*. CRC Press: US America, 2001.

Buonocore, G. G.; Del Nobile, M.; Panizza, A.; Battaglia, G.; and Nicolais, L.; Modeling the lysozyme release kinetics from antimicrobial films intended for food packaging applications. *J. Food. Sci.* **2003**, *68*(4), 1365–1370.

Buonocore, G. G.; Conte, A.; Corbo, M. R.; Sinigaglia, M.; and Del Nobile, M. A. Mono-and Multilayer Active Films Containing Lysozyme as Antimicrobial Agent. *Innov. Food Sci. Emerg. Technol.* **2005**, *6*, 459–464.

Butler, B. L.; Vergano, P. J.; Testin, R. F.; Bunn, J. M.; and Wiles, J. L.; Mechanical and barrier properties of edible chitosan film as affected by composition and storage. *J. Food Sci.* **1996**, *61*(5), 953–961.

Campos, C. A., Gerschenson, L. N. and Flores, S. K.; Development of edible films and coatings with antimicrobial activity. *Food Bioprocess Technol.* **2010**, *4*, 849–875.

Chung, D.; Papadakis, S. E.; and Yam, K. L.; Evaluation of a polymer coating containing triclosan as the antimicrobial layer for packaging materials. *Int. J. Food Sci. Technol.* **2003**, *38*, 165–169.

Dawson, P. L.; Acton, J. C.; Padgett, T. M.; Orr, R. V.; and Larsen, T.; Incorporation of antimicrobial compounds into edible and biodegradable packaging film. *Res. Dev. Activities Rep. Military Food Packag. Syst.* **1996**, *42*, 203–210.

Day, B. P. F.; Chapter 1 – Active packaging of food. In: *Smart Packaging Technologies;* Kerry, J., Butler, P., Eds.; Wiley: San Francisco, USA, **2008**.

Fabech, B.; Hellstrom, T.; Henrysdotter, G.; Hjulmand-Lassen, M.; Nilsson, J.; Rudinger, L.; Sipilainen-Malm, T.; Solli, E.; Svensson, K.; Thorkelsson, A. E.; and Tuomaala, V.; Chapter 1 – What is Active and Intelligent Packaging? In: *Active and Intelligent Food Packaging: A Nordic Report on the Legislative Aspects;* Nordic Council of Ministers: Copenhagen, **2000**; pp 21–29.

Gontard, N.; Active Packaging. In: *Proceedings of Workshop sobre Biopolimeros;* Sobral, P. J. do A., Chuzel, G., Eds.; Univ. de Sao Paulo/FZEA: Pirassununga, Brazil, **1997**; pp 23–27.

Han, J. H.; Antimicrobial food packaging. *Food Technol.* **2000**, *54*(3), 56–65.

Han, J. H.; Antimicrobial food packaging. In: *Novel Food Packaging Techniques;* Ahvenainen, R., Ed.; Woodhead Publishing Limited: England, **2003**; pp 50–95

Han, Jaejjoon. Antimicrobial Packaging System for Optimization of Electron Beam Irradiation of Fresh Produce. Ph.D. Thesis, Texas A&M University, **2006**.

Hook, P.; and Heimlich, J. E.; A History of Packaging. Ohio State University Fact Sheet. http://ohioline.osu.edu/cd-fact/0133.html.

Hotchkiss, J. H. Food-packaging interactions influencing quality and safety. *Food Addit. Contam.* **1997**, *14*, 601–607.

Hutton, T.; Food packaging: An introduction. Key topics in food science and technology – Number 7.Campden and Chorleywood Food Research Association Group: Chipping Campden, Gloucestershire, UK, **2003**; p 108.

Jiménez, A.; Fabra, M. J.; Talens, P.; and Chiralt, A;. Edible and biodegradable starch films: A review. *Food Bioproc. Technol.* **2012**, *5*, 2058–2076.

Kandemir, N.; Yemenicioğlu, A.; Mecitoğlu, C.; Elmaci, Z. S.; Arslanoğlu, A.; Gŏksungur, Y.; and Baysal, T.; Antimicrobial films from *A. pullulans* exopolysaccharides. *Food Technol. Biotechnol.* **2005**, *43*(4), 343–350.

Kerry, J. P.; Chapter 20 – Application of Smart Packaging Systems for Conventionally Packaged Muscle-Based Food Products in Part IV: Emerging Packaging Techniques and Labelling. In: *Advances in Meat, Poultry and Seafood Packaging;* Kerry, J. P., Eds.; Woodhead Publishing Limited: UK, **2012**; p 525.

Kerry J. P.; O'Grady, M. N.; and Hogan, S. A.; Past, current and potential utilisation of active and intelligent packaging systems for meat and muscle-based products: A review. *Meat. Sci.* **2006**, *74*, 113–130.

Khairuddin, N.; and Muhamad, I. I.; Preliminary Study of Antimicrobial (AM) of Starch-Based Film Incorporated with Nisin, Lysozymes and Lauric Acid. *Imposium Kimis Analisis Malaysia Ke-18,* Johor Bahru, Malaysia, Sept 12–14, **2005**.

Khairuddin, N.; Radzi, A. R. M.; and Muhamad, I. I.; Study of an Antimicrobial Starch-Based Active Packaging Systems. *Kustem 4th Annual Seminar on Sustainability Science and Management 2005,* KUSTEM: Kuala Terengganu, Terengganu, **2005**; pp 44–47.

Kim, D. H.; Na, S. K.; and Park, J. S.; Preparation and characterization of modified starch-based plastic film reinforced with short pulp Fiber. II. Mechanical Properties. *J. Appl. Polym. Sci.* **2003**, *88*, 2108–2117.

Kotler, P.; and Keller, K.; *Marketing Management,* 12th ed.; Pearson: Upper Saddle River, NJ, **2006**.

Labuza, T. P.; and Breene, W. M.; Applications of "active packaging" for improvement of shelf-life and nutritional quality of fresh and extended shelf-life foods. *J. Food Proc. Preserv.* **1988**, *13*, 1–69.

Liu, L. H.; Zabaras D.; Bennett, L. E.; Agues, P.; and Woonton, B. W.; Effects of UV-C, red light and sun light on the carotenoid content and physical qualities of tomatoes during post-harvest storage. *Food Chem.* **2009**, *115*, 495–500.

Mann, C. M.; and Markham, J. L.; A new method for determining the minimum inhibitory concentration of essential oils. *J. Appl. Microbiol.* **1998**, *84*, 538–544.

Ouattara, B.; Simard, R.; Piette, G.; Begin, A.; and Holley, R.; Diffusion of acetic and propionic acids from chitosan-based antimicrobial packaging films. *J. Food Sci.* **2000**, *65*(5), 768–772.

Pinheiro, A. C.; Bourbon, A. I.; Quintas, M. A. C.; Coimbra, M. A.; and Vicente, A. A.; K-carrageenan/chitosan nanolayered coating for controlled release of a model bioactive compound. *Innov. Food. Sci. Emerg. Technol.* **2012**, *16*, 227–232.

Pira International; Migration of Bisphenol a from Polycarbonate Plastic Food Contact Materials and Articles, **2004**. Retrieved on 16 June 2008 from http://www.food.gov.uk/science/research/researchinfo/contaminantsresearch/contactmaterials/a03prog/a03projlist/a03036proj/.

Restuccia, D.; Spizzirri, U. G.; Parisi, O. I.; Cirillo, G.; Curcio, M.; Iemma, F.; Puoci, F.; Vinci, G.; and Picci, N.; New EU regulation aspects and global market of active and intelligent packaging for food industry applications. *Food Contr.* **2010**, *21*, 1425–1435.

Risch, S. R.; Food packaging history and innovations. *J. Agric. Food Chem.* **2009**, *57*, 8089–8092.

Roberstson, G. L.; *Food Packaging Principle and Practice,* 2nd ed.; CRC Press: New York, **2006**.

Rooney, M. L.; Oxygen scavenging from air in package headspace by singlet oxygen reactions in polymer media. *J. Food Sci.* **1981**, *47*, 291–294, 298.

Rooney, M. L.; Personal communication. CSIRO: Australia. **1996**.

Sanchez-Garcia, M. D.; Lopez-Rubio, A.; and Lagaron, J. M.; Natural micro and nanobiocomposites with enhanced barrier properties and novel functionalities for food biopackaging applications. *Trend. Food Sci. Technol.* **2010**, *21*, 528–536.

Shen, X. L.; Wu, J. M.; Chen, Y.; and Zhao, G.; Antimicrobial and physical properties of sweet potato starch films incorporated with potassium sorbate or chitosan. *Food Hydrocoll.* **2010**, *24*, 285–290.

Sousa-Gallagher, M. J.; Mahajan, P. V.; and Mezdad, T.; Engineering packaging design accounting for transpiration rate: Model development and validation with strawberries. *J. Food Eng.* **2013**, *119*(2), 370–376.

Suppakul, P.; Miltz, J.; Sonneveld, K.; and Bigger, S. W.; Antimicrobial properties of basil and its possible application in food packaging. *J. Agric. Food Chem.* **2003**, *51*, 3197–3207.

Teixeira, V.; Opportunities and challenges in nanotechnology-based food packaging industry. *International Conference on Food and Agricultural Applications of Nanotechnologies*, Hotel Colina Verde: Sao pedro, SP Brasil, **2010**.

Vartiainen J.; Skytta, E.; Enqvist, J.; and Ahveinainen-Rantala, R.; Antimicrobial and barrier properties of LDPE films containing Imazalil and EDTA. *J. Plast. Film. Sheet.* **2003a**, *19*, 249–262.

Vartiainen J.; Skytta, E.; Enqvist, J.; and Ahveinainen-Rantala R.; Properties of antimicrobial plastics containing traditional food preservatives. *Packag. Technol. Sci.* **2003b**, *16*, 223–229.

Zhong, Y.; Song, X.; and Li, Y.; Antimicrobial, physical and mechanical properties of kudzu starch–chitosan composite films as a function of acid solvent types. *Carbohydr. Polym.* **2011**, *84*, 335–342.

CHAPTER 7

ADVANCES IN POSTHARVEST DISEASES MANAGEMENT IN FRUITS

EVA ARREBOLA[1]

[1]Department of Mycology, Instituto de Hortofruticultura Subtropical y Mediterranea IHSM-UMA-CSIC La Mayora

CONTENTS

7.1   INTRODUCTION

Fruits are the most consumable fresh product in the world, investigations about its qualities have shown that fruits are generally high in fiber, water, vitamin C, and sugars, although this latter varies widely from traces as in lime, to 61 percent of the fresh weight of the date (Hulme, 1970). Fruits also contain various phytochemicals that do not yet have an RDA/RDI (Recommended Dietary Allowance/Reference Daily Intake respectively) listing under most nutritional factsheets and disease prevention. Regular consumption of fruit is associated with reduced risks of cancer, cardiovascular diseases (especially coronary heart disease), stroke, Alzheimer disease, cataracts, and some of the functional declines associated with aging (Liu, 2003). Known the benefits of fruit intake, is almost mandatory its consumption, and therefore are fresh products most demanded by the population. There are three important aspects that conforms the quality of fresh product: (1) sensorial quality, that includes aroma, firmness, color; (2) safety, including pathogens and deteriorative microorganisms; and (3) nutritional value, that includes content and bioavailability of bioactive compounds. These quality properties cannot be improved after harvest, only maintained.

Focusing in the second and consequently third point we could distinguish two aspects of fruit safety, one as food safety from clinic point of view, its means human pathogen free, and another one is as fruit decay prevention to maintain the food physical and chemical characteristics. The current chapter is dedicated to overview the protocols used to forestall the fruit decay.

The battle against postharvest decay of fruits has been fought for decades but has not been won. Even the average consumer, who shops for quality fresh fruits must often discard spoiled product, recognizes the persistent problem of postharvest decay. Although the development of modern fungicides and improved storage technologies have greatly extended the shelf life of fruits after harvest, postharvest losses vary from an estimated 5 percent to more than 20 percent depending on the commodity, and could be as high as 50 percent in developing countries (Janisiewicz and Korsten 2002). Some of postharvest diseases more important that product big losses are: brown rot, which is the main postharvest disease of stone fruits and is caused by *Monilinia laxa* (Aderh and Rulh) and *Monilinia. fructicola* (Wint.) Honey (De Cal et al. 2009). Postharvest losses are typically more severe than preharvest losses, sometimes reaching high values (59%) (Larena et al., 2005). Stone fruit infection by *Monilinia* spp. mainly occurs in the field at bloom, the onset of pit hardening and between 7 and 12 days before harvest and these infections remains latent until harvest (Emery et al. 2000). However, when infection levels are high, fruit rot can appear early during the growing season and cause significant yield losses (Luo and Michailides 2001). Therefore, postharvest treatment to control brown rot may provide curative effect for both established infections and potential conidia present on the surface of the fruit. *Penicillium italicum* and *Penicillium digitatum* are the most common postharvest pathogens of citrus fruits. *P. digitatum* works by producing ethylene to accelerate ripening. It covers the fruit with green conidia, causing the fruit to shrivel and dry out. *P. italicum* causes slimy rot and produces blue-green conidia. Both diseases resemble each other in color characteristics, style of decay and infection symptoms; they fall under a general category called green and blue mold, when caused by *P. digitatum* and *P. italicum* respectively. These fungi like cold temperatures, live long time and are quite durable, including even stale adverse conditions. Sometime, fruits infected by *P. italicum* will adhere to each other to create synnemata. *Penicilium* growth typically occurs as a result of wound infections in product. *Colletotrichum gloeosporioides* is the causal agent of the disease known as anthracnose and is the most important fungal pathogen affecting tropical flowers, vegetables, and fruits. There are more than 103 plant species recorded as host of *C. gloeosporioides* (Chung et al. 2010). It is the most important disease of the crop in all mango production areas in the world (Mora et al., 2002). It is the most severe with high humidity and abundant rainfall and the symptoms can be observed on leaves, flowers, fruits, and branches of all ages. Symptoms in fruits nearing maturity the infections are quiescent and cause irregular dark spots that quickly rot the pulp of the fruit when it reaches senescence (Holguín et al. 2009, Ojeda et al., 2012). *Botrytis cinerea* is a necrotrophic fungus that affects many plants species, although its most notable hosts may be wine grapes. In viticulture, it is commonly known as botrytis bunch rot, in horticulture, it is usually called graymold or gray mold. The fungus gives rise to two different kinds of infection on grapes. The first, gray rot, is the result of consistently wet or humid conditions and typically results in the loss of the affected bunches. The second, noble rot, occurs when drier conditions follow wetter and can result in distinctive sweet dessert wines (Choquer et al., 2007).

Postharvest losses have been reduced mainly through postharvest fungicides and to a lesser degree, through postharvest management practices to reduce inoculum or effective management of the cold chain system. Actually, is possible to observe a wide variety of treatments employed to the fruit decay preservation. It may be physical or chemical through the biological treatments. Most postharvest treatments involve the alteration of the natural conditions of the fruit in order to prolong its postharvest life. Some physical postharvest techniques are, for examples, high O_2 atmospheres and irradiation which cause damage to some vital molecules of food deteriorative microorganisms, in addition to altering some biochemical processes in the fruit (Charles et al., 2009); heat treatment affect a wide range of fruit ripening processes such as ethylene synthesis, respiration, softening, and cell-wall metabolism (Zhang et al., 2009). Several studies have been related heat tolerance with the increase of heat shock proteins (HSPs), antioxidant enzymes and phytochemicals such as carotenoids and phenolic compounds (Ghasemnezhad et al., 2008). In general, it has been shown that as a secondary response, some postharvest treatments could induce some mechanism that affect the metabolic activity of the treated product, such as the triggering of antioxidant mechanism of the fruit (Gonzalez-Aguilar et al., 2010). Hormetic doses of UV-C can prolong the postharvest life and maintain the quality of fruit has been reported. These effects include delay of senescence process and fruit ripening (Gonzalez-Aguilar et al. 2007b), induction of natural defence and elicitors against fungi and bacteria (Alothman et al. 2009a). Resistance to infection by pathogens is correlated with the induction of plant defence mechanism (Gonzalez-Aguilar et al. 2007a) and DNA damage (Charles et al. 2009). This is manifested through the stimulation of antifungal chemical species such as phytoalexins, flavonoids, and degrading fungal cell-wall enzymes (El-Ghaouth et al. 1998). The induction of plant defence system can also trigger the accumulation of these compounds and other phytochemicals such as carotenoids and vitamin C which exhibit antioxidant potential, improving the nutritional status of the fruit (Alothman et al. 2009).

The chemical control has been traditionally dominated by fungicides, for example imazalil is currently the most commonly used fungicide that is effective for controlling postharvest fungal pathogens in citrus, it is used as a wax emulsion for application by spray (Alteri et al. 2013). Benzimidazole fungicide thiabendazole (TBZ) was largely used to prevent potato dry rot until resistant strains of *Fusarium* spp. were appearing in the late 1980s. In order to improve the control of the disease and maintain, when possible, TBZ efficacy through preventing the development of resistance, postharvest application of fungicides mixtures were used observing a synergistic effect in rot control. Application of TBZ+futriafol, TBZ+2-aminobutane, TBZ+imazalil resulted in variable efficacy (Bojanowski et al. 2013). Fludioxonil+cyprodinil have also been reported to be effective against potato dry rot (Daami-Remadi et al. 2010). Nowadays a new generation of fungicides are being synthesized and trend to be used in combination with physical and/or biological techniques.

Several natural compounds exhibit positive effects to conserve fruit quality. Among these the most studied are volatile compounds, acetic acids, jasmonates, glucosinolates, chitosan, active principles of some plants and some plants extracts (Tripathi and Dubey, 2004).The ability of these compounds to control microorganisms was the major reason behind their use as food additives to maintain quality of whole fresh and fresh-cut fruit and are generally recognized as safe (GRAS) (Smith et al 2005). Volatile compounds, such as 2-hexanal, are strongly antifungal in nature and their activity has been reported against *Botrytis cinerea* inhibiting hyphas growth (Utto et al. 2008). Jasmonates play an important role as signal molecule in plant defence response against pathogen attack, induce the synthesis of antioxidants such as vitamin C, phenolic compounds and increase the activity of enzymatic antioxidant system (Chanjirakul et al. 2006). Some edible coating, such as chitosan, hasantibrowning characteristics and can maintain tissue firmness and reduce microbial decay of harvested fruits for extended periods. Additionally, Jitareerat et al. (2009) found an increase in total carotenoids and vitamin C content in fruits treated with chitosan. Chitosan has the potential for inducing defence-related enzymes such as polyphenol oxidase and PAL, but the possible mechanism of action is little understood (Gonzalez-Aguilar et al. 2010). Edible coating can be used alone or in combination with controlled/modified atmosphere. Appropriate edible coating formulation may reduce gas exchange rates and water loss, as well as representing an excellent way of incorporating additives to control reactions that are detrimental to quality (Olivas and Barbosa-Canovas 2008). The effects of controlled and modified atmosphere technologies are well documented to extend the postharvest life of fruits. These effects include reduction of respiration rate, inhibition of ethylene production and action, retardation of ripening and maintenance of nutritional and sensory quality (Yahia 2009). These techniques alter the O_2, CO_2 and/or C_2H_4 concentrations in the atmosphere surrounding the commodity to produce an atmosphere composition different from that of normal air. The positive effects of the controlled/modified atmosphere depend on several factors such as type of fruit and cultivar, concentrations of O_2 and CO_2 and others gases in the atmosphere, temperature, and duration of storage (Singh and Pal 2008). Very low levels of O_2 and high CO_2 favor fermentative processes which lead to ethanol and acetaldehyde production and consequent off-flavor development (Yahia 2009).

Actually, the use of antagonistic microorganisms to control fruit disease is considered an attractive alternative to synthetic chemical compounds. Biopesticides are considered safer for the environment and human health than conventional synthetic pesticides. Bacteria such as *Bacillus* spp. or *Pseudomonas* spp. (Casals et al. 2012, Al-Mughrabi 2010),yeasts as *Candida oleophila*(Sui et al. 2012) or *Hansienaspora uvarum* (Liu et al. 2010), and fungus as *Muscodor albus* (Mlikota et al. 2006) are the most utilized in postharvest biocontrol because of very effective substances production against pathogens fungi. Compounds as lipopeptides, volatiles, and secondary metabolites have shown a strong inhibition of pathogenic fungi growth during *in vitro* and *in vivo* experiments in laboratory. However, the principal and major barrier to solve is to maintain the effectiveness of the treatment when the trial is performed in industrial scale. Scientifics around the world are dedicating time and effort to develop of competitive biopesticides and offer a real alternative to synthetic compounds.

7.2 PHYSICAL POSTHARVEST TREATMENTS

The industry of fruit, including exportation of fresh product and processed fruits such as juices and fruit-cut, is constantly growing. New techniques for maintaining quality and inhibiting pathogen microorganisms' growth are required in all steps of the distribution chain until the consumer. In this section, an overview of advances in physical treatment used in fruit handling is shown.

7.2.1 ULTRAVIOLET IRRADIATION TREATMENT

Ultraviolet (UV) radiation has been investigated as a physical treatment for decontaminating product surface, increasing phytochemical content and enhancing physiological mechanisms in fruit in order to increase general quality and storability. Decreasing the chilling injury of certain fruits is one example of the positive results obtained with the use of UV-C radiation (280–100 nm), while the effect of UV radiation treatment on red table grapes and wine grapes is one of the best known examples of a method that increases phytochemical content. In fact, the application of UV-C and UV-B (315–280 nm) radiation increases resveratrol derivatives significantly, by three and two time, respectively (Cantos et al. 2000).

Effective control of several types of plant pathogens has been achieved with postharvest UV treatments. These pathogens include fungi and bacteria that colonize the surface and internal tissues of horticultural product. Marquenie et al. (2002) found that the viable conidia of *Botrytis cinerea* decrease linearly with the applied UV-C dose. Moreover, various studies have shown that UV-C treatment applied at appropriate doses reduced decay in tomato and strawberry (Maharaj et al. 1999, Erkan et al. 2008).

Several factors have been identified as affecting the disease control performance of UV treatment, including the type of pathogen, the type of product, the maturity, cultivars, andseasonal variations (Vigneault et al. 2012). Total dosage (energy per unit area) is a main factor determining fruit response to UV-C.Lado and Yousef (2002) reported that UV-C radiation from 0.5 to 20 kJ/m^2 inhibited microbial growth by inducing the formation of pyrimidine dimers which alter the DNA helix and block microbial cell replication. Therefore, cells which are unable to repair radiation-damaged DNA die and sublethally injured cells are often subject to mutations. The intensity of the radiation (dose per unit time) may also determine treatment outcome. Lambert's law may be used to predict the radiation intensity resulting from a light source, or combination of light sources, on order to design a system capable of supplying the required radiation intensity within the desired variation limits over the entire treatment surface (Vigneault et al 2012). The intensity of the radiation is a key factor affecting the efficacy of the treatment and that the effects are distinct in different fruits. High UV-intensity may in some cases increase the outcome of the treatment and also reduce the treatment times. Excessively high intensities could result in oxidative reactions and alterations of nutritional and/or organoleptic fruit quality (Cote et al. 2013).

Although the UV dose is critical for generating the desired response, the storage temperature has been proven to be the most important postharvest condition affecting the response of product to UV treatment. In 1993, Mercier et al. demonstrated that a temperature differences as small as a few degrees Celsius is enough to generate major differences in horticultural product responses (Mercier et al. 1993). Moreover recent studies has shown that postharvest treatment of either ClO$_2$ or fumaric acid combined with UV-C can be useful for maintain the quality of strawberries; including the sensory evaluation score (Kim et al. 2010). An alternative in the use of UV is its application in low doses (0.25–8.0 kJ/m2) which still affect the DNA of microorganisms and in addition can modulate induced defence in plants.

The hormesis concept involves stimulation of a beneficial plant response by low or sublethal doses of an elicitor/agent, such as a chemical inducer or a physical stress. Hormetic UV treatment is distinguished from conventional UV treatment. In conventional treatment the UV is directed at microorganisms present on the surfaces of an object, whereas in hormetic UV treatment the object itself is the target of the incident UV. The objective of the treatment is to elicit an antimicrobial response in the fruit tissues (Ribeiro et al. 2012). Stimulation of antifungal chemical species such as phytoalexins (scoparone and resveratrol), flavonoids and degrading fungal cell-wall enzymes (chitinases and glucanases), are produced by the use of very low UV doses and the time scale for the induction of such events is measured over hours or even days (Ribeiro et al. 2012). Pombo et al. (2011) found that postharvest UV-C treatment, a few hours prior to *B. cinerea* inoculation, reduces the percentage of fruit infection in strawberry during storage. In addition, irradiation of fruit with UV-C increase the expression and activity levels of several enzymes (PAL, peroxidase, PPO, chitinases, and b-1,3-glucanases) which are involved in defence mechanism against pathogens.

7.2.2 GAMMA RADIATION POSTHARVEST METHOD

Gamma radiation is one of the three types of natural radioactivity with alpha and beta. Gamma rays are the most energetic form of electromagnetic radiation, with very short wavelength of less than one-tenth of nanometer. To be effective, gamma or e-beam treatments require time, contact, and temperature. Some microorganisms are very difficult to kill, others die easily. It is easier to kill one organism than many the amount and type of organic material that protects the microorganism. Gamma irradiation is a physical means of preservation because it kills bacteria, fungus, and yeast by breaking down DNA, inhibiting microorganism multiplication, Energy of gamma rays passes through the cells disrupting the pathogens the cause diseases and any other agents that are in the treated fruit surface. Depending of dose rate, distribution, and quality, e-beam also could produce important physiological disorders. The radiolytic products of water contained in the microorganisms yielding hydroxyl radicals, ionizing the internal environment and by production of free radicals.

One major advantage offered by irradiation is that it requires a minimal handling of food item, thus enabling decontamination without inducing any mechanism damage; it also reduces the time it take for the product to reach consumers. Mostafavi et al. (2012) showed in his studies on apples the advantage of use gamma radiation in *Penicillium expansum* control. Data showed by the author reveals that all pathogen-treated apples could prevent lesion diameter and that there were no significant differences between dose range of 300, 600, 900, and 1200 Gy after 3, 6, and9months. However, lesion diameter of nonirradiated apples was significantly increased after three months during cold storage. Therefore low irradiation doses (300 and 600 Gy) combined with cold storage which significantly retained phenol contents, antioxidant activity, firmness, weight loss and total soluble solids compared other treatment. Gamma application could be recommended as suitable method to reduce apple quality losses and for better postharvest preservation during storage period (Mostafavi et al., 2012). On the other hand, Costa-Guimaraes et al. (2013) has concluded that the use of irradiation on postharvest raspberry is a viable technique for fruit export industry. The irradiation reduces weight loss and filamentous fungi and yeast count, and the dose of 2 kGy is highly effective in controlling microorganism growth. Unfortunately it is also the dose at which raspberries lost most of their quality and firmness. Further experiments to optimize gamma irradiation conditions for postharvest treatment should be performed.

7.2.3 ULTRASONIC TREATMENT

Ultrasound is an oscillating sound pressure wave with a frequency greater than upper limit of the human hearing range. Ultrasound is thus not separated from audible sound based on differences in physical properties, only the fact that humans cannot hear it. Although this limit varies from person to person, it is approximately 20 kilohertz (kHz) in healthy, young adults. Ultrasound devices operate with frequencies from 20 kHz up to several gigahertz.

There are numerous reports on the effects of ultrasound treatment on food processing and preservation (Knorr et al. 2004, Patist and Bates 2008). However, most of these studies have been focused on the efficacy of inactivation of microorganisms and enzymes, extraction of antioxidant compounds and acceleration of heat transfer (Knorr et al. 2004). Little information has been provided on the influence of ultrasound treatments on decay incidence and quality maintenance in fruit after harvest. Previous studies have suggested that the efficacy of ultrasonic treatment can be influenced by several factors such as ultrasound power, treatment time and temperature (Valero et al 2007, Ma et al 2008, Tiwari et al 2009). Studies conducted in strawberries decay control by ultrasound has resulted that treatments using 40 or 59 kHz significantly inhibit the increase of decay index (Cao et al 2010b). Strawberry is a highly perishable fruit and the shelf life usually ends due to microbial infection, increasing the decay incidence with storage time. Cao et al. (2010b) observed that an ultrasonic treatment of 40 kHz for 10 min. reduce the decay index in 43.7% regarding to control after 8 days of storage.

It is known that ultrasound causes cytolytic effects and is effectively for inactivation of microorganisms, but the cell disruption rate depends upon the conditions and the type of microorganism. Scherba et al. (1991) examined the effects of ultrasound on *Staphylococcus aureus*, *Bacillus subtilis* and *Trichophytonmentagrophytes*. All of the microorganisms were affected by the ultrasound with the effect increasing with the storage and intensity. In Cao et al (2010b) study, the numbers of microorganism on strawberries were significantly decreased by ultrasound treatment from the first moment. The microbial population on strawberry fruit was reduced with a decreased frequency of ultrasound, but the number increased during the whole cold storage. However, total microorganisms counts on strawberries treated with 40 kHz ultrasound was significantly lower than that of the treated and nontreated fruit during storage, indicating that the ultrasound treatment effectively reduce the spoilage from microorganisms. Further experiments of Cao et al. (2010a) optimized the ultrasound treatment for strawberry using response surface methodology (RSM) as informatics tool for determining the effects of operational factors and their interactions (Bas and Boyaci 2007). The ultrasonic treatment conditions for strawberry fruit was considered optimum if decay incidence and microbial population including bacteria, yeast, and fungi reached minimum values and the responses for fruit quality parameters reached maximum values simultaneously. The optimum ultrasonic treatment conditions for the storage of strawberry fruit were an ultrasonic power of 250 W and treatment time of 9.8 min, as predicted by the computing program (Cao et al. 2010a).

7.2.4 MICROWAVE APPLICATION ON POSTHARVEST TREATMENT

Microwave (MW) are form of electromagnetic radiation with wavelengths ranging from as long as one meter to as short as one millimeter, or equivalently, with frequencies between 300 MHz (0.3GHz) and 300 GHz. The prefix "micro" in microwave is not meant to suggest a wavelength in the micrometre range. It indicates that microwaves are "small" compared to wave used in typical radio broadcasting, in that they have shorter wavelength. Microwave belong to the nonionizing radiation due to it do not contain energy enough to chemically change substances by ionization as gamma radiation. The microwave damage come from dielectric heating induced in the agent. Therefore, MW heating may be an alternative method to conventional heating in fruit industry. Microwave treatments with 2,450 MHz household oven using between 0.4 and 0.45 kW have been previously study to control postharvest diseases as *B. cinerea*, *P. expansum* and *Rhizopus stolonifer* in peaches (Karabulut and Baykal 2002; Zhang et al. 2004). Villa-Rojas et al. (2011) used a microwave oven during the studies of control of ripening and decay of strawberries. Microwave heating and refrigerated storage were found to slow down the ripening process and postharvest decay of strawberries and allow them to retain their quality attributes for a long of 12 days of assay. MW-assisted heat treatment resulted in significantly higher color lightness than conventional heat treatment as well as lower during firmness loss during storage. Both 514 and 763 W output powers were effective, the latter requiring a shorter treatment time (1 min 50 s) (Villas-Rojas et al. 2011).

Further experiment performed by Sisquell et al. (2013b) were used an industrial microwave tunnel to apply MW treatments in continuous for controlling postharvest diseases, including brown rot disease caused by *Monilinia* spp. on peaches and nectarines. The authors assayed 5 kW during 120 s observing no brown control, however when increased the MW power to 10 kW for 100 s obtained complete brown rot reduction, but a new increase to 17.5 kW for 50 s did not shown significant differences with the previous treatment. In MW heating, the rate of temperature rise in a commodity depends on the power, frequency, heating time and the material's dielectric loss factor, so higher temperature in commodities can be achieved by long heating duration and high power (Wang and Tang 2001). Maturity level and the varieties of treated stone fruit had not a significant effect on efficacy of both MW treatments at 17.5 kW for 50 s and 10kW for 95 s (Sisquella et al. 2013b). Similar results were obtained when MW heating with household oven for 2 min was applied to control natural infection of *P. expansum* and *B. cinerea* in peaches (Karabulut and Baykal 2002). However, at industrial scale using a MW tunnel, the treatment time to control natural infection of *Monilinia* ssp. has been reduced to 50 s or 95 s depending to the power level used (Sisquella et al. 2013b). The authors also observed that both MW treatment studied did not affect negatively fruit firmness in the varieties tested and even a delay of the softening was observed in "Roig d'Albesa" peaches and "PP-100" nectarines.

So far, the MW treatment for control of postharvest diseases has had very good results and is promising as a method industrial level. However, the different temperature reached at the end of the treatment depending on fruit weight could affect MW effectiveness and the internal fruit appearance.

7.2.5 HEAT TREATMENT

Hot water was original heat treatment (HT) used for the control of fungal decay, but its use has been extended even to insect control. The two main types of commercial hot water treatment are immersion and shower (spray). Hot water dips are effective for fungal pathogen control, because fungal spores and latent infections are either on the surface or in the first few cell layers under the peel of the fruit. Postharvest dips to control decay are often applied for only a few minutes at temperatures higher than heat treatment for disinfection, since it is only the surface of the commodity that requires heating (Vigneault et al. 2013). Many fruit tolerate exposure to water temperatures of 50–60°C for up to 10 min (Ranganna et al 1998), and shorter exposure at these temperatures can control many postharvest plant pathogens; however, the duration of the treatment must be precisely controlled to avoid heat damage. Afterwards, a hot water spray treatment has been developed (Fallik 2004) and designed to be part of a sorting line. The process consists of moving the commodity with brush rollers through a pressurized spray of hot water. This machine is used to clean fruit and reduce the number of pathogens. It is technically more difficult to achieve uniformity in such a treatment than in dipping process, especially when the shape of the product is heterogeneous or the product has hidden surface that the spray may not reach (Vigneault et al. 2013).

Vapour heat is a method of heating commodities that uses air saturated with water vapor at temperatures of 40 to 50°C to kill insect eggs and larvae as quarantine treatment before market shipments. This method is disputed because the temperature of the fruit surface is even higher than that of the surrounding water vapor media during the treatment (Vigneault et al. 2013). Otherwise, hot air has been used for both fungus and insects control (Yahia and Ortega-Zaleta 2000, Hoa et al. 2006). Forced hot air was preferred by many countries for development of quarantine procedures (Hallman 1996). Lurie (1998) reviewed hot air heat treatment and concluded that exposure to high temperatures in forced or static air can decrease fungal infection. However, the processing times are relatively long, running from 12 to 96 h at temperatures ranging from 38 to 46°C. Vicente et al. (2002) studied the bacterial and fungal reduction in strawberries cv. Selva treated in an air oven at 45°C for 3 h and then stored at 0°C for 0, 7, and 14 days. The authors observed that in the absence of storage, heat-treated fruits showed lower decay at 20°C than control without treatment. The percentage of decayed fruits was lower in heat-treated than control fruits after 7 days at 0°C followed by 72 h at 20°C. The treatment decreased the initial bacterial population, but did not modify the amount of fungus initially present. After 7 days of cold storage, the colony forming units (CFU) number for bacteria was lower than in control fruits. This difference was still significant after 48 h at 20°C. In case of fungi, HT that were stored for 7 or 14 days at 0°C and then transferred to 20°C for 48 h showed lower CFU value than controls (Vicente et al. 2002).

Aside from controlling pathogenic organisms, HT may also reduce disease by changing the physical characteristics of the interior pericarp. For example, HT not only delayed fruit ripening and coloration, inhibited fruit softening and slowed ethylene production (Lara et al. 2009), it also increase fruit hardness, as well as increasing the content of polyphenolics, flavonoids, and dopamine (Zhang et al. 2011, Perotti et al. 2011). Some scientists discovered HT induces phenylalanine ammonia lyase and accumulates phenylpropanoid compounds, which play a vital role in heat-induced fruit resistibility (Tu et al. 2010). Proteome analysis reveals the induction of heat shock proteins, allergen proteins, dehydrin, and other proteins involved in the stress response (Lara et al. 2009). The function of phenylpropanoid compounds in plant defence range from preformed or inducible physical and chemical barriers against infection (Dixon et al. 2002), these compounds has been found in the lignification of monolignols to lignin (Dennes et al. 2011). Yu et al. (2013) has observed that HT decreased the level of precursors needed in lignin synthesis (ferulic acid, sinapic acid, cinnamic acid and caffeic acid), but increase lignin content. The accumulation of lignin induced by HT creates a thickening of cell walls which forms an effective physical barrier to pathogen, delaying the invasion of disease organisms. For the same reason HT increase fruit firmness increasing the shelf life of the fruit during storage (Yun et al 2013).

Actually, the combination of HT with generally recognizes as safe substances or low-toxicity chemical have showed a synergistic relationship. Mari et al. (2007)

showed better control when hot water treatment was combined with sodium bicarbonate to control brown rot caused by *Monilinia* spp. on stone fruit than either hot water or sodium bicarbonate used alone. Calcium and salicylic acid combined with water at 45°C for 15 min had least decay in strawberry cv Camarosa than control fruit (Shafiee et al. 2010). Sisquella et al. (2013a) concluded after their studies in brown rot on peach and nectarines that immersion for 40 s in 200 mg L-1 of peracetic acid at 40°C may provide an alternative treatment to control recent infections of *Monilinia* spp. whatever their concentration, without generally affecting fruit quality.

The combination of two physical postharvest treatments has been studied by Shao et al (2012). In apple fruit, hot air treatment at 38°C for 4 days has been considered to be optimal temperature and duration to preserve postharvest storage quality. HT can delay ripening and maintain firmness of apple fruit to improve consumer acceptability while also controlling the development of common postharvest diseases caused by *Penicilium expansum*, *Botrytis cinerea* and *Colletotrichum acutatum*(Shao et al. 2007). On the other hand, chitosan (CTS) (poly-b-[1,4] N-acetyl-D-glucosamine) polymer is industrially produced by chemical deacetylation of the chitin found in arthropod exoskeletons. CTS is a safe biopolymer that can be used to form an edible semipermeable film on the surface of fruit to extend storage life and reduce several forms of decay caused by fungi during storage (Bautista-Baños et al. 2006). The results of combination of these two postharvest treatment showed that apples treated with hot air at 38°C for 4 days and then coated with CTS before storage at 0°C for 8 weeks, followed by further commercial shelf storage at 20°C for 7 days, decreased mold growth and decay development and it is an effective method to maintain fruit quality (Shao et al. 2012).

7.3   EDIBLE COATINGS

Application of edible coating forms a transparent film on the fruit surface, acts as barrier to water and gases (O_2 and CO_2), imparts gloss, better color, reduces fruit weight loss, extends storage life and minimize microbial spoilage. Edible coating offers an attractive alternative to film packaging due to their environment-friendly characteristics. Semipermeable coatings increase a fruit's skin resistance to gas diffusion, modify its internal atmosphere composition and depress its respiration rate. It can also create a modified atmosphere (MA) around the fruit similar to controlled atmosphere (CA), but the changes in concentrations of O_2 and CO_2 are dependent on temperature and relative humidity (RH) (Singh et al. 2013).

Various types of edible coatings such as carnauba, wax, shellac, zein, cellulose derivatives, chitosan and its derivatives, and other composite mixtures containing sucrose esters of fatty acids and sodium salts of carboxymethylcellulose have been tested to extend storage life of fruit with variable results. For example a cellulose-based polysaccharide coating improves aroma volatiles, ethanol, and acetaldehydes in mango fruit, while carnauba wax does not affect aroma volatile concentration. However, it depends of the mango genotype."Kensington Pride" mango fruit coated with mango carnauba wax exhibit significantly improves fruit aroma volatiles including total aroma volatiles, monoterpenes, sesquiterpenes, and aldehydes while maintaining comparative aromatic and lactose concentration and reducing total alcohol levels compared with untreated mango fruits (Singh et al. 2013).

Among the most commonlyused coatings that currently being optimized, we could find alginate-based films. Alginate is a natural polysaccharide extracted from brown sea algae (*Phaeophyceae*), and it is composed of two uronic acids: b-D-mannuronic acid and a-L-guluronic acid. Sodium alginate is composed of block polymers of sodium poly(L-guluronate), sodium poly(D-mannuronate) and alternating sequences of both sugars. Alginate is known as a hydrophilic biopolymer that has a coating function because of its well-studied unique colloidal properties which include its use for thickening, suspension forming, gel forming and emulsion stabilizing (Acevedo et al. 2010). Sodium alginate has been effective on maintaining postharvest quality of tomato and peach (Zapata et al. 2008, Maftoonazad et al. 2008). Díaz-Mula et al. (2012) treated sweet cherry fruits harvested at commercial maturity stage with edible coating based on sodium alginate at several concentrations (1%, 3% or 5% w/v). The coatings (preferably 3 and 5%) were effective on delaying the evolution of the parameters related to postharvest ripening and decay. In addition, the edible coating showed a positive effect on maintaining higher concentration of total phenolics and total antioxidant activity. The results revealed that untreated sweet cherry fruit could be storage 8 days at 2°C plus 2 days at 20°C, while alginate-coated cherries could be storage 16 days at 2°C plus 2 days at 20°C (Díaz-Mula et al. 2012). Moreover, the effect of alginate-based edible coating as carrier of anti-browning agents (ascorbic and citric acids) on fresh-cut mangoes "Kent" stored at 4°C was evaluated by Robles-Sánchez et al. (2013). The authors observed that ascorbic acids as anti-brown agent carried in alginate-based edible coating contribute not only to color retention in fresh-cut mango, but also improve the antioxidant potential of fruit. According with results obtained, the fresh-cut mangoes can be storage 12 days at 4°C without quality loss.

Pectin is a purified carbohydrate product obtained by aqueous extraction of appropriate edible plant material usually citrus fruits or apples. Due to their hydrophilic nature, pectin edible coatings exhibit low water vapor barrier, but good O_2 and CO_2 barrier properties. Low methoxy (LM) pectins are often used as edible coatings because of their ability to form strong gels upon reaction with multivalent metal cations, like calcium. The incorporation of calcium in the polysaccharide edible coatings reduces their water vapor permeability, making the coatings water-insoluble (Lacroix and Tien 2005). Furthermore, calcium salts have been employed as a tissue structural preservative, since this ion acts as a cross-linking agent, forming complexes with the cell wall and middle lamella pectin, improving structural integrity and promoting greater tissue firmness (Martín-Diana et al. 2007). Ferrari et al. (2013) has studied the effect of osmotic dehydration in combination with pectin edible coating on quality and shelf life of fresh-cut melon. The authors osmodehydrated the melon pieces in 0.5 percent calcium lactate solution with or without 40°Bx of sucrose. Then the samples were coated with 1 percent pectin and storage at 5°C for 14 days. The shelf life of untreated fresh-cut melon was limited to 9 days due to microbial growth and sensory rejection, while treated samples showed a shelf life of 14 days. Higher preservation of firmness in coated fruit pieces was attributed to the action of calcium salt on melon structure, causing a strengthening of cell wall. Calcium lactate also inhibited microbial growth along storage improving microbiological stability of fresh-cut melon. The combination of osmotic dehydration and pectin coating was a good preservation for fresh-cut melon, since it improve fruit sensory acceptance, promoting the reduction of product respiration rate, as well as the maintenance of quality parameters during 14 days (Ferrari et al. 2013).

Edible coatings that have not antimicrobial properties by itself, as chitosan, could be combined with microencapsulated antimicrobial complex to acquire inhibitory characteristics of postharvest decay. There are several categories of antimicrobials that can be potentially incorporated into edible coatings, including organic acids (acetic, benzoic, lactic, sorbic, propionic), fatty acid esters, polypeptides, plant essential oils, nitrites, and sulfites, among others. Essential oils are designed as GRAS and are outstanding alternative to chemical preservatives. Mantilla et al (2013) used sodium alginate, pectin, and calcium chloride to prepare three different solutions to combine with b-cyclodextrin and *trans*-cinnamaldehyde, to assess fresh-cut pineapple. The addition of the antimicrobials to edible coating was performed by using the layer-by-layer technique. The method consists to immerse the fruits into a series of different solutions that contain oppositely charged polyelectrolytes. Pineapples were washed, disinfected, and treated by layer-by-layer method. After, the fruit was stored in sealed containers for 15 days at 4°C. Microbiological analysis presented by demonstrated the effectiveness of the coating as a carrier of antimicrobial compounds and the effectiveness of the compound against microbial growth. The effectiveness against psycothrophics, yeast, andmolds was particularly significant. The best coating formulation in terms of the preservation of quality attributes of fresh-cut pineapple was 1 g/100 g of alginate, 2 g/100 g of antimicrobial compound (trans-cinnamaldehyde) and 2 2/100 g of pectin (Mantilla et al.2013).

Chitosan is a linear polysaccharide composed of randomly distributed b-(1-4)-linked D-glucosamine and N-acetyl-D-glucosamine units. Chitosan-based edible coating is derived from natural sources by deacetylation of chitin and has been studied for efficacy in inhibiting decay and extending shelf life on fruit. Chitosan and its derivatives have been shown to inhibit the growth of a wide range of fungi (Aider 2010). They can also trigger defensive mechanisms in plant and fruits against infections caused by several pathogens. A particular case is the treatment of strawberries with chitosan studied by Wang and Gao (2013). Decay of strawberries coated with chitosan was reduced significantly in comparison with untreated fruits. Early signs of mold development in strawberries appeared at day 3 of 10°C and at day 6 of

5°C storage for untreated berries. At day 9 of 5°C storage, the percentages of decay were 21.5 and 11.7 percent for untreated fruits and chitosan coated (0.5g/100mL) samples, respectively. After 9 days of storage at 10°C, the percentage of decay for chitosan coated berries (0.5, 1.0 and 1.5 g/mL) were 25.3 percent, 15.7 percent and 6.9 percent respectively, while the untreated control berries was 95.2 percent. Strawberries treated with chitosan also maintain better fruit quality with higher levels of phenolics, anthocyanins, flavonoids, antioxidant enzyme activity and oxygen radicals than untreated fruits (Wang and Gao 2013). The combination of chitosan and beeswax was further study by Velickova et al. (2013) to prolong the shelf life of strawberries stored at 20°C and 35–40 percent of relative humidity. Strawberries were coated with four different coating formulations: chitosan as monolayer, three layers coating consisting of separated beeswax-chitosan-beeswax, three layers coating where chitosan was crosslinked with sodium tripolyphosphate and composite. Fruits treated with chitosan and composite coating gave better visual appearance and taste and was therefore more preferable by 90 percent of the judges than three-layer coating (Velickova et al. 2013). Moreover, the effectiveness of chitosan and short hypobaric treatments, alone or in combination, to control storage decay of sweet cherries, was investigated over 2 years by Romanazzi et al. (2003). In single treatment, chitosan was applied by postharvest dipping or preharvest spraying at 0.1, 0.5 and 1 percent concentrations; hypobaric treatment at 0.5 and 0.25 atm were applied for 4 h. In combined treatments, sweet cherries were dipped in 1 percent chitosan and then exposed to 0.5 and 0.25 atm, or sprayed with chitosan (0.1, 0.5 and 1%) 7 days before harvest and exposed to 0.5 atm soon after harvest. Untreated fruit kept at normal pressure (near 1 atm) were used as controls. Rot incidence was evaluated after 14 days storage at 0±1°C, followed by a 7 days shelf life. In both years, chitosan and hypobaric treatments applied alone significantly reduced brown rot, graymold and total rots, the latter also including blue mold, *Alternaria*, *Rhizopus*, and green rots. A combined treatment with 1 percent chitosan and 0.5 atm was the best in controlling decay, showing in the first year, a synergistic effect in the reduction of brown rot and total rots. These results indicated to the authors that the combination of chitosan and hypobaric treatments was a valid strategy for increasing the effectiveness of the treatments in controlling postharvest decay of sweet cherries (Romanazzi et al. 2003).

Today, the research direction in the edible coatings is the development of novel coatings and new formulations with better properties to introduce in the market. A novel edible coating that is being optimized to be applicable to chilgoza (pine nuts) is the gum *Cordia*. *Cordia myxa* is a small deciduous tree, which grows nearly all over the Indo-Pak subcontinent. The ripe fruits contain an anionic polysaccharide having good adhering property. Chilgoza (*Pinus gerardiana*), specie of genus *Pinus* (pine nuts) grows in Pakistan, Afghanistan, and India. It is a rich source of unsaturated fatty acids but the un-shelled nuts are highly susceptible to rancidity. Haq et al. (2013) investigated the efficacy of gum *Cordia* in comparison with carboxy methyl cellulose (CMC) as edible coating to retard this oxidation. Gum *Cordia* and CMC with or without natural antioxidants were used during the study. Methanolic extract of *C. myxa* and a-tocopherol were selected as antioxidants. Pine nuts stored 112 days at 35°C showed a significant improvement of shelf life when was treated with gum *Cordia* containing the extract of *C. myxa* (Haq et al. 2013).

On the other hand, Zambrano-Zaragoza et al. (2013) has performed an interesting study which objective was to prepare solid lipid nanoparticles (SLNs) based on the hot lipid dispersion method using a rotor-stator device in order to obtain a submicronic system and evaluate the effect of SLN-xanthan gum coating on the guava shelf life. Candeuba wax was used as a component of SLN. The coating was formulated with xanthan gum (4 g/L) and polyethylene glycol (5 g/L) in order to form a continuous film retaining the SLN. The best results were obtained with SLN concentrations of 60 and 65 g/L since at these concentrations, guavas showed the lowest range of weight loss and preserved the best quality compared with the fruits processed above 70 g/L. Higher SLN concentration caused physiological damage and also delay the maturation (Zambrano-Zaragoza et al. 2013). A different coating formulation has been studied by Das et al. (2013) to enhance tomatoes shelf life stored at room temperature. Tomatoes coated with rice starch-based edible coating formulation containing coconut oil and tea leaf extract were studied for the effect of coating on biochemical changes and during storage for 20 days. Coconut oil and tea extracts also exhibited microbe-barrier property. The edible coating formulation of starch+glycerol+lipid+antioxidant extract was found to form a rigid and continuous fruit coating that was able to extend the ripening period of tomatoes on storage at room temperature along with a good microbe-barrier property (Das et al. 2013).

7.3.1 ATMOSPHERE MODIFICATION

During the 1980s, large amount of information were published on the optimal gas composition for a wide variety of fresh product. However, the development and testing of controlled atmosphere (CA) technology for new product or the optimization of gas composition for product already stored in a CA environment, continued in order to address the specificity of each new cultivar. From an engineering point of view, the recent development of technologies to generate atmospheres of any composition, from N_2 generators to C_2H_4 and CO_2 gas scrubbers opened up opportunities for using purging systems and better controlled gas concentrations (Vigneault et al. 2013).

Typically, it is recommended that purging be done to reduce O_2 levels to less or equal to5kPa and that the remaining reduction be achieve naturally through respiration by product. To achieve better control of gas concentrations, the product must be at its required storage temperature when purging is completed (Silva et al. 2006). Significant technologies advances in recent years have shown that elevated O_2 concentrations can be used to preserve and maintain the postharvest quality of whole and fresh-cut fruit. Improvements in the sensory quality, microbial safety and shelf life of fresh-cut product, as well as inhibition of decay in soft fruit, have been achieved by using high O_2 concentration (Schotsmans et al. 2007). The atmosphere modification can be done for different time periods. Thus short-term or pulse treatment involves a rapid variation of either the O_2 or CO_2 concentration, or both, under constant temperature conditions, which generates produce stress. Some types of product respond to these stresses by maintaining or increasing their phytochemical content. This treatment has been used as an alternative to chemical treatment to prevent physiological disorders and delay ripening in fruit (Polenta et al. 2005). When the modification of the surrounding gas concentration is achieved in 4 to 6 days it is a medium-term treatment. From an engineering perspective, medium-term treatments are easier to generate than pulse treatment since their duration is much longer, less precision is required for processing time, and the maximum time periods for reaching the optimal gas concentrations are longer (hours instead seconds) (Vigneault et al. 2013). Long-term treatment or storage is applied when gas composition is altered in fruit environment until be consumed. Currently, CA and modified atmosphere with low temperatures are used to control fungal and bacterial growth. It has been reported that molds and Gram negatives aerobic bacteria are highly sensitive to CO_2, while low O_2 levels inhibit of most aerobic microorganisms, although low O_2 appears to have limited fungistatic properties and is less effective than elevated CO_2 (Conte et al. 2009). Serradilla et al. (2013) has studied three different combinations of O_2 and CO_2 (3, 5 or 8 % O_2 combined with 10% CO_2) and their influence in microbiological quality on "Ambrunés" sweet cherries. The authors concluded that sweet cherries stored under CA showed an acceptable microbial quality at least during 30 days of storage and subsequent shelf life, since it delayed the spoilage due to microbial growth. In general, the most pronounced effect of CA was observed after 15 days of storage and 5% O_2+ 10% CO_2 and 8% O_2+ 10% CO_2, these treatments presented the lowest counts, showing significant differences with the untreated control (Serradilla et al. 2013). Researches in controlled and modified atmosphere are accompanied of the development of new technologies. Chong et al. (2013) has developed a mathematic model to investigate the effect of storage cut on gas separation performance of hollow fiber membrane modules using binary and ternary mixed gas systems.

Oxygen vs carbon dioxide modification can be combined with variations of pressure. Pressure treatment is a physical method that can result in high quality, microbiologically safe food with an extended shelf life. Pressure treatment offers homogeneity as it is applied instantaneous and uniformly around each single product or throughout an entire mass of food, independently of its size, shape, or composition. However, considering the large range of pressures that can be applied to product, pressure treatment need to be divided in two main categories: low and high pressure techniques. A low pressure treatment (0 to 1 MPa) can be hypobaric or hyperbaric and it can be applied to fresh product. High pressure treatments (above 100

MPa) are generally applied to processed food. The pressure too high can irreversibly damage pressure-sensitive product such as fresh horticultural crops and too low can produce any significant effect on microorganisms and enzymes inactivation (Vigneault et al. 2013). Hypobaric storage at 0.04 to 0.05 MPa reduce decay of loquat fruit by 87 percent compared with a control stored at 2°C (Gao et al. 2006). Pressure of 0.05 MPa applied for 4 h was the most effective treatment against mold and rots on sweet cherries and strawberries, and 0.025 MPa applied for 24 h decreased the incidence of mold on table grapes (Romanazzi et al. 2001). On the other hand, the effects of hyperbaric treatment on the respiration rate and quality attributes of avocado were investigated (Vigneault et al. 2013). The fruits were stored at ambient temperature using different pressure levels. Hyperbaric pressure was found to slow down the ripening process resulting in extended storage life. Surface color change and weight loss were reduced when using a higher hyperbaric pressure. In fact, the surface color changes and the weight loss of the products stored at 0.9 MPa and ambient temperature were similar to those stored at 5°C (Vigneault et al. 2013).

7.3.2   PLANT DEFENCE INDUCTION

Plants possess a range of defences that can be actively expressed in response to pathogens and parasites of various scales, ranging from microscopic viruses to insect herbivores. The timing of these defence responses is critical and can be difference between able to cope or succumbing to the challenge of a pathogen. Systematic acquired resistance (SAR) and induced systemic resistance (ISR) are two forms of induced resistance; in both SAR and ISR, plant defences are preconditioned by prior infection or treatment that results in resistance (or tolerance) against subsequent challenge by a pathogen. Great strides have made over the identification of a number of chemical and biological elicitors, some of which are commercially available for use in conventional agriculture (Vallad and Goodman 2004).

7.3.3   CALCIUM AND SALICYLIC ACID, EXOGENOUS APPLICATION OF PLANT DEFENCE INDUCERS

As a versatile signaling ion with special function acting at multiple sites in diverse networks of signaling cascade, calcium (Ca^{2+}) serves as a major regulatory ion in horticultural crops. These pathways receive signals from a wide array of biotic and abiotic sources and cause changes in gene expression (Dodd et al. 2010). Calcium is also considered to be an important mineral element regulates fruit quality, specially, maintenance of fruit firmness, a decrease in postharvest decay and incidence of physiological disorders such as water core, bitter pit and internal breakdown (Lurie 2009, Aghdam et al. 2012). Besides, salicylic acid (SA), an endogenous plant growth regulator, has been found to generate a wide range of metabolic and physiological responses in plants thereby affecting their growth and development. SA as a natural and safe phenolic compound exhibits a high potential in controlling postharvest losses of horticultural crops.

The effect of postharvest dips of $CaCl_2$ and/or the auxin naphthalene acetic acid (NAA) on quality parameters, cell-wall composition and expression of cell wall-modify genes of Chilean strawberry fruits has been studied by Figueroa et al. (2012). After treatments, the fruit was cold stored and samples were evaluated at 0, 5, and 8 days, plus 2 days at room temperature. No differences in physiology characteristics were found after cold storage between treatments. Respecting to cell wall integrity, $CaCl_2$ treated fruits showed high contents of ionically-bound pectins and NAA-treated fruits exhibited low levels of endoglucanase gene (*EG1*) during cold storage. After cold storage, $CaCl_2$ in combination with NAA, produced a reduction in transcriptional levels of polygalacturonase (*PG1*), pectate lyase (*PL*) and *EG1* genes. In contrast, an increase in the expression levels of pectin methylesterase (*PE1*) and xyloglucan endotransglycosylate/hydrolase (*XTH1*) genes was observed after cold storage in $CaCl_2$ and NAA-treated fruits, respectively (Figueroa et al. 2012). In addition, a study to examine the activity of *PG* and *PE* enzymes during storage of dragon fruit (*Hylocereus polyrhizus*) harvested at 28 and 34 days after anthesis and postharvest treated with 0, 2.5, 5.0 and 7.5 g/L $CaCl_2$ was performed by Awang et al. (2013). The *PG* activity was lower in younger fruit and vice-versa for *PE* activity. Increasing concentration of $CaCl_2$ effectively reduced the activity of both enzymes thus slowing down the softening process; these results are agreement with Figueroa's finding. Another study that confirms the influence of calcium in fruit defence was done by Yang et al. (2013). The authors studied seven transcription factors dependent of calcium (calcium/calmodulin-regulation) in tomato fruit exposed to different stresses such as chilling, wounding, and *Botrytis cinerea* infection. Gene expression studies revealed that the seven transcription factor differentially respond to different stress signals. *SlSR2* was the only gene upregulated by all the treatments, the rest of studied genes were upregulated by salicylic acid (Yang et al. 2013). A combined action of calcium with other kind of treatment shows a synergic effect in comparison with every treatment alone. This was proved by Yu et al. (2012) who evaluated the efficacy of the biocontrol yeast *Cryptococcuslaurentii* and $CaCl_2$ in supressing the blue and gray mold rots in pear fruit wounds. Their results showed that combined treatment was much better approach for inhibition of *Penicillium expansum* and *Botrytis cinerea* infection than *C. laurentii* or $CaCl_2$ alone. Calcium chloride neither affected the growth of *C. laurentii in vitro* or *in vivo*, nor directly inhibited the mold rots in pear fruit. However, $CaCl_2$ was shown to elicit the fruit resistance to mold rots when the time interval between $CaCl_2$-treatment and pathogen-inoculation was increased up to 24 h, being associated with an activation of peroxidase activity of pear fruit. The authors concluded that the mechanism by which $CaCl_2$ reinforced the biocontrol efficacy of *C. laurentii* has been mainly due to its ability to induce the fruit natural resistance (Yu et al. 2012).

Salicylic acid or acetyl salicylic acid (ASA), a synthetic analog of SA, when applied exogenously induced the expression of pathogenesis-related (*PR*) genes and also conferred resistance against various pathogens. Exogenous application of SA at nontoxic concentrations to susceptible fruits could enhance resistance to pathogens and control postharvest decay (Asghari and Aghdam 2010). Salicylic acidin concentration dependent manner from 1 to 2 mM effectively reduced fungal decay in Selva strawberry fruit (Barbalar et al. 2007). In Hayward kiwifruit postharvest decay was significantly affected by methyl salicylate (MeSA) vapor at the end of storage period. Decay incidence in fruit treated with 32 ml/L was 6.3% whereas it was 34.2 percent in control fruits (Aghdam et al. 2009). Recently, persimmon fruits cv. Karaj were treated at harvest with SA at 1 and 2 mM to study its possible storage extension in comparison with untreated fruit (Khademi et al 2012). The most noticeable effect of postharvest SA application on stored persimmon fruit was the reduction of disease incidence at 2mM concentration, while 1 mM SA failed to control diseases. Furthermore the results showed that SA did not affect fruit commercial qualities such as fruit firmness, titratable acidity or color, however was unable to suppress ethylene production (Khademi et al 2012). SA treatment at 2 mM concentration reduce postharvest disease incidence of persimmon fruit by inducible defence mechanism, being suitable for increasing postharvest life of the fruit. In addition, fungal growth inhibitory characteristics have been found in SA by Panahirad et al. (2012). The results obtained by the authors from *in vitro* experiments showed that SA significantly reduced *Rhizopus stolonifer* growth at all concentrations used, at 5 mM, the growth was completely inhibited. *In vivo* experiments performed in peach showed relatively high level of inhibition, though the postimmersion treatment as compared with the preimmersion treatment resulted in higher inhibitory effects (Panahirad et al.2012). Cao et al. (2013) has obtained proofs of the defence induction from treated Chinese jujube fruit by SA application. Disease incidence and lesion area in the jujube fruit inoculated with *Alternaria alternate* were significantly inhibited by 2 and 2.5mM SA dipping. Naturally infected decay rate and index in jujubes were also significantly reduced by SA dipping during long-term storage at 0°C. SA enhanced activities of the main defence-related enzymes including phenylalanine ammonia-lyase, peroxidase, chitinase and b-1,3-glucanase in the fruit during storage. SA strongly decreased catalase activity but increased superoxide dismutase activity and ascorbic acid content in jujubes. Therefore, the beneficial effects of SA on fruit protection may be due to its ability to activate several highly coordinated defence-related systems in jujubes, instead of its fungicidal activity (Cao et al. 2013). SA is also used in combination with others compounds to improve the beneficial action in fruit. Shirzadeh and Kazemi (2012) combined SA, calcium nitrate and mentha essential oil treatment to study its effectiveness on postharvest quality and storage behavior of apples cv. Granny Smith. Apples were dipped in concentrations, 2.5 mM of SA, 1 percent of calcium nitrate and 150ppm of essential oil for 10 min and stored at 0–2°C for 20 weeks for shelf life analysis. The fruit quality parameters were measured at 20, 80, and 140th days of postharvest life. The results obtained by the authors indicated the potential improvement of shelf life of apples by SA and calcium nitrate pretreatment of the fruit, since this treatment did further improve product quality (Shirzadeh and Kazemi 2012).

7.3.4   OTHER CHEMICALS WHICH INDUCE PLANT DEFENCES

Although calcium and salicylic acid are the most used compounds to induce fruit defences by exogenous application in postharvest treatments, there are other compounds which cause the same effect. b-Aminobutyric acid as defence inducer has been studied by Zhang et al. (2013). The authors analyzed the effect of b-amino-butyric acid (BABA) on control of anthracnose caused by *Colletotrichum gloeo-sporioides* in mango fruit. The results obtained by the authors show that BABA treatments effectively suppressed the expansion of lesion in mango fruit inoculated with *C. gloeosporioides* during storage at 25°C, with the greatest efficacy being obtained using 100 mM BABA. However, BABA at 25–400 mM did not exhibit direct antifungal activity against *C. gloeosporioides*. Furthermore, BABA treatment at 100 mM enhanced the activities of b-1,3-glucanase, chitinase and phenylalanine amonio lyase. b-Aminobutyric acid treatment also contributed to the accumulation of hydrogen peroxide, while decreasing the rate of superoxide radical production. Concurrently, BABA increased the activity of superoxide dismutase while inhibiting catalase and ascorbate peroxidase activities. These results indicated to the authors that increased disease resistance of mango fruit after BABA treatment during storage might be attributed to an elicitation of defence response involving in the enhancement of defence-related enzyme activities and modulation of antioxidant system activities (Zhang et al. 2013).

Another example is a natural elicitor from *Arcitum lappa* designated burdock frutooligosaccharide (BFO). The effects of BFO in controlling postharvest disease in grape, apple, banana, strawberry, and pear were investigated by Sun et al. (2013). The disease index, decay percentage and area under the disease progress curve indicated that BFO has general control effects on postharvest disease of fruits. Kyoho grapes were studied by the authors to elucidate the mechanism of BFO in boosting the resistance of grapes to *Botrytis cinerea*infection. BFO treatment induced up-regulation of the atrionatriuretic peptide receptor A (*NPR1*), pathogenesis-related protein 1 (*PR1*), phenylalanine amonio lyase (*PAL*) and stilbene synthase (*STS*) genes, and inhibited the total phenol content decrease, which activated chitinase and b-1,3-glucanase. These results indicated that the salicytic acid-dependent signaling pathway was induced. The delayed color change and peroxidase and polypheno-loxidase activity suggested that BFO delayed grape browning. The reduced respiration rate, weight loss, and titratable acidity prolonged the shelf life of postharvest grapes. Therefore BFO is a promising elicitor in postharvest disease control (Sun et al. 2013).

Nevertheless, not all resistance inducers in the market can act against the fungal growth for postharvest diseases control, and therefore an evaluation of these chemical is necessary. Moscoso-Ramirez and Palou (2013) have evaluated seven different chemical for green and blue mold control on oranges (cultivar "Valencia" or "Lanelate"). The authors performed preventive and curative postharvest treatment with selected chemical on oranges artificially inoculated with *Penicillium digitatum* and *Penicillium italicum*. *Invivo* primary screening to select the most effective chemicals and concentration were done with benzothiadiazole, b-aminobutyric acid, 2,6-dichloroisonicotinic acid, sodium silicate, salicylic acid, acetylsalicylic acid and harpin. 2,6-dichloroisonicotinic acid at 0.03mM, salicylic acid at 0.25 mM, b-aminobutyric acid at 0.3 mM and benzothiadiazole at 0.9 mM were selected and tested afterwards as dips at 20°C for 60 s or 150 s with oranges artificially inoculated before or after the treatment and incubated for 7 days at 20°C. Although it was an effective treatment, sodium silicate at 1 M was discarded because of potential phytotoxicity to the fruit rind. Preventive or curative postharvest dips at room temperature had no effect or only reduced the development of green and blue molds very slightly. Therefore, the authors could not recommended these treatments for inclusion in postharvest decay management programs for citrus packinghouse (Moscoso-Ramirez and Palou, 2013).

The use of combined treatments or to establish a protocol of action for disease control in postharvest storage is the actual trends in this field. Preharvest dips of mango fruit in plant defence inducing chemicals (PDIC) integrated with postharvest treatments with inorganic salts and hot water were evaluated for management of anthracnose on artificially inoculated mango fruits by Dessalegn et al. (2013). Either of the PDICs salicylic acid or potassium phosphate at 100 mg/L combined with a fruit dip for 3 min in 3 percent aqueous sodium bicarbonate at 51.5°C significantly reduced disease development as compared to other treatments and untreated control. This combination kept anthracnose severity, as lesion development, below 5 percent during much of the 12 days experimental period and had the maximum proportion of marketable fruit (93.3%). The mean disease severity on untreated control fruit exceeded 30%, disease incidence reached 100 percent and marketability dropped to 0%. The treatments also maintained quality of mango; pH, total soluble solids, titratable acidity, firmness, and color of mango fruit significantly differed from those of the control. Heating calcium chloride (3%) to 51.5°C did not significantly improve this effect on severity of mango anthracnose even when combined with preharvest PDICs. The integrated measures involving sodium bicarbonate offer effective options for the management of mango fruit rot due to anthracnose (Dessalegn et al. 2013).

7.3.5   BLUE LIGHT AS METHOD OF CONSERVATION

Light has been shown to have a profound effect on growth and metabolism in several species of plant pathogens and food contaminants (Schmidt-Heydt et al. 2010), raising the possibility of using light as a strategy for decay control. The circadian clock modulates plant responses to environmental stress and the role of the light signaling in plant-pathogen interaction is becoming increasing evident (Roden and Ingle 2009). Phospholipases D and A$_2$ (*PLD* and *PLA$_2$*) and their enzymatic substrates and produces are involved in plant immunity responses and are key elements in the lipid signaling pathway (Munnik et al. 1998, Ryu et al. 2008). Red and white light inhibit *PLD* activity in etiolated oat seedlings, whereas far-red light counteracts this effect, suggesting the involvement of phytochrome (Kabachevskaya et al. 2007). In citrus, diurnal expression of different genes encoding *PLD* enzymes has been demonstrated: *CsPLDa* was regulated by light in citrus leaf blade but not in fruit peel, whereas *CsPLDg* was regulated by light in both tissues (Malladi and Burns 2008). In citrus leaf blade, both *CssPLA$_2$a* and *CsPLA$_2$b* gene expression was redundantly mediated by blue, green, red, and red/far-red light, but blue light was a major factor affecting *CssPLA$_2$a* and *CsPLA$_2$b* expression (Liao and Burns 2010).

Blues light also has marked effect on growth and development of plants, including flowering, stomatal movement and phototropic responses (Christie 2006). Furthermore, direct effects of blue light on fungal development occur in a variety of fungal species. For example, blue light promoted sporulation of *Paecilomyces fumosoroseus* and *Trichoderma atrovirie* (Sánchez-Murillo et al. 2004, Casas-Flores et al. 2006), sexual transmission in *Cryptococcus neoformans* (Idnurm and Heitman 2005), supporting the global effects of blue light on development of many fungi. Alferez et al. (2012) reported that blue light with emission between 410 and 540 nm with peak emission at 460 nm, reduced *Penicillium digitatum* infection in harvested tangerines. Alternating cycles of 12 h blue light/12 h dark greatly reduced fungal colonization in "Fallglo" tangerines as compared to continuous dark conditions. Furthermore, the percentage of infected wounds was similarly reduced in fruit held under the 12 h blue light/12 h dark cycle (Alferez et al. 2012). The authors observed a correlation of the inhibitory effect of blue light and the increment of PLA$_2$ genes transcription. Therefore, the blue light decay reduction is not simply due to inhibition fungal growth, since *P. digitatum* growth and sporulation were unaffected by blue light. However, blue light treatment induced *PLA$_2$* gene expression and reduce rate of infection, whereas inhibition of *PLD* expression by red light was correlated with infection. These data strongly suggested to Alferez et al. that induction of lipid signaling cascade by light inhibits fungal colonization (Alferez et al. 2012). Further experiments showed that blue light effectively suppress mycelial growth and postharvest symptoms development caused by *P. citri* and *P. italicum* beside *P. digitatum* on tangerine and sweet orange. However, there was a little to no effect on decay caused by *Lasiodiplodia theobromae* and *Colletotrichum gloeosporioides*. The condtion of *P. digitatum* and *P. italicum* on flavedo was suppressed by blue light but the fungal growth was not significantly affected in tissues beneath the fruit surface where blue light had limited penetration. Blue light reduce *Penicillium* postharvest decay reducing fungal growth of *P. italicum* and *P. citri*, reduce cell-wall digestion enzyme activity of *P. digitatum* and induce flavedo octanal production (Liao et al

2013), which is an aldehyde volatile with potent antimicrobial characteristics for several pathogens (Utama et al. 2002).

7.3.6 *PHYSICAL POSTHARVEST TREATMENT THAT INDUCE PLANT DEFENCE*

Many physical postharvest treatments induce natural defence mechanisms as stress inflicted on the fruit, therefore, the treatment starts defence pathways stimulated by mechanical response. In other cases the treatment stimulates the plant defences by itself. Zeng et al. (2010) published an interesting work about the evaluation of disease resistance and reactive oxygen species metabolism in harvest navel oranges (*Citrussinensis* L. *Osbeck*) treated with chitosan. Fresh navel oranges were treated with 2 percent chitosan or 0.5 percent glacial acetic acid (control) solution 1 min, and some of them were inoculated with *Penicillium italicum* and *Penicillium digitatum*. Then, the fruit were stored at 20°C and 85–90 percent relative humidity. Treatment with 2% chitosan significantly reduced the disease incidence and the lesion diameter compared with control fruit. This treatment effectively enhanced the activities of peroxidase and superoxide dismutase, and levels of glutathione and hydrogen peroxide, inhibited the activities of catalase and the decreases of ascorbate content during navel oranges fruits storage. Ascorbate peroxidase activity in the orange fruit was induced slightly by the chitosan treatment during 14-21 days storage. However, glutathione reductase activity in the fruit was not enhanced by the chitosan treatment. The results obtained by the authors indicated that chitosan treatment could induce the navel orange fruit disease resistance by regulating the hydrogen peroxide levels, antioxidant enzymes and ascorbate-glutathione cycle (Zeng et al. 2010). Furthermore, the effects of chitosan and oligochitosan on resistance induction of peach fruit against brown rot caused by *Monilinia fructicola* were investigated by Ma et al. (2013). Both chitosan and oligochitosan showed significant effect on controlling this disease. Moreover, chitosan and oligochitosan delayed fruit softening and senescence. The two antifungal substances enhanced antioxidant and defence-related enzymes, such as catalase, peroxidise, b-1,3-glucanase and chitinase and they also stimulated the transcript expression of peroxidise and glucanase. These finding suggested to the authors that the effects of chitosan and oligochitosan on disease control and quality maintenance of peach fruit is associated with their antioxidant property and the elicitation of defence responses in fruit (Ma et al. 2013), as also has been observed in the previous study. In fact, the efficacy of practical grade chitosan when used in solution and water-soluble commercial chitosan formulation were compared to each and with other plant resistance inducers such as acetic, glutamic formic and hydrochloric acids by Romanazzi et al. (2013), in controlling postharvest diseases of strawberry. The commercial chitosan formulation and other resistance inducers based on benzothiadiazole, oligosaccharides, soybean lecithin, calcium and organic acids, and *Abies sibirica* and *Urtica dioica* extracts were also tested. The commercial chitosan formulation was as effective as the practical grade chitosan solutions in the control of gray mold and *Rhizopus* rot of strawberries immersed in these solutions and kept for 4 days at 20±1°C. Moreover, the treatment with commercial and experimental resistance inducers reduced gray mold, *Rhizopus* rot and blue mold of strawberries stored 7 days at 0±1°C and then exposed to 3 days shelf life. The highest disease reduction was obtained with commercial chitosan formulation, followed by benzothiadiazole, calcium, and organic acids (Romanazzi et al. 2013).

In other occasions, a physical action inflicted on fruit can shoot the plant defences. Hypobaric treated (50kPa, 4h) strawberries had reduced rot incidence from natural infection during subsequent storage for 4 days at 20°C and later inoculation with *Botrytis cinerea* and *Rhizopus stolonifer* spores. Biochemical analyses of strawberries performed by Hashmi et al. (2013), suggested that activities of defence-related enzymes were increased with the hypobaric treatment; phenylalanine ammonia-lyase (PAL) and chitinase peaked 12h after treatment, while peroxidase increase immediately. Polyphenol oxidase activity remained unaffected during after storage for 48h at 20°C. In addition, the effect of low oxygen treatment (10% at 101 kPa, 4h) was also investigated to determine if the lower partial pressure of oxygen generated during hypobaric treatment contributed to the observed effect. However the low oxygen treatment did not influence rot development, suggesting that the treatment effects were pressure rather than oxygen related. These results have suggested that hypobaric treatment causes reduced decay incidence due to stimulation of defence-related enzymes (Hashmi et al. 2013). Another example of physical treatment which can induce proteins related with plant defence is heat treatment. The temperature increase have a direct effect slowing germ tube elongation or of inactivating or outright killing germinating spores, thus reducing the effective inoculum size and minimizing rots. Heat treatment can also indirectly affect decay development via physiological responses of the fruit tissues, or enhancing wound healing. Heat treatment can induce chitinase and b-1,3 glucanase, stabilize membrane, elicit antifungal compounds, or inhibit the synthesis of cell-wall hydrolytic enzymes and delay the degradation rate of preformed antifungal compounds that are present in unripe fruit (Schirraa et al. 2000, Pavoncello et al. 2001).

7.4 CHEMICAL POSTHARVEST TREATMENTS

When referring to chemical treatments it is mainly fungicides application but also can include others chemicals such as ozone, chloride dioxide gas or sulfur dioxide. Fungicides are biocidal chemical compounds or biological organism used to kill or inhibit fungi or fungal spores. The most used fungicides in postharvest treatment are contact compound which are not taken into the fruit tissues however some kind of fungicide residues have been found for human consumption. Some fungicides are dangerous for human health, such as vinclozolin, which has now been removed from use. The actual trend is to use natural compounds which are safe for human health and environmental. These chemical derived from plant extracts or essential oils such as Cinnamon, Jojoba, Neem, Oregano, Rosemary, Tea, and many more that are been discovered every day.

7.4.1 GASES USED IN POSTHARVEST TREATMENTS

Ozone is 3,000 times more efficient than chlorine in destroying microbes. Ozone or trioxygen (O_3) is formed from dioxygen (O_2) by action of ultraviolet light but is much less stable than O_2, breaking down in the O_2 + oxygen radical (O^-). For this reason ozone is a powerful oxidant which causes damage to microorganism and it is used as biocidal.However ozone can also be dangerous to plants and animals above 100ppm due to can damage mucus and respiratory tissues.

Ozone treatment can delay the fruit ripening in 20 or 30 percent which increase the shelf life of treated food. This is mainly due to ozone reaction with ethylene ($H_2C=CH_2$). Additionally the high oxidative power of ozone reacts with carbon dioxide (CO_2) and water (H_2O) as follow:

$$H_2C=CH_2 + O_3 \text{''} C_2H_4O + O_2$$

$$C_2H_4O + 2O_2 \text{''} 2CO_2 + 2H_2O + O^-$$

It should be note that ethylene oxide (C_2H_4O), product of the first reaction, is an effective inhibitor of microorganism, which effect could be added to ozone effect.

In 1997 Food and Drug Administration of United State (US-FDA) declared ozone to be generally recognized as safe (GRAS) for food contact application in the United States (Graham et al. 1997, US-FDA 1997). For the postharvest treatment of fresh fruit, ozone can be used as a relatively brief prestorage or storage treatment in air or water, or as a continuous or intermittent component of the atmosphere throughout storage or transportation. Palou et al. (2002) published a research focused on the effect of ozone on stone fruit and table grapes and the continuously supplied gas at 0.3 ppm. This treatment inhibited aerial mycelial growth and sporulation on "Elegant Lady" peaches wound inoculates with *Monilinia fructicola*, *Botrytis cinerea*, *Mucor piriformis*, or *Penicillium expansum* and stored for 4 weeks at 5°C and 90% of relative humidity. Aerial growth and sporulation, however, were resumed afterward in ambient atmosphere. Gray mold nesting among "Thompson Seedless" table grapes was completely inhibited under 0.3 ppm ozone when fruit were stored for 7 weeks at 5°C. Gray mold incidence, however, was not significantly reduced in spray inoculated fruit (Palou et al. 2002).

Another powerful oxidant gas used in food industry is chloride dioxide (ClO_2) and it is used as disinfectant agent. Trinetta et al. (2010) has published a study about the potential use of ClO_2 in the food disinfection. The authors' objective is to offer tomatoes free of human pathogens but the treatment also affect to natural microbe population and consequently to the shelf life of tomatoes. Trinetta evaluated high-concentration-short-time chloride dioxide gas treatments effects on *Salmonella*-inoculated "Roma" tomatoes and determined the optimal treatment conditions for microbial inactivation and shefl-life extension. Effects of ClO_2 concentration (2, 5, 8, and 10 mg/L) and exposure time (10, 30, 60, 120, and 180 s) on inoculated tomatoes with *Salmonella enterica* strains serotype Montevideo, Javiana, and Baildon were studied. After ClO_2 treatments, tomatoes were stored at room temperature for 28 days. Inherent microbial population, changes in tomato color and chloride dioxide gas residuals were evaluated by the authors. The statistic analysis of obtained results showed that both ClO_2 concentration and exposure time were determinant for *Salmonella* inactivation. Surviving *Salmonella* population of 3.09, 2.17 and 1.16 log CFU/cm² were obtained treating tomatoes with 8 mg/L ClO_2 for 60 s, 10 mg/L ClO_2 for 120 s and 10 mg/L ClO_2 for 180 s, respectively (initial inoculum was 6.03 ± 0.11 log CFU/cm²). The selected treatments significantly reduce background microflora while fruit color and residual contents were not significantly different as compared with control. These results did suggest the authors to propose high-concentration-short-time chloride dioxide gas treatments as an effective method in pathogen inactivation during tomatoes packaging (Trinetta et al. 2010).

Sulfurdioxide (SO_2) is a natural and toxic gas that could be found during volcanic activity and is a potent global warming gas. Further oxidation of SO_2, usually in the present of a catalyst such as NO_2 forms H_2SO_4, and thus acid rain. However, it is a potent disinfectant gas used in food industry to rid of pathogens the export fruits. The effectiveness of SO_2 concentration x time (Ct) treatments on gray mold control produced by *Botrytis cinerea* was determined in the laboratory and validated, prior to refrigerating the fruit, using pallet scale SO_2 fumigation treatment on the following blueberry cultivars: "Brigitta", "Legacy", "Liberty" and "O'Neal". In inoculated "Brigitta" and "Liberty" blueberries, gray mold prevalence varied from 97.2 percent to 97.5 percent in nontreated fruit and this value was reduced from 7.9 percent to 6.1 percent in blueberries that were exposed to a SO_2 Ct product of 400 (ml/L)/h. The relationship between SO_2 Ct products and gray mold prevalence under laboratory conditions was best explained by exponential models, which had a determination coefficient (R^2) that ranged from 0.88 to 0.96. The estimated EC_{90} values varied between 245 and 400 (ml/L)/h was validated using a pallet scale application treatment to obtain the best control and minimal variation (Rivera et al. 2013).

The most extend gas used in postharvest industry is 1-Methylcyclopropene (1-MCP). The mode of action of this gas consist that 1-MCP involves its tightly binding to the ethylene receptors in plants, thereby blocking the effects of ethylene and delay the fruit ripening. The substance is applied in gas form (as a fumigant) in the storage room. In long-term postharvest cold storage, fruit are placed in boxes (usually plastic or wooden bins) and stacked in a specific pattern. The top of the boxes are frequently covered with a thin plastic sheet for the purpose of reducing fruit moisture loss. Wooden boxes, card lining and other plant porous materials used in bins have 1-MCP adsorption capacity. Plastic covers affect the airflow and with that the 1-MCP transport. The influence of box materials and plastic cover on the distribution of 1-MCP in cold storage was studied by Ambaw et al. (2013) using validates computational fluid dynamics models. Reynolds Average Navier-Stokes equations with the SST k-ω turbulence model were used to calculate the airflow. Diffusion, convection, and adsorption of 1-MCP were modeled to obtain 3D spatial and temporal distribution of 1-MCP inside a storage container, boxes, andfruit. Time dependent profiles of calculated 1-MCP concentrations in air in the container agreed well with measurement data. The plastic cover imposed no effect on the adsorption of 1-MCP. Wooden boxes notably absorbed 1-MCP from the treatment atmosphere and may reduce the efficacy and uniformity of the treatment (Ambaw et al. 2013).

7.4.2 FUNGICIDES APPLIED FOR CONTROL OFPOSTHARVEST DISEASES

The application of fungicides is one of the methods that are used to suppress the fungal growth. Inefficient fungicide treatment of fruits could result in lower biological efficacy, environmental contamination, loss of fungicide, fungicide residue above the allowable limit and decay due to fruit injury; therefore the fungicide application should be properly designed. A computational fluid dynamics (CFD) models was used by Delele et al. (2012) to study the operation of postharvest storage fungicide fogging systems. The modeling was based on an Eulerian-Lagrangian multiphase flow model. The effect of air circulation rate, circulation intervals, bin design, stacking pattern and room design on deposition of fungicide was investigated by the authors. Air circulation rates of 0 m³/h (no circulation), 2,100 and 6,800 m³/h were used. Interval circulation of air was also investigated. Obviously, the highest fungicide deposition was observed during fogging without circulation while the lowest deposition corresponded to fogging with the highest circulation rate. For the considered on/off combination times, the effect of circulation interval on overall average deposition and uniformity was not significantly different from the case of fogging with continuous air circulation. Bin with higher vent hole ratio and the presence of air deflector increased the amount and uniformity deposition. Good agreement was found between measured and predicted results of deposition of fungicide particles (Delele et al 2012). An alternative fungicide application is by wax emulsion for application by spray; actually it is the currently most used method to apply Imazalil (IMZ) for controlling postharvest fungal pathogens in citrus. Alteri et al. (2013) have published a study where describe tests with a pilot plant utilized to develop a method called Imazalil thim film treatment (ITFT). The method was designed to reduce the amount of fungicide used and problems related to wastewater disposal. The efficacy of three methods of fungicide application was tested by the authors, spraying, dipping, and ITFT. The decay of fruit and IMZ residues in fruits were evaluated over a 60 days storage period at 5°C and after one week of shelf life at

20°C. At the end of the shelf life period, the incidence of curative fruit decay after ITFT treatment (10.4%) which did not significantly differ from that of spraying (11.5%), but it was significantly higher than that of dipping (3%). During storage, the level of IMZ decreased for all treatments following first-order destruction rate kinetics (with a half-life to 19.4 days). For the dipping and ITFT treatments, IMZ residues were significantly higher than the residues in the spraying treatment. Further, significant differences in the IMZ residues were found between the dipping and ITFT treatments (Alteri et al. 2013).

On the other hand, the fungicides application time can change depending of kind of fruit and chemical compound. For example, postharvest anthracnose and stem-rot, caused by *Colletotrichum gloeosporioides* and *Lasidiplodia theobromae*, respectively can be severe problems to certain ethylene-degreened early season citrus cultivars. Preharvest application of fungicides can be an effective approach for control of these diseases. Zhang and Timmer (2007) studied the potential of five fungicides: benomyl, thiophanate methyl, azoxystrobin, fludioxonil, and pyraclostrobin applied 2, 14, 21, and 28 days before harvest for control of postharvest anthracnose, stem rot and green mold (*Penicillium digitatum*) on early season Florida Fallglo and Sunburst tangerine hybrids in 2003 and 2004. Most fungicides significantly reduced anthracnose incidence on Fallglo when applied 2 days before harvest in both years and 14 and 21 days before harvest in 2003. At other application dates, none of the fungicides was effectives. On Fallglo fruit in 2004, the five fungicides reduced postharvest anthracnose by 37.4–62.6 percent when sprayed 2 days before harvest. Little anthracnose was observed on Sunburst fruit in either year. On both cultivars and in years, benomyl and thiophanate methyl consistently and significantly reduced stem-rot incidence. Other fungicides were less effective or ineffective in controlling stem-rot. Benomyl and thiophanate methyl also reduced green mold incidence by 58.9–100 percent on Sunburst tangerines in 2004, but were ineffective in 2003 due to the presence of resistant strains. Thiophanate methyl appeared to be an excellent alternative to benomyl for citrus postharvest diseases control since benomyl has been removed from the market (Zhang and Timmer 2007). Another alternative is to apply the fungicide directly to harvested fruit during postharvest process. Imazalil (IMZ) is widely used in citrus packhouses to manage green mold, caused by *Penicillium digitatum*. Njombolwana et al. (2013) has evaluated the green mold control efficacy of IMZ applied in a wax coating, and the combination of aqueous dip and coating IMZ applications. Single application of IMZ at 3 mg/mL in carnauba wax coating at rate of 0.6, 1.2, and 1.8 L/ton of fruit gave better protective (mean 13% infection) than curative (mean 70% infection) control of the sensitive isolate. Imazalil residue levels increased (0.85 to 1.75 mg/g) with increasing coating load. However, the resistant isolate could not be controlled (>74% infection). Dip only treatment (IMZ sulfate at 500 mg/mL for 45 s and 90 s) gave good curative control of the sensitive isolate at residue loading of 0.12–0.73 mg/g. Wax coating only treatment (IMZ at 3 mg/mL at 1.8 L wax ton) gave good protective control and improved sporulation inhibition at residue loading of 1.32-7.09 mg/g. The maximum residue limit of 5 mg/g was exceeded at high wax loads on navels and clementines. Double application with dip (45 s in IMZ sulfate at 500 mg/mL) followed by 2 mg/mL IMZ in wax coating at 0.6, 1.2 and 1.8 L wax /ton resulted in residue loading of 1.42–2.83 mg/g, increasing protective control as well as curative control. In all treatments, poor curative and protective control of the resistant isolates was observed. Finally, the authors concluded that double application of Imazalil demonstrated superior green mold control by giving good curative and protective control and sporulation inhibition (Njombolwana et al. 2013).

As has already been comment before, the great problem of fungicides application is not that can leave residues on the fruits or the contaminant waste from postharvest treatments. The big problem is the inadvertent selection of resistant fungal strains of fungicides used. For example, anthracnose disease caused by *C. gloeosporioides* is a worldwide problem and is especially important in Southeast Asia owing to the severe economic damage they cause to tropical fruits that are grown for local consumption and export. Benzimidazoles are systemic fungicides widely used for controlling anthracnose disease in Taiwan. Thirty-one isolates of *C. gloeosporioides* from mango and strawberry grown in Taiwan were examined for their sensitivity to benzimidazole fungicides (Chung et al. 2010). The response of the isolates grown on benzimidazole-amended culture media were characterized as sensitive, moderately resistant, resistant, or highly resistant. Analysis of point mutations in b-tubulin gene by DNA sequencing of PCR-amplified fragment revealed a substitution of GCG for GAG at codon 198 in resistant and highly resistant isolates and substitution of TAC for TTC at codon 200 in moderately resistant isolates. A set of specific primers, TubGF1, and TubGR, was designed to amplify a portion of the b-tubulin gene for the detection of benzimidazole-resistant *C. gloeosporioides*. *Bsh*1236I restriction maps of the amplified b-tubulin gene showed that the resistant isolate sequence, but not the sensitive isolate sequence was cut. The PCR restriction fragment length polymorphism (PCR-RFLP) was validated to detect benzimidazole-resistant and benzimidazole-sensitive *C. gloeosporioides* isolates recovered from avocado, banana, carambola, dragon fruit, grape, guava, jujuve, lychee, papaya, passion fruit and wax apple. Therefore, this method can be used as valuable tool for monitoring the occurrence of benzimidazole-resistant *C. gloeosporioides* and for assessment of the need for alternative management practise (Chung et al. 2010). Another example is possible to detect with thiabendazole (TBZ). This fungicide is commonly used as a postharvest treatment for control of blue mold in apples caused by *Penicillium expansum*. Different point mutations in the b-tubulin gene conferring benzimidazole resistance have been reported in plant pathogens, as stated above, but molecular mechanism of TBZ resistance in *P. expansum* from apples has been studied by Yin and Xiao (2013) very recently. The determination of TBZ resistance level showed that all 102 TBZ-resistant (TBZ-R) isolates were highly resistant. Sequencing of the majority of the b-tubulin gene showed that 76 TBZ-R isolates harbored the E198V mutation and 26 harbored the F167Y mutation, and all the sensitive isolates did not posses any of the mutations, indicating that these two point mutations in the b-tubulin gene were correlated with TBZ resistance in *P. expansum* from apple. There was no association between levels of TBZ resistance and types of mutations (E198V or F167Y) in the b-tubulin gene. A multiplex allele-specific PCR assay was developed by the authors to detect these two mutations simultaneously. Microsatellite-primed PCR derived presence-absence matrix used to assess the genetic relationship among 56 isolates suggested that the resistance mutations originated several times independently and that there was no correlation between the type of point mutation and the genetic background of isolates (Yi and Xiao 2013).

7.4.3 ESSENTIAL OILS AND PLANT EXTRACTS

An essential oil is a concentrated hydrophobic liquid containing volatile aroma compounds from plants. Oil is "essential" in the sense that it carries a distinctive scent, or essence, of the plant. Essential oils and their components are gaining increasing interest by consumers due to its eco-friendly and biodegradable properties (Tzortzakis 2007). Application of essential oils is an attractive method to control postharvest diseases in postharvest systems because their bioactivity in the vapor phase and the limitation of aqueous sanitation for many commodities, make them useful as possible fumigants. Hussein et al. (2012) evaluated the antifungal properties of several essential oils and plant extracts. Harmal seeds (*Peganum harmala* L.), cinnamon bark (*Cinnamomum cassia* L.) and sticky fleabane leaves (*Inula viscosa* L.) combined with food preservatives such as sodium benzoate, sodium molybdate, ammonium heptamolybdate tetrahydrate, potassium carbonate and sodium bicarbonate, against *Penicillium digitatum* and *Penicillium italicum*. Both disc agar diffusion and broth dilution methods were used to evaluate the antifungal activity of the plant extracts and food preservatives *in vitro*. Results revealed that methanolic fraction of cinnamons' bark and sticky fleabane leaves showed the highest efficacy. Minimal inhibitory concentration (MIC) values of 150 and 37.5 mg/mL were obtained with cinnamons fraction against *P. digitatum* and *P. italicum*, respectively. Sodium benzoate was the most effective chemical against *P. digitatum* and *P. italicum* with obtained MIC of 37.5 and 75 mg/mL, respectively. Mixtures of sodium benzoate and fractions of either cinnamon or sticky fleabane leaves reflected synergistic effects against *P. italicum* and antagonistic effects against *P. digitatum* (Hussein et al. 2012). Although the results are very promising *in vitro*, the most important test is the application of treatment with essential oils on fruit and examine the decay incidence and the organoleptic characteristics of treated fruits. Four plant essential oils (cinnamon oil, linlool, *p*-cymene and peppermint leaf oil) and the plant oil-derived biofungicides Sporan (rosemary and wintergreen oils) and Sporatec (rosemary, clove and thyme oils) were evaluated as postharvest fumigants by Mehra et al. (2013) to manage fungal decay under refrigerated holding conditions. Hand-harvest Tifblue rabbiteye blueberry fruit were inoculated at the stem end with conidia suspension of

Alternaria alternata, Botrytis cinerea, Colletotrichum acutatum or sterile deionized water (check inoculation) and subjected to biofumigation treatment under refrigeration (7°C) for 1 week. Sporatec volatiles reduced disease incidence significantly in most cases. Whereas others treatments had no consistent effect on postharvest decay. Sensory analysis of uninoculated and biofumigated berries was performed utilizing a trained sensory panel and biofumigation was found to have significant negative impacts on several sensory attributes such as sourness, astringency, juiciness, bitterness, and blueberry-like flavor. Biofumigated fruits were also analyzed for antioxidant capacity and individual anthocyanins, and no consistent effects on these antioxidant-related variables were found in treated berries. Because of limited efficacy in reducing postharvest decay, negative impacts on sensory qualities and failure to increase antioxidant levels, the authors concluded that the potential for postharvest biofumigation of blueberries under refrigerated holding conditions appears limited (Mehra et al. 2013).

The efficiency of essential oils in postharvest fungal control directly depends of target fungus, treated fruit and oil used, considering all three as one system. Complete inhibition of mycelial growth of *Colletotrichum gloeosporioides in vitro* was observed with citronella or peppermint oils at 8 ml/plate and thyme oil at 5 ml/plate. Thyme oil at 66.7 ml/L significantly reduced anthracnose from 100 percent (untreated control) to 8.3 percent after 4 days and 13.9 percnet after 6 days in artificially wounded and inoculated avocados cultivars "Fuerte" and "Hass". Gas chromatography (GC/MS) analysis revealed thymol (53.19% RA), menthol (41.62% RA) and citronellal (23.54% RA) as the dominant compounds in thyme, peppermint, and citronella oils respectively. The activities of defence enzymes including chitinase, 1,3-b-glucanase, phenylalanine ammonia-lyase and peroxidise were enhanced by thyme oil (66.7 ml/L) treatment and the level of total phenolics in thyme oil treated fruit was higher than that in untreated (control) fruit. In addition, thyme oil treatment enhanced the antioxidant enzymes such as superoxide dismutase and catalase (Sellamuthu et al. 2013). A new approach to the control of postharvest decay, while maintaining fruit quality, has been implemented by the application of carboxymethyl cellulose coating enriched with extract of *Impatiensbalsamina* L. stems, a commonly used traditional Chinese medicine, amended coating to "Newhall" navel orangen. After harvest, navel oranges were dipped in amended coating and then the samples were stored at 5°C and 90–95 relative humidity after being dried naturally. The data suggested that the coating treatment reduced the decay rate and weight loss of fruit from 10.2 to 6.1 percent and from 6.33 percent to 2.91 percent, respectively after 100 days storage. None deleterious effects on fruit quality such as soluble solids content, titratable acidity and ascorbic acid content were observed. Moreover, the activities of scavenger antioxidant and defence enzymes, including peroxidase, superoxide dismutase, chitinase and b-1,3-glucanase were also increased by the enriched coating treatment. All these results suggest that the film may be an effective and safe alternative preservative for navel oranges (Zeng et al. 2013). Previous studies of Zeng et al. (2012) already had been revealed the potential use of essential oils, specifically clove extract, in postharvest handling of oranges fruits. After harvest, the authors dipped navel oranges in 100 mg/mL clove extract for about 3 min and then the samples were stored at 7°C and 90–95 percent of relative humidity. The results showed that the clove extract treatment significantly reduced the fruit's decay rate and weight loss compared with those of the control group. Meanwhile, total soluble solid content, titratable acidity and ascorbic acid content deterioration were substantially lower in the clove extract treated group than those of the control group. Moreover, the activities of superoxide dismutase, peroxidise, and chitinase were effectively enhanced by the clove extract treatment, while no difference in b-1,3-glucanase activity was found between the two groups. Based on this study, Zeng et al. (2012) proposed the use of plant extract/essential oils as alternative and effective treatment for extend the shelf life of stored navel oranges.

7.4.4 SALTS USED IN POSTHARVEST TREATMENTS

The used of inorganic salts have been increasing in last years, mainly because they are compounds easily available in the market. Furthermore, they show positive reaction in control of important postharvest diseases and they also are compatibles with treatments of sanitation and/or diseases control. For example, in the current year (2013) is possible to find a high number of publications regarding to salts application in postharvest disease control, some of them are shown in the present section.

Cerioni et al. (2013a) studied the application of potassium phosphite (KP) alone or in combination with others chemicals to control *Penicillium*mold in citrus. To control green or blue mold, fruit were inoculated with *P. digitatum* and *P. italicum* and then immersed 24 h later in KP, calcium phosphite (CaP), sodium carbomate, sodium bicarbonate or potassium sorbate for 1 min at 20 g/L for each at 25°C or 50°C. Mould incidence was lowest after potassium sorbate, CaP, or KP treatment at 50°C. CaP was often more effective than KP but left a white residue on fruit. KP was significantly more effective when fruit were stored at 10 or 15°C after treatment compared with 20°C. Acceptable levels of control were achieved only when KP was used in heated solution or with fungicides. KP was compatible with imazalil (IMZ) and other fungicides and improved their effectiveness, but KP increased thiabendazole and IMZ residues slightly. Phosphite residues did not change during storage for 3 weeks, except they declined when KP was applied with IMZ. KP caused no visible injuries or alteration in the rate of color change of citrus fruit in air or ethylene at 5 ml/L (Cerioni et al. 2013a).Another example of combined treatment studied by Cerioni et al. (2013b)was the application of potassium sorbate, sodium bicarbonate and potassium phosphite combined with heat and hydrogen peroxide in the presence of $CuSO_4$ to control major lemon postharvest diseases was investigated on inoculated fruit. Green and blue molds, which both require wounds for infections to occur, were controlled by combination of hydrogen peroxide followed by inorganic salts, even when the temperature solutions were 25°C. Control of sour rot was poor with salts solutions alone but significantly improved in treatments including hydrogen peroxide followed by potassium sorbate or sodium bicarbonate at 50°C. *Phomopsis* stem-end rot was effectively controlled by potassium sorbate and potassium phosphite at 20°C, and diplodia stem-rot was partially controlled only by potassium sorbate. Applications of either potassium sorbate or hydrogen peroxide followed by potassium phosphite were the most promising treatment, primary because they controlled most of the diseases without the need to heat the solutions. These treatments controlled postharvest citrus diseases to useful levels and could be suitable alternative to conventional fungicides, or could be applied with them to improve their performance or to manage fungicide resistant isolates (Cerioni et al. 2013b). On the other hand, Fallanaj et al. (2013) has evaluated the effect of electrolyzed salt solution using thin-film diamond-coated electrodes on *Penicillium* spp. population of citrus fruit wash water and fruit decay. Although different organic and inorganic salts were tested, electrolyzed water (EW) in combination with sodium bicarbonate (SBC) was the most effective treatment in inhibiting spore germination of *Penicillium* spp. and among the best in reducing *Penicillium* rot, with no deleterious effect on the fruits. Commercial trials conducted in packinghouses in Sicilia (insular Italy) confirmed that the electrolyzed SBC solution was more effective than the electrolyzed tap water in reducing the population of *Penicillium* spp. Indeed, in the presence of SBC a 93 percent reduction of the pathogen population was observed 1 h after the beginning of the electrolysis process, whereas in the absence of salt similar results were observed only after 7 h. In addition, rot incidence in fruit exposed to electrolyzed SBC solution was reduced by up to 100 percent as compared to 70 percent in absence of the salt. These results demonstrated that among the range of salts tested, the combination of electrolysis and SBC has a synergistic effect (Fallanaj et al. 2013).

7.4.5 BIOLOGICAL CONTROL IN POSTHARVEST DISEASES

In comparison to the field environment, postharvest environments are well defined; abiotic and biotic factors can be determined with relative ease and manipulated to an antagonist's advantage. Although the mechanism(s) of biocontrol have not yet been fully explained and, to date, there have been only a few attempts to exploit these mechanisms to improve postharvest biocontrol, reports on the mechanism of biocontrol in postharvest diseases suggest that competition for nutrients and space plays a major role in most cases. Rapid colonization of fruit wounds by the antagonist is critical for decay control and manipulations leading to improved colonization enhance biocontrol. Within microbial communities, interactions are density dependent, and then one type of interaction can occur at any one time, depending on the growth phase of different microorganisms, population density, and species diversity. Three different types of interactions, competition for nutrients, competition for space, and inhibition by secondary metabolites, were observed with preharvest sprays of *Bacillus subtilis* to control *Colletotrichum gloeosporioides* on avocado (Korsten et al., 1997). The main approaches used to improve biocontrol of postharvest diseases are manipulation of environment, use of mixed cultures of antagonists, physiological, and genetic manipulation of antagonists, combining field and postharvest applications, manipulation of formulations and integration with other methods (Janisiewicz and Korsten, 2002).

7.4.6 BACTERIAL ANTAGONISTS USED IN POSTHARVEST TREATMENTS

The principal bacterial genus used in postharvest biocontrol is *Bacillus*, followed closely by *Pseudomonas*, however, this one is more used in preharvest treatment. A very nice example of antagonists' test was published by Recep et al. (2009). Although his studies were applied to potato and this chapter is dedicated to fruit treatments, worth discussing their findings in this section. In Recep et al. study, a total of 17 Plant Growth Promoting Rhizobacteria (PGPR) strains, consisting of eight different species (*Bacillus subtilis*, *Bacillus pumilus*, *Burkholderia cepacia* (also known as *Pseudomonas cepacia*), *Pseudomonas putida*, *Bacillusamyloliquefaciens*, *Bacillus atrophaeus*, *Bacillus macerans* and *Flavobacterbalastinium*), were tested for antifungal activity *in vitro* (on Petri plate) and *in vivo* (on potato tuber) conditions against *Fusarium sambucinum*, *Fusarium oxysporum* and *Fusarium culmorum* cause of dry rot disease of potato. All PGPR strains had inhibitory effects on the development of at least one or more fungal species *in vitro*. The strongest antagonism was observed in *B. cepacia* strain with inhibition zones between 3.5 and 4.7 cm. All PGPR strains were then tested on tuber of two potato cultivars "Agria" and "Granola" under storage conditions. Only *B. cepacia* had significant effects on controlling potato dry rot caused by three different fungi species on the two potato cultivars (Recep et al. 2009). This report demonstrates the difficulties that may be found in biocontrol treatments. Moreover show inhibitory capacity, must have good adaptation to system fruit-pathogen-antagonist, adding adaptation to the storage conditions as well. However, the postharvest biocontrol is a strong and firm commitment in current researches. Wang et al. (2013a) analyzed the efficiency of *Bacillus subtilis* SM21 on controlling *Rhizopus* rot caused by *Rhizopusstolonifer* in postharvest peach fruit and the its possible mechanism of control. The *invitro* test showed significant inhibitory effect of *B. subtilis* SM21 on mycelial growth of *R. stolonifer* with inhibition rate close to 49%. Furthermore, *in vivo* application of SM21 strain resulted in a reduction of lesion diameter and disease incidence by 37.2 percent and 26.7 percent on the second day of inoculation compared with the control. Analysis of treated peaches showed that *B. subtilis* SM21 enhanced activities of chitinase and b-1,3-glucasane, and promoted accumulation of H_2O_2. Total phenolic content and 2,2-diphenyl-1-picrylhydrazyl (DPPH) radical-scavenging were also increased by this treatment. Moreover, transcription of seven defence related genes (*NPR1, PR, CHI, GNS, PAL, LOX1*, and *CAT1*) was much stronger in fruit treated with *B. subtilis* SM21 or those both inoculated with *R. stolonifer* and treated with *B. subtilis* compared with fruit inoculated with *R. stolonifer* alone. These results suggested to the authors that *B. subtilis* SM21 could control *Rhizopus* rot by directly inhibiting growth of the pathogen, and indirectly inducing disease resistance in the fruit (Wang et al. 2013a).

Treatments with antagonist agents or biopesticides are increasing. In fact, treatments with *Bacillus* have been compared with other more traditional postharvest treatment such as chitosan application. Casals et al. (2012) published the finding from application of chitosan or *Bacillus subtilis* (strain CPA-8) on peaches preexposed to 50°C for 2 h and 95–99 percent of relative humidity. These two treatments were evaluated for their ability to prevent *Monilinia fructicola* infections and their ability to complement heat treatment. Two chitosan concentrations (0.5% or 1%) were applied at three temperatures (20, 40 and 50°C) for 1 min to wounded and unwounded fruit that were artificially inoculated with *M. fructicola*. One percent chitosan applied at 20°C had a preventive effect against further *M. fructicola* infection on heat-treated fruit that has been previously inoculated: brown rot incidence was reduced to 10 percent in comparison with the control (73%). However, chitosan applied to wounded fruit had a poor preventive effect. The antagonist, *B. subtilis* CPA-8, had a preventive effect in controlling *M. fructicola* infection: the incidence of brown rot was reduced to less 15 percent for both varieties evaluated ("Baby Gold 9" and "Andros" peaches), in comparison with the control fruit (higher than 98%). In contrast, when fruit were stored at 0°C, this preventive effect was not detected. These finding indicated that heat-treated fruit could be protected from subsequent fruit infection after heat treatment by use of chitosan or *B.subtilis*, thereby providing packinghouses with an effective biologically based, combined approach to the management of postharvest brown rot (Casals et al. 2012). This author also has shown the current trend of combine biological treatments with others compatible methods of storage. The efficacy of *Bacillus amyloliquefaciens* DGA14, sodium carbonate (SC), sodium bicarbonate (SBC) and sodium hypochlorite (SH), applied alone or in various combinations, was evaluated *in vitro* and *in vivo* by Alvindia (2013) to control banana crown rot. The antagonist had variable responses in 1 percent (w/v) salts such a normal cell growth in SBC, reduction of cell cultivability by 1000-fold in SC and total inhibition by SH. *B. amyloliquefaciens* as stand-alone treatment controlled pathogens *in vitro* by 70 percent while SC, SBC, or SH by 20–45 percent. *Bacillus* and SC together restricted the mycelium growth of pathogen by 88%, *Bacillus* + SBC by 83 percent and *Bacillus* + SH by 52 percent. The efficacy of *B. amyloliquefaciens* against pathogens *in vitro* was enhanced by 21 percent and 15 percent with the amelioration of SC and SBC, respectively. Postharvest application showed that fruit dipped for 30 min in antagonist + SBC reduced crown rot by 93 percent, antagonist + SBC by 70 percent and antagonist + SH by 9 percent. The author finding indicated that the combination of *B. amyloliquefaciens* DGA14 with SBC managed crown rot disease comparable with synthetic fungicides without negative effects on fruit quality 14 days after treatment (Alvindia 2013). Recently, this combined treatment of *B. amyloliquefaciens* and sodium bicarbonate along with hot water treatment have been tested to control green and blue mold, and sour rot in mandarin by Hong et al. (2014). Population of antagonist (strain HF-01)were stable in the presence of 1 or 2 percent SBC treatment, and spore germination of *Penicilium digitatum, Penicillium italicum* and *Geotrichum citri-aurantii* in potato dextrose broth was greatly controlled by hot water treatment of 45°C for 2 min. Individual application of SBC at low rates and hot water treatment, although reducing disease incidence after 8 weeks or 4 weeks of storage at 6°C or 25°C respectively, was not as effective as the fungicide treatment. The treatment comprising *B. amyloliquefaciens*combined with 2 percent SBC or/and hot water (45°C for 2 min) was as effective as the fungicide treatment and reduced decay to less than 80% compared with the control. *B. amyloliquefaciens* alone or in combination with 2 percent SBC or/and hot water significantly reduced postharvest decay without impairing fruit quality after storage at 25°C for 4 weeks or at 6°C for 8 weeks. These results suggest that the combination of *B. amyloliquefaciens* HF-01, SBC and hot water could be a promising method for the control of postharvest decay on citrus while maintaining fruit quality after harvest (Hong et al. 2014).

7.4.7 EUKARYOTIC ANTAGONISTS USED IN POSTHARVEST TREATMENTS

Yeasts are eukaryotic microorganisms classified in the kingdom Fungi, with 1,500 species currently described (Kurtzman and Fell 2006) (estimated to be 1 percent of all fungal species). Yeasts are unicellular, although some species with yeast forms may become multicellular through the formation of strings of connected budding cells known as pseudohyphae or false hyphae. Yeast size can vary greatly depending on the specie, normally measuring 3–4 mm in diameter, although some yeasts can reaches over 40 mm (Walker et al. 2002). Most yeasts reproduce asexually by mitosis, and many do so by an asymmetric division process called budding.

Yeasts are interesting to be used in biological control methods because they are relatively easy to produce and maintain and they have several characteristics that can be manipulated in order to improve its use and efficiency (Pimenta et al. 2009). In particular, the high efficiency of yeasts applied as biocontrol agents is related to: (i) their adaptation to both the immediate environment and the nutritional conditions prevailing at the wound site, (ii) their capacity to grow at low temperatures and (iii) their ability to colonize wounds (Janisiewicz et al. 2010). Robiglio et al. (2011) isolated 75 yeast cultures from healthy pears from two Patagonian cold-storage packinghouses. *Aureobasidium pullulans, Cryptococcus albidus, Cryptococcus difluens, Pichiamembranifaciens, Pichia philogaea, Rhodotorula mucilaginosa* and *Saccharomycescerevisiae* yeast species were identified. Additionally, 13 indigenous isolates of *Penicillium expansum* and 10 isolates of *Botrytis cinerea* were obtained from diseased pears, characterized by aggressiveness and tested for sensibility to postharvest fungicides. The authors preselected the yeasts for their ability to grow at low temperature. In a first biocontrol assay using the most aggressive and the most sensitive isolate of each pathogen, two epiphytic isolates of *A. pullulans* and *R. mucilaginosa* were the most promising isolates to be used as biocontrol agents. They reduced the decay incidence by *P. expansum* to 33% and the lesion diameter in 88 percent after 60 days of incubation in cold. Foreign commercial yeast used as a reference in assays, only reduced 30 percent of lesion diameter in the same conditions. However, no yeast could reduce the incidence of *B. cinerea* decay. The control activity of the best two yeasts was compared with the control caused by fungicides in a second bioassay, obtaining higher levels of protection against *P. expansum* by the yeasts (Robiglio et al. 2011). Successful results have been also obtained by Luo et al. (2013) using *Pichia membranaefaciens* against *Penicillium digitatum* and *Penicillium italicum* to control green and blue mold respectively. The lesion diameter caused by both molds on citrus fruits was remarkably reduced when the fruit was point-inoculated or dipped in a suspension of *P. membranaefaciens* at 10^8 cfu/mL. The application of *P. membranaefaciens* on citrus fruit enhanced the activity of superoxide dismutase, ascorbate peroxidase and glutathione reductase, as well as the levels of hydrogen peroxide, the superoxide anion and glutathione, but inhibited the decreasing ascorbic acid content. These results indicated to the authors that the yeast treatment induced the synthesis of antioxidant enzymes which might have antagonist effects against *Penicillium* infection in citrus fruit (Luo et al. 2013). Another specie of *Pichia*, *P. caribbica*, has been tested to control *Rhizopus stolonifer* in peach fruit. Results obtained from bioassays have shown the decay incidence and lesion diameter of *Rhizopus* decay of peaches treated by *P. caribbica* were significantly reduced compared with the untreated fruit, and the higher the concentration of *P. caribbica*, the better the efficacy of biocontrol. Also a rapid colonization of the yeast in the peach wounds stored at 25°C was observed. Additionally, the peroxidase, catalase, and phenylalanine ammonia-lyase activities were significantly induced by yeast treatment compared by untreated fruit (Xu et al. 2013).

The biocontrol conducted by *P. membranaefaciens* can be enhanced by combination with other organism. In fact, a synergistic effect has been detected in the combination of *P.membranaefaciens* and lyophilized culture of *Lentinula edodes* (Wang et al. 2013b). The combined treatment induced higher phenolic accumulation and the up-regulation of superoxide dismutase, catalase, ascorbate peroxidise, chitinase and b-1,3-glucanase activities in fruit than *L. edodes* or yeast alone, and resulted in a lower lesion diameter and disease incidence. The use of *L edodes* may be an effective method to improve the biological activity of *P. membranefaciens*, and induce host defences appear to contribute to the control mechanism (Wang et al. 2013b). Culture filtrate of basidiomycete *L. edodes* was also used to enhance the biocontrol activity of the yeast *Crytococcus laurentii* strain LS28. *In vitro*, *L. edodes* cultures filtrates improve the growth of *C. laurentii* and the activity of its catalase, super-oxide dismutase and glutathione peroxidise, which play a key role in antioxidant scavenging. In addition, *L. edodes* also delayed *P. expansum* conidia germination. The biocontrol effect of LS28 used together with *L. edodes* in wounded apples improved the inhibition of *P. expansum* growth and patulin production in comparison with LS28 alone, under both experimental and semicommercial conditions (Tolaini et al. 2010). On the other hand, the effectiveness of *C. laurentii* can be also enhances by low concentration of pyrimethanil (40 mg/mL). *C. laurentii* at 10^7 CFU/mL combined with pyrimethanil reduce blue mold rot in pear *in vivo* more efficiently than each one alone. However, there was no additive activity when pyrimethanil was combined with other biocontrol yeasts such as *Rhodosporidium paludigenum* or *Rhodotorula glutinis*. Combination of pyrimethanil and *C. laurentii* at low concentration also inhibited blue mold rot when *P. expansum* was inoculated into fruit wounds 12 h before treatment and fruit was stored at low temperature (4°C). Pyrimethanil at 0.04 to 400 mg/mL did not influence the survival of *C. laurentii in vitro*, and it only slightly reduced the population growth of *C. laurentii* after 48 h of inoculation in the pear fruit wounds. There was no significant difference in quality parameters including total soluble solids, titratable acidity and ascorbic acid of pear fruit wounds among all treatments after 5 days of treatment at 25°C (Yu et al 2013).

7.5   CONCLUSIONS

After the information showed in this chapter, the main conclusions that can be drawn is that there is a wide range of postharvest methods for the maintenance of the shelf life of the fruit, and the efficiency of each method depends directly on the type of fruit and the pathogens for which is necessary to protect. Each treatment has its own characteristics and has to be considered when choosing one or the other. In addition, each exposed treatment presents some limitations that currently the researcher wants to counteract by combining various postharvest methods or even of combination of preharvest treatment and postharvest methods. Anyway, the current search focuses on the elimination of the use of chemicals that can be dangerous for the consumer either by direct consumption or generate hazard waste. For example, the combination of physical methods, such as heat treatment, waxes, and cold storage could present synergistic effect for fruit protection. Also edible coating method could be complemented with plant extracts with biocidal capacity and/or inorganic salts which could induce plat defences. This and other combinations are the principal trends studying for today.

KEYWORDS

- **Postharvest disease**
- **Physical treatments**
- **Irradiation**
- **Ultrasonic treatment**
- **Essential oils**
- **Fruits and vegetables**

REFERENCES

Acevedo, C. A.; López, D. A.; Tapia, M. J.; Enrione, J.; Skurtys, O.; Pedreschi, F.; Brown, D. I.; Creixell, W.; and Osorio, F.; Using RGB image processing for designating an alginate edible film. *Food Bioproc. Technol.* **2010**, *5*, 1511–1520.

Aghdam, M. S.; Hassanpouraghdam, M. B.; Paliyath, G.; and Farmani, B.; The language of calcium in postharvest life of fruit, vegetables and flowers. *Sci. Hortic-Amsterdam.* **2012**, *144*, 102–115.

Aghdam, M. S.; Mostofi, Y.; Motallebiazar, A.; Ghasemneghad, M.; and Fattahi-Moghaddam, J.; *Book of Abstracts*, 6th International Postharvest Symposium. Antalya, Turkey, **2009** April 8–12,.

Aider, M.; Chitosan application for active bio-based films production and potential in the food industry: Review. *LTW-Food Sci. Technol.* **2010**, *43*, 837–842.

Alferez, F.; Liao, H. L.; and Burns, J. K.; Blue light alters infection by *Penicilliumdigitatum* in tangerines. *Postharvest Biol. Technol.* **2012**, *63*, 11–15.

Al-Mughrabi, K. I.; Biological control of Fusarium dry rot and other potato tuber diseases using *Pseudomonas fluorescens* and *Enterobacter cloacae*. *Biol. Control* **2010**, 53, 280–284.

Alothman, M.; Bhat, R.; and Karim, A.; Effects of radiation processing on phytochemicals and antioxidants in plants produce. *Trends. Food. Sci. Tech.* **2009**, *10*(4), 512–516.

Alteri, G.; Di Renzo, G. C.; Genovese, F.; Calandra, M.; and Strano, M. C.; A new method for the postharvest application of imazalil fungicide to citrus fruit. *Biosyst. Eng.* **2013**, *115*, 434–443.

Alvindia, D. G.; Enhancing the bioefficacy of *Bacillus amyloliquefaciens* DGA14 with inorganic salts for the control of banana crown rot. *Crop Prot.* **2013**, 51, 1–6.

Ambaw, A.; Verboven, P.; Defraeye, T.; Tijskens, E.; Schenk, A.; Opara, U. L.; andNicolai, B. M.; Effect of box material on the distribution of 1-MCP gas during cold storage: A CFD study. *J. Food Eng.* **2013**, 119, 150–158.

Asghari, M.; and Aghdam, M. S.; Impact of salicylic acid on post-harvest physiology of horticultural crops. *Trends. Food. Sci. Technol.* **2010**, *21*, 502–509.

Awang, Y. B.; Chuni, S. H.; Mohamed, M. T. M.; Hafiza, Y.; and Mohamad, R. B.; Polygalacturonase and pectin methylesterase activities of $CaCl_2$ treated red-fleshed dragon fruit (*Hylocereus polyrhizus*) harvested at different maturity. *Am. J. Agric. Biol. Sci.* **2013**, *8*(2), 167–172.

Babalar, M.; Asghari, M.; Talaei, A.; and Khosroshahi, A.; Effect of pre-and postharvest salicylic acid treatment on ethylene production, fungal decay and overall quality of Selva strawberry fruit. *Food Chem.* **2007**, *105*, 449–453.

Bas, D.; and Boyaci, I. H.; Modeling and optimization I: Usability of response surface methodology. *J. Food Eng.* **2007**, *78*, 836–845.

Bautista-Baños, S.; Hernández-Lauzardo, A. N.; Velázquez-Del, V. M. G.; Hernández-López, M.; Ait, B. E.; Bosquez-Molina, E.; and Wilson, C. L.; Chitosan as a potential natural compound to control pre and postharvest diseases of horticultural commodities. *Crop Prot.* **2006**, *25*, 108–118.

Bojanowski, A.; and Avis, T. J.; Pelletier, S.; Tweddell, R. J.; Management of potato dry rot. *Postharvest Biol. Technol.* **2013**, *84*, 99–109.

Cantos, E.; García-Viguera, C.; de Pascual-Teresa. S.; Tomás-Barberán, F. A. Effect of postharvest ultraviolet irradiation on resveratrol and other phenolics of cv. Napoleon table grapes. *J. Agric. Food Chem.* **2000**, *48*, 4606–4612.

Cao, S.; Hu, Z.; and Pang, B.; Optimization of postharvest ultrasonic treatment of strawberry fruit. *Postharvest Biol. Technol.* **2010a**, *55*, 150–153.

Cao, S.; Hu, Z.; Pang, B.; Wang, H.; Xie, H.; and Wu, F.; Effect of ultrasound treatment on fruit decay and quality maintenance in strawberry after harvest. *Food. Contr.***2010b**, *21*, 529–532.

Cao, J.; Yan, J.; Zhao, Y.; and Jiang, W.; Effects of postharvest salicylic acid dipping on Alternaria rot and disease resistance of jujube fruit during storage. *J. Sci. Food Agric.* **2013**, *93*, 3252–3258.

Casals, C.; Elmer, P. A. G.; Viñas, I.; Teixido, N.; Sisquella, M.; and Usall, J.; The combination of curing with either chitosan or *Bacillus subtilis* CPA-8 to control brown rot infection caused by *Monilinia fructicola*. *Postharvest. Biol. Technol.* **2012**, *64*, 126–132.

Casas-Flores, S.; Rios-Momberg, M.; Rosales-Saavedra, T.; Martínez-Hernández, P.; Olmedo-Monfil, V.; and Herrera-Estrella, A.; Cross talk between a fungal blue-light perception system and the cyclic AMP signaling pathway. *Eukaryot. Cell.* **2006**, *5*, 499–506.

Cerioni, L.; Rapisarda, V. A.; Doctor, J.; Filckert, S.; Ruiz, T.; Fassel, R.; and Smilanick, J. L.; Use of phosphite salts in laboratory and Semicommercial test to control citrus postharvest decay. *Plant. Dis.* **2013a**, *97*(2), 201–212.

Cerioni, L.; Sepulveda, M.; Rubio-Ames, Z.; Volentini, S. I.; Rodríguez-Montelongo, L.; Smilanick, J. L.; Ramallo, J.; and Rapisarda, V. A.; Control of lemon postharvest diseases by low-toxicity salts combined with hydrogen peroxide and heat. *Postharvest. Biol. Technol.* **2013b**, *83*, 17–21.

Conte, A.; Scrocco, C.; Lecce, L.; Mastromatteo, M.; and Del Nobile, M. A;Ready to eat sweet cherries: study on different packaging systems. *Innov. Food Sci. Emerg.* **2009**, 10, 564–557.

Cote, S.; Rodoni, L.; Miceli, E.; Concellón, A.; Civello, P. M.; and Vicente, A. R.; Effect of radiation intensity on the outcome of postharvest UV-treatment. *Postharvest Biol. Technol.* **2013**, 83, 83–89.

Costa-Guimaraes, I.; Tavares-Menezes, E. G.; Silva de Abreu, P.; Costa-Rodríguez, A.; Siriano-Borges, P. R.; Batista, L. R.; Cirilo, M. A.; and de Oliveira-Lima, L. C.; Physicochemical and microbiological quality of raspberries (*Rudus idaeus*) treated with different doses of gamma irradiation. *Food. Sci. Technol. Campinas.* **2013**, *33*(2), 316–322.

Chanjirakul, K.; Wang, S. Y.; Wang, C. Y.; and Siriphanich, J.; Effect of natural volatile compounds on antioxidant capacity and antioxidant enzymes in raspberries. *Postharvest Biol. Technol.* **2006**, *40*(2), 106–115.

Charles, M. T.; Tano, K.; Asselin, A.; and Arul, J.; Physiological basis of UV-C induced resistance to *Botrytis cinerea* in tomato fruit. Constitutive defence enzymes and inducible pathogenesis-related proteins. *Postharvest Biol. Technol.* **2009**, *51*(3), 414–424.

Chong, K. L.; Peng, N.; Yin, H.; Lipscomb, G. G.; and Chung, T. S.; Food sustainability by designing and modelling a membrane controlled atmosphere storage system. *J. Food Eng.* **2013**, *114*, 361–374.

Choquer, M.; Fournier, E.; Kunz, C.; Pradier, J. M..; Simon, A.; and Viaud, M.;*Botrytis cinerea* virulence factors: new insights into a necrotrophic and polyphageos pathogen. *FEMS Microbiol. Lett.* **2007**, 277(1), 1–10.

Christie, J. M. Phototropin blue-light receptors. *Annu. Rev. Plant Biol.* **2006**, 58, 21–45.

Chung, W. H.; Chung, W. C.; Peng, M. T.; Yang, H. R.; and Huang, J. W.; Specific detection of benzimidazole resistance in *Colletotrichum gloeosporioides* from fruit crops by PCR-RFLP. *New. Biotechnol.* **2010**, 27, 17–24.

Daami-Remadi, M.; Ayed, F.; Jabnoun-Khiareddine, H.; Hibar, K.; andEl Mahjoub, M.; *In vitro, in vivo* and *in situ* evaluation of fungicides tested individually or in combination for control of the Fusarium dry rot of potato. *Int. J. Agric. Res.* **2010**, 5, 1039–1047.

Das, D. K.; Dutta, H.; and Mahanta, C. L.; Development of a rice starch-based coating with antioxidant and microbe-barrier properties and study of its effect on tomatoes stored at room temperature. *LWT.-Food. Sci. Technol.* **2013**, 50, 272–278.

De Cal, A.; Gel, I.; Usall, J.; Viñas, I.; and Melgarejo, P.; First report of brown rot caused by *Monilina fructicola* in peach orchads in Ebro Valley, Spain. *Plant. Dis.* **2009**, 93, 763.

Delele, M. A.; Vorstermans, B.; Creemers, P.; Tsige, A. A.; Tijskens, E.; Schenk, A.; Opara, U. L.; Nicolaï, B. M.; and Verboven, P.; Investigating the performance of thermonebulisation fungicide fogging system for loaded fruit storage room using CFD model. *J. Food. Eng.* **2012**, 109, 87–97.

Denness, L.; McKenna, J. F.; Segonzac, C.; Wormit, A.; Madhou, P.; Bennett, M.; Mansfield, J.; Zipfel, C.; and Hamann, T.; Cell wall damage-induced lignin biosynthesis is regulated by a reactive oxygen species – and jasmonic acid-dependent process in *Arabidopsis. Plant Physiol.* **2011**, 156, 1364–1374.

Dessalegn, Y.; Ayalew, A.; and Woldetsadik, K.; Integrating plant defense inducing chemical, inorganic salt and hot water treatment for the management of postharvest mango anthracnose. *Postharvest Biol. Technol.* **2013**, 85, 83–88.

Díaz-Mula, H. M.; Serrano, M.; and Valero, D.; Alginate coating preserve fruit quality and bioactive compounds during storage of sweet cherry fruit. *Food Bioprocess. Technol.* **2012**, 5, 2990–2997.

Dixon, R. A.; Achnine, L.; Kota, P.; Liu, C. J.; Reddy, M. S. S.; and Wang, L. J.; The phenylpropanoid pathway and plant defence – a genomics perspective. *Mol. Plant Pathol.* **2002**, 3, 371–390.

Dodd, A. N.; Kudla, J.; and Sanders, D.; The language of calcium signaling. *Annu. Rev. Plant. Biol.* **2010**, 61, 41–42.

El-Ghaouth, A.; Wilson, C. L.; and Wisniewski, M.; Ultrastructural and cytochemical aspects of the biological control of *Botrytis cinerea* by *Candida saitoana* in apple fruit. *Phytopathol.* **1998**, 88(4), 282–291.

Emery, K. M.; Michailides, T. J.; and Screm, H.; Incidence of latent infection of immature peach fruit by *Monilinia fructicola* and relationship to brown rot in Georgia. *Plant Dis.* **2000**, 84, 853–857.

Erkan, M.; Wang, S. Y.; and Wang, C. Y.; Effect of UV-treatment on antioxidant capacity, antioxidant enzyme activity and decay in strawberry fruit. *Postharvest Biol. Technol.* **2008**, 48, 163–171.

Fallanaj, F.; Sanzani, S. M.; Zavanella, C.; and Ippolito, A.; Salt addition improves the control of citrus diseases using electrolysis with conductive diamond electrodes. *J. Plant Pathol.* **2013**, 95(2), 373–383.

Fallik, E.; Prestorage hot water treatments (immersion rinsing and brushing). *Postharvest. Biol. Technol.* **2004**, 32, 125–134.

Ferrari, C. C.; Sarantópoulos, C. I. G. L.; Carmello-Guerreiro, S. M.; and Hubinger, M. D.; Effect of osmotic dehydration and pectin edible coating on quality and shelf life of fresh-cut melon. *Food Bioproc. Technol.* **2013,** *6,* 80–91.

Figueroa, C. R.; Opazo, M. C.; Vera, P.; Arriagada, O.; Díaz, M.; and Moya-León, M. A.; Effect of postharvest treatment of calcium and auxin on cell wall composition and expression of cell wall-modifying genes in the Chilean strawberry (*Fragaria chiloensis*) fruit. *Food Chem.* **2012,** *132,* 2014–2022.

Gao, H. Y.; Chen, H. J.; Chen, W. X.; Yang, J. T.; Song, L. L.; Zheng, Y. H.; andJiang, Y. M.; Effect of hypobaric storage on physiological and quality attributes of loquat fruit at low temperature. Proceeding of the IVth IC. In MQIC. Purvis, A. C., et al., Eds.; *Acta. Hortic.* **2006,** *712,* 269–274.

Ghasemnezhad, M.; Marsh, K.; Shilton, R.; Babalar, M.; and Woolf, A.; Effect of hot water treatments on chilling injury and heat damage in satsuma "mandarins": Antioxidant enzymes and vacuolar ATPase, and pyrophosphatase. *Postharvest. Biol. Technol.* **2008,** *48(*3), 364–371.

Gonzalez-Aguilar, G. A.; Villa-Rodriguez, J. A.; Ayala-Zavala, J. F.; and Yahia, E. M.; Improvement of the antioxidant status of tropical fruits as secondary response to some postharvest treatments. *Trends. Food. Sci. Tech.* **2010,** *21,* 475–482.

Gonzalez-Aguilar, G. A.; Villegas-Ochoa, M. A.; Martínez-Téllez, M. A.; Gardea, A. A.; and Ayala-Zavala, F.; Improving antioxidant capacity of fresh-cut mangoes treated with UV-C. *J. Food Sci.* **2007a,** *72*(3), S197–S202.

Gonzalez-Aguilar, G. A.; Zabaleta-Gatica, R.; and Tiznado-Hernandez, M. E.; Improving postharvest quality of mango "Haden" by UV-C treatment. *Postharvest. Biol. Technol.* **2007b,** *45*(1), 108–116.

Graham, D. M.; Pariza, M.; Glaze, W. H.; Newell, G. W.; and Erdman, J. W.; Borzelleca, J. F.; Use of ozone for food processing. *Food. Technol.* **1997,** *51,* 72–76.

Hallman, G. J.; Mortality of third instar caribbean fruit fly (*Diptera: Tephritidae*) reared in diet or grapefruit and immersed in heated water or grapefruit juice. *Fla. Entomol.* **1996,** *79,* 168–172.

Hashmi, M. S.; East, A. R.; Palmer, J. S.; and Heyes, J. A.; Hypobaric treatment stimulates defence-related enzymes in strawberry. *Postharvest. Biol. Technol.* **2013,** *85,* 77–82.

Hoa, T. T.; Clark, C. J.; Waddell, B. C.; and Woolf, A. B.; Postharvest quality of Dragon fruit (*Hylocereus undatus*) following disinfesting hot air treatments. *Postharvest. Biol. Technol.* **2006,** 41, 62–69.

Holguín, M. F.; Huerta, P. G.; Benítez, C. F.; and Toledo, A. J.; Epidemiología de la antracnosis *Colletotrichum gloeosporioides* (Penz.) Penz. And Sacc. en mango (*Mangifera indica* L.) cv. Ataulfo en el Soconusco, Chiapas, México. Revista Mexicana de Fitopatología. **2009,** *27,* 93–105.

Hong, P.; Hao, W.; Luo, J.; Chen, S.; Hu, M.; and Zhong, G.; Combination of hot water, Bacillus amyloliquefaciens HF-01 and sodium bicarbonate treatment to control postharvest decay of mandarin fruit. *Postharvest Biol. Technol.* **2014,** *88,* 96–102.

Hulme, A. C.; *The Biochemistry of Fruits and Their Products,* 1th ed.; Academic Press: London & New York, **1970.**

Hussein, E. I.; Kanan, G. J. M.; Al-Batayneh, K. M.; Alhussaen, K.; Al-Khateeb, W.; Qar, K.; Jacob, J. H.; Muhaidat, R.; and Hegazy, M. I.; Evaluation of food preservatives, low toxicity chemicals, liquid fraction of plant extracts and their combinations as alternative options for controlling citrus post-harvest green and blue moulds *in vitro*. *Res. J. Med. Plant.* **2012,** *6*(8), 551–573.

Idnurm, A.; and Heitman, J.; Light control growth and development via a conserved pathway in the fungal kingdom. *PloS. Biol.* **2005,** *3,* 615–626.

Janisiewicz, W. J.; and Korsten, L.; Biological control of postharvest diseases of fruits. *Annu. Rev. Phytopathol.* **2002**, *40*, 441–441.

Janisiewicz, W. J.; Kurtzman, C. P.; and Buyer, J. S.; Yeasts associated with nectarines and their potential for biological control of brown rot. *Yeast.* **2010**, *27*, 389–398.

Jitareerat, P.; Paumchai, S.; Kanlayanarat, S.; and Sangchote, S.; Effect of chitosan on ripening enzymatic activity, and disease development in mango (*Mangifera indica*) Fruit. *New Zealand J. Crop Hort.* **2007**, *35*(2), 211–218.

Kabachevskaya, A. M.; Liakhnovich, G. V.; Kisel, M. A.; and Volotovski, I. D.; Red/far red light modulated phospholipase D activity in oat seedlings: relation of enzyme photosensitivity to photosynthesis. *J. Plant. Physiol.* **2007**, *164*, 108–110.

Karabulut, O. A.; and Baykal, N.; Evaluation of the use of microwave power for the control of postharvest diseases of peaches. *Postharvest Biol. Technol.* **2002**, *26*, 237–240.

Khademi, O.; Zamani, Z.; Mostofi, Y.; Kalantari, S.; and Ahmadi, A.; Extending storability of persimmon fruit cv. Karaj by postharvest application of salicylic acid. *J. Agric. Sci. Technol.* **2012**, *14*(5), 1067–1074.

Kim, J. Y.; Kim, H. J.; Lim, G. O.; Jang, S. A.; and Song, K. B.; Research note. The effects of aqueous chloride dioxide or fumaric acid treatment combined with UV on postharvest quality of "Maehyang" strawberries. *Postharvest Biol. Technol.* **2010**, *56*, 254–256.

Knorr, D.; Zenker, M.; Heinz, V.; and Lee, D. U.; Applications and potential of ultrasonics in food processing. *Trend. Food Sci. Technol.* **2004**, *15*, 261–266.

Korsten, L.; De Villiers, E. E.; Wehner, F. C.; and Kotzé, J. M.; Field sprays of *Bacillussubtilis* and fungicides for control of preharvest fruit diseases of avocado in South Africa. *Plant Dis.* **1997**, *81*, 455–459.

Kurtzman, C. P.; and Fell, J. W.; Yeast systematic and phylogeny-Implications of molecular identification methods for studies in ecology. In *Biodiversity and Ecophysiology of Yeasts. The Yeast Handbook.* Springer: Heidelberg, **2006**.

Lacroix, M.; and Tien, C.; Edible films and coating from non-starch polysaccharides. In *Innovation in food packaging;* Han, J. H., Ed.; Elsevier Academic: San Diego, **2003**, pp 338–361

Lado, B.; and Yousef, A.; Alternative food preservation technologies: efficacy and mechanisms. *Microbes Inf.* **2002**, *4*, 433–440.

Lara, M. V.; Borsani, J.; Budde, C. O.; Lauxmann, M. A.; Lombardo, V. A.; Murray, R.; Andreo, C. S.; and Drincovich, M. F.; Biochemical and proteomic analysis of "Dixiland" peach fruit (*Prunus persica*) upon heat treatment. *J. Exp. Bot.* **2009**, *60*, 4315–4333.

Larena, I.; Torres, R.; De Cal, A.; Liñan, M.; Melgarejo, P.; Domenichini, P.; Bellini, A.; Mandrin, J. F.; Lichou, J.; Ochoa de Eribe, X.; and Usall, J.; Biological control of postharvest brown rot (*Monilinia* spp.) of peach by field application of *Epicoccumnigrum*. *Biol. Contr.* **2005**, *32*, 305–310.

Liao, H. L.; Alferez, F.; and Burns, J. K.; Assessment of blue light treatment on citrus postharvest diseases. *Postharvest. Biol. Technol.* **2013**, *81*, 81–88.

Liao, H. L.; and Burns, K.; Light controls phospholipases $A_2\alpha$ and β gene expression in *Citrus sinensis*. *J. Exp. Bot.* **2010**, *61*, 2469–2478.

Liu, R. H.; Health benefits of fruit and vegetables are from additive and synergistic combinations of phytochemicals. *Am. J. Clin. Nutr.* **2003**, *78*(3), 517S.

Liu, H. M.; Guo, J. H.; Cheng, Y. J.; Luo, L.; Liu, P.; Wang, B. Q.; Deng, B. X.; and Long, C. A.; Control of gray mold of grape by *Hansienaspora uvarum* and its effectson postharvest quality parameters. *Ann. Microbiol.* **2010**, *60*, 31–35.

Luo, Y.; and Michailides, T. J.; Factors affecting latent infection of prune fruit by *Monilinia fructicola*. *Phytopathol.* **2001**, *91*, 864–872.

Luo, Y.; Zhou, Y.; and Zeng, K.; Effect of Pichia membranaefaciens on ROS metabolism and postharvest disease control in citrus fruit. *Crop. Prot.* **2013**, *53*, 96–102.

Lurie, S.; Postharvest heat treatment. *Postharvest Biol. Technol.* **1998**, *14*, 257–269.

Lurie, S.; Stress physiology and latent damage. In *Postharvest Handing: A Systems Approach*; Florkowski, W. J.; Shewfelt, R. L.; Brueckner, B.; and Prussia, S. E. Eds.; Academic Press, **2009**; pp 443–459.

Ma, Y. Q.; Chen, J. C.; Liu, D. H.; and Ye, X. Q.; Effect of ultrasonic treatment on the total phenolic and antioxidant activity of extracts from citrus peel. *J. Food Sci.* **2008**, *73*,115–120.

Ma, Z.; Yang, L.; Yan, H.; Kennedy, J. F.; and Meng, X.; Chitosan and oligochitosan enhance the resistance of peach fruit to brown rot. *Carbohyd. Polym.* **2013**, *94*, 272–277.

Maftoonazad, N.; Ramaswamy, H. S.; and Marcotte, M.; Shelf life extension of peaches through sodium alginate and metylcellulose edible coatings. *Int. J. Food Sci. Technol.* **2008**, *43*(6), 951–957.

Maharaj, R.; Arul, J.; and Nadeau, P.; Effect of photochemical treatment in the preservation of fresh tomato (*Lycopersicon esculentum* cv. Capello) by delaying senescence. *Postharvest Bio. Technol.* **1999**, *15*, 13–23.

Malladi, A.; and Burns, J. K.; CsPLDα and CsPLDγ1 are differential induce during leaf and fruit abscission and diurnally regulated in *Citrus sinensis*. *J. Exp. Bot.* **2008**, *59*, 3729–3739.

Mantilla, N.; Castell-Perez, M. E.; Gomes, C.; and Moreira, R. G.; Multilayered antimicrobial edible coating and its effect on quality and shelf-life of fresh-cut pineapple (*Ananas comosus*). *LWT-Food Sci. Technol.* **2013**, *51*, 37–43.

Mari, M.; Torres, R.; Casalini, L.; Lamarca, N.; Mandrin, J. F.; Lichou, J.; Larena, I.; De Cal, A.; Melgarejo, P.; and Usall, J.; Control of postharvest brown rot on nectarine by *Epicoccum nigrum* and physio-chemical treatments. *J. Sci. Food Agric.* **2007**, *87*, 1271–1277.

Marquenie, D.; Lammertyn, J.; Geeraerd, A.; Soontjens, C.; Van Impe, J.; Nicolai, B.; and Michiels, C.; Inactivation of conidia of *Botrytis cinerea* and *Monilinia frutigena* using UV-C and heat treatment. *Int. J. Food. Microbiol.* **2002**, *74*, 27–35.

Martín-Diana, A. B.; Rico, D.; Frías, J. M.; Barat, J. M.; Henehan, G. T. M.; and Barry-andRyan, C.; Calcium for extending the shelf life of fresh whole and minimal processed fruits and vegetables: a review. *Trend. Food Sci. Technol.* **2007**, *18*(4), 210–218.

Mehra, L. K.; MacLean, D. D.; Shewfelt, R. L.; Smith, K. C.; and Scherm, H.; Effect of postharvest biofumigation on fungal decay, sensory quality, and antioxidant levels of blueberry fruit. *Postharvest Biol. Technol.* **2013**, *85,* 109–115.

Mercier, J.; Arul, J.; Ponnampalam, R.; and Boulet, M.; Induction of 6-methoxymellein and resistance to storage pathogens in carrot slices by UV-C. *J. Phytopathol.* **1993**, *137*, 44–54.

Mlikota Gabler, F.; Fassel, R.; Mercier, J.; and Smilanick, J. L.; Influence of temperature, inoculation interval, and dose on biofumigation with *Muscodor albus* to control postharvest gray mold on grapes. *Plant Dis.* **2006**, *90,* 1019–1025.

Mora, A. A.; and Téliz, O. D.; Enfermedades del mango. In *Mango: Manejo y Comercialización. Colegio de Postgraduados en Ciencias Agrícolas;* Mora, A. A.; Téliz, O. D. Rebouças, S. J. A. Eds.; México 2002, 55–171.

Moscoso-Ramirez, P. A.; and Palou, L.; Evaluation of postharvest treatments with chemical resistance inducers to control green and blue molds on oranges fruits. *Postharvest Biol. Technol.* **2013**, *85*, 132–135.

Mostafavi, H. A.; Mirmajlessi, S. M.; Mirjalili, S. M.; Fathollahi, H.; and Askari, H.; Gamma radiation effects on physico-chemical parameters of apple fruit during commercial post-harvest preservation. *Radiat. Phys. Chem.* **2012**, *81*, 666–671.

Munnik, T.; Irvine, R. F.; and Musgrave, A; Phospholipid signalling in plants. *Biochim. Biophys. Acta.* **1998**, *1389*, 222–272.

Njombolwana, N. S.; Erasmus, A.; and Fourie, P. H.; Evaluation of curative and protective control of *Penicilium digitatum* following imazalil application in wax coating. *Postharvest Biol. Technol.* **2013**, *77,* 102–110.

Ojeda, A. M.; Mora-Aguilera, J. A.; and Villegas-Monter, A.; Nava-Diaz, C.; Hernandez-Castro, E.; Otero-Colina, G.; Hernandez-Morales, J. Temporal analysis and fungicide management strategies to control mango anthracnose epidemics in Guerrero, Mexico. *Trop. Plant. Pathol.* **2012**, *37*(6), 375–385.

Olivas, G. I.; and Barbosa-Canovas, G.; Edible films and coatings for fruits and vegetables. In *Edible Film and Coating for Food Applications*; Embuscado, M. E., and Hubber, K. C., Eds.; Springer: New York, **2008**; 211–244.

Palou, L.; Crisosto, C. H.; Smilanick, J. L.; Adaskaveg, J. E.; and Zoffoli, J. P.; Effects of continuous 0.3 ppm ozone exposure on decay development and physiological responses of peaches and table grapes in cold storage. *Postharvest. Biol. Technol.* **2002**, *24*, 39–48.

Panahirad, S.; Zaare-Nahandi, F.; Safaralizadeh, R.; and Alizadeh-Salteh, S.; Postharvest control of *Rhizopus stolonifer* in peach (*Prunus persica* L. Batsch) fruits using salicylic acid. *J. Food Saf.* **2012**, *32*, 502–507.

Patist, A.; and Bates, D.; Ultrasonic innovation in the food industry : from the laboratory to commercial production. *Innov. Food Sci. Emerg. Technol.* **2008**, *9*, 147–154.

Pavoncello, D.; Lurie, S.; Droby, S.; and Porat, R.; A hot water treatment induces resistance to *Penicillium digitatum* and promotes the accumulation of heat shock and pathogenesis-related proteins in grapefruit flavedo. *Physiol. Plantarum* **2001**, *111*, 17–22.

Perotti, V. E.; Del Vecchio, H. A.; Sansevich, A.; Meier, G.; Bello, F.; Cocco, M.; Garrán, S. M.; Anderson, C.; Vazquez, D.; and Podestá, F. E.; Proteomic, metabolic, and biochemical analysis of heat treatment Valencia oranges during storage. *Postharvest Biol. Technol.* **2011**, *62*, 97–114.

Pimenta, R. S.; Morais, P. B.; Rosa, C. A.; and Corrêa, A.; Utilization of yeasts in biological control programs. In *Yeast Biotechnology, Diversity and Applications*; Satyanarayana, T.; Kunze, G., Eds.; Springer Science, 2009.

Polenta, G.; Budde, C.; and Murray, R.; Effects of different pre-storage anoix treatment on ethanol and acetaldehyde content in peaches. *Postharvest Biol. Technol.* **2005**, *38*, 247–253.

Pombo, M. A.; Rosli, H. G.; Martínez, G. A.; and Civello, P. M.; UV-treatment affects the expression and activity of defense genes in strawberry fruit (Fragaria x ananassa, Duch.). *Postharvest Biol. Technol.* **2011**, *59*, 94–102.

Ranganna, B.; Raghavan, G. S. V.; and Kushalappa, A. C.; Hot water dipping to enhance storability of potatoes. *Postharvest Biol Technol.* **1998**, *13*, 215–223

Recep, K.; Fikrettin, S.; Erkol, D.; and Cafer, E.; Biological control of the potato dry rot caused by Fusarium species using PGPR strains. *Biol. Contr.* **2009**, *50*, 194–198.

Ribeiro C.; Canada, J.; and Alvarenga, B.; Prospects of UV radiation for application in postharvest technology. *Emir. J. Food. Agric.* **2012**, *24*(6), 586–597.

Rivera, S. A.; Zoffoli, J. P.; and Latorre, B. A.; Determination of optimal sulfur dioxide time and concentration product for postharvest control of gray mold of blueberry fruit. *Postharvest Biol. Technol.* **2013**, *83*, 40–46.

Robiglio, A.; Sosa, M. C.; Lutz, M. C.; Lopes, C. A.; and Sangorrín, M. P.; Yeast biocontrol of fungal spoilage of pears stored at low temperature. *Int. J. Food Microbiol.* **2011,** *147,* 211–216.

Robles-Sánchez, R. M.; Rojas-Graü, M. A.; Odriozola-Serrano, I.; González-Aguilar, G.; and Martin-Belloso, O.; Influence of alginate-based edible coatig as carrier of antibrowning agents on bioactive compounds and antioxidant activity in fresh-cut Kent mangoes. *LWT-Food Sci. Technol.* **2013,** *50,* 240–246.

Roden, C.; and Ingle, R. A.; Light, rythms, infection: the role of light and the circadian clock in determining the outcome of plant-pathogen interaction. *Plant. Cell.* **2009,** *21,* 2546–2552.

Romanazzi, G.; Feliziani, E.; Santini, M.; and Landi, L.; Effectiveness of postharvest treatment with chitosan and other resistance inducers in control of storage decay of strawberry. *Postharvest Biol. Technol.* **2013,** *75,* 24–27.

Romanazzi, G.; Nigro, F.; and Ippolito, A.; Short hypobaric treatments potentiate the effect of chitosan in reducing storage decay of sweet cherries. *Postharvest Biol. Technol.* **2003,** *29,* 73–80.

Romanazzi, G.; Nigro, F.; Ippolito, A.; and Salerno, M.; Effect of short hypobaric treatment on postharvest rots of sweet cherries, strawberries and table grapes. *Postharvest Biol. Technol.* **2001,** *22,* 1–6.

Ryu, S. B.; Lee, H. Y.; and Hwang, I. W.; Transgenic plants with increased resistance to biotic and abiotic stresses and accelerated flowering time due to overexpression of a secondary phospholipase A_2 (sPLA$_2$). PCT International Application Patent WO 2008100112, 2008.

Sánchez-Murillo, R. I.; de la Torre-Martínez, M.; Auirre-Linares, J.; and Herrera-Estrella, A.; Light-regulated asexual reproduction in *Paecilomyces fumosoroseus. Microbiology.* **2004,** *150,* 311–319.

Scherba, G.; Weigel, R. M.; and O'Brien, W. D.; Quantitative assessment of the germicidal efficacy of ultrasonic energy. *Appl. Environ. Microb.* **1991,** *57,* 2079–2084.

Schirraa, M.; D'hallewina, G.; Ben-Yehoshuab, S.; and Fallikb, E.; Host-pathogen interaction modulated by heat treatment. *Postharvest Biol. Technol.* **2000,** 21(1), 71–85.

Schmidt-Heydt, M.; Bode, H.; Raupp, F.; and Geisen, R.; Influence of light on ochratoxin biosynthesis by *Penicillium. Mycotoxicol. Res.* **2010,** *26,* 1–8.

Schotsmans, W.; Molan, A.; and MacKay, B.; Controlled atmosphere storage of rabbiteye blueberries enhances postharvest quality aspects. *Postharvest Biol. Technol.* **2007,** *44,* 277–285.

Sellamuthu, P. S.; andSivakumar, D.; Soundy, P.; and Korsten, L.; Essential oil vapours suppress the development of anthracnose and enhance defence related and antioxidant enzyme activities in avocado fruit. *Postharvest. Biol. Technol.* **2013,** *81,* 66–72.

Serradilla, M. J.; Villalobos, M. C.; Hernández, A.; Martín, A.; Lozano, M.; and Córdoba, M. G.; Study of microbiological quality of controlled atmosphere packaged "Ambrunés" sweet cherries and subsequent shelf-life. *Int. J. Food Microbiol.* **2013,** *166,* 85–92.

Shafiee, M.; Taghavi, T. S.; and Babalar, M.; Addition of salicylic acid to nutrient solution combined with postharvest treatment (hot water, salicylic acid, and calcium dipping) improved postharvest fruit quality of strawberry. *Sci. Hortic-Amsterdam.* **2010,** *124,* 40–45.

Shao, X. F.; Tu, K.; Tu, S.; and Tu, J.; A combination of heat treatment and chitosan coating delays ripening and reduces decay in "Gala" apple fruit. *J. Food Qual.* **2012,** *35,* 83–92.

Shao, X. F.; Tu, K.; Zhao, Y. Z.; Chen, L.; Chen, Y. Y.; and Wang, H.; Effects of post-harvest heat treatment on fruit ripening and decay development in different apple cultivars. *J. Hortic. Sci. Biotechnol.* **2007,** *82,* 297–303.

Shirzadeh, E.; and Kazemi, M.; Effect of salicylic acid and essential oils treatment on quality characteristics of apple (*Malus domestica* var Granny Smith) fruits during storage. *Asian J. Biochem.* **2012,** *7*(3), 165–170.

Silva, F.; Goyette, B.; Bourgeois, G.; and Vigneault, C.; Comparing forced air cooling and water cooling for apples. *Int. J. Food Agric. Environ.* **2006**, *4*, 33–36.

Singh, S. P.; and Pal, R. K.; Controlled atmosphere storage of guava (*Psidium guajava* L.) fruit. *Postharvest Biol. Technol.***2008**, *47*(3), 296–306.

Singh, Z.; Singh, R. K.; Sane, V. A.; and Nath, P.; Mango-postharvest biology and biotechnology. *CRC-Cr Rev. Plant Sci.* **2013**, *32*, 217–236.

Sisquella, M.; Casals, C.; Viñas, I.; Teixidó, N.; and Usall, J.; Combination of peracetic acid and hot water treatment to control postharvest brown rot on peaches and nectarines. *Postharvest. Biol. Technol.***2013a**, *83*, 1–8.

Sisquella, M.; Viñas, I.; Teixidó, N.; Picouet, P.; and Usall, J.; Continuous microwave treatment to control postharvest brown rot in stone fruit. *Postharvest Biol. Technol.* **2013b**, *86*, 1–7.

Smith, R. L.; Cohen, S. M.; Doull, J.; Feron, V. J.; Goodman, J. I.; Marnett, L. J. et al. GRAS flavouring substances 22. *Food. Technol.-Chicago.***2005**, *59*(8), 24–62.

Sui, Y.;Liu, J.; Wisniewski, M.; Droby, S.; Norelli, J.; and Hershkovitz, V.; Pretreatment of the yeast antagonist, *Candida oleophila*, with glycine betaine increase oxidative stress tolerance in the microenvironment of apple wounds. *Int. J. Food Microbiol.* **2012**, *157*, 45–51.

Sun, F.; Zhang, P.; Guo, M.; Yu, W.; and Chen, K.; Burdock fructooligosaccharide induces fungal resistance in postharvest Kyoho grapes by activating the salicylic acid-dependent pathway and inhibiting browning. *Food Chem.***2013**, *138*, 539–546.

Tiwari, B. K.; O'Donnell, C. P.; Muthukumarappan, K.; and Cullen, P. J.; Ascorbic acid degradation kinetic of sonicated oranges juices during storage and comparison with thermally pasteurized juice. *LWT. Food. Sci. Technol.* **2009**, *42*, 700–704.

Tolaini, V.; Zjalic, S.; Reverberi, M.; Fanelli, C.; Fabbri, A. A.; Del Fiore, A.; De Rossi, P.; and Ricelli, A.; *Lentinula edodes* enhances the biocontrol activity of *Cryptococcuslaurentii* against *Penicillium expansum* contamination and patulin production in apple fruits. *Int. J. Food Microbiol.* **2010**, *138*, 243–249.

Trinetta, V.; Morgan, M. T.; and Linton, R. H.; Use of high-concentration-short-time chloride dioxide gas treatment for the inactivation of *Salmonella enterica* spp. inoculated onto Roma tomatoes. *Food Microbiol.* **2010**, *27*, 1009–1015.

Tripathi, P.; and Dubey, N. K.; Exploitation of natural products as an alternative strategy to control postharvest fungal rotting of fruit and vegetables. *Postharvest Biol. Technol.* **2004**, *32*(3), 235–245.

Tu, K.; Shao, X. F.; Tu, S. C.; Su, J.; and Zhao, Y.; Effects of heat treatment on wound healing in gala and Red fuji Apple fruits. *J. Agric. Food Chem.* **2010**, *58*, 4303–4309.

Tzortzakis, N. G.; Maintaining postharvest quality of fresh produce with volatile compounds. *Innov. Food Sci. Emerg. Technol.* **2007**, *8*, 111–116.

Utama, I. M. S.; Wills, R. B. H.; Ben-Yehoshua, S.; and Kuek, C.; *In vitro* efficacy of plant volatiles for inhibiting the growth of fruit and vegetable decay microorganisms. *J. Agric. Food Chem.* **2002**, *50*, 6371–6377.

Utto, W.; Mawson, A. J.; and Brolund, J. E.; Hexanal reduces infection of tomatoes by *Botrytis cinerea* whilst maintaining quality. *Postharvest Biol. Technol.* **2008**, *47*(3), 434–437.

Valero, M.; Recrosio, N.; Saura, D.; Munoz, N.; Martin, N.; and Lizama, V.; Effect of ultrasonic treatment in orange juice processing. *J. Food Eng.* **2007**, *80*, 509–516.

Vallad, G. E.; and Goodman, R. M.; Systematic acquired resistance and induced systematic resistance in conventional agriculture. *Crop. Sci.* **2004**, *44*, 1920–1934.

Velickova, E.; Winkelhausen, E.; Kuzmanova, S.; and Alves, V. D.; Impact of chitosan-beeswax edible coating on the quality of fresh strawberries (*Fragaria ananassa* cv Camarosa) under commercial storage conditions. *LWT-Food. Sci. Technol.* **2013**, *52*, 80–92.

Vicente, A. R.; Martínez, G. A.; Civello, P. M.; and Chaves, A. R.; Quality of heat-treated strawberry fruit during refrigerated storage. *Postharvest Biol. Technol.* **2002**, *25*, 59–71.

Villas-Rojas, R.; López-Malo, A.; and Sosa-Morales, M. E.; Hot water bath treatment assisted by microwave energy to delay postharvest ripening and decay in strawberries (*Fragaria* x *ananassa*). *J. Sci. Food Agric.* **2011**, *91*, 2265–2270.

Vigneault, C.; Leblanc, D. I.; Goyette, B.; and Jenni, S.; Invited review: Engineering aspects of physical treatment to increase fruit and vegetable phytochemical content. *Can. J. Plant Sci.* **2012**, *92*, 373–397.

Walker, K.; Skelton, H.; and Smith, K.; Cutaneous lesions showing giant yeast forms of Blastomyces dermatitidis. *J. Cutan. Pathol.* **2002**, *29*(10), 616–618.

Wang, S. Y.; and Gao, H.; Effect of chitosan-based edible coating on antioxidants, antioxidant enzyme system, and postharvest fruit quality of strawberries (*Fragaria* x *aranssa* Duch.). *LWT-Food Sci. Technol.* **2013**, *52*, 71–79.

Wang, S.; and Tang, J.; Radio frequency and microwave alternative treatment for insect control in nuts: a review. *Int. Agric. Eng. J: CIGR.* **2001**, *10*(3–4), 105–120.

Wang, X.; Wang, J.; Jin, P.; and Zheng, Y.; Investigating the efficacy of *Bacillus subtilis* SM21 on controlling *Rhizopus* rot in peach fruit. *Int. J. Food Microbiol.* **2013a**,*164*, 141–147.

Wang, J.; Wang H. Y.; Xia, X. M.; Li, P. P.; and Wang, K. Y.; Synergistic effect of *Lentinula edodes* and *Pichia membranefaciens* on inhibition of *Penicillium expansum* infection. *Postharvest Biol. Technol.* **2013b**, *81*, 7–12.

Xu, B.; Zhang, H.; Chen, K.; Xu, Q.; Yao, Y.; and Gao, H.; Biocontrol of postharvest Rhizopus decay of peaches with Pichia caribbica. *Curr. Microbiol.* **2013**, 67, 255–261.

Yahia, E. M., Ed.; *Modified AND Controlled Atmospheres for the Storage, Transportation and Packaging of Horticultural Commodities,* 1st ed.; CRC Press/Taylor and Francis Group: Boca Raton, FL, 2009.

Yahia, E. M.; and Ortega-Zaleta, D.; Mortality of eggs and third instar larvae of *Anastrepha ludens* and *A. oblique* with insecticidal controlled atmosphere at high temperatures. *Postharvest Biol. Technol.* **2000**, *20*, 295–302.

Yang, T.; Peng, H.; Whitaker, B. D.; and Jurick, W. M.; Differential expression of calcium/calmodulin-regulates SISRs in response to abiotic and biotic stresses in tomato fruit. *Physiol. Plantarum.* **2013**, *148*(3), 44–455.

Yin, Y. N.; and Xiao, C. L.; Molecular characterization and a multiplex allele-specific PCR method for detection of thiabendazole resistance in *Penicillium expansum* from apple. *Eur. J. Plant. Pathol.* **2013**, *136*, 703–713.

Yu, T.; Yu, C.; Lu, H.; Zunun, M.; Chen, F.; Zhou, T.; and Sheng, K.; Effect of *Crytocccuslaurentii* and calcium chloride on control of *Penicillium expansum* and *Botrytis cinerea* infections in pear fruit. *Biol. Cont.* **2012**, *61*, 169–175.

Yu, C.; Zhou, T.; Sheng, K.; Zeng, L.; Ye, C.; Yu, T.; and Zheng, X.; Effect of pyrimethanil on Cryptococcus laurentii, Rhodosporidium paludigenum, and Rhodotorula glutinis biocontrol of Penicillium expansum infection in pear fruit. *Int. J. Food Microbiol.* **2013**, *164*, 155–160.

Yun, Z.; Gao, H.; Liu, P.; Liu, S.; Luo, T.; Jin, Q.; Xu, J.; Chang, Y.; and Deng, X.; Comparative proteomic and metabolomic profiling of citrus fruit with enhancement of disease resistance by postharvest heat treatment. *BMC. Plant. Biol.* **2013**, *13*, 44.

Zambrano-Zaragoza, M. L.; Mercado-Silva, E.; Ramirez-Zamorano, P.; Cornejo-Villejas, M. A.; Gutiérrez-Cortez, E.; and Quintanar-Guerrero, D.; Use of solid lipid nanoparticles (SLNs) in edible coatings to increase guava (*Psidium guajava* L.) shelf-life. *Food Res. Int.* **2013**, *51*, 946–953.

Zapata, P. J.; Guillén, F.; MArtínez-Romero, D.; Castillo, S.; Valero, D.; andSerrano, M.; Use of alginate or zein as edible coatings to delay postharvest ripening process and to maintain tomato (*Solanum lycopersicon* Mill) quality. *J. Sci. Food Agric.* **2008**, *88*, 1287–1293.

Zhang, Z. K.; Bi, Y.; Ge, Y. H.; Wang, J. J.; Deng, J. J.; Xie, D. F.; and Wang, Y.; Multiple preharvest treatment with acibenzolar-S-methyl reduce latent infection and induce resistance in muskmelon fruit. *Sci. Hortic-Amsterdam.* **2011**, *130*, 126–132.

Zhang, H. Y.; Fu, C. X.; Zheng, X. D.; Xi, Y. F.; Jiang, W.; and Wang, Y. F.; Control of postharvest Rhizopus rot of peaches by microwave treatment and yeast antagonist. *Eur. Food Res. Technol.* **2004**, *218*, 568–572.

Zhang, Z.; Nakano, K.; and Maezawa, S.; Comparison of the antioxidant enzymes of broccoli after cold or heat shock treatment at different storage temperature. *Postharvest Biol. Technol.* **2009**, *54*(2), 101–105.

Zhang, J.; and Timmer, L. W.; Preharvest application of fungicides for postharvest disease control on early season tangerine hybrids in Florida. *Crop. Prot.* **2007**, *26*, 886–893.

Zhang, Z.; Yang, D.; Yang, B.; Gao, Z.; Li, M.; Jiang, Y.; and Hu, M.; β-Aminobutyric acid induces resistance of mango fruit to postharvest anthracnose caused by *Colletotrichumgloeosporioides* and enhances activity of fruit defense mechanisms. *Sci. Hortic-Amsterdam.* **2013**, *160*, 78–84.

Zeng, K.; Deng, Y.; Ming, J.; and Deng, L.; Induction of disease resistance and ROS metabolism in navel oranges by chitosan. *Sci. Hortic-Amsterdam.* **2010**, *126*, 223–228.

Zeng, R.; Zhang, A.; Chen, J.; and Fu, Y.; Postharvest quality and physiological responses of clove bud extract dip on "Newhall" navel oranges. *Sci. Hortic-Amsterdam* **2012**, *138*, 253–258.

Zeng, R.; Zhang, A.; Chen, J.; and Fu, Y.; Impact of carboxymethyl cellulose coating enriched with extract of *Impatiens balsamina* stems on preservation of "Newhall" navel oranges. *Sci. Hortic-Amsterdam.* **2013**, *160*, 44–48.

CHAPTER 8

INTEGRATED POSTHARVEST PEST MANAGEMENT IN FRUITS AND VEGETABLES

TAMOGHNA SAHA[1*], NITHYA C.[1] AND S. N. RAY[1]

[1]Department of Entomology, Bihar Agricultural University, Sabour, Bhagalpur-813210, Bihar; Email: tamoghnasaha1984@gmail.com

CONTENTS

8.1 INTRODUCTION

Horticulture plays a significant role in Indian Agriculture. Fruits and vegetables are highly valued in human diet mainly for vitamins and minerals. In India, the fruits have been given a place of honor on being offered to God at every festival and have also been mentioned in our epics like Mahabharata, Ramayana, and writings of Sushrutha and Charaka. Fruits and vegetables contribute approximately 91 percent of vitamin C, 48 percent of vitamin A, 27 percent of vitamin B6, 17 percent of niacin, 16 percent of magnesium, 19 percent of iron and 9 percent of calories to the human diet. Other important nutrients supplied by fruits and vegetables, include iron, riboflavin, zinc, calcium, potassium, and phosphorus (USDA, 1983).

A considerable amount of fruits and vegetables produced in India is lost due to improper postharvest operations; as a result there is a considerable gap between the gross production and net availability (http://tnau.ac.in).The estimated postharvest losses of fruits and vegetables lie in the range of 20–40 percent (Wills et al. 2004). The presence or potential presence of arthropods on and in horticultural products shipped from areas where the pest is present to areas where is it not has been the source of many interstate, inter, and intra country trade issues for many years (Table 8.1). The first line of defense against insects and disease is good management during production. Planting resistant varieties, the use of irrigation practices that do not wet the leaves or flowers of plants, avoiding over-fertilization with nitrogen, and pruning during production to reduce canopy overgrowth can all serve to reduce produce decay before and after harvest. A second important defense is careful harvesting and preparation for market in the field. Thirdly, sorting out damaged or decaying produce will limit contamination of the remaining, healthy produce. Yet, even when the greatest care is taken, sometimes produce must be treated to control insects or decay-causing organisms.

Postharvest losses management can include some treatments conventional chemicals, irradiation, and organic treatments. The list of insect causes postharvest losses are given in table no. 1. Management options are chemical treatments such as fumigants, pesticide dips and detergents are considered conventional since they do not fit the organic standards of the USA and many other countries. Among the many chemical fumigants identified for controlling postharvest arthropods are phosphine (Fields and White, 2002) and sulfuryl fluoride (SF) (Schneider et al., 2003).Although irradiation is considered a physical treatment, it is not commonly considered an 'organically compliant' measure, even though it does not render the commodity or affected arthropods radioactive nor does it leave any detectable residues deemed to be harmful to human or animal health (Wall, 2008 and Anon, 2006).

Many of the organic treatments developed rely upon the use of physical treatments such as the use of temperature extremes (eg, short term high and longer term low temperatures) as well as the use of controlled atmospheres (CAs) (Follett and Neven, 2006 and Neven et al., 2009). High temperature treatments can include, but are not exclusive to, hot-water dips, hot forced air, vapor heat, microwave, and radio frequency. Low temperature treatments generally use long-term cold storage above freezing, but there have been limited uses for freezing and flash freezing to control postharvest pests (Johnson and Valero, 2003). The application of CAs has also gained much interest, especially the use of CA in combination with temperature extremes (Neven, 2008). In addition to physical treatments, a number of chemicals that meet organic requirements have been investigated, such as neem (Obeng, 2007), vegetable oils (Rajendran and Sriranjini, 2008) and biofumigants (Lecay et al., 2009).

TABLE 8.1 Some examples of insects considered as postharvest pests in some country

Common Name	Scientific Name	Common Hosts	Distribution
Codling moth	*Cydia pomonella* (L.)	Apples, pears, quince, walnut, prunus spp.	Worldwide, except Japan
Sweet potato weevil	*Cylas formicarius* (Fab.)	Sweet potato	Asia, Africa, Australia, Hawaii
Mango seed weevil	*Sternochetus mangiferae* (Fabricius)	Mango	Asia, Africa, Australia, Hawaii
Red legged earth mite	*Dysmicoccus brevipes* (Ckll.)	Pineapple	Asia, Africa, Australia, South America
Fruit flies			
Apple maggot fly	*Rhagoletis pomonella* (Walsh)	Apple, blueberry	U.S.A, Canada
Caribbean fruit fly	*Anastrepha suspense* (Loew)	Tropical and subtropical fruits	Caribbean, Southern Florida
Mexican fruit fly	*Anastrepha ludens* (Loew)	Citrus, Mango, some other tropical and subtropical fruits	Mexico, Central America
European cherry fruit fly	*Rhagoletis cingulata* (Loew)	Cherry, *Lonicera* spp.	Europe
Mediterranean fruit fly	*Ceratitis capitata* (Wiedemann)	Deciduous subtropical and tropical fruits	South Europe, Africa, Central America, Hawaii
Melon fruit fly	*Bactrocera cucurbitae* (Coquillett)	Cucurbits, tomato, several other fleshy fruits	Hawaii, Asia, Papua New Guinea
Oriental fruit fly	*Bactrocera dorsalis* (Hendel)	Mostly fleshy fruits and vegetables	Asia, Hawaii
Queensland fruit fly	*Bactrocera tryoni* (Froggatt)	Deciduous subtropical and tropical fruits	Australia, Pacific Island

(Sourse: Elhadi, 2006)

From the foregoing discussion, it is clear that emphasis should be given to formulate national policy to minimize postharvest losses of fruits and vegetable, and the government would take initiatives and allocate resources to improve the postharvest handling conditions, and thereby improve the socioeconomic status of the stakeholders in the fruits and vegetables supply chain.

8.2 NATURE AND CAUSES OF POSTHARVEST LOSSES

Losses occur after harvesting is known as postharvest losses. It starts first from the field, after harvest, in grading and packing areas, in storage, during transportation and in the wholesale and retail markets. A number of losses occur because of poor facilities, lack of know-how, poor management, market dysfunction or simply the carelessness of farmers.

(a) Extend of postharvest loss

The estimation of postharvest loss is crucial to make available more food from the existing level of production.

A current joint study conducted by the management consultancy firm, McKinsey and Co. and (The Confederation of Indian Industry (CII), at least 50 percent of the production of fruits and vegetables in the country is lost due to insect pests, wastage, and value destruction. The wastage cost is estimated to be Rs.23, 000 crores each year. Swaminathan Committee (1980) reported that the postharvest handling accounts for 20–30 percent of the losses at different stages of storage, grading, packing, transport, and finally marketing as a fresh produce or in the processed form. Chadha (2009) reported that India loses about 35–45 percent of the harvested fruits and vegetables during handling, storage, transportation etc. leading to the loss of Rs. 40,000 crores per year.

(b) Important sites of postharvest losses

Important sites where postharvest losses (Table 2 and Table 3) are noticed in India are:

- Farmer's field (15–20%)
- Packaging (15_2004)
- Transportation (30–40%)
- Marketing (30–40%)

TABLE 8.2   Estimated loss of fruits

Crop	Estimated loss (%)
Papaya	40–100%
Grapes	27%
Banana	20–28%

TABLE 8.2   *(Continued)*

Crop	Estimated loss (%)
Citrus	20–95%
Avocado	43%
Apple	14%

TABLE 8.3   Estimated loss of vegetables

Onion	25–40%
Garlic	08–22%
Potato	30–40%
Tomato	5–347%
Cabbage & cauliflower	7.08–25.0%
Chili	4–35.0%
Radish	3–5%
Carrot	5–9%

Sources: www. Postharvest.ucdavis.edu, www.postharvest.ifsa.ufl.edu

It is true that postharvest losses occur in every country, but the magnitude and major causes of losses and the effective remedial methods differ greatly from one country to another, one season or even one day to another. This book chapter addresses the most current state of postharvest and quarantine treatments in development and currently in use on horticultural products. The current implementation these treatments is also addressed.

8.3   POSTHARVEST TREATMENTS

8.3.1   HEAT AND COLD

Cold treatments can control some insect pests, and are currently used for the control of fruit flies (Table 8.4). Treatment requires 10 days at O°C (32 OF) or below, or 14 days at 1.7°C (35 °F) or below, so treatment is only suited to commodities capable of withstanding long-term low-temperature storage such as apples, pears, grapes, kiwifruit, and persimmons. For produce packed before cold storage treatment, package vents should be screened to prevent the spread of insects during handling. Control of storage insects in nuts and dried fruits and vegetables can be achieved by freezing.

Cold storage (less than 5°C or 41 ° F), heat treatments or the exclusion of oxygen (0.5% or lower) using nitrogen, packaging in insect-proof containers is needed to prevent subsequent insect infestation.

TABLE 8.4   Cold Treatment Protocol for Mediterranean fruit fly (*Ceratitis capitata*)

Temperature	Exposure period (days)
0°C (32°F) or below	10
0.6°C (33°F) or below	11
1.1°C (34°F) or below	12
1.7°C (35°F) or below	14
2.2°C (36°F) or below	16

Source: USDA APHIS PPQ Treatment Manual, (www.aphis.usda.gov/ppq/manuals/online-manuals.htm)

Some fungi and bacteria in their germination phase are susceptible to cold, and infections can be reduced by treating produce with a few days of storage at the coldest temperature the commodity can withstand without any damage (0°C for apple, pear, grape, kiwifruit, persimmon, and stone fruits). *Rhizopus stolonifer* and *Aspergillus niger* (black mold) can be killed when germinating by 2 or more days at 0 °C (32 °F) (Adaskaveg et al, in Kader, 2002), and pathogen growth can be nearly stopped by storage at temperatures below 5°C (41 °F).

Hot water dips or heated air can also be used for direct control of postharvest insects. In mangoes an effective treatment is 46.4°C for 65–90 min depending on size. Fruit should not be handled immediately after heat treatment. Whenever heat is used with fresh produce, clean, cool water showers or forced cold air should be provided to help return the fruits to their optimum temperature as soon as possible after completion of the treatment.

8.4   HOT WATER TREATMENTS

Hot water immersion treatments have been used to disinfest fruit flies (Figure8.2) from different fruits such as mangoes, bananas, papaya, and guavas. The results of hot-water treatment and storage temperature (4°C, 13°C or 22°C) on the quality and impedance of outer and inner mesocarp of mango fruit (Figure 8.1) were evaluated in

FIGURE 8.1   Fruit fly infested mango.

FIGURE 8.2   Adult oriental fruit fly.

two experiments during storage by Nayanjage et al. (2001). Fruit were subjected to equivalent heat unit at 36.5°C for 60 min plus 46.5°C for 90 min by hot-water treatments on the assumption of cumulative heat effects and base temperature of 12–13°C. Fruit reflectance decreased whereas chroma and huge angle increased over storage time and also with increase in storage temperature. The yellow color increased with a rise in storage temperature in hot-water treated mangoes. Impedance of hot-water treated fruit was poorly correlated with soluble solids content and chroma, but well correlated with reflectance of fruit pulp at 22°C.

Evaluated of mature, green mangoes subjected to a simulated in hot-water immersion at 50 °C for 60 min and subsequent storage at 5°C and 20°C by Talcott et al. (2005). Fruit held at 5°C were transferred to 20°C after 8 days of storage to complete ripening, where by symptoms of chilling injury were observed. Storage temperature during ripening and not the hot-water treatment was the major factor contributing to changes in polyphenolic content, with an antioxidant capacity unaffected by either postharvest treatments or ripening. Carotenoid concentrations were highest in hot-water treated fruit stored at 20°C, whereas storage at 5°C initially delayed ripening. Despite appreciable differences in fruit quality during quarantine treatment or low-temperature storage, only minor differences in antioxidant phytochemicals were observed. Therefore, immersion in water at 46.1°C for 110 min was suggested to provide postharvest and quarantine security against Mexican fruit fly (*Anastrepha ludens*) for fruit weighing between 700 and 900 gm without adversely affecting fruit market quality (Shellie and Mangan, 2002).

The fruit of guava infested with third instars of the Caribbean fruit fly (*Anastrepha suspense*) that were immersed in hot water at 46.1°C for 35 min and hydrocooled until fruit center temperatures returned to 24°C had only slightly reduction in quality of fruit held at 10°C (Gould and Sharp, 1992). The hot-water exposure time had a greater effect on banana fruit ripening than water temperature. Untreated bananas ripened after 13–15 day and ripening was delayed by 2–7 days when fruit were exposed to hot water for 15–20 min. Hot water treatments did not inhibit pulp softening but peels tended to be firmer for bananas immersed in 49–51°C water than control fruit. There was no difference in soluble solids content or titrable acidity between heat-treated bananas and controls.

Hot water dips of sweet cherry fruit of sufficient duration were effective in controlling codling moth but cause significant loss of quality. The market quality and condition of grapefruit were compared after three heat treatments for quarantine control of Caribbean fruit fly (*Anastrepha suspense*) (McGuire, 1991). The immersion at a constant 48°C significantly increased weight loss and promoted injury and decay, while reducing firmness and color intensity after 4 weeks of storage. No loss in quality resulted from treatment by forced hot air. Four heat treatments for quarantine and postharvest control of Caribbean fruit flies (*Anastrepha suspense*) in mango did not affect fruit quality but variably controlled two postharvest diseases (McGuire, 1992). Immersion of fruit in water at a constant temperature of 46 °C for 90 to 115 min significantly reduced anthracnose (*Collecttrichum gloeosporioides*) on three cultivars by 60–87 percent and stem end rot (*Diplodia natalensis*) by 61-88%. The author has recommended immersion in water at a constant temperature of 46 °C for the disinfestation of mangoes because it controls diseases without reducing market quality.

8.4.1 HOT WATER DIPS

Hot water treatments were originally used for fungal control. Their use has, however, been extended to curing and insect disinfestations. Postharvest hot-water dips are often applied for a few minutes in order to control decay.

The use of warm water dips has been tested on sweet lime fruits. Fruit were wounded and inoculated with spores of Penicillium intalicum, following which they were retained at room temperature for 4–5 h and subsequently dipped in warm water at 25, 45, and 55°C for 2 and 5 min. Dipping in warm water at 25 and 45°C for 5 min was observed to significantly reduce decay during storage (Mahmoodabadi et al., 2000).

Procedures for hot-water dipping for the disinfestations of a number of tropical and subtropical fruits from various species of fruity fly have also been developed (Lurie, 1998).

8.5   HOT AIR TREATMENT

Early, mid and late-season grape fruit were treated with hot air at 46, 48, and 50°C for 3,5 or 7 hrs to determine the effects of time and temperature on market quality. Early and late-season fruit were found to be more easily damaged by the higher temperatures than mid-season fruits (McGuire and Reeder, 1992). Increased time at lower temperatures had less of a deleterious effect on weight loss, loss of firmness and color and susceptibility to scalding injury and fungal decay than did shorter times at higher temperatures.

High Temperature Forced Air (HTFA) treatments have been developed for disinfesting Mediterranean fruit fly, melon fly, and oriental fruit fly from papayas (Anonymous, 1992). An approved treatment for disinfestations of grapefruit of Mexican fruit fly consists of a HTFA procedure that required a stepped heating profile of 40°C for 120 min followed by 50°C for 90 min and 52°C until fruit center temperature reach 48°C. This treatment was further shortened by using hot forced air procedure that required the maintenance of fruit center temperature at 44°C for 100 min, providing that the fruit reaches the temperature in 90 min.

The postharvest quality of nine early, mid, and late cultivars of California was determined following forced air heat treatment was used for quarantine disinfestations of Mediterranean fruit fly (Obenland et al., 1999). The treatment consisted of heating the nectarines to a seed surface temperature of 47.2°C for 2 min over a period of 4 hrs. Nectarines were evaluated after storage at 0°C for 14 days and 20°C for 4 days, which simulated shipping and sale of the fruit. Surface appearance indexed by color measurements and visual ratings remained generally unchanged by heat treatment and would have had no influence on the marketability of the fruit. A detectable change in the internal flesh was shown in a number of cultivars but was imperceptible when rated by visual evaluation. Heat treated fruits were firmer than unheated fruit but retained the ability to soften. Only the minor differences in soluble solid content and titratable acidity heat treated and untreated control fruits were detected. This study conducted that the forced-air heat treatment of the Mediterranean fruit fly for quarantine required for international trade did not adversely affect the quality and marketability of grown nectarines.

Grapefruit infested artificially with late third instars of Mexican fruit fly (*Anastrepha ludens*) and treated with forced hot air (with an incremental air temperature increase) until the fruit center temperatures were 48°C, with humidity controlled to maintain a dew point that was 2°C lower than the fruit surface temperature, had no fruit damage resulting from scald by condensation and by desiccation (Mangan and Ingle, 1994). The treatment significantly not hamper fruit appearance or flavor quality ratings, although rating for flavor and over all preference were lower for treated late season commercially stored fruit.

Vapour heat treatments have been developed for mangoes, grapefruit, papayas, litchis, tomatoes, bell peppers, eggplants, Chinese peas, cucumbers, green beans, lima beans and low wax beans. Ruby Red grapefruit tolerated a high temperature, forced air, vapor heat treatment of 43.5 °C for 260 min, a treatment applied for control of the Caribbean fruit fly (*Anastrepha suspense*), where fruit did not develop symptoms of quality of deterioration during subsequent storage (Miller and McDonald, 1991).

Vapor heat at 43.5 °C and 100 % relative humidity for 5 h reduced pill pitting fivefold compared with control fruit after 5 weeks of storage (4 weeks at 10 °C + 1 week at 21 °C) in freshly harvested Florida grape fruit and did not cause peel discoloration or rind breakdown (Miller et al., 1991). There is no difference in volume between treated and untreated fruit after 1 week of storage or in weight loss, peel color, total soluble solid content acidity and pH after 5 weeks. Fruit were slightly less firm after vapor heat treatment and remained less firm throughout storage, compared with control fruit. This study suggested that the vapor heat treatment tested is a potentially viable postharvest treatment for control of the Caribbean fruit fly (*Anastrepha suspense*) because it is not phytotoxic to grapefruit and has been reported effective in disinfestation of this pest in grapefruit

Treatments for reducing/eliminating heat injury

Several treatments have been tested for reducing or eliminating heat injury. Conditioning fruit at 40 °C for up to 16 h before hot-water treatment accelerated fruit ripening, as reflected in higher total soluble solids and lower titrable acidity levels (Jacobi et al., 2001). A conditioning treatment of either 22 °C or 40 °C before hot-water treatment could prevent the appearance of cavity at all maturity levels. The 40 °C conditioning temperature was found to more effective in increasing fruit heat tolerance than the 22 °C treatment. Therefore, for maximum fruit quality, it was recommended that mature fruit are selected and conditioned before hot-water treatment to reduce the risk of heat damage.

Immersion of guavas for 35 min in water at 46.1 °C slowed softening, sweetening, and color development of fruit, delayed ripening by 2 days, increased susceptibility to chilling injury, decay, and weight loss in storage but overall loss of quality is minimal (McGuire, 1997).

Chilling injury symptoms where avocados were held at 37–38 °C for 17–18 hrs and then air-cooled at 20 °C for 4 hrs before storage at 1.1 °C for 14 days but nonheated fruit developed serious surface discoloration and pitting (Sanxter et al., 1994). Chilling injury symptoms were reduced further when the heated fruit were stored in perforated polythene bags during 1.1 °C storage and no treatment surpassed the quality of fruit in untreated controls.

Hot water has been shown to cause severe injury to grape fruit. Grape-fruit pre harvest treated with gibberellic acid or not treated, were postharvest treated with vapor heat or hot-water such as that the surfaces of fruit were exposed to the same rate of temperature increases and treatment durations and quality attributes were then compared with ambient air and ambient water control after storage (Miller and McDonald, 1997). After 4 weeks storage at 10 °C plus 1 week at 20 °C, scald affected 5 % of hot water treated and 20 % of vapor heat treated fruit. No scald develop on control fruit and at the end of storage, mass loss for hot water treated and vapor heat treated fruit was approximately 5%. Gibberellic acid and heat treatments reduced decay relative to the control, and gibberellic acid treated fruit remained greener during storage than control fruit. This finding indicate that vapor heat and hot-water treatments at the temperatures and duration to control the Caribbean fruit fly (*Anastrepha suspense*) will likely cause peel injury to grapefruit produced in Florida, regardless of treatment with gibberellic acid.

Controlled modified atmosphere treatments

For commodities that tolerate high CO_2 levels, 15 to 20 % CO_2 enriched air canbe used as a control decay-causing pathogens such as *Botrytis cinera*on strawberry,blueberry, blackberry, fresh fig and table grapes during transport.

Modified atmospheres and controlled atmosphere with very low O_2 or very high CO_2 pressures, with or without addition of other gases such as CO have been tested as postharvest or quarantine systems (Hinsch et al., 1992). Atmospheres that can control insects in short period of time should contain very low concentrations of O_2 (less than 1%) or very high concentrations of CO_2 (up to 50-80 %). These insecticidal atmospheres can eliminate many insects within a period of 2-4 days at room temperatures (Yahia, 1998;Ke *et al.*, 1995). A potential problem that hinders the possibility for developing this system for fresh horticultural commodities is the possible fermentation of the tissue.

LowO_2 and/or high CO_2 have been used to kill certain insects in commodities that can tolerate these conditions. The effectiveness of insecticidal atmospheres depends upon the temperature, relative humidity, duration of exposure and life stage of the insect. The following are some examples from Mitcham et al (1997): Insecticidal atmospheres (0.5% or lower O_2 and/or 40 % or higher CO_2) have been shown to bean effective substitute for methyl bromide fumigation to disinfest dried fruits, nuts, and vegetables.

Papaya fruit exposed to a continuous flow of an atmosphere containing < 0.4% O_2 (balance N_2) at 0–5 days at 20 °C had decay and some fruit had developed off-flavors after 3 days in low O_2 plus 3 days in air at 20 °C (Yahia et al., 1992). The intolerance of the fruit to low O_2 correlates with an increase in the activity of pyruvate decarboxylase and lactate dehydrogenase, but not with the activity of alcohol dehydrogenase. Therefore, authors suggested that insecticidal O_2 (< 0.4%) atmospheres can be used as a postharvest insect control treatment in papaya for periods < 3 days at 20 °C without the risk of significant fruit injury.

The first and third instars of the green headed leafroller (*Planolortnx excessana*) and the first and fifth instars of the brown headed leaf roller (*Ctenopseustis obliquana*) and the light brown apple moth (*Epiphyas postvitana*) are completely killed in 2 months when apples are stored at 0.5 °C in 3% O_2 and 3% CO_2.

The eggs of the apple rust mite (*Aclilus schlechtel1dali*) and the European red mite (*Panonychusulmi*) are killed in 5.3 months when apples are stored at 2.8 °C in an atmosphere of 1% O_2 and 1% CO_2.

Codling moth larvae (*Cydia pomollella*) are killed in 3 months when apples (Fig 3 and 4) are stored at 0 °C, 1.5–2% O_2 and less than 1% CO_2.

FIGURE 8.3   Codling moth larvae in apple.

FIGURE 8.4   Apple infested by codling moth.

Sweet potato weevil (*Cylas formicarius elegantulus*) (Figure 6) has been controlled at ambient temperature in stored tropical sweet potatoes (Figure 5) by treatment with low oxygen and high carbon dioxide atmospheres. At 25°C (76 °F), storage in 2 to 4% O_2 and 40 to 60% CO_2 results in 100% mortality of adult weevils in 2 to 7 days.

FIGURE 8.5 Sweet potato infested by potato weevil.

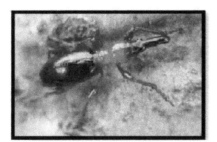

FIGURE 8.6 Adult sweet potato weevil.

In kiwifruit, the adult two-spotted spider mite (*Tetranychus Urticae*) is killed by 40°C, 0.4% O_2 and 20% CO_2 in only 7 h. When persimmons are stored at 20°C, 0.5% O_2 and 5% CO_2, the third instar of leaf rollers (*Plnotortrix excessalla*) is killed in 4 days and the larvae and adult mealy bug (*Pseudococcus longispinus*) is killed in 7 days.

Codling moth (*Cydia pomonella*) in stone fruits can be controlled at 25°C (76 °F) by using atmospheres of 0.5% oxygen and 10% carbon dioxide for 2 to 3 days (adult or egg) or 6 to 12 days (pupa). Normal color and firmness changes during ripening are not affected by treatment (http://postharvest.ucdavis.edu/files/93615.pdf).

Treatments with 45% CO_2 at 0 °C (32 °F) are being developed for several surface pests, including omnivorous leafroller (*Platynota sultana*), western flower thrips (*Frankliniella occidentalis*), and Pacific spider mite (*Tetranychus pacificus*) on harvested table grapes. This treatment requires 13 days at 0 ° to 2 °C (32 ° to 36 °F) and could be conducted in a marine container during transport. However, neither this nor other insecticidal CA treatments are approved yet as quarantine treatments (Mitcham et al., 2002).

Pesticide

A wide variety of chemicals are available for postharvest pest management. They are used in various ways— as dips, sprays, dusts, or applied on a pad of absorbent paper. Always follow label instructions and be aware that recommendations for use may differ by state and commodity, when using chemical pest controls, you need to consider cost, availability, regulations for proper use and residue tolerances. Recently many chemical controls have been banned due to concerns over residues and the possible consequences for human health. Others such as benomyl are no longer registered for postharvest applications; whenever possible it is a good idea to try to reduce your reliance on chemical controls.

When using chemicals in solution in the field on packinghouse, make sure you get good coverage by applying to the run-off stage. Always use potable water for spraying—recent *Cyclospora* outbreaks in raspberries in Guatamala were traced to contaminated water used to apply pesticides. The low cost, simple equipment illustrated here can ensure postharvest chemicals are applied as intended. The tray has perforations on the base to allow the solution to drain and the produce to dry before further handling.

Some plant materials are useful as natural pesticides. The pesticidal properties of the seeds of the neem tree (as an oil or aqueous extract) are becoming more widely known and used throughout the world. Native to India, neem (*Azadirachta indica*) acts as a powerful pesticide on food crops but appears to be completely nontoxic to humans, mammals, and beneficial insects (NRC, 1992). Any "natural pesticide" must be shown to be safe for humans before its approval by regulatory authorities.

The FPA has also approved a neem-based biological pesticide developed by Tata Oil Mills Ltd (TOMeo) for use of a wide range of food crops, fruits, and grains during production. We expect to hear more about this bio-pesticide in the next few years and eventually to be able to use it safely in postharvest horticulture applications.

Irradiation

Use of irradiation with dose of up to 1.0 kilogray (KGy) has been approved by United States Food and Drug Administration (Anonymous, 1987). Irradiation can kill insects may also damage the commodity and therefore postharvest treatments have been proposed by using lower doses that sterilize the insects rather than killing them. The most feasible and potentially useful application of irradiation in fruits and vegetables are probably for disinfestation as a postharvest treatment. Moy (1993) has observed that all stages of fruit fly will become sterile upon being irradiated at a minimum dose of 0.15 KGy, the dose level approved by United States Department of Agriculture in January 1989 for treating Hawaiian papayas as a quarantine procedure.

Irradiation as a postharvest treatment is easy to apply, quick, and generally safe. The research has demonstrated that irradiated insects are unable to either continue development or successfully reproduce. In most of the irradiated insects do not directly die as a result of radiation treatment was an initial concern for regulatory agencies. In addition to that, there was a general lack of consistent biochemical markers that could be used to indicate whether an insect had received an appropriate dose of radiation to render it biologically inactive. With continued research and improved dosimetry and documentation, regulatory agencies became more comfortable with the 'wriggler' issue (Lisa, 2010).

Low doses (0.15-1.00 KGy) of gamma irradiation have been reported to disinfest fruit flies from several fruits such as mangoes, papayas, bananas, litchis, cucumbers, tomatoes, bell peppers and egg plants and also on other insects such as mango seed weevils. A generic dose of 150 Gy has been proposed for use on tephritid fruit flies. Contrary to the 150 Gy dose, approved irradiation quarantine treatment doses for Mediterranean fruit fly, melon fly and oriental fruit fly in Hawaii are 210-250 Gy (Follet, 2004). Follet (2004) showed that 250-300 Gy can control Hawaii sweet potato pests. Approved irradiation quarantine treatment doses for melon fly (*Bactrocera cucurbitae*, Coquillet), Mediterranean fruit fly (*Ceratitis capitata*, Wiedemann) and Oriental fruit fly infesting fruits and vegetables for export from Hawaii to Continental United States are 210, 225, and 250 Gy, respectively (Follet and Armstrong, 2004).

Sweet potato roots of two Hawaii grown clones were treated with 100-600 Gy x-ray irradiation and evaluated for quality before and after cooking. Irradiation had the greatest effect on sucrose concentrations, which increased linearly in response to dose as starch concentrations decreased. A sensory panel perceived sweet potato roots treated with 600 Gy irradiation showed as sweeter than control roots.

However, grapefruit from gibberellic acid treated trees were irradiated at 0.3 or 0.6 KGy, and evaluated for quality after treatment and simulated commercial storage (Miller and McDonald, 1996). There was a general decline in the sensory preference for juice flavor and pulp flavor and texture as irradiation dose increased. The author has concluded that gibberellic acid treated grapefruit will tolerate irradiation dosage of 0.3 KGy without serious damage. However, at a dosage of 0.6 KGy, serious peel damage detrimental to fruit quality will likely to develop during storage.

The response of apple and winter pear fruit to irradiation postharvest treatment. Irradiation at doses between 0.30 and 0.90 KGy reduced apple firmness, doses of < 0.30 KGy had no effect and firmness loss due to irradiation was cultivar dependent (Drake et al., 1999). However, Bosc pear lost pears lost firmness due to irradiation, and the lost firmness due to irradiation and the firmness loss was dose dependent. Both Anjou and Bosc pear ripened normally after irradiation exposure. There was an increase in scald for Anjou pear that was dose dependent, and disease dependence f Fuji and Granny smith apples caused by *Panicum expansum* was reduced from about 80% of wounds, with lesions to 30% after irradiation exposure. Irradiation had no effect on number of lesions caused by either *Botrytis cinerea*. No effect was observed on decay of Anjou pear fruit naturally infected with *P. expansum* and *B. cinerea*.

Radiofrequency

The efficacy of using radiofrequency (RF) at 27.12 MHz as a postharvest quarantine treatment was evaluated against fifth instars of the codling moth (*Cydia pomonella*) in apples. Results demonstrated that the energy fields between the RF units electrodes were neither predictable nor uniform. Moving fruit submerged in water during RF exposure may improve uniformity, but pulp temperature varied considerably among fruit, among sites on the same fruit and at different depths within the same site (Hansen *et al.*, 2004).

Biological control and plant growth regulators

Mainly two biological control products (antagonistic organisms) are currently used as complementary tools (to chemical and/or heat treatments) for the management of postharvest decays together with other strategies as part of an integrated pest management program for a few fruits and vegetables (see Table 5).

Certainly two plant growth regulators can be used to delay senescence of citrus fruits and consequently delay their susceptibility to decay (see Table 5).

TABLE 8.5 Commercially available biological control materials and plant growth regulators (PGR) registered as postharvest treatments:

Category	Organisms/ Product	Year introduced	Crop	Decay organisms or function	Methods of application	Residue tolerance (ppm)
Biocontrol	*Pseudomonas syringae*	1995	Citrus	*Penicillium digitatus*	Dip or spray	exempt
	Bio-Save)			*P. italicum,*		
				Geotrichum citriaurantii		
			Cherries	*Penicillium expansum Botrytis cinerea*	Drench	exempt
			Apples pears	*Penicillium expansum Botrytis cinerea*	Dip or drench	exempt
				Mucor piriformis		
			Potatoes	*Fusarium sambucinum*	Dip or spray	exempt
				Helminthosporium solani		

TABLE 8.5 *(Continued)*

Biocontrol	Candida oleophila (Aspire)	1995	Pome fruits	Decay pathogens	Any type of application	exempt
			Citrus	Decay pathogens	Any type of application	exempt
PGR	Gibberellic acid (Pro Gibb)	1955	Citrus	Delays senescence (delays onset of decay)	Storage wax	exempt
PGR	2,4-D (Citrus Fix)	1942	Citrus	Delays senescence of buttons (delays onset of decay)	Storage wax	5

Source: Adaskaveg et al, in Kader (2002)

Irradiation and heat

Marsh grape fruit was treated with vapor heat (2 hrs at 38 °C) and treatment with fungicide of thiabendazole (4 gm/l) and thiabendazole (1 gm/l) plus imazalil (1 gm/l) prior to irradiation at 0.5 or 1.0 kGy [100]. Vapour heat reduced the severity and incidence of peel injury by about 50% without adversely affecting other quality parameter; fungicide did not reduce the peel injury.

Modified atmosphere (MA) and controlled atmosphere (CA) at high temperatures

CA and MA at high temperatures can eliminate many insects within a period of 2 to 4 hrs at > 40 °C (70). Potential problems that hinder the possibility for developing this system for fresh horticultural commodities are possible fermentation (due to the use of anaerobic gas mixtures) and heat injury. However, the advantage of using this system at high temperatures is to accomplish the mortality of insects in a short period. The potential development of postharvest insect control system using CA at room or higher temperatures in several crops has been tested for several horticultural commodities. Mangoes were found to be sufficiently resistant, and therefore this technique can be developed commercially for this fruit (Yahia, 1998; Yahia and Tizando, 1993; and Yahia and Vazquez, 1993).

Several treatments combining MA or CA with heat (34, 70, and 71) have been studied for different crops. Reducing the availability O_2 during heating stress hinders the insects' ability to support elevated metabolic demands due to heat load. There is also evidence that a heat treatment under an oxic condition reduces the production of heat shock proteins in insects (Thomas and Shellie, 2000), and elevated CO_2 atmospheres may interfere with the insects ability to produce ATP (Friedlander, 1983).

Yellow and red white fleshed peach and nectarine of mid and late-season maturity classes were subjected to combined CA temperature treatments using heating rates of either 12 °C/h (slow rate) or 24 °C/h (fast rate) with a final chamber temperature of 46 °C, while maintaining a CA of 1% O_2 and 15 % CO_2 (Obenland *et al.*, 2005). Soluble solids, acidity, weight loss and color were either not affected or changed to a very small degree as a result of the treatment. A sensory panel preferred the taste of untreated fruit over treated fruit but the rating of treated and untreated and treated fruit were generally similar and it is unclear whether an average consumer would detect the difference.

The response of Kiwifruit to high temperature CA treatments for control of two-spotted spider mite (*Tetranychus urticae* Koch) was investigated by Lay Yee and Whiting (La Yee and Whiting, 1996). Kiwifruit were subjected to 40 °C for 7 or 10 hrs in 20% CO_2 or in air. Fruit were cooled with following treatment in ambient water or ambient air, stored at 0 °C in air for 8 weeks, afterwards held at 20 °C overnight and assessed for quality. Comparative to untreated control, no significant damage was found with fruit subjected to 40 °C air treatments. There were no significant damage was observed with fruit treated for 7 hrs with 20% CO_2 followed by hydrocooling. The treatment without hydrocooling and 10 hr treatments with hydrocooling showed only slight damage, while the 10 hr CA without hydro-cooling had moderate fruit damage.

Quarantine treatments, CA using at high temperature were also tested for codling moth (Fig 7) in sweet cherries (Neven, 2005), the treatments used at 45 °C for 45 min and 47 °C for 25 min at 1% O_2, 15% CO_2 and − 2 ° C dew point environments. These treatments have been recorded to provide control of all life stages of codling moth while preserving commodity market quality.

FIGURE 8.7   Codling moth adult.

Heated cherries and MeBr- fumigated cherries were lesser firm after 14 days of cold storage than nonheated, control fruit. The stems of MeBr fumigated cherries were less green than heated or nonheated cherries. Cherries exposed to 45 °C had lower titrable acidity than nonheated cherries, fumigated cherries or cherries exposed to 47 °C, cherry quality after 14 days of cold storage was no affected by hydro-cooling before heating or by the method of cooling after heating (hydro-cooling, forced air cooling). Cherry stored for 14 days at 1 °C in 6% O_2 with 7% CO_2 had similar market quality as cherries stored in air at 1 °C. This study has suggested that Bing sweet cherry can tolerate heating in an atmosphere of low O_2 containing elevated CO_2 as doses that may provide quarantine security against codling moth (*Cydia pomonella*) and western cherry fruit fly (*Rhagoletis cingulata*). Some studies indicated that the slower the rate of heating and lower the final treatment temperature, he longer the total treatment to control the insect pests (Neven *et al.*, 1996; Neven and Rehfield, 1995).

Hot water and radiofrequency

Hansen *et al.* (2005) reported that hot-water treatments and radiofrequency for postharvest control of codling moth in Bing sweet cherries were examined as a potential alternative method. Codling moth infested each Bing sweet cherries submerged in water at 38 °C for 6 min pretreatment, exposed to various temperatures generated by radiofrequency and held at different temperatures and durations; 50 °C for 5 min, 51.6 °C for 4 min, 53.3 °C for 5 min. Insect mortality was recorded 2 hrs after treatment and 7 and 14 days of storage at 1 °C. Not a single larva survived at 50 °C and 51.6 °C treatments. The color of noninfested cherry fruits was darker at temperature increased. Stem color was severely affected after 7 days of storage, even in warm water bath of 38 °C for 6 min, as was fruit firmness the same treatment. The quality of fruit losses increased after 14 days of storage, compared with after 7 days of storage.

Comparative study between different treatments

Know *et al* (2004) reported that Comparative effect of gamma irradiation and MeBr fumigation were determined for fresh chestnut on mortality of pests and quality stability. Chestnut was exposed to both irradiation at 0 to 10 kGy and MeBr fumigation in commercial conditions, and then stored at 5 °C for 6 months. Pest with postharvest importance for chestnut revealed 100% mortality by MeBr at the third day after fumigation and by irradiation at 0.5 kGy in about 4 weeks. Sprouting was controlled for 6 months with treatments with 0.25 kGy or more and of MeBr, but rotting rate dramatically increased from 2 months after fumigation. I rradiation over 1.0 kGy as well as fumigation caused significant changes in color of stored chestnut. However the cumulative mortality of chestnut pests, these author recommended irradiation at 0.5 kGy as an alternative to MeBr fumigation for both quarantine and sprout control purposes. In an additional study by Know (2005) gamma irradiation (up to 10 kGy) was recorded as an alternative to MeBr fumigation for the pest control of chestnuts. Mortality rate 100% was achieved by MeBr on the third day after fumigation, and by irradiation at 0.5 kGy in about 3 weeks. Respiration rate of samples one day after treatments increased in proportion to irradiation dose. Respiration pattern of MeBr group was equal to that of the 10 kGy group. Both MeBr and irradiation at 0.25 kGy or higher showed that 100% inhibition of sprouting during storage at 5 °C for 6 months. Flesh firmness was significantly reduced by MeBr or irradiation over 5 kGy 1 day after treatments. MeBr fumigation resulted in appreciable decrease in flesh weight, reducing sugar and ascorbic acid contents compared with irradiated samples. Know (2005) has indicated that irradiation at 0.5 kGy was effective as an alternative to MeBr in controlling pests while maintaining overall quality of fresh chestnut during storage.

Hot air treatment (48.5 °C for 3-4 hrs) of papaya fruit, a developed postharvest treatment for fruit flies, did not significantly reduced incidence of postharvest diseases when compared with fungicide or hot-water treatments, on the other hand when combined with thiabendazole (4 gm ai/l) or hot-water immersion (49 °C for 20 min), the incidence of postharvest diseases was reduced (Nishijima et al., 1992). However incidences of disease were not significantly affected by the sequence of hot air or hot-water application along with pitting and scalding symptoms increased when hot water preceded hot air treatment, but these symptoms did not occur when hot air preceded ho air treatment.

Fruit fly quarantine treatments on Arkin carambolas fruit by hot-water immersion at 43.3 °C–43.6 °C for 55–70 min, 46.0–46.3 °C for 35–45 min, or 49.0–49.3 °Cfor 25 or 35 min, or vapor heat at 43.3–43.6 °C for 90–120 min, 46.0–46.3 °C for 60–90 min, or 49.0–49.3 °C for 45–60 min (Hallman, 1991). The 49.0–49.3 °Ctreatment showed in excessive damage to carambolas 2–4 days after treatment, but there were not statistically significant differences in the variables measured among the other treatments and control.

The fruits of litchi were found to either 15 days at 1.1 °C or to gamma irradiation at 100, 200, or 300 Gy (McGuire, 1997). Cold treated Mauritius fruit losses some color intensity externally and internally, and the pale flesh had a greener hue. Mauritius fruit were also more susceptible to decay following irradiation at 300 Gy and 6 days of storage at 5 °C. The cultivars lost firmness after this treatment. The pericarp of irradiated Mauritius fruit becomes more orange, whereas the flesh of the cultivars becomes greener. Loss of quality was minimal either cold or irradiation treatment, and they were recommended to be acceptable for lychness requiring quarantine treatment for eradication of exotic pests.

The fruit of canistel were found to cold storage (1 or 3 °C for 17 days) and hot-water immersion (46 °C for 90 min or 48 °C for 65 min) treatments known to kill immature Caribbean fruit flies (Anastrepha suspense Loew) in other fruit (Hallman, 1995). Canistel quality did not loss in cold storage compared with fruit stored at normal 10 °C, and unripe canistels submerged in both hot-water treatments developed dark blotches on the peel and a 2-3 mm thick layer under the peel that did not soften.

Immersion in hot water and irradiation postharvest treatments were used to disinfest lychee of fruit flies and other pests before export from Hawaii to the United States mainland (Follet and Sanxter, 2003) compared the quality of fruit exposed to each of these treatments. One day after harvest, Kaimana lychee fruit were subjected to hot-water immersion at 49 °C for 20 min, irradiation treatment at a minimum absorbed doses of 400 Gy or left untreated as control. Afterwards fruit were stored at 2 or 5 °C in perforated plastic bags and quality attributes were evaluated after 8 days. Lychee fruit treated with hot-water immersion were darker (lower lightness) and less intensely colored than irradiated and untreated fruit at both storage temperatures. Fruit stored at 2 °C were darker (lower lightness) than fruit stored at 5 °C had greater weight loss. External appearance of fruit treated with hot-water immersion was rated as unacceptable.

Rambutan was irradiated with 250 Gy is an APHIS approved quarantine treatment, but a hot forced air treatment has also been proposed for eliminating fruit fly pests, and therefore evaluated the effect of each of these treatments on the quality of rambutan (Follet and Sanxter, 2000).After two days of harvest rambutan fruit were subjected to hat air at a seed surface temperature of 47. 2 °C, irradiation treatment at 250 Gy, or left untreated as controls. Then fruits were stored at at 10 °C in perforated plastic bags and quality characters were evaluated after 4, 8, and 12 days.

8.6   CONCLUSION

Postharvest treatment is very important for facilitating national and international trade of horticultural commodities. These treatments have increased significantly since last few years because horticultural crops help to increase in trade and restrict the use of fumigants in horticultural crops. A number of deficiencies currently exist in the postharvest management and processing of fruits and vegetables in India. Action must be taken in order to upgrade systems, in order to reduce the levels of postharvest losses in India.

Most importantsafety and significant research programmes have been reported in developing physical treatments and systems, because interest in reducing health and environmental problems. Presence of pests in horticultural foodstuffs has been the center of attention of many procedures and treatments to prevent continuous damage to the foodstuffs or accidental movement of pests from one area to another. Use of chemicals in postharvest management, are harmful to environment as well as human body. Therefore it is desirable to explore some alternative methods are being developed to be more environmentally safe and have no impact on human health. Proper infrastructure, logistics, and management and human resources are essential to improving postharvest management and marketing of fruits and vegetables. Several other treatments are still in progress and may require moreresearch to gain prevalent acceptance.

KEYWORDS

- **Postharvest pests**
- **Postharvest loss**
- **Fruits and vegetables**
- **Postharvest treatments**
- **Radio frequency**
- **Quality**
- **Infestation**

REFERENCES

Adaskaveg, J. E.; Forster, H.; and Sommer, N.F.; Principles of postharvest pathology and management of decays of edible horticultural crops. In: *Postharvest Technology of Horticultural Crops,* 3rd ed.; Kader, A. A., Ed.; University of California, ANR Publication 3311, **2002**; pp 196–195.

Anonymous; FDA-US Food and Drug Administration. Irradiation in the production processing and handling of food, final rules. *Federal Register.* **1987**, *51,* 13375–13399

Anonymous; Food irradiation update. *Food Technol.***2006**, *60*(10), 73–75.

Anonymous. U.S.; Department of Agriculture–Animal and Plant Health Inspection Service. T108 (a) (1) and (2) Fumigation plus refrigeration. In Plant protection and quarantine treatment manual. US Department of Agriculture, Hyattsville, Maryland, USA, **1992**; pp 5,63–5,64.

Chadha, K. L.; *Handbook of Horticulture;* IARI Publications: New Delhi, **2009**.

Drake, S. R.; Sanderson, P. G.; and Neven, L. G.; Response of apple and winter pear fruit quality to irradiation as a quarantine treatment. *J. Food Process. Pres.***1999**, *23*(4), 203–216.

Elhadi, M. Y.; Effect of insect quarantine treatments on the quality of horticultural crops. *Stew. Postharvest Rev.* **2006**, *1,* 1–18.

Fields, P. G.; and White, N. D. G.; Alternatives to methyl bromide treatments for stored-product and quarantine insects. *Ann. Rev. Entomol.* **2002**, *47,* 331–359.

Follett, P. A.; Irradiation to control insects in fruits and vegetables for export from Hawaii. *Radiat. Phys. Chem.* **2004**, *71* (1–2), 163–166.

Follett, P. A.; andArmstrong, J. W.; Revised irradiation doses to control Melon fly, Mediterranean fruit fly and Oriental fruit fly (Diptera: Tephritidae) and a genetic dose for tephritid fruit flies. *J. Econ. Entomol.* **2004**, *97*(4), 1254–1262.

Follett, P. A.; and Neven, L. G.; Current trends in quarantine entomology. *Ann. Rev. Entomol.* **2006**, *51*, 359–385.

Follett, P. A.; and Sanxter, S. S.; Comparison of rambutan quality after hot forced air and irradiation quarantine treatments. *Hortic. Sci.* **2000**, *35*(7), 1315–1318.

Follett, P. A.; and Sanxter, S. S.; Lychee quality after hot water immersion and X-ray irradiation quarantine treatments. *Hortic. Sci.* **2003**, *38*(6), 1159–1162.

Friedlander, A.; Biochemical reflections on a non chemical control method: The effect of controlled atmosphere on the bio-chemical process in stored products insects. In: *Proceeding, Third*

International Working Conference on Stored Product Entomology; Kannas State University: Manhattan (KS), **1983**; pp 471–486.

Gould, W. P.; and Sharp, J. L.; Hot water immersion quarantine treatment for guavas infested with Caribbean fruit fly (Diptera: Tephritidae). *J. Am. Soc. Hortic. Sci.* **2002**, *127*(3), 430–434.

Hallman, G. J.; Cold storage and hot water immersion as quarantine treatments for canistel infested with Caribbean fruit fly. *Hortic. Sci.* **1995**, *30*(3), 570–572.

Hallman, G. J.; Quality of carambolas subjected to postharvest hot water immersion and vapour heat treatments. *Hortic. Sci.* **1991**, *26*(3), 286–287.

Hansen, J. D.; Drake, S. R.; Hedit, M. L.; Tang, J.; and Wang, S.; Radiofrequency treatments for postharvest codling moth in fresh apples. *Hortic. Technol.* **2004**, *14*(4), 533–537.

Hansen, J. D.; Drake, S. R.; Hedit, M. L.; Tang, J.; Wilkins, M. A.; and Wang, S.; Evaluation of radiofrequency- hot water treatments for postharvest control of codling moth in sweet cherries. *Hortic. Technol.* **2005**, *15*(3), 613–616.

Hinsch, R. T.; Harris, C. M.; Hartsell, P. L.; and Tebbets, J. C.; Fresh nectarine quality and methyl bromide residues after in-package quarantine treatments. *Hort. Sci.* **1992**, *27*(12), 1288–1291.

Jacobi, K. K.; MacRae, E. A.; and Hetherington, S. E.; Postharvest heat disinfestations treatments of mango fruit. *Sci. Hortic.* **2001**, *89*(3), 171–193.

Johnson, J. A.; and Valero, K. A.; Use of commercial freezers to control Cowpea Weevil, *Callosobruchus maculatus* (Coleoptera: Bruchidae) in organic garbanzo beans. *J. Econ. Entomol.* **2003**, *96*(6), 1952–1957.

Ke, D.; Yahia, E.; Hess, B.; Zhou, L.; and Kader, A.; Ethanolic fermentation of pears as influenced by ripening stageand controlled atmosphere storage. *J. Am. Soc. Hortic. Sci.* **1994**, *119*(5), 976–982.

Know, J. H.; Effects of gamma irradiation and methyl bromide fumigation on the quality of fresh chestnut during storage. *Food Sci. Biotech.* **2005**, *14*(2), 181–184

Know, J. H.; Known, Y. J.; Byun, M. W.; and Kim, K. S.; Comparativeness of gamma irradiation with fumigation for chestnuts associated with quarantine and quality security. *Radiat. Phys. Chem.* **2004**, *71*(1–2), 43–46.

La Yee, M.; and Whiting, D. C.; Response of Kiwifruit to high temperature controlled atmosphere treatments for control of two spotted spider mite (*Tetranychus urticae*). *Postharvest Biol. Tech.* **1996**, *7*(1–2), 73–83.

Lacey, L. A.; Horton, D. R.; Jones, D. C.; Headrick, H. L.; and Neven, L. G.; Efficacy of the bio-fumigant fungus *Muscodoralbus* (Ascomycota: Xylariales) for control of Codling Moth, *Cydiapomonella* (L.)(Lepidoptera:Tortricidae) in simulated storage conditions. *J. Econ. Ento.* **2009**, 102,43–49.

Lisa, G. N.; Postharvest management of insects in horticultural products by conventional and organic means, primarily for quarantine purposes. *Stewart. Postharvest Rev.* **2010**, *6*(1), 1–11.

Lurie, S.; Postharvest heat treatments of horticultural crops. *Hortic. Rev.* **1998**, *22*, 91–121.

Mahmoodabadi, K.; Rahemi, M.; and Banihashemi, Z.; Postharvest curing of sweet lime by heat treatments to reduce fruit decay by *Penicillium italicum*. *Iran, J. Plant Path.* **2000**, *36*, 245–259.

Mangan, R. L.; and Ingle, S. J.; Forced hot air quarantine treatment for grapefruit infested with Mexican fruit fly (Diptera: Tephritidae). *J. Econ. Entomol.* **1994**, *87*(6), 1574–1579.

McGuire, R. G.; Concomitant decay reductions when mangoes are treated with heat to control infestations ofCaribbean fruit flies. *Plant. Dis.* **1992**, *79*(9), 946–949.

McGuire, R. G.; Market quality of grapefruits after heat quarantine treatments. *Hortic. Sci.* **1991**, *26*(11), 1393–1395.

McGuire, R. G.; Market quality of guavas after hot-water quarantine treatment and application of carnauba waxcoating. *Hortic. Sci.* **1997**, *32* (2), 271–274.

McGuire, R. G.; Response of lychee fruit to cold and gamma irradiation treatments for eradication of exotic pests. *Hortic. Sci.* **1997**, *32*(7), 1255–1257.

McGuire, R. G.; and Reeder, W. F.; Predicting market quality of grapefruit after hot air quarantine treatment. *J. Am. Soc. Hortic. Sci.* **1992**, *117*(1), 90–95.

Miller, W. R.;and McDonald, R. E.; Comparative response of postharvest GA treated grapefruit to vapour heat and hot water treatments. *Hortic. Sci.* **1997**, *12*(01), 15–19.

Miller, W. R.; and McDonald, R. E.; Postharvest quality of GA-treated Florida grapefruit after gamma irradiation with TBZ and storage. *Postharv. Biol. Technol.* **1996**, *7*(3), 253–260.

Miller, W. R.; and McDonald, R. E.; Quality of stored Marsh and Ruby Red grapefruit after high-temperature, forced air treatment. *Hort. Sci.* **1991**, *26*(9), 1188–1191.

Miller, W. R.; McDonald, R. E.; Hallman, G.; and Sharp, J. L.; Condition of Florida grapefruit after exposure to vapour heat treatment. *Hortic. Sci.* **1991**, *26*(1), 42–44.

Mitcham, E. J.; Zhou, S.; and Kader, A. A.; Potential for CA for postharvest insect control in fresh horticultural perishables: An update of summary tables compiled by Ke and Kader, **1992**; pp 78–90 In: CA'97 Proceedings. Volume 1: CA Technology and Disinfestation Studies. Department of Pomology Postharvest Hort. Series No. 15, Thompson, J. F., Mitcham, E. J., Eds.; **1997**.

Mitcham, E. J.; Mitchell, F. G.; Arpaia, M. L.; and Kader, A. A.; Postharvest Treatments forinsect control. In: *Postharvest Technology of Horticultural Crops,* 3rd ed.; Kader, A. A., Ed.; University of California, ANR Publication 3311, **2002**; pp 251–257.

Moy, J. H.; Efficacy of irradiation vs thermal methods as postharvest treatments for tropical fruits. *Radiat. Phys. Chem.* **1993**, *42*(1–3), 269–272.

Neven, L. G.; Combined heat and controlled atmosphere quarantine treatments for control of Codling Moth, *Cydiapomonella*, in sweet cherries. *J. Econ. Entomol.* **2005**, *98*(3), 709–715.

Neven, L. G. Organic quarantine treatments for tree fruits. *Hortic. Sci.* **2008**, *43*(1), 22–26.

Neven, L. G.; and Rehfield, L. M.; Comparison of prestorage heat treatments on the fifth instar codling moth Lepidoptera: Totricidae) mortality. *J. Econ. Entomol.* **1995**, *88*(5), 1371–1375.

Neven, L. G.; Rehfield, L. M.; and Shellie, K. C.; Moist and vapour forced air treatments of apple and pears: Effect on the morality of fifth instar codling moth (Lepidoptera: Totricidae). *J. Econ. Entomol.* **1996**, *89*(3), 700–704.

Neven, L. G.; Yahida, E.; and Hallman, G.; Effects on insects. In: *Modified and Controlled Atmospheres for the Storage, Transportation, and Packaging of Horticultural Commodities;* Yahida, E., Ed.; Taylor and Francis, LLC: Boca Raton, Florida, **2009**; pp 233–316.

Nishijima, K. A.; Miura, C. K.; Armstrong, J. W.; Brown, S. A.; and Hu, B. K. S.; Effect of forced, hot air treatment of papaya fruit on fruit-quality incidence of post harvest diseases. *Plant Dis.* **1992**, *76*(7), 723–727.

Nyanjange, M. O.; Wainwright, H.; and Bishop, C. F. H.; Effect of hot water treatments and storage temperatures on ripening and the use of electrical impedance as an index for assessing post harvest changes in mango fruits. *Ann. Appl. Biol.* **2001**, *139*, 21–29.

Obeng-Ofori, D.; The use of botanicals by resource poor farmers in Africa and Asia for the protection of stored agricultural products. *Stewart. Postharvest Rev.* **2007**, *3*(6), 10.

Obenland, D. M.; Arpaia, M. L.; and Aung, L. H.; Quality of nectarine cultivars subjected to forced air heat treatment for Mediterranean fruit fly disinfestations. *J. Hortic. Sci. Biotech.* **1999**, *74*(5), 553–555.

Obenland, D.; Neipp, P.; Mckey, B.; and Neven, L.; Peach and nectarine quality following treatment with high temperature forced air combined with controlled atmosphere. *Hortic. Sci.* **2005**, *40*(5), 1425–1430.

Rajendran, S.; and Sriranjini, V.; Plant products as fumigants for stored–product insect control. *J. Stor. Prod. Res.* **2008**, *44*(2), 126–135.

Sanxter, S. S.; Nishijima, K. A.; and Chan, H. T.; Heat treating Sharwil avocado for cold tolerance in quarantine cold treatments. *Hortic. Sci.* **1994**, *29*(10), 1166–1168.

Schneider, S. M.; Rosskopf, E. N.; Leesch, J. G.; Chellemi, D. O.; Bull, C. T.; and Mazzola, M.; United States Department of Agriculture – Agricultural research service research on alternatives to methyl bromide: Pre-plant and post-harvest. *Pest. Manage. Sci.* **2003**, *59* (6–7), 814–826.

Shellie, K. C.; and Mangan, R. L.; Hot water immersion as a quarantine treatment for large mangoes: Artificial versus cage infestation. *J. Am. Soc. Hortic. Sci.* **2002**, *127*(3), 430–434.

Talcott, S. T.; Moore, J. P.; Lounds-Singleton, A. J.; and Pereival, S. S.; Ripening associated phytochemical changes in mango (*Mangifera indica*) following thermal quarantine and low temperature storage. *J. Food. Sci.* **2005**, *70*(5), 337–341.

Thomas, D. B.; and Shellie, K. C.; Heating rate and induced thermotolerance in Mexican fruit fly (Diptera:Tephritidae) larvae, a quarantine pest of citrus and mangoes. *J. Econ. Entomol.* **2000**, *93*(4), 1373–1379.

USDA; Composition of fruits and fruit juices raw processed, prepared. US Department of Agriculture, Agricultural Handbook 8-9 (http://www.nal.usda. gov/finc/ foodcomp). **1983**.

Wall, M.; Quality of postharvest horticultural crops after irradiation treatment. *Stewart Postharv. Rev.* **2008**, *4*(2), 1.

Wills, R.; McGlasson, B.; Graham, D.; and Joyce. D.; *Postharvest: Introduction to the Physiology and Handling Fruits, Vegetables and Ornamentals*, 4th ed.; University of New South Wales Press Ltd.: Sydney 2052, Australia, **2004**; p 262.

Yahia, E. M.; Modified/Controlled atmospheres for tropical fruit. *Hortic. Rev.* **1998**, *22*, 123–183.

Yahia, E. M.; and Tizando, M.; Tolerance and response of avocado fruit to insecticidal oxygen atmospheres. *Hortic Sci.* **1993**, *28*(10), 1031–1033.

Yahia, E. M.; and Vazquez, I.; Responses of mango to insecticidal oxygen atmospheres. *Hortic Sci.* **1993**, *26*(1), 42–48.

Yahia, E. M.; Rivera, M.; and Hernandez, O.; Response of papaya to short term insecticidal oxygen atmosphere. *J. Am. Soc. Hortic. Sci.* **1992**, *117*(1), 96–99.

E REFERENCES

http://tnau.ac.in

http://postharvest.ucdavis.edu/files/93615.pdf

www.Postharvest.ucdavis.edu

www.postharvest.ifsa.ufl.edu

www.aphis.usda.gov/ppq/manuals/online-manuals.htm

CHAPTER 9

FLOWER SENESCENCE

DONAL BHATTACHARJEE[1]; PRAN KRISHNATHAKUR[1];
JEEBIT SINGH[1] and R. S. DHUA[1]

[1]Department of Postharvest Technology of Horticultural Crops; Bidhan
Chandra Krishi Viswavidyalaya, Mohanpur-741252, Nadia, West Bengal,
India

CONTENTS

Senescence is the process of aging, both in whole organisms (organismal senescence) and individual cells (cellular senescence) within these organisms. In the strictest scientific sense, senescence means the period of decline that follows the development phase in an organism's life. This word is derived from the Latin word *senex* meaning "old age". In the longevity and healthy aging fields, senescence is the decline in health and function associated with aging. The process of senescence is incredibly complex, and accompanied with a myriad of chemical and physical reactions. As organisms age, they slowly break down, experiencing tissue death and more general malfunction, whether they are plants, animals, fungi, or one-celled organisms. Without the processes of senescence, organisms would be immortal and the Earth would be choked with living organisms as a result. From a biological standpoint, senescence promotes evolution, and keeps Earth from being too crowded.Senescence occurs when cells lose the ability to divide because of DNA damage or a shortening of telomeres, they go through a transformation that results in decline or destruction. The cells either self-destruct (called apoptosis) or go into a period of decline (called senescence). The end result is cell death. Cell death is a normal part of biological functioning and occurs regularly in all living organisms. The science of biological aging is known as biogerontology.

9.1   FLOWER SENESCENCE

The first thought that usually comes to mind when looking at flowers is their esthetic value. Longer lasting flowers are valued not only by consumers of floriculture products, but also by retailers, who have additional time to transport, store, and market them. However, this is not the only reason flowers are important to humans. Flowers are, in many cases, instrumental in attracting insects for pollination. Consequently, understanding and manipulating flower senescence to create longer-lived flowers should therefore result in both esthetic and economic benefits to the flower industry and agriculture as a whole.

Senescence is a tightly regulated process affecting single cells,, organs or the whole plant and eventually leading to death within a developmental program, which allows the plant or the species to survive (Leopold, 1961; Quirinoet al., 2000; Lim et al., 2007). Senescence is the final phase in the ontogeny of the organ in which a series of normally irreversible events is initiated that leads to cellular breakdown and death of the organ. During senescence nutrients are recovered through the degradation of macromolecules and transported to other plant parts working like a sink. This recycling mechanism plays a relevant role in determining yield and quality of crop plants.

Watada et al. (1984) defined senescence as those processes that follow physiological maturity and lead to the death of the whole plant, organ tissue or cell. Senescence of higher plants is classified into three major types (Leopold, 1961; Beevers, 1976; Medawar, 1957; Leopold, 1975): (a) Population senescence (e.g., annual plants), (b) organism or individual plant senescence, and (c) determinate organ senescence (e.g., leaves, fruits, flowers, petals, etc.). Determinate organs such as flowers, most probably, undergo a genetically programmed senescence, which is independent of specific "aging genes" and is devoid of any special genetical mechanism of senescence. Sacher (1973) described determinate organ senescence on the basis of general gene action of normal plant development. Experimental evidence is lacking to demonstrate the presence of specific "aging genes". Ample evidence however, points to the increased activity of several enzymes, which function during plant development, may be playing an important role during the organ senescence.

The flower is a complex organ composed of different tissues that senesce at different rates, allowing studies of inter-organ communication (Borochov and Woodson, 1989). Flower senescence refers most of the time to the parts of flower that are regarded as attractive (Stead, 1992). Understanding and controlling flower longevity is therefore important commercially. The limited life span and the existence of an irreversible program largely independent of environmental factors make the flower a useful system for studying the senescence process in plant (Rogers, 2006). In addition, the onset of senescence is often readily visible (sometimes as a color changes (Macnishet al. 2010a), coordinated within a single large organ comprising relatively uniform cells, and may be manipulated by simple triggers (pollination, ethylene, and photoperiod). The overall picture of floral senescence that has emerged from recent studies is one of a controlled disassembly of the cells of the corolla, probably by a mechanism homologous with apoptosis (van Doorn, 2011), and transport of the resulting nutrients to other parts of the inflorescence or beyond. Senescence in flower organs has been pointed out as a form of programmed cell death (PCD), a genetically controlled cell death involving a suicidal pathway regulated at different levels and associated with well-defined and characteristic events (Danon et al., 2000; Rubinstein, 2000; Zhou et al., 2005). Flower senescence is certainly programmed and irreversibly leads to cell death making the distinction between senescence and PCD unnecessary for this system (van Doorn and Woltering, 2004a; Rogers, 2006).

Flowers are the structures responsible for sexual reproduction, and thus play a relevant role in the perpetuation of the earth's most dominant group of plants (Rubinstein, 2000). Plants have often showy and gaudy flowers and produce scents to attract insects and other animals. Most insect-pollinated plants pay their pollinators in energy rich nectar whilst others, including the early spider orchid (*Ophrys-sphegodes*), mimic females and promise their pollinators a sexual partner (Ledford, 2007). Despite its irreplaceable ecological role the flower is energetically expensive to maintain beyond its useful life (Ashman and Schoen, 1994). Flower senescence allows the removal of the structures no longer necessary and the redistribution of the nutrients to the growing ovary or to other plant organs. The length of time a flower remains open and functional varies among species from one day to several weeks depending on ecological adaptations. The effect of pollination in promoting senescence does not occur in all species and floral longevity can evenbe increased following pollination-induced flower closure or a pollination-induced change in color (van Doorn, 1997). The pollination effect seems to be mediated by ethylene.

Moreover, ethylene was shown to be involved in senescence of many flowers, which therefore are classified as ethylene-sensitive (Woltering and van Doorn, 1988; van Doorn, 2001). Studies on flower senescence have been often carried out using species like *Petunia hybrida*, *Dianthus caryophyllus*, and *Mirabilis jalapa* in which flower wilting, withering, or abscission are regulated by ethylene (Xu et al., 2007; Hoeberichts et al., 2007; O'Donoghue et al., 2009). However, in several species senescence is not associated with the production of this hormone and ethylene sensitivity is null or very low (Woltering and van Doorn, 1988; van Doorn, 2001). Ethylene-insensitive species include important bulb plants such as *Lilium*, *Narcissus*, *Tulipa*, *Iris*, and *Hemerocallis*. The regulation of flower senescence in ethylene-insensitive flowers has not been clearly understood so far and a role for hormones other than ethylene has been hypothesized but not yet fully demonstrated (Panavas et al., 1998; van der Kop et al., 2003; Hunter et al., 2004). Although the signals and the translation pathway leading to flower senescence have not been well understood, the events associated with the progression of flower degeneration have been at least partially elucidated in both ethylene-sensitive and ethylene-insensitive species. During senescence of both ethylene-sensitive and ethylene-insensitive flowers, marked changes occur in the biochemical and biophysical properties of the cell membranes (Leverenz et al., 2002).

Another category consists of flowers with petals that show wilting because of water stress in the entire cut flowering stem. In many rose cultivars, for example, the petals of uncut flowers abscise without senescence symptoms. In contrast, the petals of these flowers often wilt due to an occlusion in the xylem, which blocks the flow of water to the petals. Such petal wilting is not due to senescence and will not be discussed here.

Loss of membrane permeability is a typical feature used to determine the onset of senescence through the measurement of the ion leakage (Panavas et al., 1998; Stephenson and Rubinstein, 1998; Hossain et al., 2006). The loss of cellular organization, the degradation of cytoplasm and organelles were observed in several species (Smith et al., 1992; O'Donoghue et al., 2002; Wagstaff et al., 2003). Ultrastructural studies of senescing petals showed the presence of vesicles, cytoplasm, organelles, and electron dense structures within the vacuole leading to the conclusion that the autophagic machinery could be implicated (Phillips and Kende, 1980; Smith et al., 1992). Recent work strengthens the hypothesis that autophagy may be involved in petal senescence (Shibuya et al., 2009a, b).

Numerous genes become (transiently) up-regulated during senescence, whilst many other genes are down-regulated. Most of the changes in gene expression are related to remobilization of macromolecules, and transport of the mobile compounds out of the petal. A considerable part of differential gene expression is related to plant defence. No genes have as yet been identified that are specific for cellular death. Degradation of macromolecules in the senescent cell is mainly due to autophagic processes in the vacuole, in addition to protein degradation in mitochondria, nuclei, and the cytoplasm, fatty acid breakdown in peroxisomes, and degradation of nucleic acids in the nuclei.

It is as yet unknown how the onset of senescence is regulated. This is true for the onset of the increase in ethylene production in systems that are controlled by ethylene as well as for the regulation in species where senescence is not under the control of ethylene. Although many transcription factors and signal transduction factors are differentially expressed during petal senescence, as yet the functional role is known of only a few (e.g. EIL3 and Aux/IAA).

It is also not clear how the petal cells die. Death is preceded by the disappearance of most of the cellular contents, and might therefore be due to ongoing remobilization, but might also occur independently of the remobilization process. Several factors that control programmed death of animal cells, such as caspases, proteins released by mitochondria, and proteins that control the release of compounds through membranes, have no apparent counterpart in plant cells, indicating that the regulation of cell death during petal senescence is quite different from the regulation of cell death in animal cells. Tonoplast rupture (or lack of tonoplast semipermeability) is widely held to be the immediate cause of developmental cell death in plants, but the data that support this contention in petals are still weak. The mechanism that accounts for the rupture is completely unknown.

The ovary is central in the regulation of ethylene-sensitive senescence, at least in carnation flowers. It mediates the hastening effect of pollination on senescence. After pollination the ovary produces ethylene which induces a rapid increase in petal ethylene production. In the absence of pollination, the ovary is also causal in petal senescence, although the mechanism by which ethylene is then produced in the ovary is not clear.

Numerous mRNAs are produced during senescence and such production requires intact mechanisms of both translation and posttranslational RNA modification (van Doorn, 2008). As shown by several molecular studies, the up-regulation of genes involved in cell death regulation and cell dismantling is essential for senescence to occur (Buchanan-Wollaston, 1997). In spite of the key role for mRNA synthesis during developmental cell death and senescence, both RNA and DNA are degraded during these processes. The breakdown of the nucleic acids can take place within the nucleus, mitochondria, and plastids by localized enzymes. Nuclear DNA fragmentation in plant cells was observed in nuclei during physiological senescence of seed tissues (Kladnik et al., 2004; Lombardi et al., 2007a), leaves (Gunawardena et al., 2004; Lee and Chen, 2002), flowers (Orzaez and Granell, 1997; Langson et al., 2005; Hoeberichts et al., 2005) and other plant organs (Groover et al., 1997; Gladish et al., 2006).

9.2 FACTORS AFFECTING SENESCENCE

Senescence might be triggered by several factors including biotic and abiotic stresses. Flowers being most colorful and susceptible part of plant are easily affected by several environmental factors such as insect-mediated pollination, seasonal changes, lack of water, and various stresses such as invasion by a pathogen or attack by a predator (Taylor and Whitelaw, 2001). Temperature and water relations are the main environmental determinants of petal senescence rate in cut flowers. Pollination activates a series of postpollination developmental events contributing to reproduction and ovary growth, pigmentation changes and petal senescence (Stead, 1992; Woltering et al., 1994). The postpollination events trigger increase in hydrolytic enzymes and degradation of macromolecules. Pollination-induced senescence is accelerated in 60 different genera, most of which are suspected of being sensitive to ethylene.

9.3 EVENTS ASSOCIATED WITH SENESCENCE

1. Ultrastructural Changes

The maintenance of the cellular organization and compartmentalization is an essential prerequisite so that the cell can survive and carry out its metabolic functions. During senescence cells undergo structural and biochemical changes and gene regulation. The activation of the catalytic machinery allows the recovery of the nutrients accumulated during the growth phase. The earliest and most significant change in leaf cell structure is the breakdown of the chloroplast, the organelle that contains up to 70 percent of the leaf proteins (Lim et al., 2007). The roles of petals and leaves are very different, as are their development and the signaling mechanisms that trigger their senescence (Price et al., 2008). However, leaf and flower senescence share several structural and biophysical features. Moreover, a large proportion of remobilization-related genes are up-regulated in both leaf and petals (Price et al., 2008).

The delicacy of petal cells and their rapid collapse during senescence is an excellent tool for studying ultrastructural changes during senescence. The petal has a complex structure including different cell types and tissues in which the senescence process is not spatially homogeneous. Van Doornet al. (2003) used Iris as a model for examining ultrastructural and molecular changes during opening and senescence, and found dramatic changes in ultrastructure that were clearly related to eventual senescence well before any of the normal hallmarks of senescence (petal inrolling, wilting) had occurred. One of the earliest changes in cell ultrastructure, at least during senescence of Iris petals, is the closure of the plasmodesmata. Plasmodesmata, if open, allow transfer of relatively small molecules, such as sugars, hormones, and RNA molecules, between neighboring cells. If the plasmodesmata are closed, this transport is halted. In the outer petals of Iris the plasmodesmata of mesophyll cells close 2 days before those of the epidermis cells. The mesophyll cells also die 2 days earlier than the epidermal cells. Compared with epidermis cells, the mesophyll cells also exhibited all other ultrastructural changes, indicative of processes leading to cell death, nearly 2 days earlier. This suggests a role for plasmodesmata in the early regulation of senescence. In several species, mesophyll cells die well before epidermal cells and visible signs of petal senescence have been associated with mesophyll collapse (O'Donoghue et al., 2002; van Doorn et al., 2003; Wagstaff et al., 2003). In *Gypsophila paniculata,* there was no difference in the timing of cell death betweenmesophyll and epidermal cells, but visible senescence closely correlates with collapse ofepidermal cells (Hoeberichts et al., 2005). In Alstroemeria, petal senescence starts in the petalmargin (Wagstaff et al., 2003), while in tobacco flowers it begins in the proximal part wherethe abscission will eventually take place (Serafini-Fracassini et al., 2002). In carnation, theultrastructural changes were shown to be asynchronous in the mesophyll tissue as well as inthe parenchyma surrounding the vascular tissues (Smith et al., 1992). In some flowers it hasbeen reported that the vascular tissues are the very last to degrade (Matile and Winkenbach,1971; Wagstaff et al., 2003). This is probably correlated with the need to recover andreallocate nutrients until the late stage of flower senescence. The difference in timing of celldeath and degradation makes the understanding of the mechanisms underlying flower senescence difficult.

Several genes have been shown to be up-regulated during flower senescence. DNA fragmentation or laddering has been detected during petal senescence in a range of species, including pea, petunia, freesia, alstroemeria, gypsophila, sandersonia, and gladiolus (Orz'aez and Granell 1997; Xu and Hanson 2000; Yamada et al. 2001, 2003; Wagstaff et al. 2003; Hoeberichts et al. 2005). In gypsophila, sandersonia, iris, and alstroemeria, terminal deoxynucleotidyltransferase-mediated dUTP nick-end labeling (TUNEL) staining, light microscopy and transmission electron microscopy showed that nuclear degradation was already under way before the flowers were fully open. Although the epidermal cells remained intact, the mesophyll cells degenerated completely before visible senescence (Bailly et al. 2001; O'Donoghue et al. 2002; van der Kop et al. 2003; Wagstaffet al. 2003; Hoeberichts et al. 2005).

The tonoplast of *Ipomoea* (= *Convulvulus*) *tricolor* petal cells; numerous vesicles were found in the vacuole. These vesicles are probably due to microautophagy and/or macroautophagy. The vacuoles also contained structures reminiscent of mitochondria. The vacuole therefore seems a main site of membrane and organelle degradation (Winkenbach, 1970b; Matile and Winkenbach 1971; Phillips and Kende, 1980; Smith et al., 1992). Petal cells in *Ipomoea* (Winkenbach, 1970a), carnation (Smith et al., 1992), *Hemerocallis* (Stead and van Doorn, 1994), and Iris (van Doorn et al., 2003) exhibit a decrease in the number of small vacuoles, and an increase in vacuolar size. The ongoing increase in vacuole volume in Iris was accompanied by the loss of a considerable part of the cytoplasm and with the disappearance of most organelles. In Iris, one of the earliest ultrastructural senescence symptoms (after closure of the plasmodesmata) was the loss of most of the endoplasmic reticulum (ER) and attached ribosomes. Several other organelles, such as Golgi bodies, then also become less frequent. Many mitochondria became degraded, but some remained until vacuolar collapse. Similar changes have been reported in carnation (Smith et al., 1992).

The nucleus has been found to remain until a late stage of senescence. During cell dismantling the nucleus is one of the last organelles to be degraded, even if it undergoes ultrastructural changes (van Doorn and Woltering, 2008). In senescing petals of carnation the nuclei were small, sometimes horseshoe shaped, and with compact nucleoli (Smith et al., 1992). Wagstaff et al., (2003) reported a reduction of nuclear size in Alstroemeria petals and sepals from just before flower opening until the oldest stage, when the nucleus was the only recognizable structure in the cell. In tobacco petals, the nuclei showed nuclear blebbing,stained with DAPI or toluidine blue, similar to that observed during apoptosis in animal cells (Serafini-Fracassini et al., 2002). Nuclei in young cells have a rather even distribution of DNA (chromatin). Changes in nucleus include also chromatin condensation and fragmentation. If fragmentation results in the formation of DNA masses that do not stay inside the same nuclear envelope, it may be referred to as nuclear fragmentation (Yamada et al., 2006b). During cell senescence, the chromatin clumps into patches that are found either throughout the nucleus or mainly at the nuclear periphery. In several species the nucleus showed brighter fluorescence after staining the DNA with a fluoro probe, indicating DNA condensation. This is often accompanied by a decrease in nuclear diameter (Yamada et al., 2006a, b). In the petals of *Ipomoea*, a doubling of the number of DNA masses ("nuclei") per petal was found, together with a continuing decrease in the diameter of these masses (Yamada et al., 2006a). As the DNA masses contained an outer membrane, the results indicate nuclear fragmentation. Evidence for nuclear fragmentation was also found in the petals of *Petunia hybrida* and *Argyranthemumfrutescens*, but not in petals of *Antirrhinum majus* (Yamada et al., 2006b).

Other organelles also disappear and the volume of cytoplasm becomes very low, leaving only a small strip next to the cell wall. The role of the vacuole as a lytic compartment has been often emphasized (Matile and Winkenbach, 1970; Phillips and Kende, 1980; Smith et al., 1992; Wagstaff et al., 2003). Decaying mitochondria, ribosomes, and membranes have been found in the vacuole of senescing cells. In *Ipomeapurpurea* invaginations of the tonoplast were reported, resulting in the sequestration of cytoplasmic material into the vacuole. At least some mitochondria, and in carnation some plastids, stayed until a late phase of senescence (Winkenbach, 1970b; Smith et al., 1992; van Doorn et al., 2003). When the cytoplasm of the cell is almost or completely empty, electron microscopy (EM) work shows that the vacuolar membrane (tonoplast) collapses and then disappears. This occurs earlier than the collapse and disappearance of the plasma membrane (Winkenbach, 1970b; Smith et al., 1992; van Doorn et al., 2003). Although it cannot be excluded that this is an artifact of EM preparation, the earlier collapse of the tonoplast might be a real effect, similar to that in other plant cells, for example during the PCD of tracheal elements (Obara et al., 2001), and in several other examples of PCD/senescence (Jones, 2001; Jones and Dangl, 2004). One of the last events in senescing cells is vacuole rupture, which is probably the direct cause of cellular death (van Doorn and Woltering, 2008). Cell depletion and the loss of wall functions promote the cellular crushing and lead to visible signs of petal senescence.

2. Biochemical Changes

At the biochemical level, senescence is associated with changes in membrane fluidity and leakage of ions in several different flowers (Thompson et al., 1997). When plant tissues undergo senescence, cellular membranes progressively lose their organization and changes in the bilayer structure occur leading to impairment of functions. During the course of petal senescence, a decrease in the level of many macromolecular components like starch (Ho and Nichols, 1977; Horie, 1961), cell wall polysaccharides, proteins, and nucleic acids were recorded. The main constituents of the sugar pool of mature petals of carnation and rose were the reducing sugars, rather than nonreducing sucrose, majority of transformation in sugars being accompanied by starch. A significant increase in the pH of the vacuole of some senescing flower petals was ascribed to proteolysis and an increase in the level of basic amino acid, asparagine, in the old petals, followed by accumulation of free ammonia.

The loss of membrane lipid phosphate is one of the best documented indices of membrane lipid metabolism during senescence (Thompson et al.. 1998). The increasing activity of phospholipases and acyl hydrolases leads to the replacement of phospholipids with neutral lipids. Although, the release of electrolytes from senescing tissues is considered a hallmark of membrane degradation and senescence, changes in lipid content and membrane structure occur before the leakage of nutrients is evident (Thompson et al., 1998; Leverentz et al,. 2002). The decline of the phospholipids/neutral lipids ratio accompanies the increase insterol/phospholipids ratio contributing to the reduction of membrane fluidity. Another senescence-associated event that leads to a loss of membrane permeability is the oxidation of existing membrane components (Rubinstein, 2000). The oxidation of membrane fatty acid takes place through two distinct mechanisms:

(a) The enzymatic oxidation mediated by lipoxygenase (LOX) and

(b) The nonenzymatic autoxidation.

In several species, an increase of LOX activity was detected during flower senescence. Lipid peroxidation is generally measured as thiobarbituric acid reactive substance (TBARS). In daylily, a short living ethylene-insensitive flower, lipid peroxidation increases from 24 h before flower opening reaching a peak at 18 hrs after flower opening. Lipid peroxidation is followed by an increase in LOX activity before flower opening (Panavas and Rubinstein, 1998b). The increase in lipid peroxidation and LOX activity was estimated also in ethylene-sensitive species such as rose and carnation (Fobel et al., 1987; Bartoli et al., 1995; Fukuchi-Mizutani et al., 2000). On the contrary, in *Alstroemeria peruviana* LOX activity declined during senescence and loss of membrane function was not related to LOX activity or accumulation of lipid hydroperoxyde (Leverentz et al., 2002). In the ethylene-sensitive category, LOX activity may promote senescence through oxidative membrane damage. However, in some ethylene-sensitive plants such as Phalaenopsis, the LOX seems not to play any apparent role. On the other hand, in the ethylene-insensitive category, LOX promotion of senescence has been proposed in daylily and implicated in gladiolus species. Thus, Alstroemeria represents a distinct pattern of floral senescence that is both ethylene and LOX independent.

3. Metabolic Changes

Internal metabolic changes in respiration of cut-flowers have been demonstrated. The rate of respiration in many cut-flowers reaches its peak at the time of opening of flowers which decreases as the flowers mature and senesce. Later is a second dramatic increase in respiration over a relatively short period followed by a final decline (Mayak and Halevy, 1980). The second peak in the respiration drift signifies the last phase of senescence. It has been considered to be an analog to the climacteric rise in respiration of many fruits. The second peak in respiration of flowers may be employed to assess the effectiveness of senescence-retarding substances, provided this hike in respiration reflects internal metabolic changes associated with aging chemicals that delays the occurrence of the second respiratory peak, have been reported to increase vase-life of flowers (Ballantyne, 1966; Kende and Hanson, 1976).

The gradual decline in respiration of senescing flowers may be the result of short supply of readily oxidizable substrates. The translocation of respirable substrates within the flower from the petals to the ovary has been demonstrated in senescing flowers (Coorts, 1973; Rogers 1973). The hydrolysis of carbohydrates, proteins are degraded to a mixture of smaller polypeptides and amino acids (Parups, 1971). The conversion of the excess ammonia into amides such as asparagine and glutamines is the probable mechanism of detoxification (Sacalis, 1975). The onset of hydrolysis of cellular components, such as proteins and carbohydrates, is initiated in response to the depletion of free sugars used up in respiration, to supply alternative respirable substrates. A decreased ATP production resulting from the progressive uncoupling of electron transport and oxidative phosphorylation in leaves may occur before the total depletion of sugars. The partial energy shortage created may activate alternative or complementary pathways with the formation of oxidative products that trigger senescence.

4. Changes in Pigments

Discoloration or fading of color is a common symptom of many senescing flowers. The carotenoids and anthocyanins, two major classes of pigments responsible for different color of flowers, change significantly during the development and maturation of plant organs. Simpson et al. (1975) monitored changes in the composition of carotenoid in *Strelitziareginae* flowers throughout the development of plastids from small, colorless plastids through the green chloroplasts to large spindle-shaped chromoplasts. They noted an increase in the concentration of oxygenated carotenoids with age. Differential changes in the anthocyanin content of senescing flowers have been noted. Certain flowers such as the short lived chicory fade and turn white upon senescing. Such type of fading has been attributed to the decolourization by an enzyme complex having catecholase property.

Stewart et al. (1975) noted that changes in the color of senescing petals are significantly influenced by change in the pH of the vacuole. Co-pigmentation with other flavonoids and related compounds is the decisive factor in determining the intensity of the color in most flowers. The color and pH changes associated with the senescence of flowers may proceed at different rates in adjacent cells indicating that contiguous cells differ from each other in the rates of their proteolysis and aging.

5. Oxidative Events

Due to various oxidative reactions, reactive oxygen species (ROS) are known to be involved in normal death of plant cells including petals. ROS is a collective term that includes both oxygen radicals and certain nonradicals that are oxidizing agents and/or are easily converted into radicals (HOCl, HOBr, O_3, $ONOO^-$, $^1O_2$ and H_2O_2) (Halliwell, 2006). ROS have been shown to have a role in stress-induced and normal senescence in animals and plants and are possible determinants of membrane degradation (Panavas and Rubinstein, 1998). Cells have evolved strategies to utilize ROS as biological signals that control various genetic stress programs (Laloi et al., 2004). As H_2O_2 are both an important signaling molecule and a toxic by-product of cell metabolism (Gechev and Hille, 2005). The ROS is produced from hydrogen peroxide, thus the hydrogen peroxide level regulating enzymes showed differential expression during senescence (Halliwell and Gutteridge, 1989). Under normal conditions, ROS are rapidly metabolized with the help of constitutive antioxidant enzymes and other metabolites via nonenzymatic pathways such as antioxidant vitamins, proteins, and nonprotein thiols (Kotchoni and Gachomo, 2006). Following the observation that high levels of H_2O_2 may be correlated with flower aging, several studies were made in order to analyze the level and the activity of such scavenging enzymes during flower senescence (Bartoli et al., 1995; Panavas and Rubinstein, 1998; Bailly et al.; 2001; Hossain et al., 2006). H_2O_2 is reduced by ascorbate peroxidase (APX) with the consequent oxidation of ascorbate to dehydroascorbate while catalase converts H_2O_2 into H_2O and O_2.

Glutathione reductase (GR) reduces glutathione; one of the most important antioxidants. Superoxide dismutase (SOD) is effective in ROS scavenging but produces H_2O_2 which has to be degraded to avoid cell damage. The increase in SOD activity over the senescing period could be due to the overexpression of genes induced by H_2O_2 accumulation (Hossain et al., 2006). Natural antioxidants, such as ascorbate, glutathione and α-tocopherol, are also present in flower petals (Rubinstein, 2000). In chrysanthemum petals, the activity of antioxidant enzymes increased during the early stage of senescence. The content of α-tocopherol and total thiols increased at the onset of flower senescence and decreased in browning petals (Bartoliet al., 1997). In petals, the activity of antioxidative enzymes and the level of nonenzymatic antioxidant have been assessed only in vitro and in whole petals not in various cells and cell compartments. However, there is at present no conclusive evidence that ROS are either an early signal or a direct cause of cellular death during petal senescence (van Doorn and Woltering, 2008).

6. Protein Degradation and Protease Activity

During the time course of senescence targeted protein degradation is a critical part, therefore the activity of proteolytic enzymes is an essential element in these processes. Proteases degrade proteins by hydrolyzing internal peptide bonds and are one of the best characterized cell death proteins in plants. Protein degradation plays a crucial role in recycling intracellular damaged proteins or proteins that are no longer necessary and to ensure a constant turnover of macromolecules. Selective protein breakdown takes place in the proteasome, plastids, mitochondria, nucleus, and vacuole and carries out a number of different roles during plant development. At a cellular level it prevents deleterious effects due to the presence of abnormal and nonfunctional proteins, regulates the activity of metabolic pathways through the modulation of key enzymes and regulates the levels of receptors (Callis, 1995).

In flowers the transition from sink to source involves the activation of catabolic pathways allowing the recovery of organic and inorganic compounds. Proteins are involved as executioners, regulators, or substrates during the remobilization processes. The role of protein synthesis and breakdown during flower senescence has been widely studied by Lay-Yee et al., 1992; Jones et al., 1994; Stephenson and Rubinstein, 1998; Wagstaff et al.,2002. The decrease in protein level has been reported for several plant species and it can be due to a decrease in synthesis as well as an increase in degradation (van Doorn and Woltering, 2008). The decrease in protein level has been reported for several plant species and it can be due to a decrease in synthesis as well as an increase in degradation (van Doorn and Woltering, 2008). The decrease in protein content may begin before flower opening (Wagstaff et al., 2002; Azeez et al., 2007) or after flower opening (Stephenson and Rubinstein, 1998; Pak and van Doorn, 2005).

Among all the proteases, the cysteine proteases are the most frequent and well characterized (Stephenson and Rubinstein, 1998). In support of their role in flower senescence several cysteine proteases have been shown to be up regulated and further cloned from petals of Petunia (Jones et al., 2005), Alstroemeria (Wagstaff et al., 2002), Sandersonia (Eason et al., 2002) and Narcissus (Hunter et al., 2002). A rapid decrease in protein level was reported after compatible pollination in *Petunia inflata* (Xu and Hanson, 2000). In *Petunia* x *hybrid* protein content decreased starting from 4 days after flower opening (Jones et al., 2005). Proteindegradation has often been reported to be followed by an increase in protease activity. Using differential screening of a cDNA library, Guerrero et al. (1998) reported a cDNA clone encoding a daylily thiol protease (SEN11), whose expression is strongly regulated in flower tepal senescence. Earlier reported thiol protease SEN102 (Valpuesta et al., 1995) and SEN11 transcripts were not detectable in flower buds at the opening stage, but a significant increase in transcripts in both wilting petals and sepals was reported. The expression pattern of these genes coding for proteases suggests their involvement in the protein hydrolysis in tepals at the late senescence stage. Azeez et al. (2007) reported the expression of specific serine proteases during senescence-associated proteolysis in Gladiolus flowers. During the senescence, serine protease activity increases up to two-third of total proteases activity.

7.   Nucleic Acid Synthesis and Breakdown

During senescence of floral parts, the degradation of DNA and RNA is the most common feature (Thomas et al., 2003). Numerous mRNAs are produced during senescence and such production requires intact mechanisms of both translation and posttranslational RNA modification (van Doorn and Woltering, 2008). It was revealed by several molecular studies that the up-regulation of genes involved in cell death and cell dismantling is essential for senescence to occur. Using suppressive subtractive hybridization and microarray analysis, Breeze et al. (2004) found that in Alstroemeria petals both cell wall related genes and genes involved in metabolism were present in higher proportions in the earlier stages, while genes encoding metal binding proteins were the major component of senescence enhanced libraries. In senescing wallflower petals a large proportion of up-regulated genes were related to their mobilization process (Price et al., 2008). A limiting factor in understanding the mechanism underling flower senescence could be the heterogeneity of the petal tissues with different cells dying at different times. The up-regulation or down-regulation of genes cannot be ascribed to cells at a specific stage of development or senescence but only to the petal tissue. In spite of the key role of mRNA synthesis during developmental cell death and senescence, both RNA and DNA are degraded during these processes. The breakdown of the nucleic acids can take place within the nucleus, mitochondria, and plastids by localized enzymes. In animal cells caspase-activated DNase (CAD) preexists in living cells as an inactive complex with an inhibitory subunit. The cleavage of the inhibitory sub unit mediated by caspase-3 results in the release and activation for the catalytic subunit responsible of the characteristic DNA laddering (Hengartner, 2000). Although in plants there is no evidence for a protease activated DNase, the degradation of DNA and RNA is also a common feature of the dying cells. In *Petunia inflata* the increase in corolla size and fresh/dry weight ratio until 24 hrs after pollination was coupled with the continuous decrease in RNA content (Xu and Hanson, 2000). Molecular changes, therefore, follow compatible pollination despite a healthy appearance of the petal. TdT-mediated dUTP-biotin nick end labeling (TUNEL) has been a widely used method to assess DNA fragmentation in plants (Orzaez and Granell, 1997; Lee and Chen, 2002; Gunawardena et al., 2004; Hoeberichts et al., 2005; Lombardi et al., 2007). In *Pisumsativum* petals DNA fragmentation was observed during normal senescence and TUNEL positive nuclei were not detected in cells treated with inhibitors of ethylene (Orzaez and Granell, 1997). In *Gypsophylapaniculata* the DNA degradation became apparent well before visible signs of petal senescence and TUNEL positive nuclei appeared before the increase in ethylene production. The occurrence of TUNEL positive nuclei was stimulated by ethylene treatments but was not prevented by silver thiosulfate (Hoberichts et al., 2005).

Nuclease activity results in DNA degradation and sometimes in the appearance of the nucleosomal ladder. Often the first step of DNA degradation leads to the formation of high molecular weight fragments (50–300 kb) (Nagata et al., 2003). Afterwards, through the activation of specific DNase, the internucleosomal cleavage produces 180 bp multiple fragments that look like a ladder when placed on a gel. However, DNA fragmentation into oligonucleosomal units should not be considered as an indicator of PCD without parallel evaluation of morphological changes especially in quickly dying cells. However, DNA laddering has been detected in several plant tissues and cells undergoing natural or induced death. Balk et al. (2003) identified two different mitochondrial associated pathways for DNA degradation in Arabidopsis. One was reported to be associated with the formation of high molecular weight DNA fragments and chromatin condensation. The second one appeared to mediate DNA laddering. This double mechanism resembles the DNA degradation pathway duringapoptosis in animals in which the intermembrane- space protein AIF mediates high molecularweight fragmentation and chromatin condensation while the caspase-activated DNase CAD leads to DNA laddering.

Panavas et al. (1999) cloned a cDNA from daylily petals, with similarity to fungal S1- and P-type endonucleases that degrade both single-stranded DNA and RNA and only a single copy of the petal-specific gene was reported. The transcript level of this putative nuclease increases at flower opening and continues to increase during senescence. The advanced stages of petal senescence in Petunia were found to be associated with DNA laddering and increased nuclease activity (Xu and Hanson, 2000). Five DNases with specific activity against ssDNA were identified in petals of Petunia and all of them were shown to be increased during the senescence of pollinated flowers. Langston et al. (2005), using the same flower species, characterized a single cobalt dependent senescence-specific nuclease PhNUC1, which is a glycoprotein and is characterized by higher activity during both pollination-induced and age-related senescence. Activity of PhNUC1 was induced in nonsenescing corollas by treatment with ethylene. The increased PhNUC1 activity was delayed in ethylene-insensitive flowers (*35S:etr1-1*) suggesting ethylene-dependent regulation of DNA fragmentation and nuclease activity. In another study using flow cytometry and fluorescence microscopy, Yamada et al. (2006a) reported DNA degradation, chromatin condensation and nuclear fragmentation during PCD in petals of Ipomoea. Yamada et al. (2006b) studied flower PCD in the petals of Antirrhinum, Argyranthemum, and Petunia, using DNA degradation and changes in nuclear morphology as parameters. The petals of all three flowers showed loss of turgor (wilting) and DNA degradation. Two distinct types of nuclear morphology were observed during PCD in these petals. One type was characterized by chromatin clumping into spherical bodies inside the nuclear membrane lacking nuclear fragmentation during PCD as in Antirrhinum. Nuclear fragmentation did not occur in Antirrhinum, whereas nuclear fragmentation was reported in Argyranthemum and Petunia (Yamada et al., 2006b).

9.4   HORMONAL REGULATION OF FLOWER/PETAL SENESCENCE

Plant hormones were reported to play an important role in senescence process. The specific regulation of flower senescence is very complex process and is governed by endogenous levels and sensitivity of hormones. It is difficult to say that all the hormones are actively involved in floral programmed cell death (PCD) in all species. There should be some quantitative and qualitative species-specific differences present to mediate programmed cell death. In most of the flowers, it is an increase in ethylene following pollination that triggers the changes associated with petal senescence. The roles of individual hormones are discussed below.

1.   Ethylene

Senescence in several plants is accelerated by the naturally occurring plant hormone ethylene. Ethylene is a gas produced by plants and exposure to ethylene causes premature senescence of both the leaves and flowers. In flowers, senescence is triggered by hormonal changes that follow pollination. Like fruits, most flowers also show a climacteric rise in ethylene following pollination (O'Neill, 1997). In flowers like Petunia that are self-incompatible, the senescence-related changes in petals are observed after compatible pollination and not after pollination by incompatible pollen (Singh et al., 1992).

The role of ethylene in flower development has been studied to a great extent in Petunia, Geranium, orchids, and carnation. In Petunia, out of four ACC oxidase (ACO) gene members ACO1 is expressed specifically in senescing corollas and in other floral organs following ethylene exposure while ACO3 and ACO4 were expressed in developing pistil tissue (Tang et al., 1994). The timing and tissue specificity of the increased expression of ACO transcripts correlated with pollination-induced ethylene production in styles and stigma followed subsequently by corollas leading to senescence (Tang et al., 1996). Pollination in orchids was associated with an increase in ACO transcript in stigma and petals (Nadeau et al., 1993). Ethylene biosynthetic genes were also differentially expressed during carnation flower senescence (ten Have et al., 1997).

Ethylene receptors and downstream genes of ethylene signaling cascade were also shown to play an active role in the progression of senescence. In carnation, Shibuya et al. (2002) showed that DC-ERS2 and DCETR1 are ethylene receptor genes responsible for ethylene perception and that their expression during flower senescence is regulated in a tissue specific manner and independently of ethylene. In roses, senescence of flowers seems to be related to elevated levels of ETR3 indicating that flower senescence may be triggered by the perception of endogenous ethylene by ethylene receptors (Muller et al., 2000). Expression of ethylene receptor gene ERS1 and signaling regulator gene CTR1 from Delphinium florets was found to increase after treatment of florets with ethylene (Kuroda et al., 2004).

In *Rosa hybrida* higher expression of two CTR genes were reported during ethylene induced flower senescence. RhCTR1 levels were up-regulated in senescing flowers, while RhCTR2 was constitutively expressed during flower senescence (Muller et al., 2002). In Dianthus, a putative EIN3-like protein, DC-EIL1 transcript level decreased in flower tissue, especially in petals, during natural senescence and in response to ethylene and ABA treatment (Waki et al., 2001). Similarly, in rose the gene RhEIN3 was constitutively and stably expressed during flower development in response to ethylene and ABA (Muller et al., 2000). These findings suggest that control of ethylene sensitivity during senescence occurs upstream of EIN3 and its homologues.

Several ethylene mutants were studied in order to elucidate the role of ethylene in regulation of flower Senescence. Transgenic carnation flower with antisense ACO showed delayed senescence which enhanced the shelf life from five days to nine days (Woltering et al., 1997). The expression of the Arabidopsis etr1-1 gene in transgenic carnations Petunia and tomato delayed flower senescence, resulting in a significant increase in vase life (Bovy et al., 1990) (Wilkinson et al., 1997).

2.   Abscisic acid

Exogenous application of ABA to certain flowers hastens flower senescence (Borochovet al., 1989). In daylilies, which are ethylene-insensitive, ABA is thought to be the primary hormonal regulator of flower senescence, and many of the senescence-related changes are brought about by ABA. These include ion leakage, changes in lipid peroxidation, protease activity and expression of novel DNases and RNases (Panavas et al., 1998). In ethylene-sensitive flowers, like carnation, ABA accelerates flower senescence by increasing the endogenous production of ethylene (Ronen et al., 1981). Hunter et al. (2004) reported that the ABA content increased in tepals of senescent flowers. The increased ABA content coincided with the appearance of visual signs of senescence in tepals. Exogenous ABA enhanced the premature accumulation of senescence-associated transcripts in the tepals indicating that ABA induced the transcripts independently of ethylene.

3.   Cytokinins

In contrast to ethylene and ABA, cytokinins delay senescence in floral tissues. The inverse relationship between cytokinin content and senescence occurs in some flowers, like Petunia (Lara et al., 2004), roses (Mayaket al., 1972) and carnation (Van Staden et al., 1980).

In a study, Chang et al. (2003) confirmed the role of cytokinins in flower senescence. The transgenic plants over expressing IPT gene under the SAG promoter exhibited significant delays in flower senescence resulted in the increased cytokine in content and less ethylene sensitivity. Increased endogenous ethylene production was measured after pollination but this increase was delayed in IPT flowers. Flowers from IPT plants were less sensitive to exogenous ethylene, and showed delayed ethylene responsiveness, corolla senescence and a senescence-related cysteine protease. These plants confirm that the regulation of flower senescence involves the interaction of cytokinins, ethylene, and ABA.

4. Auxins, gibberelic acids and jasmonic acid

The effect of auxins and gibberelic acids is not well characterized in flower senescence. Applications of auxins to cut flowers stimulate senescence of some ethylene-sensitive flowers (Stead et al., 1992). Gibberellic acid is known to delay senescence in some cut flowers by acting as an antagonistic to ethylene. Saks et al. (1992) showed that exogenous application of gibberellic acid delayed the senescence of cut carnation flowers with reduced ethylene production. Recently, Setyadjit*et al.* (2006) treated an ethylene-sensitive flower Grevillea "Sylvia" and this treatment did not enhance the longevity of inflorescences and ACC level. At higher level gibberellic acid concentrations enhanced flower abscission rather than senescence. Jasmonic acid also shows stimulating effect on flower senescence. Jasmonic acid along with several other metabolites promotes senescence of orchid species, presumably by elevating ACC, thereby stimulating ethylene production. However, in orchid petals for 50 h after pollination-induced senescence neither lipoxygenase activity nor jasmonic acid content changed (Porat*et al.*, 1995).

9.4.1 NONHORMONAL REGULATION OF FLOWER SENESCENCE

The flower senescence is also known to be regulated by several signaling pathways including G protein and calcium signaling, polyamines, and sugars. A heterotrimeric G proteins linked phospholipase *a* and*c* (PLA and PLC) increase in petals of roses and Petunia before the onset of senescence (Borochov et al., 1994). PLC is known to hydrolyze the membrane component phosphatydiylinositoldiphosphate to yield inositol triphosphate and diacylglycerol (DAG). In addition, DAG level increased before to the burst of ethylene in Petunia flower senescence (Borochov et al., 1997). Secondary messenger calcium levels increase in leaves concomitant with senescence (Huang et al., 1997) and in the orchid Phalaenopsis, exogenous calcium increased flower sensitivity to applied ethylene accelerating senescence. The possibility thus exists that the calcium wave is the signal for onset of senescence. This increased level of calcium may result in the calcium-dependent phosphorylation of proteins that are necessary for the up-regulation of suicide proteins.

In plants, polyamines (PAs) have a well-established role in the stimulation of cell division, in growth, and in the delay of senescence, hence called as "juvenility" factors. The senescence of carnation petals is inhibited by spermine, which may be due to a corresponding inhibition of ethylene synthesis. Addition of an inhibitor of polyamine synthesis leads to elevated levels of transcripts for ACC synthase and ACC oxidase as well as to increased ethylene production (Lee et al., 1997). Fracassini et al. (2002) showed the role of a polyamine in corolla senescence of tobacco. Results show that bis-derivatives decreased with the progression of senescence, while monoderivatives increased during early senescence. These derivatives were present in different amounts in the proximal and distal parts of the corolla. In excised flowers, exogenous spermine delayed senescence and PCD, and caused an increase in polyamine levels.

vanDoorn (2004) reviewed evidence for sugar starvation to be a cause of petal senescence. Changes in ultrastructure, metabolism, and gene expression in senescent petals are remarkably similar to those in sugar-starving organs. The delay in protein degradation and delayed expression of a number of genes was reported after sugar feeding (Eason et al., 1997). Thus, low sugar levels may elevate ethylene production and trigger petal cell death in flowers. In recent study, Hoeberichtset al. (2007) reported that soluble sugars, like sucrose, in the carnation petal act as a repressor of senescence at the transcriptional level. It was observed that sucrose acts more efficiently than the silver thiosulfate (STS an ethylene receptor inhibitor) in ethylene-signaling inhibition. For example the senescence-associated increase in *Dc-EIL3* expression was delayed by STS and prevented by sucrose treatment. Sucrose starvation, therefore, might negatively affect ethylene signaling subsequent to SAGs expression.

9.4.2 SENESCENCE-ASSOCIATED GENES (SAGS) IN FLOWER

Several genes associated with floral senescence have been isolated from a number of flowers using different tools In addition to genes involved in ethylene biosynthesis; there are also reports of expression of some other genes such as a glutathione S-transferase (Meyer et al., 1991). The GST1 from carnation was enhanced by ethylene and contained a 22 bp ethylene response enhancer element that was bound by a 32 kDa protein, CEBP-1 (Maxsonet al., 1996). The differential display has been used in orchid flowers to isolate three MADS box genes of the AP1/AGL9 subfamily (Yu et al., 2000). The ubiquitin pathway is also reported to be involved in the degradation of petal proteins during floral senescence in daylily (Courtney et al., 1994), and a proteasome inhibitor, Z-leu-leu-Nva-H, was shown to delay senescence in Iris tepals (Pak et al., 2005). Yanet al. (2006) showed that a tonoplast-localized cytochrome P450 expressed at a higher level in senescing petals of Petunia than in vegetative tissue. Up regulation occurs in response to compatible pollination, ethylene treatment, or jasmonate treatment. Recently Shutianet al. (2007) reported a MAP kinase p56 involved in self-incompability induced early PCD signaling cascade in incompatible pollen. The p56 activation is necessary for progression of PCD in pollen and use of specific inhibitor of MAPK, U0126, inhibits pollen tube growth.

In Arabidopsis delayed abscission mutants also showed delayed flower senescence. For example delayed floral organ abscission mutants (dab-1, dab-2, dab-3)

(Patterson et al., 2004) and inflorescence deficient in abscission (ida), in whichfloral organs remain attached to the plant body after the shedding of mature seeds, showed delayed flower senescence (Butenko et al., 2003). MADS-box genes, like AGL15 and AGL18 (AGAMOUS-like 15 and 18) have also been shownto play an important role in flower senescence. Plants that constitutively overexpress AGL15 exhibit delayed flowering, and delayed floral organ abscission and senescence (Fernandez et al., 2000). Another MADS box gene AGL18 overexpression showed prolonged longevity and retention of perianthorgans. Thus AGL15 and AGL18 have thecapacity to delay floral organ senescence as well as flowering time when expressed at elevated levels (Adamczyket al., 2007).

Using subtractive hybridization 54 genes were isolated from Daffodil, including genes encoding a few regulatory proteins and several cysteine and serine proteases (Hunter et al., 2002). van Doornet al. (2003) through microarray identified a number of genes that were highly expressed during senescence. These included a number of genes with unknown function and sequences encoding a Grap2 and cyclin-D interacting protein, a MADS-domain transcription factor, a casein kinase, and a nucleotide-gated ion channel interacting protein might be important elements in the regulation of senescence. Breeze et al. (2004) identified 109 genes associated with flower senescence in Alstroemeria, out of which 93 were up-regulated and 16 were down-regulated. The up-regulated genes encoded e.g., a zinc finger protein, a Xa21 receptor-type protein kinase, and an aspartic proteinase. Among the down-regulated genes were sequences encoding a gibberellins induced protein and a cytochrome P450.

Tetsuyaet al. (2006) using differential screening reported several SAGs in morning glory flower senescence. One of them was protein kinase with up regulated expression which may be involved in signaling cascade during senescence as leucine-rich repeat transmembrane protein kinase has been reported to become up regulated, rather specifically, during leaf senescence in Arabidopsis. Another Arabidopsis homologue ataxin2 was also reported during flower senescence. Ataxin2 caused premature cell death in yeast due to defects in actin filament formation (Satterfield et al., 2002). This actin depolymerization is also reported in cell death in pollen tubes (Thomas et al., 2006).

Xinjiaet al. (2007) reported that the opening and senescence of *M. jalapa*flowers appears to be under photoperiodic control and using differential screening reported expression of several light-responsive and circadian clock-related transcription factors. The transcription factors significantly up regulated during senescence were homologues of bZIP proteins. A remarkable abundance of transcripts of a gene encoding a RING zinc finger ankyrin protein increased 40,000-fold as the flowers senescent. It might be possible that RING zinc finger protein, that shows such dramatic up-regulation during senescence, may play a key role in the control of the process and further detailed study in flower senescence is needed. Hoeberichtset al. (2007) using microarray analysis of senesced carnation flowers reported differential expression of several genes especially up-regulation of numerous ethylene response genes. A total of 268 genes were analyzed and they mainly grouped in genes associated with protein degradation, cell wall degradation, ROS, lipid degradation, defense, and signal transduction.

In general, thousands of genes are regulated in a tissue at a particular time, and hundreds of them may change expression during PCD/senescence. Several associated changes caused by senescence have been difficult to identify because sets of genes are typically involved. Instead of targeting individual genes, it has been necessary to identify groups of genes that coordinate and regulate senescence process synergistically. The large numbers of genes were identified in last few years using several differential tools and showed up and down-regulation pattern during the flower senescence. In fact, a number of SAGs identified through differential analysis have yet to be characterized (Marianne et al., 2007). Another most important aspect is that a combination of physiological, biochemical, genetic, and molecular assets of SAG products will be required to fully elucidate the regulation of flower senescence.

KEYWORDS

- **Ultrastructural changes**
- **Membrane fluidity**
- **Metabolic changes**
- **Discolouration**
- **Proteolytic enzymes**
- **Hormonal regulation**

REFERENCES

Adamczyk, Lehti-Shiu Melissa D.; Fernandez Donna, E.; The MADS domain factors AGL15 and AGL18 act redundantly as repressors of the floral transition in *Arabidopsis*. *Plant J.* **2007**, *50*, 1007–1019.

Ashman, T.-L.; and Schoen, D. J.; How long should flowers live? *Nature*. **1994**. *371*, 788–791.

Azeez, A.; Sane, A. P.; Bhatnagar, D.; and Nath, P.; Enhanced expression of serine proteases during flower senescence in *Gladiolus*. *Phytochemistry.* **2007**, *68*, 1352–1357.

Bailly, C.; Corbineau, F.; and van Doorn W. G.; Free radical scavenging and senescence in *Iris* tepals. *Plant Physiol. Biochem*. **2001**, *39*, 649–656.

Balk, J.; Chew, S. K.; Leaver, C. J.; and McCabe, P. F.; The inter membrane space of plant mitochondria contains a DNase activity that may be involved in programmed cell death. *Plant J.* **2003**, *34*, 573–583.

Bartoli C. G.; Simontacchi, M.; Guiamet, J.; Montaldi, E.; and Puntarulo, S.; Antioxidant enzyme and lipid peroxidation during aging of *Chrysanthemum morifolium* RAM petals. *Plant Sci.* **1995,** *104,* 161–168.

Bartoli, C. G.; Simontacchi, M.; Montaldi, E. R.; and Puntarulo, S.; Oxidants and antioxidants during aging of chrysanthemum petals. *Plant Sci.* **1997,** *129,* 157–165.

Beevers, L.; *Senescence in Plant Biochemistry.* 3rd ed.; Bonner, J., and Varner, J. E., Eds.; Academic Press: New York: pp 771–794.

Borochov, A.; Cho, M. H.; Boss, W. F.; Plasma membrane lipid metabolism of Petunia petals during senescence. *Physiol. Plant.* **1994,** *90,* 279–284.

Borochov, A.; Spiegelstein, H.; and Philosoph-Hadas, S.; Ethylene and flower petal senescence: Interrelationship with membrane lipid catabolism. *Physiol. Plant.* **1997,** *100,* 606–612.

Borochov, A.; Woodson W. R.; Physiology and biochemistry of flower petal senescence. *Hortic. Rev.* **1989,** *11,* 15–43.

Bovy, A. G.; Angenent, G. C.; Dons, H. J. M.; and van Altvorst, A. C.; Heterologous expression of the Arabidopsis etr1-1allele inhibits the senescence of carnation flowers. *Mol. Breed.* **1999,** *5,* 301–308.

Breeze, E.; Wagstaff, C.; Harrison, E.; Bramke, I.; Rogers, H.; and Stead, A.; Gene expression patterns to define stages of post-harvest senescence in *Alstroemeria* petals. *Plant Biotechnol. J.* **2004,** *2,* 155–168.

Breeze E.; Wagstaff, C.; Harrison, E.; Bramke, I.; Rogers, H.; Stead, H.; Thomas, B.; and Buchanan- Buchanan-Wollaston, V.;. The molecular biology of leaf senescence. *J. Exp. Bot.* **1997,** *48,* 181–199.

Butenko, M. A.; Patterson, S. E.; Grini, P. E.; Stenvik, G. E.; Amundsen, S. S.; Mandal, A.; and Aalen, R. B.; Inflorescence deficient in abscission controls floral organ abscission in Arabidopsis and identifies a novel family of putative ligands in plants. *Plant Cell.* **2003,** *15,* 2296–2307.

Callis, J.; Regulation of protein degradation. *Plant Cell.* **1995,** *7,* 845–857.

Chang, H.; Jones, M.; Banowetz, G. M.; and Clark, D. G.; Overproduction of cytokinins in Petunia flowerstransformed with PSAG12-IPT delays corolla senescence and decreases sensitivity to ethylene. *Plant Physiol.* **2003,** *132,* 2174–2183.

Courtney, S. E.; Rider, C. C.; Stead, A. D.; Changes in protein ubiquitination and the expression of ubiquitinencoding transcripts in daylily petals during floral development and senescence. *Physiol. Plant.* **1994,** *91,* 196–204.

Danon, A.; Delorme, V.; Mailhac, N.; and Gallois, P.; Plant programmed cell death: a common DNA during apoptosis. *Cell. Death. Differ.* **2000.** *10,* 108–116.

Eason, J. R.; de Vre L. A.; Somerfield, S. D.; Heyes, J. A.; Physiological changes associated with *Sandersoniaaurantiaca* flower senescence in response to sugar. *Postharvest Biol. Technol.* **1997,** *12,* 43–50.

Eason, J. R.; Ryan, D. J.; Pinkney, T. T.; and O'Donoghue, E. M.; Programmed cell death during flower senescence: Isolation and characterization of cysteine proteases from *Sandersoniaaurantiaca. Funct. Plant Biol.* **2002,** *29,* 1055–1064.

Fernandez, D. E.; Heck, G. R.; Perry, S. E.; Patterson, S. E.; Bleecker, A. B.; and Fang, S. C.; The embryo MADS domain factor AGL15 acts postembryonically: Inhibition of perianth senescence and abscission via constitutive expression. *Plant Cell.* **2000,** *12,* 183–198.

Fobel, M.; Lynch, D. V.; and Thompson, J. E.; Membrane deterioration in senescing carnation flowers. *Plant Physiol.* **1987.** *85,* 204–211.

Fukuchi-Mizutani, M.; Ishiguro, K.; Nakayama, T.; Utsunomiya, Y.; Tanaka, Y.; Kusumi, T.; and Ueda, T.; Molecular and functional characterization of a rose lipoxygenasecDNA related to flower senescence. *Plant Sci.* **2000**, *160*, 129–137.

Fukuchi-Mizutani , M.; Ishiguro, K.; Nakayama, T.; Utsunomiya, Y.; Tanaka, Y.; Kusumi, T.; Ueda Gechev, T. S.; and Hille, J.; Hydrogen peroxide as a signal controlling plant programmed cell death. *J. Cell. Biol.* **2005**. *168*, 17–20.

Gechev, T. S.; and Hille, J.; Hydrogen peroxide as a signal controlling plant programmed cell death. *J. Cell. Biol.* **2005**, *168*, 17–20.

Gladish, D. K.; Xu, J.; and Niki, T.; Apoptosis-like programmed cell death occursinprocambium and ground meristem of pea (*Pisumsativum*) root tips exposed to sudden flooding. *Ann. Bot.* **2006**, *97*, 895–902.

Groover, A.; DeWitt, N.; Heidel, A.; and Jones, A.; Programmed cell death of plant trachearyelements differentiating in vitro. *Protoplasma*. **1997**, *196*, 197–211.

Guerrero, C.; de la Calle, M.; Reid, M. S.; and Valpuesta, V.; Analysis of the expression of two thiolprotease genes from daylily (*Hemerocallis spp.*) during flower senescence. *Plant Mol. Biol.* **1998**, *36*, 656–671.

Gunawardena, A. H. L. A. N.; Greenwood, J. S.; Dengler, N. G.;. Programmed cell death remodels lace plant leaf shape during development. *Plant Cell*. **2004**, *16*, 60–73.

Halliwell, B.; and Gutteridge, J. M. C.; *Free Radicals in Biology and Medicine*. Oxford, UK: Clarendon Press; **1989**. 450–499

Halliwell, B.; Reactive species and antioxidants. Redox biology is a fundamental theme of aerobic life. *Plant Physiol.* **2006**. *141*, 312–322.

Hengartner, M. O.; The biochemistry of apoptosis. *Nature*. **2000**, *407*, 770–776.

Hoeberichts, F. A.; de Jong, A. J.; and Woltering, E. J.; Apoptotic-like cell death marks the earlystage of gypsophila (*Gyp; sophila paniculata*) petal senescence. *Postharvest Biol. Technol.* **2005**, *35*, 229–236.

Hoeberichts, F. A.; van Doorn, W. G.; Vorst, O.; Hall, R. D.; and van Wordragen M. F.; Sucrose prevents up-regulation of senescence-associated genes in carnation petals. *J. Exp. Bot.* **2007**, *58*, 2873–2875.

Hossain, Z.; Mandal, A. K. A.; Datta, S. K.; Biswas, A. K.; Decline in ascorbateperoxidaseactivity – a prerequisite factor for tepal senescence in gladiolus. *J. Plant Physiol.* **2006**. *163*, 186–194.

Huang, F. Y.; Philosoph-Hadas, S.; Meir, S.; Callahan, D. A.; Sabato, R.; Zelcer, A.; Hepler, P. K.; Increases in cytosolic Ca2C in parsley mesophyll cells correlated with leaf senescence. *Plant Physiol.* **1997**, *115*, 51–60.

Hunter, D. A.; Ferrante, A.; Vernieri, P.; and Reid, M. S.; Role of abscisic acid in perianth senescence of daffodil (*Narcissus pseudonarcissus* 'Dutch Master'). *Physiologia Plantarum*. **2004**. *121*, 313–321.

Hunter, D. A.; Steele, B. C.; Reid, M. S.; Identification of genes associated with perianth senescence in Daffodil (*Narcissus pseudonarcissusL*. "Dutch Master") *Plant Sci.* **2002**, *163*, 13–21.

Jones, M. L.; Chaffin, G. S.; Eason, J. R.; and Clark, D. G.; Ethylene-sensitivity regulates proteolytic activity and cysteine protease gene expression in petunia corollas. *J. Exp. Bot.* **2005**. *56*, 2733–2744.

Jones, R. B.; Serek, M.; Kuo, C. L.; and Reid, M. S.; The effect of protein synthesis inhibition on petal senescence in cut bulb flowers. *J. Am. Soc. Hortic. Sci.* **1994**, *119*, 1243–1247.

Jones, A. M.; Programmed cell death in development and defense. *Plant Physiol.* **2001**, *125*, 94–97.

Kladnik, A.; Chamusco, K.; Dermastia, M.; and Chourey, P.; Evidence of programmed celldeath in post-phloem transport cells of the maternal pedicel tissue in developing caryopsis of maize. *Plant Physiol.* **2004,** *136,* 3572–3581.

Kotchoni, S. O.; and Gachomo, E. W.; The reactive oxygen species network pathway: an essential prerequisite for perception of pathogen attack and the acquired disease resistance in plant. *J. Biosci.* **2006.** *31,* 389–404.

Kuroda, S.; Hirose, Y.; Shiraishi, M.; Davies, E.; and Abe, S.; Co-expression of an ethylene receptor gene, *ERS1,* and ethylene signaling regulator gene, *CTR1,* in Delphinium during abscission of florets. *Plant Physiol. Biochem.* **2004,** *42,* 745–751.

Laloi, C.; Apel, K.; and Danon, A.; Reactive oxygen signalling: the latest news. *Curr. Opin. Plant Biol.* **2004.** *7,* 323–328.

Langston, B. J.; Bai, S.; Jones, M. L.; Increase in DNA fragmentation and induction of asenescence-specific nuclease are delayed during corolla senescence in ethylene-insensitive (etr1-1) transgenic petunias. *J. Exp. Bot.* *56,* 15–23.

Lara, M. E. B.; Garcia, M. C. G.; Fatima, T.; Ehness, R.; Lee, T. K.; and Proels, R.; Extracellular invertase is an essential component of cytokinin-mediated delay of senescence. *Plant Cell.* **2004,** *16,* 1276–1287.

Lay-Yee, M.; Stead, A. D.; and Reid, M. S.; Flower senescence in daylily (*Hemerocallis*). *Physiologia. Plantarum.* **1992,** *86,* 308–314.

Ledford, H.; The flower of seduction. *Nature.* **2007.** *445,* 816–817.

Lee, M.; Lee, S. H.; and Park, K. Y.; Effects of spermine on ethylene biosynthesis in cut carnation (*Dianthus caryophyllus*L.) flowers during senescence. *J. Plant Physiol.* **1997,** *151,* 68–73.

Lee, R. H.; and Chen, S. C. G.; Programmed cell death during rice leaf senescence is non-apoptotic. *New Phytologist.* **2002.** *155,* 25–32.

Leopold, A. C.; Senescence in plant development. *Science.* **1961,** *134,* 1727–1732.

Leopold, A. C.; Aging, senescence, and turn over in plants. *Bioscience.* **1975,** *25,* 659.

Leverentz, M. K.; Wagstaff, C.; Rogers, H. J.; Stead, A. D.; Chanasut, U.; Silkowski, H.; Thomas, B.; Weichert, H.; Feussner, I.; and Griffiths, G.; Characterization of a novel lipoxygenase-independent senescence mechanism in Alstroemeria peruviana floral tissue. *Plant Physiol.* **2002.** *130,* 273–283.

Lim, P. O.; Kim, H. J.; Nam, H. G.; Leaf senescence. *Ann. Rev. Plant Biol.* **2007,** *58,* 115–136.

Lincoln , J. E.; Richael, C.; Overduin, B.; Smith, K.; Bostock, R.; and Gilchrist, D. G.; Expression of the anti-apoptotic baculo virus p35 gene in tomato blocks programmed cell death and provides broad-spectrum resistance to disease. *Proc. Natl. Acad. Sci.* USA. **2002.** *99,* 15217–15221.

Lombardi, L.; Ceccarelli, N.; Picciarelli, P.; Lorenzi,; R.; DNA degradation during programmed cell death in *Phaseoluscoccineus* suspensor. *Plant Physiol. Biochem.* **2007a,** *45,* 221–227.

Macnish, A.; Jiang, C.-Z.; Negre-Zakharov, F.; and Reid, M. S.; Physiological and molecular changes during opening and senescence of *Nicotianamutabilis* flowers. *Plant Sci.* **2010a.** *179,* 267–272.

Marianne Hopkins; Taylor Catherine; Liu Zhongda; Ma Fengshan; McNamara Linda; Wang Tzann-Wei; and Thompson John, E.; Regulation and execution of molecular disassembly and catabolism during. *Senescene. New. Phytol.* **2007,** *175,* 201–214.

Matile, P.; and Winkenbach, F.; Function of lysosomes and lysosomal enzymes in the senescing corolla of the morning glory (*Ipomoea purpurea*). *J. Exp. Bot.* **1971,** *22,* 759–771.

Maxson, J. M.; and Woodson, W. R.; Cloning of a DNA-binding protein that interacts with the ethylene responsive enhancer element of the carnation *GST1* gene. *Plant Mol. Biol.* **1996,** *31,* 751–759.

Mayak, S.; and Halevy, A. H.; Interrelationships of ethylene and abscisic acid in the control of rose petalsenescence. *Plant Physiol.* **1972,** *50,* 341–346.

Medawar, P. B.; *The Uniqueness of the Individual,* Basic Books Inc., New York; **1957,** pp. 191.

Meyer, R. C.; Goldsbrough, P. B.; and Woodson, W. R.; An ethylene-responsive flower senescence-related gene from carnation encodes a protein homologous to glutathione *S*-transferases. *Plant Mol. Biol.* **1991,** *17,* 277–281.

Muller, R.; Lind-Iversen, S.; Stummann, B. M.; Serek, M.; Expression of genes for ethylene biosynthetic enzymes and an ethylene receptor in senescing flowers of miniature potted roses. *J. Hortic. Sci. Biotechnol.* **2000,** *75,* 12–18.

Muller, R.; Owen, C. A.; Xue, Z. T.; Welander, M.; Stummann, B. M.; Characterization of two CTR-like protein kinases in *Rosa hybrida*and their expression during flower senescence and in response to ethylene. *J Exp Bot.* **2002,** *53,* 1223–1225.

Muller, R.; Stummann, B. M.; and Serek, M.; Characterization of an ethylene receptor family with differential expression in rose (*Rosa hybrida* L.) flowers. *Plant Cell Rep.* **2000,** *19,* 1232–1239.

Nadeau, J. A.; Zhang, X. S.; Nair, H.; and O'Neill, S. D.; Temporal and spatial regulation of 1-aminocyclopropane-1-carboxylate oxidase in the pollination-induced senescence of orchid flowers. *Plant Physiol.* **1993,** *103,* 31–39.

Nagata, S.; Nagase, H.; Kawane, K.; Mukae, N.; and Fukuyama, H.; Degradation of chromosomal DNA during apoptosis. *Cell. Death. Differ.* **2003,** *10,* 108–116.

O'Donoghue, E. M.; Somerfield, S. D.; and Heyes, J. A.; Organization of cell walls in *Sandersoniaaurantiaca* floral tissue. *J. Exp. Bot.* **2002,** *53,* 513–523.

O'Donoghue, E. M.; Somerfield, S. D.; Watson, L. M.; Brummell, D. A.; and Hunter, D. A.; Galactose metabolism in cell walls of opening and senescing Petunia petals. *Planta.* **2009,** *229,* 709–721.

Obara, K.; Kuriyama, H.; and Fukuda, H.; Direct evidence of active and rapid nuclear degradation triggered by vacuole rupture during programmed cell death in *Zinnia. Plant Physiol.* **2001,** *125,* 615-626.

O'Neill, S. D.; Pollination regulation of flower development. *Annu. Rev. Plant Physiol. Plant Mol. Biol.* **1997,** *48,* 547–574.

Orz'aez, D.; and Granell, A.; The plant homologue of the defender against apoptotic death gene is down-regulated during senescence of flower petals. *FEBS Lett.* **1997.** *404,* 275–278.

Pak, C.; van Doorn, W. G.; Delay of Iris flower senescence by protease inhibitors. *New. Phytologist.* **2005,** *165,* 473–480.

Panavas, T.; Pikula, A.; Reid, P. D.; Rubinstein, B.; and Walker, E. L.; Identification of senescence-associated genes from daylily petals. *Plant Mol. Biol.* **1999,** *40,* 237–248.

Panavas, T.; Rubinstein, B.; Oxidative events during programmed cell death of daylily (*Hemerocallis hybrid*) petals. *Plant Sci.* **1998,** *133,* 125–138.

Panavas, T.; Walker, E. R.; and Rubinstein, B.; Possible role of abscisic acid in senescence of daylily petals. *J. Exp. Bot.* **1998,** *49,* 1987–1997.

Patterson, S. E.; and Bleecker, A. B.; Ethylene-dependent and -independent processes associated with floral organ abscission in Arabidopsis. *Plant Physiol.* **2004,** *134,* 194–203.

Phillips, H. L.; and Kende, H.; Structural changes in flowers of Ipomoea tricolor during flower opening and closing. *Protoplasma.* **1980;** *102,* 199–215.

Porat, R.; Borochov, A.; Halevy, A. H.; Pollinationinduced senescence in Phalaenopsis petals: Relationship of ethylene sensitivity to activity of GTP-binding proteins and protein phosphorylation. *Physiol Plant* **1994**, *90*, 679–684.

Porat, R.; Halevy, A. H.; Serek, M.; and Borochov, A.; An increase in ethylene sensitivity following pollination is the initial event triggering an increase in ethylene production and enhanced senescence of Phalaenopsis orchid flowers. *Physiologia. Plantarum.* **1995**, *93*(4), 778–784.

Price, A. M.; Orellana, D. F. A.; Salleh, F. M.; Stevens, R, Acock, R.; Buchanan-Wollaston, V.; Stead, A. D.; and Rogers, H. J.; A comparison of leaf and petal senescence in wallflower reveals common and distinct patterns of gene expression and physiology. *Plant Physiol.* **2008**, *147*, 1898–1912.

Quirino, B. F.; Noh, Y.-S.; Himelblau, E.; Amasino, R. M.;. Molecular aspects of leafsenescence. *Trend. Plant Sci.* **2000**, *5*, 278–282.

Rogers, H. J.; Programmed cell death in floral organs: how and why do flowers die? *Ann. Bot.* **2006**, *97*, 309–315.

Ronen, M.; and Mayak, S.; Interrelationship between abscisic acid and ethylene in the control of senescenceprocesses in carnation flowers. *J. Exp. Bot.* **1981**, *32*, 759–765.

Rubinstein, B.; Regulation of cell death in flower petals. *Plant Mol. Biol.* **2000**, *44*, 303–318.

Sacher, J. A.; Senescence and postharvest physiology. *Ann. Rev. Plant Physiol.* **1973**, *24*, 197.

Saks, Y.; Van Staden, J.; and Smith, M. T.; Effect of gibberellic acid on carnation flower senescence evidence that the delay of carnation flower senescence by gibberellic acid depends on the stage of flowerdevelopment. *Plant Growth. Regul.* **1992**, *11*, 45–51.

Satterfield, T. F.; Jackson, S. M.; and Pallanck, L. J.; A *Drosophila* homolog of the polyglutamine disease gene *SCA2* is a dosage-sensitive regulator of actin filament formation. *Genetics.* **2002**, *162*, 1687–1702.

Serafini-Fracassini, D.; Del Duca, S.; Monti, F.; Poli, F.; Sacchetti, G.; Bregoli, A. M.; Biondi, S.; and DellaMea, M.; Transglutaminase activity during senescence and programmed cell death in the corolla of tobacco (*Nicotianatabacum*) flowers. *Cell Death Diff.* **2002**, *9*, 309–321.

Setyadjit, Irvin DE; Joyce, D. C.; Simons, D. H.; Vase treatments containing gibberellic acid do not increaselongevity of cut Sylvia' inflorescences. *Australian J. Exp. Agric.* **2006**, *46*, 1535–1539.

Shibuya, K.; Nagata, M.; Tanikawa, N.; Yoshioka, T.; Hashiba, T.; and Satoh, S.; Comparison of mRNA levels of three ethylene receptors in senescing flowers of carnation (*Dianthus caryophyllus* L.) *J. Exp. Bot.* **2002**, *53*, 399–406.

Shibuya, K.; Yamada, T.; Suzuki, T.; Shimizu, K.; Ichimira, K.; InPSR26, a putative membrane protein, regulates programmed cell death during petal senescence in Japanese morning glory. *Plant Physiol.* **2009a**, *149*, 816–824.

Shibuya, K.; Yamada, T.; and Ichimira, K.; Autophagy regulates progression of programmed cell death during petal senescence in Japanese morning glory. *Autophagy.* **2009b**, *5*, 546–547.

Shutian, Li; SamajJozef; Franklin-Tong Vernonica, E.; A MAPK signals to programmed cell death induced by self-incompatibility in Papaver pollen. *Plant Physiol.* **2007**, *145*, 236–245.

Singh, A.; Evenson, K. B.; and Kao, T. H.; Ethylene synthesis and floral senescence following compatible and incompatible pollinations in *Petunia inflata. Plant Physiol.* **1992**, *99*, 38–45.

Smith, M. T.; Saks, Y.; and van Staden; Ultrastructural changes in the petal of senescing flowers of *Dianthus caryophyllus* L. *Ann. Bot.* **1992**, *69*, 277–285.

Stead, A. D.; and van Doorn, W. G.; Strategies of flower senescence—a review. In: *Molecular and Cellular Aspects of Plant Reproduction,* Scott, R. J.; and Stead, A. D.; Eds.; Cambridge University Press: Cambridge; **1994**, pp 215–238.

Stead, A. D.; Pollination-induced flower senescence: A review. *Plant Growth. Regul.* **1992**, *11*, 13–20.

Stephenson, P.; and Rubinstein, B.; Characterization of proteolytic activity during senescence in daylily. *Physiologia Plantarum.* **1998**, *104*, 463–473.

Tang, X.; Gomes, A. M. T. R.; Bhatia, A.; and Woodson, W. R.; Pistil-specific and ethylene-regulated expression of 1- aminocyclopropane-1-carboxylate oxidase genes in Petunia flowers. *Plant Cell.* **1994**, *6*, 1227–1239.

Tang, X.; and Woodson, W. R.; Temporal and spatial expression of 1-aminocyclopropane-1-carboxylate oxidase mRNA following pollination of immature and mature Petunia flowers. *Plant Physiol.* **1996**, *112*, 503–511.

Taylor, J. E.; and Whitelaw, C. A.; Signals in abscission. *New Phytol.* **2001**, *151*, 323–339. 138.

ten Have, A.; Woltering, E. J.; Ethylene biosynthetic genes are differentially expressed during carnation (*Dianthus caryophyllus* L.) flower senescence. *Plant Mol. Biol.* **1997**, *34*, 89–97.

Tetsuya Yamada; Ichimura Kazuo; Kanekatsu Motoki; van Doorn outer, G.; Gene expression in opening and senescing petals of morning glory (*Ipomoea nil*) flowers. *Plant Cell. Rep.* **2007**, *26*, 823–835.

Thomas, H.; Ougham, H. J.; Wagstaff, C.; and Stead, A. D.; Defining senescence and death. *J. Exp. Bot.* **2003**, *54*, 1127–1132.

Thomas, S. G.; Huang, S.; Li, S.; Staiger, C. J.; and Franklin-Tong, V. E.; Actin depolymerization is sufficient to induce programmed cell death in self-incompatible pollen. *J. Cell Biol.* **2006**, *174*, 221–229.

Thompson, J. E.; Froese, C. D.; Hong, Y.; Hudak, K. A.; and Smith, M. D.; Membrane deterioration during senescence. *Can. J. Bot.* **1997**, *75*, 867–879.

Thompson, J. E.; Froese, C. D.; Madey, E.; Smith, M. D.; and Hong, Y.; Lipid metabolism during plant senescence. *Prog. Lipid Res.* **1998**. *37*, 119–141.

Valpuesta, V.; Lange, N. E.; Guerrero, C.; and Reid, M. S.; Upregulation of a cysteine protease accompanies the ethylene-insensitive senescence of daylily (*Hemerocallis*) flowers. *Plant Mol. Biol.* **1995**, *28*, 575–582.

van der Kop, D. A. M.; Ruys, G.; Dees, D.; van der Schoot, C.; de Boer, A. D.; and van Doorn, W. G.; Expression of defender against apoptotic death (DAD-1) in Iris and Dianthus petals. *PhysiologiaPlantarum.* **2001**, *117*, 256–263.

van Doorn, W. G.; Effects of pollination on floral attraction and longevity. J. Exp. Bot. **1997**, *48*, 1615–1622.

van Doorn, W. G.; Categories of petal senescence and abscission: a re-evaluation. Ann. Bot. **2001**, *87*, 447–456.

van Doorn, W. G.; Is petal senescence due to sugar starvation? Plant Physiol. **2004**, *134*, 35–42.

van Doorn, W. G.; Is the onset of senescence in leaf cells of intact plants due to low or high sugar levels? J. Exp. Bot. **2008**, *59*, 1963–1972.

van Doorn, G. W.; Classes of programmed cell death in plants, compared to those in animals. J. Expt. Bot. **2011**, *62*, 4749–4761.

van Doorn, W. G.; and Woltering, E. J.; Senescence and programmed cell death: substance or semantics? J. Exp. Bot. **2004**, *55*, 2147–2153.

van Doorn, W. G.; and Woltering, E. J.; Physiology and molecular biology of petal senescence. J. Exp. Bot. **2008**, *59*, 453–480.

van Doorn, W. G.; Balk, P. A.; van Houwelingen, A. M.; Hoeberichts, F. A.; Hall, R. D.; Vorst, O.; van der Schoot, C.; and vanWordragen, M. F.; Gene expression during anthesis and senescence in iris flowers. Plant Mol. Biol. **2003**, *53*, 845–863.

Van Staden, J.; Dimalla, G. G.; The effect of silver thiosulfate preservative on the physiology of cut carnations: II. Influence of endogenous cytokinins. *Z Pflanzenphysiol.* **1980**, *99*, 19–26.

Wagstaff C, Leverentz MK, Griffiths G, Thomas B, Chanasut U, Stead AD, Rogers HJ. Cysteine protease gene expression and proteolytic activity during senescence of Alstroemeria petals. *J. Exp. Bot.* **2002**;*53*:233–240.

Wagstaff, C.; Malcolm, P.; Rafiq, A.; Leverentz, M.; Griffiths, G.; Thomas, B.; Stead, A.; and Rogers, H. J.; Programmed cell death (PCD) processes begin extremely early in Alstroemeria petal senescence. *New Phytol.* **2003**, *160*, 49–59.

Waki, K.; Shibuya, K.; Yoshioka, T.; Hashiba, T.; and Satoh, S.; Cloning of a cDNA encoding EIN3-like protein (DCEIL1) and decrease in its mRNA level during senescence in carnation flower tissues. *J. Exp. Bot.* **2001**, *355*, 377–379. [PubMed]

Watada, A. E.; Kerner, R. C.; Kader, A. A.; Romani, R. J.; and Staby, G. L.; Terminology for the description of developmental stages of horticultural crops. *Hortic. Sci.* **1984**, *19*, 20–21

Wilkinson, J. Q.; Lanahan, M. B.; Clark, D. G.; Bleecker, A. B.; Chang, C.; Meyerowitz, E. M.; Klee, H. J.; A dominant mutant receptor from Arabidopsis confers ethylene insensitivity in heterologous plants. *Nat. Biotechnol.* **1997**, *15*, 444–447.

Winkenbach, F.; The metabolism of blossoming and withering corolla of the morning glory *Ipomoea purpurea* magnificent I relationships between formative pretext l, fuel transportation, breathable and invertase. *Rep. Swiss. Bot. Soc.* **1970a**, *80*, 374–390.

Winkenbach, F.; The metabolism of blossoming and withering corolla of the morning glory Ipomoea purpurea magnificent II function and de novo synthesis of lysosomal enzymes during wilting. *Rep. Swiss. Bot. Soc.* **1970b**, *80*, 391–406.

Woltering, E. J.; and van Doorn, W. G.; Role of ethylene in senescence of petals: morphological and taxonomical relationships. *J. Exp. Bot.* **1988**, *39*, 1605–1616.

Woltering, E. J.; de Vrije, T.; Harren, F.; and Hoekstra, F. A.; Pollination and stigma wounding: Same response, different signal? *J Exp Bot.* **1997**, *48*, 1027–1033.

Woltering, E. J.; ten Have, A.; Larsen, P. B.; Woodson, W. R.; Ethylene biosynthetic genes and inter organ signalling during flower senescence. In: *Molecular and Cellular Aspects of Plant Reproduction*, Scott R. J., Stead, A. D., Eds.; Cambridge University Press: Cambridge, UK; **1994**, pp 285–307.

XinjiaXu; Gookin Tim; Jiang Cai-Zhong; Reid MichaelGenes associated with opening and senescence of Mirabilis jalapa flowers. *J. Exp. Bot.* **2007**, *58*, 2193–2201.

Xu, X.; Gookin, T.; Jiang, C. Z.; Reid, M.; Gene associated with opening and senescence of *Mirabilis jalapa* flowers. *J. Exp. Bot.* **2007**, *58*, 2193–2201.

Xu, Y.; and Hanson, M. R.; Programmed cell death during pollination-induced petal senescence in petunia. *Plant Physiol.* **2000**, *122*, 1323–1333.

Yamada, T.; Ichimura, K.; van Doorn, W. G.; DNA degradation and nuclear degeneration during programmed cell death in petals of *Antirrhinum, Argyranthemum*, and *Petunia. J. Exp. Bot.* **2006b**, *57, 3543–3552.*

Yamada, T.; Takatsu, Y.; Kasumi, M.; Ichimura, K.; van Doorn, W. G.; Nuclear fragmentation and DNA degradation during programmed cell death in petals of morning glory (*Ipomoea nil*). *Planta.* **2006a**; *224*, 1279–1290.

Yamada, T.; Takatsu, Y.; Kasumi, M.; Manabe, T.; Hayashi, M.; Marubashi, W.; and Niwa, M.; Novel evaluation method of flower senescence in Freesia (*Freesia hybrida*) based on apoptosis as an indicator. *Plant Biotechnol.* **2001,** *18,* 215–218.

Yamada, T.; Takatsu, Y.; Manabe, T.; Kasumi, M.; and Wataru, M.; Suppressive effect of trehalose on apoptotic cell death leading to petal senescence in ethylene-insensitive flowers of gladiolus. *Plant Sci.* **2003,** *164,* 213–221.

Yan Xu; Ishida Hiroyuki; Reisen Daniel; and Hanson Maureen, R.; Upregulation of a tonoplast-localized cytochrome P450 during petal senescence in *Petunia inflate. BMC. Plant Biol.* **2006,** *6,* 8.

Yu, H.; Goh, C. J. Identification and characterization of three orchid MADS-box genes of the AP1/AGL9 subfamily during floral transition. *Plant Physiol.* **2000,** *123,* 1325–1336.

CHAPTER 10

POSTHARVEST MANAGEMENT OF FRESH CUT FLOWERS

VIGYA MISHRA[1] and SHAILENDRA K. DWIVEDI[2]

[1]Amity International Centre for Post Harvest Technology and Cold Chain Management Amity University Uttar Pradesh, Sec 1235-Noida 201 301

[2]College of Horticulture, RVS Krishi Vishwavidyalaya, Mandsaur Campus, Madhya Pradesh, India, Pin- 458 001

CONTENTS

10.1   INTRODUCTION

Cut flower production has been emerged as an important industry mainly to supply to the needs of the demand in the overseas market. It is being viewed as a high growth industry in our economy. There is a tremendous transformation in our floriculture sector mainly due to the entry of corporate who are producing cut flowers to meet the emerging demand in the developed countries for floricultural products.

Fresh cut flower industry accounts for around 2/3rd of the world trade in floriculture and has been growing at around 8–10 percent annually. In the past 50 years, the cut flower market has changed dramatically, from a local market with growers located on city outskirts, to a global one. Flowers and cut foliage from all over the world are sold as bunches or combined into arrangements and bouquets in the major target markets like North America, Japan, and the European Union. The changing trend of the market has occurred with little consideration for its postharvest consequences. Flowers that used to be obtained from local growers and were retailed within days of harvest may now take as long as 3weeks to arrive at the retail florist or supermarket. Increased emphasis on holidays as occasions for sale of cut flowers has exacerbated this trend. The volume of flowers required to meet thedemand for the major occasions has led to their widespread storage. Because of their perishability, flowers produced in distant growing areas have traditionally been shipped by air transportation system to offset the disadvantages of poor temperature management and low humidity conditions) (Reid and Jiang, 2012). Therefore, the postharvest handling of fresh cut flowers is a major consideration right after harvesting up to it reaches the consumer level. The chapter is focused on the postharvest handling, storage, and transportation of a variety of cut flowers.

10.2   WHAT ARE CUT FLOWERS?

The term "cut flowers" represent the flowers starting to blossom or flower buds that are cut with branches, stems, and leaves. Cut flowers are grouped into two categories: "fresh cut flowers" and "non-fresh cut flowers" such as preserved flowers.

Fresh cut flowers are highly perishable and maintain only limited life-supporting processes by taking water up through their stems. Fresh cut flowers are used for decorative purposes such as vase arrangements and bouquets at formal events, designs for weddings and funerals, gifts on occasions such as Mother's Day, Valentine's Day etc, in times of illnessand informal displays to beautify homes and public places. More than 200 different types of fresh cut flowers are sold in the world (Reid and Jiang, 2012). Typical fresh cut flowers include roses, carnations, orchids, chrysanthemums,alstroemeria, strelitzia, tuberose, tulips, gladiolus, and lilies etc while some nontraditional or exotic cut flowers include yellow anthurium, bird of paradise, red ginger etc (Figure 10.1).

Preserved flowers are cut flowers that have been dehydrated, preserved with a chemical solution and then air- or oven-dried. They may be used in boutonnieres, corsages, wreaths, formal, and informal displays, and similar ornamental articles. Preserved flowers, known in the industry as "everlasting flowers" or "everlastings," are not as perishable as fresh cut flowers (Anonymous, 2003).

10.3 GLOBAL SCENARIO OF CUT FLOWER INDUSTRY

Globally more than 145 countries are involved in the cultivation of ornamental crops and the area under these crops is increasing steadily. The production of flower crops has increased significantly as there is a huge demand for floricultural products in the world, resulting in growing International Flower Trade. Due to globalization and its effect on income, there is growing per capita floriculture consumption in most of the countries. In case of developed countries, the consumption of flowers is closely linked with GNP per capita income and urban population.

Red Ginger (*Alpiniapurpurata*)   Torch Ginger (*Etlingeraelatior*)

Beehive Ginger (*Zingiberspectabile*)   Bird of Paradize (*Strelitziareginae*)

Anthuriumadreanum   *Heliconiaspp*

FIGURE 10.1   Some important nontraditional cut flowers (da Silva Vieira et al. 2014).

10.3.1 INDIAN SCENARIO AND TRADE

India is bestowed with diverse agroclimatic and ecological conditions, which are favorable to grow all types of commercially important flowers generally found in different parts of the world. It also enjoys the best climate in selected pockets for floriculture during winter months. India is in an enviable position to become a leader in the world floricultural trade because of the prevailing congenial location, overall favorable climate of liberalization and globalization and also specific incentives by the government and floricultural development. The production of cut flowers in India is increasing per year due to its increasing demand (Figure 10.2). According to NHB Database (2013), the production of cut flowers has increased from 47942 lakh no.s in 2009 to 76732 lakh no. in 2013.

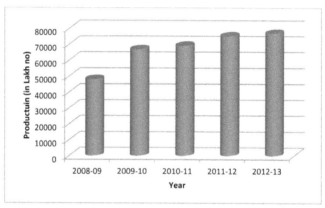

FIGURE 10.2   Production of cut flowers in India in 2012–2013 (Source: NHB Database, 2012–2013).

The leading cut flower-producing states of India include: West Bengal, Kanataka, Maharashtra, Andhra Pradesh, and Odisha (Figure 10.3). West Bengal tops the list with 33.10 percent of the total production which comprises 25429.1 lakh nos of cut flowers per year.

In India a variety of cut flowers are grown important ones being: rose, gladiolus, orchids, tuberose, carnation, anthurium etc (Table 10.1). Rose is the principal cut flower grown all over India. The maximum production of cut roses in India is in Karnataka (7132 lakh no.s) followed by Andhra Pradesh (6,899 lakh no.s) and Odisha (3,690 lakh no.s) (NHB Database, 2013). The larger percentage of the area in many states is used for growing scented rose, usually local varieties to be sold as loose flowers. For cut flower use, the old rose varieties like Queen Elizabeth, Super Star, Montezuma, Papa Meilland, Christian Dior, Eiffel Tower, Kiss of Fire, Golden Giant, Garde Henkel, First Prize etc. are still popular. In recent times the export purpose varieties like First Red, Grand Gala, Konfitti, Ravel, Tineke, Sacha, Prophyta, Pareo, Noblesse. Virsilia, Vivaldi etc. are also being grown commercially.

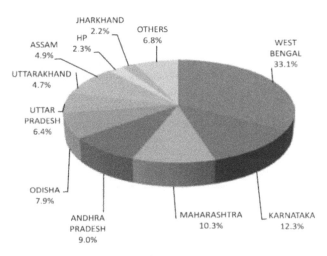

FIGURE 10.3 Leading cut flower producing states.

TABLE 10.1 Major cut flowers of India and their production in 2012–2013.

FLOWERS	QUANTITY (in lakh no.s)
ANTHURIUM	320.27
CARNATION	148.7
GERBERA	249.2
CHRYSANTHEMUM	30.58
GLADIOLUS	7067.95
ORCHIDS	0.18
ROSE	19902.76
TUBE ROSE	1560.7
TULIP	45.1
OTHERS	47406.42
TOTAL	76731.86

Gladiolus is the next most important cut flower crop in the country. Earlier it was considered a crop for temperate regions in India. However, with improved agronomic techniques and better management, the northern plains of Delhi, Haryana, Punjab, Uttar Pradesh, as well as Maharashtra and Karnataka have emerged as the major areas for production of gladiolus. Uttar Pradesh is the leading producer of gladiolus cut flower followed by Odisha.

Tuberose, a very popular cut flower crop in India is grown mainly in the eastern part of the country (i.e., West Bengal) and also in northern plains and parts of south. Leading states involved in the production of tuberose cut flower are Karnataka, Haryana, and J & K. Both single and double flower varieties are equally popular. Tuberose flowers are also sold loose in some areas for preparing garlands and wreaths.

The other main cut flower item is orchid. Its production is restricted mainly in the north-eastern hill regions, besides parts of the southern states of Kerala and Karnataka. The main species grown are Dendrobiums, Vanda, Paphiopedilums, Oncidiums, Phalaenopsis, and Cymbidiums.

10.4   QUALITY LOSS IN CUT FLOWERS

In cut flowers loss of quality of stems, leaves or flower parts may result in their rejection in the marketplace. In some ornamentals loss of quality may result from one of several causes, including wilting or abscission of leaves and/or petals, yellowing of leaves, and geotropic or phototropic bending of scapes and stems. There are many causes of quality loss important ones being:

Growth, development, and aging

In plants, death of individual organs and of the whole plant itself is an integral part of the life cycle. Even in the absence of senescence of floral organs or leaves, the continuing growth process can result in quality loss, for example in spike-type flowers that bend in response to gravity for example, *Alstroemeria* and *Zentedeschia, Narcissus*etc (Philisoph et al., 1995; Philosoph et al., 1996)

Flower senescence

The early death of flowers is a common cause of quality loss and reduced vase life for cut flowers. Flowers can be divided into several categories in terms of their senescence. Some flowers are extremely long-lived, especially in the daisy and orchid families. Others are short-lived, including many of the bulb crops like tulip, Iris, and *Narciussus*etc.

Wilting

Extended life for cut and potted ornamentals depends absolutely on a continuous and adequate supply of water. Failure of water supply, whether through obstruction of the cut stems, or through inadequate watering of pots, results in rapid wilting of shoot tips, leaves, and petals.

Leaf yellowing and senescence

Yellowing of leaves and even of other organs (buds, stems) commonly is associated with the end of display life in some cut flowers (*Alstroemeria* being an important example). Leaf yellowing is a complex process that may be caused by a range of different environmental factors.

Shattering

The process of loss of leaves, buds, petals, flowers or even branchlets, is called as 'shattering', or "abscission". It is also a common problem in cut flowers. Very often, this problem is associated with presence of ethylene in the air, but other environmentalfactors may also be involved (Reid and Jiang, 2012).

10.5 FACTORS AFFECTING POSTHARVEST QUALITY AND VASE LIFE OF CUT FLOWERS

Like other perishable horticultural crops, the life of ornamentals is also affected by physical, environmental, and biological factors. Some ornamentals, particularly cut flowers can be extraordinarily long-lived. Nevertheless, the majority of the ornamentals of commerce have relatively short lives. The delicate petals of flowers are easily damaged and are often highly susceptible to diseases. Even under optimum conditions, their biology leads to early wilting, abscission or both. Type of plant material and preharvest factors plays an important role. After harvest, temperature is of overriding importance and affects plant water relations, growth of disease, response to physical stresses, carbohydrate status and the interplay among endogenous and exogenous growth regulators. Much has been studied in the past years about the role of these factors and the response of ornamentals to them, and some of the research findings have led to technologies that can greatly improve marketing and postharvest quality of ornamentals.

The vase life (VL) of cut flowers refers to the duration from placement of stems in a vase solution to the loss of visible ornamental value and is synonymous with display life, keeping or lasting quality (Halevy and Mayak, 1981). The VL terminating criteria are mainly either water stress symptoms (van Doorn, 2011) or *Botrytis cinerea* infection symptoms (van Meeteren, 2007; Macnish *et al.*, 2010d) (Figure 10.4).

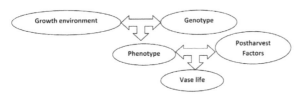

FIGURE 10.4   Factors determining vase life of cut flowers.

Four major factors are involved in determining the storage and vase life of floral crops: temperature, water relation, carbohydrate supply and plant growth regulators (Reid and Kofranek, 1980).

10.5.1 PREHARVEST FACTORS

For many years research has mainly focused on maximizing vase life (VL) during the postharvest period. However, the physiological and anatomical characteristics that ultimately determine the VL potential of the cut flower are formed during the preharvest period (genotype × growth environment interaction (Fanourakis et al., 2012b). Understanding the effect of growth environment on the VL potential and its interaction with the genetic background would contribute to the efficient utilization of an optimal combination of growth conditions and genotype, with the aim of maximizing the postharvest performance of cut flowers.

10.5.1.1 VARIETIES

Many commercial cut flowers are patented cultivars, characterized by specific attributes such as color, form, disease resistance, and size. Sometimes, breeders fail to consider other commercially important attributes. For example, some of the modern *Alstroemeria* cultivars have wonderful flowers, but their display life is short because of rapid leaf yellowing under commercial conditions. (Van Der Meulen-Muisers et al., 1997).

10.5.1.2 ENVIRONMENTAL FACTORS

The environmental factors include season, light intensity, atmospheric temperature, photoperiod etc. The effect of some important environmental factors on vase life of cut flowers has been discussed below:

10.5.1.3 LIGHT INTENSITY

Supplemental assimilation light (AL) is frequently used to control both the photoperiod and light intensity in greenhouses in northern Europe (Heuvelink et al., 2006; Hemming, 2011). It is often applied in horticulture to enhance productivity, but also to reduce the visual variation in flower quality throughout the year (Fjeld et al., 1994). A number of different AL sources is commercially available such as fluorescent tubes, high pressure sodium lamps and more recently light emitting diodes that differ in the light quality they emit (Hogewoning et al., 2010; Savvides et al., 2012). High pressure sodium lamps are the most commonly used AL sources in protected cultivation, due to their highest energy efficiency (van Ieperen and Trouwborst,

2008), while light emitting diodes are currently gaining importance (van Ieperen, 2012). The number of crops, where AL is applied, is increasing, and the applied light levels are reportedly higher compared to five years ago (Heuvelink *et al.*, 2006). Additionally, higher light intensities are realized in modern greenhouses through newly introduced designs (less constructional elements resulting in less shading), cover materials with higher light permeability and materials increasing light transmission inside the greenhouse (e.g. white ground cover; Hemming, 2011; Max *et al.*, 2012). Another emerging trend in horticulture is greenhouse cover materials that influence the directional quality of light, by scattering it, without reducing its intensity (Hemming *et al.*, 2008). Diffuse light penetrates deeper into the canopy (i.e.more uniform vertical distribution of light) as compared to direct light, enhancing production in various ornamental crops (Hemming *et al.*, 2008; Markvart *et al.*, 2010), including roses (Victoria *et al.*, 2012).

10.5.1.4   PHOTO PERIOD

Continuous light (i.e. 24 h light period) induces visible injuries (e.g. chlorosis and necrosis) in several plant species such as tomato and eggplant, while in others, including roses, it does not lead to injuries but enhances productivity (Velez-Ramirez et al., 2011). Due to its yield promoting effect, it is sometimes applied in cut rose crops (Mortensen and Gislerod, 1996). Mortensen and Fjeld (1998) showed that extending the photoperiod from 16 to 20 h did not affect the VL of cut roses. However, a further extension of the photoperiod to continuous light resulted in a shorter VL (5–47%, depending on the cultivar). An increase of the photoperiod from 12 through 16–20 h did not affect night time stomatal conductance, indicating no effect on stomatal responsiveness (Mortensen and Fjeld, 1998).

10.5.1.5   RELATIVE HUMIDITY

Relative air humidity (RH) is the ratio of the amount of water vapor in the air relative to the amount of water vapor that would be present at saturation, and it is routinely measured in many greenhouses. RH depends on both the amount of moisture available and air temperature (T_{air}). Vapour pressure deficit (VPD) combines the effects of both RH and temperature and is the difference between saturation vapor pressure and actual air vapor pressure. VPD characterizes the evaporative demand of air and it is one of the key drivers of transpiration. However, RH remains the commonly used measure in horticulture, and it is also used in the relevant literature.

High RH is common in greenhouses especially during the winter, mainly due to low ventilation (Max et al., 2009). High RH-grown cut roses had higher rates of water loss, compared to roses grown at moderate RH, as a result of less responsive stomata to both water stress and darkness. High RH has been shown to affect neither stem hydraulic conductivity nor its recovery after artificial induction of air emboli at the cut surface (Fanourakis et al., 2012a). Torre et al. (2003) reported that leaves of high RH-grown plants show reduced density of vascular tissue (expected to increase leaf hydraulic resistance (Nardini et al., 2012) and increased intercellular air spaces (expected to decrease leaf hydraulic resistance; Cochard et al., 2004; Sack and Holbrook, 2006). However, the degree and the relative importance of these responses have as yet not been quantified. Leaf hydraulic resistance data will yield a deeper understanding of the consequences of a more humid environment on the leaf water balance.

10.5.1.6   GENETIC FACTORS

The genetic makeup of a cultivar exerts a pronounced effect on vase life of cut flowers. Large differences in vase life have been observed among cultivars that were grown and tested under identical conditions. For instance, in a study on roses VL ranged between 13 and 24 days depending on the cultivar (Marissen and Beninga, 2001). A large number of studies examining genotypic variation in VL, by comparing different cultivars grown under different environmental conditions, showed considerable differences (Macnish et al., 2010a; Borda et al., 2011). However, the genetic and/or environmental factors that account for these differences cannot be distinguished. The reasons underlying genotypic variation in VL are not fully understood. It is common knowledge that the postharvest life of flowers varies enormously, from the ephemeral flowers of the daylily to the extremely long-lived flowers of some orchid genera. Less extreme, but still marked variations are also seen within genera and even species, and certainly this variation provides a great opportunity for breeders to develop longer lasting flowers. Color, form, productivity, and disease resistance continue to be the targets of breeding programs. This can be seen by comparing the postharvest life of different cultivars from the same breeder. Elibox and Umaharan (2008) reported that vase lives of *Anthurium* cultivars ranged from 14 to 49 days. A simple model, based on abaxial stomatal density and flower color accurately predicted the relative vase life ranking of different cultivars, providing an excellent tool for future breeding. Variations in other important postharvest characteristics have also been reported, for example, for ethylene sensitivity in carnations (Woltering and van Doorn 1988; Wu *et al.* 1991; Reid and Wu 1992) and in roses (Macnish et al. 2010c). A difference in vase life of modern rose cultivars has been observed from 5 to 19 days (Macnish et al., 2010c). Five out of the 38 cultivars tested were insensitive to ethylene indicating the breeding opportunities not only for extending vase life, but also eliminating the problem of ethylene-induced senescence and abscission. Mokhtari and Reid (1995) analyzed the difference in vase life between two rose cultivars, and noted several morphological and anatomical characteristics that correlated with improved water uptake and longer vase life.

Clements and Atkins (2001) characterized a single-gene recessive mutant of *Lupinusangustifolius* L. 'Danja' in which no organs abscise in response to continuous exposure to high concentrations of ethylene. A long-lived Delphinium mutant (Tanaseet al. 2009) also showed no ethylene-induced sepal abscission. These mutants indicate the opportunity for a genetic approach to prevent flower abscission and petal abscission that is a common postharvest problem in cut flowers.

The effects of some other pre harvest factors have also been reported by many workers. The positive effect of preharvest citric acid sprays on post- harvest longevity of cut flowers has been reported on tuberose (Eidyan,2010) and lilium (Darandeh and Hadavi, 2012).

10.5.1.7   TRANSGENIC STRATEGIES FOR EXTENDING THE VASE LIFE OF FRESH CUT FLOWERS

Although floriculture crops have been a target for transgenic manipulation, the primary focus of commercial activities has been changing flower color, especially to produce "blue" roses. The fact that the products of these efforts are now commercially available indicates the potential for using transgenic approaches to modify other features of floral crops. Floral crops offer several advantages for commercialization of transgenic approaches. The high value of floricultural crops, the diversity of taxa to which the same transgenic approaches can be applied, and the relatively short life cycle of these crops all argue for the value of a transgenic approach to plant improvement as opposed to the time-consuming approaches of conventional breeding. Since ornamentals are nonfood crops, registration of transgenic plants is much less cumbersome and expensive than for food crops, and consumer acceptance has already demonstrated by the transgenic "Moondust" carnations and "Applause" blue roses. Indeed, it seems that ornamentals can be an excellent pilot programfor demonstrating the value and safety of transgenic breeding in horticultural crops. Transgenic carnations expressing an antisense fragment of ACO have a longer shelf life than untransformed flowers (Savin et al. 1995). Flowers of transgenic Petunia plants transformed with high homology antisense gene fragments for broccoli ACS and broccoli ACO display delayed senescence (Huang et al. 2007). Likewise, VIGS of ACO in petunia resulted in extended flower life (Chen *et al.* 2004). These findings have limited commercial value, since inhibiting ethylene biosynthesis does not prevent perception of the exogenous ethylene that is a common contaminant in supermarkets and homes.

The success of transgenic approaches to extending flower life are an indication of future prospects for using such approaches to improve the postharvest performance of all floricultural crops, including cut flowers. By using an inducible expression system, or a drought-responsive promoter, it will be possible to reduce the negative effects of limited water supply on the postharvest life of ornamentals. Similarly, the beneficial effects of CKs in improving flower opening and delaying leaf senescence that have long been reported in ornamentals can be obtained by transgenic modification of CK biosynthesis. Gan and Amasino (1995) demonstrated that leaf senescence could be delayed in transgenic plants expressing isopentenyl-transferase (IPT), an enzyme that catalyzes the rate-limiting step in CK synthesis. Chang et al. (2003) demonstrated that overexpression of IPT under the control of the promoter from a senescence-associated gene (SAG12) in petunia resulted in a 6–10 day delay in floral senescence, relative to wild-type (WT) flowers. Flowers from IPT plants were less sensitive to exogenous ethylene and required longer treatment times to induce endogenous ethylene production, corolla senescence, and up-regulation of a senescence-related cysteine protease. The IPT transgene might also be deployed to reduce the negative effects of postharvest water stress in potted plants. Rivero et al. (2007) demonstrated, in tobacco, that suppression of drought-induced leaf senescence in transgenic plants where IPT is driven by a drought-stress promoter resulted in outstanding drought tolerance of the transgenic plants, as well as minimal yield loss when the plants were watered with only 30 percent of the amount of water used under control conditions.

10.6   HARVEST FACTORS

The most important factors for harvest are when, how and "when" the plant material will reach the optimum stage of development and "when" during the day to harvest. Each plant material has its own best harvest stage which can vary depending on the use of, and market for, the plant material. Materials for preserving usually are harvested more mature than those for fresh, wholesale markets. Some general rules of thumb for when to harvest are:

(i)   Spike-type flowers— harvest when one-fourth to one-half of the individual florets are open;

(ii) Daisy-type flowers— harvest when flowers are fully open.

The other important thing is the best time of day for harvesting flowers. The besttime is the coolest part of the day and when there is no surface water from dew or rain on the plants. Right stage, method, and time of harvesting of flowers are of considerable importance to ensure their long vase-life. The stems should be cut with sharp knives or secateurs. Hardwood stems should always be given slanting cut to expose maximum surface are to ensure rapid water absorption. The flowers of dahlia and poinsettia release latex upon cutting. To overcome such problem, stems should be given a dip in hot water (80–900C) for a few seconds. The flowers of rose, carnation, gladiolus, tuberose, daffodils, lily, iris, freesia, and tulip should be harvested at bud stage since their buds continue to open in water. The flowers of snapdragon, Harvesting of flowers at bud stage is always preferred as their buds have long vase-life, are less sensitive to ethylene, easy to handle during storage and transport and are less prone to diseases and pests.A fresh cut flower remains a living specimen regardless of its removal from the mother plant. Its potential vase life, even at its peak harvest quality is short (Blessington, 2004). Thequality of flowers at harvest is set and can only stabilize or decrease.

A significant effect of harvesting time has been found on the severity of chilling symptoms. In *Heliotropium arborescens* and *Lantana camara* cuttings, Friedman, and Rot (2005) demonstrated that cuttings harvested in the morning were more sensitive to storage at chilling temperatures than those harvested at noon. Generally cut flowers are harvested at different heights or at different nodes (in carnations) without knowing its impact on growth and flower production in successive crop. To induce early sprouting of buds and transformation of laterals, the levels of harvest plays an important role and also have an impact on number of buds sprouting at the bottom or top of the left over harvested shoots, which finally determines the number of flower stalks produced per harvested stalk. The buds sprouted at different levels have direct impact on the quality of flowerstalk and flower bud.

Chandra Sekhar et al. (2013) reported a difference in vase life of cut flower of carnation due to harvesting at different heights of flower stalk and cultivars (Table 10.2). Harvesting of flower stalk at 10 cm height recorded maximum vase life of cut flower (11.83 days). Among the cultivars, cv. Domingo registered maximum vase life of cut flower which was significantly superior to rest of the Cvs. Keiro and Dover. Flower stalks harvested at lower levels recorded maximum vase life than flower stalks harvested at higher levels. It might be due to better quality flowers produced by harvesting at lower levels which has maximum length and diameter of flower.

TABLE 10.2   Effect of harvesting at different heights on length of flower stalk on vase life of cut flower (days) in three cultivars of carnation (Chandra Sekharet al., 2013)

Height from the ground at the time of harvesting	Stem length				Vase life			
	Domingo	Keiro	Dover	Mean	Domingo	Keiro	Dover	Mean
5	79.33	71.86	73.43	74.87	11.20	10.06	8.71	9.99
10	98.30	80.33	84.36	87.66	14.00	11.50	10.00	11.83
15	92.29	79.53	81.30	84.37	12.00	10.84	10.16	11.00
20	84.80	76.70	77.83	79.77	11.80	10.76	9.00	10.52

10.7   POSTHARVEST FACTORS

Postharvest factors are grouped into (a) harvest factors, (b) cutting and conditic ing methods, and (c) test room conditions and vase life terminating symptom These factors exert a considerable effect on storage and vase life of cut flowe There are many factors that can interact to reduce fresh flower vase life. The include: carbohydrate depletion, attack by bacteria and fungi, normal maturati and aging, wilting caused by water, stress and xylem blockage, bruising and crus ing, temperature fluctuations between storage and transit, color change (bluing accumulation of ethylene, poor water quality and suboptimal cultural practices conditions. Storage at the cool temperature helps to delay the normal senescenc bacterial and fungal attack and bluing. Floral preservatives, meticulous handli and proper sanitation practices help prevent carbohydrate depletion, poor wat quality, bruising or crushing, wilting, and incursion of microorganisms. Growe must be aware of these obstacles and be prepared to address them with corre postharvest handling procedures.

10.7.1 STORAGE TEMPERATURE

The rates of development and senescence of cut flowers are strongly influenc by temperature. For Example between the normal storage temperature (0°C) a room temperature (20°C), the respiration of roses and carnation increases appro mately 25-fold. Relatively short exposures to elevated temperatures can therefc greatly reduce the overall storage or vase life of cut flowers. Proper temperatu management is obviously the primary goal in upgrading their handling. Most c flowers are cooled by simply placing them, packed or unpacked into a cool roo A need for rapid cooling of large volumes of flowers prior to truck transportati has led to the construction of forced- air pre coolers. Cut flowers originating fro tropical and sub tropicqal regions may be deleteriously affected by temperatur below about 12.5°C. The marked effects of temperature on the life of cut flowe were first quantified in 1973 (for carnations) by Maxie et al. (1973). Respirati of flowers has a very high Q10 value, higher than that of most other perishab crops. The Q10 for Narcissus, has been found to be more than 7 between 0 a 10°C (Cevallos and Reid 2001). The close link between respiration and grow and senescence in these poikilotherms means that a Narcissus flower held at 10 may lose as much vase life in 1 day as does a similar flower held for one week 0°C.

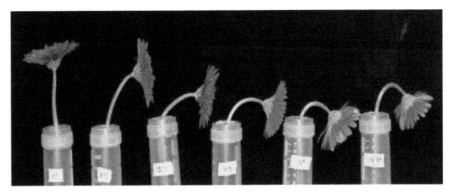

FIGURE 10.5 Effect of storage temperature on quality of fresh cut gerbera flowers after storage.

Reid (2001) reported that storage temperature adversely affected the storage quality of fresh cut gerbera flowers (Figure 10.5).

Although it is recommended that most ornamentals should be stored close to the freezing point (0°C), there are some exceptions also, including the tropical cut flowers such as anthurium, heliconias, and gingers, most foliage plants, and some important flowering plants (including *Poinsettia* and African violet). These tropical species must be transported and handled at temperatures above 10°C. Symptoms of exposure to chilling temperatures include wilting, necrosis, and browning of colored bracts and petals. Some flowers, such as Zinnia, Celosia, and Cosmos perform better when stored at temperatures above 0°C (Dole et al. 2009).

10.7.2 RESPIRATION RATE

Cut flowers have extremely high rates of respiration, and respiration increases exponentially with temperature (Table 10.3) with Q10 values that range from 1.5 to as high as 7. Different cultivars of the same species may have quite different respiration rates and may respond differently to temperature.

TABLE 10.3 Q_{10} values and respiration rates (ml CO_2Kg^{-1} h^{-1}) relating respiration to storage temperature of selected cut flower species and cultivars (Cevallos, 1998).

Flower Species/Cultivar	Q_{10} 0–10°c	Q_{10} 10–20°c	Respiration 10°c
Anemone mona Lisa	3.08	2.45	163.0
Aster' Matsumoto	3.16	2.90	117.3

TABLE 10.3 *(Continued)*

Flower Species/Cultivar	Q_{10} 0–10°c	Q_{10} 10–20°c	Respiration 10°c
Calla Lily	2.95	2.21	62.8
Carnation "Imperial White"	3.10	2.66	133.9
Carnation Ruri	4.46	2.78	145.2
Daffodil King Alfred'	6.76	2.97	151.5
Gerbera	2.94	2.88	125.4
Iris 'Telstar	3.37	2.76	83.5
Killia Daisy	2.97	2.30	86.7
Lisianthus	3.67	3.44	67.3
Jonquil "Geranium"	3.86	2.85	174.7
Narcissus "Paper white	3.48	2.61	146.6
Ranunculus	3.13	2.24	236.1
Rose "Ambiance"	4.13	3.85	109.9
Rose "Cara Mila"	5.74	3.17	134.0
Rose "Fire and Ice"	5.35	3.05	155.7
Rose "First Red"	5.67	3.00	108.0
Rose "Kardinal"	4.11	2.71	149.9
Rose"Preference"	4.82	4.76	59.1
Rose "Raphaella"	2.40	1.54	128.3
Rose "Tineke"	2.39	2.78	73.0
Snapdragon	2.65	2.53	210.5
Tulip	3.32	2.98	180.0

A close association between flower respiration during storage and vase life after storage shows the potential usefulness of controlled (CA) or modified (MA) atmospheres, in which the O_2 content of the storage atmosphere is reduced, sometimes with an increase in the CO_2 content. Such atmospheres have been tested with cut flowers, results have been disappointing (Reid 2001). Commercial trials have failed to demonstrate benefits, and where such benefits have been claimed, the absence of valid controls has clouded the credibility of the results. Recent studies have focused on the use of sealed packages for single flowers or small bouquets but they only apply to specific species or even varieties of flowers and are therefore of very little general utility. The reasons for this have not been examined, but it seems possible that the difference in surface/volume ratio between bulky fruits and thin petals may be part of the reason for the difference. O_2 diffusion is likely to be limiting in bulky fruit, so that the terminal oxidases are limited for O_2 at much higher external O_2 than their actual Km would suggest. Curiously, too, the flowers failed to show the rise in respiration at very low O_2 levels (the Pasteur effect) that results from the onset of anaerobic respiration. Increased CO_2 production under anaerobic conditions has been attributed to increased glycolysis and the increased decarboxylation of phosphoenyl pyruvate. Although we have no explanation for the absence of a Pasteur effect, it is clear that low $O2$ results in anaerobic respiration, since flowers stored under anaerobic conditions can smell alcoholic and may collapse shortly after placing at room temperature in air (Macnish et al. 2009).

10.7.3   WATER RELATION

Cut flowers differ markedly from other perishable commodities in their water relation. In most cases they are very susceptible to desiccation, due both to transpiration from leaves and to their high surface area to volume ratio. Water lost during the postharvest period can normally be replaced from vase solution when the commodity enters the retail distribution system. Adequate water relation in harvested cut flowers is an obvious and important element of their postharvest management. Water balance is determined by the differential between water supply and water loss, and optimal postharvest handling includes managing both sides of this relationship. The primary tool in reducing water loss is temperature control. The water content of saturated air rises in an exponential fashion (doubling for every 11°C). Depending on the humidity, therefore, water loss can rise with temperature in a similar fashion. Sealed bags, or perforated polyethylene wraps can maintain higher humidity and thus reduce water loss after harvest, but at higher temperatures the likelihood of condensation and attendant proliferation of diseases is greatly accentuated. Intuitively, providing adequate water to a cut flower should be an easy matter, since the vase solution has direct access to the xylem, without the need for transport from the soil and across the tissues of the root. In practice, water uptake is frequently impeded, by the desiccation that occurs during extended dry handling of the flowers, by air emboli that form when the water column in the xylem is broken, and very commonly by microbial occlusion and/or the formation of physiological plugs, tyloses, and gels (van Doorn and Reid 1995). Differences among species and even between varieties of the same species are a function of the structure of the xylem, the size of emboli and cavitations, embolism repair ability (Brodersen et al. 2010) and the likelihood of colonization of the stem by microbes.

Desiccation. Desiccation is one of the most Important postharvest problems in the handling of cut flowers, the prime example being "bent neck" a disorder of cut roses where the water needs of the foliage and flower are provided at the expanse of the relatively unsclerified stem tissue just below the flower (Reid and Kofranek, 1980) Present information on the movement of water in the stems of cut flowers is sketchy But it is known that it can be strongly affected by the composition of the vase solution. Although floral tissues, devoid of functional stomates, lose relatively little water themselves, water loss can occur rapidly through the stomata of stems and leaves during postharvest handling. Surprisingly, the opening and vase life of flowers, at least in roses (Macnish *et al.* 2009) and gypsophila (Rot and Friedman 2010), is not affected unless desiccation is in excess of 15% of the fresh weight. In their study, Rot and Friedman used the apoplastic fluorescent dye 8-hydroxypyrene-1,3,6-trisulfonic acid (HPTS) to measure water uptake by florets and whole stems. These dye studies verified the effects of anionic detergents (such as Triton X-100) in improving water uptake in dehydrated flowers (Jones et al. 1993).

Emboli. Formed immediately on cutting the stem, as tension in the xylem water column is released, emboli can result in a temporary reduction in water uptake that may become permanent if the rate of transpiration exceeds the water conductance of the embolized stem. In some taxa, such as Heliconia spp. that have very long vessels, the embolism can result in permanent failure of conducting elements. More often, the emboli are resorbed by the xylem (apparently by water influx from surrounding living cells, with individual droplets expanding over time, filling vessels, and forcing the dissolution of entrapped gas (Brodersen et al. 2010). Detergent dips or vase solutions, low pH, and hydrostatic pressure all overcome emboli in the stem, as, of course does recutting under water—one of the very traditional practices in the floral trade. We have tested deep water treatments, in which the stem is immersed in water, containing a biocide, that is at least 50 cm as much as 1 m deep. This pretreatment improved rehydration and vase life of recalcitrant cut flowers, such as heliconia, ginger, and a range of woody species (Reid and Kofranek, 1980).

10.7.4 PH OF VASE SOLUTION

Acidic solutions, for example, move much more readily through the stems of cut flowers than solution which are neutral or alkaline. Plugging of the cut surface of the stem, whether by microbes contaminating the vase solution, particulate or Colloidal material in the water or exudations from the cells surrounding the conducting tissues in the stem, are considered to be a major limitation to the vase life of many cut flowers. Because of the dramatic effects of water quality (PH, dissolved solids, gases, particulate, and colloidal matter) on the vase life of cut flowers, provision of good quality water is an important part of postharvest handling of cut flowers. Many operators therefore use ion exchange resin deionizers to provide water of adequate quality for postharvest use. Considerable improvement in hydration of cut flowers can be achieved with any water supply by simply adding sufficient acid to reduce the pH to 3–3.5. In practice citric acid is a good additive because it produces this pH without danger of the pH falling lower if too much is used. The salts of 8-hydroxy-quinoline and aluminum sulfate are commonly used in commercial flower preservatives. As biocides but their effectiveness probably also relates to the low pH of their solutions. With improved temperature management, the handling of cut flowers could change In the future to a dry handling system where the flower is not put in to water from the time of harvest until it enters the retail distribution system. This goal cannot be achieved without modifications of systems of harvesting, handling, and storage so That water loss is reduced to a minimum. Specifically, reduction of the time from harvest until the commodity reaches the proper storage temperature by using forced-air precooling will play a major role. It will also be important to prevent desiccation during storage; high humidity storage systems, humidifier, and containers with vapor barriers may all be important, if combined with adequate control of fungal diseases (Reid and Kofranek, 1980).

10.7.5 MICROORGANISMS

A rapidly respiring and wounded stem placed in water quickly depletes the oxygen in the vase solution, providing perfect growing conditions for microbes (yeasts and bacteria) that benefit from the cellular contents released from the cells damaged during cutting. Occlusion by microbes (Figure 1.5) and the extracellular polysaccharide that they elaborate is by far the most common cause of poor water relations in cut flowers (Macnish et al., 2008). The standard treatment for avoiding these events is the use of bactericides ($HClO_4$, $Al_2(SO_4)_3$, and quaternary ammonium compounds are among the most popular). Reduction in pH of the solution [citric acid, $Al_2(SO_4)_3$] is also helpful in reducing bacterial growth but is insufficient on its own, since acidophilic yeasts and bacteria can quickly colonize a vase or bucket solution. Not all bacteria are deleterious in the vase solution. Zagory and Reid (1986), demonstrated that some of the microbial species isolated from vase solutions have no affect (or may even augment) the life of carnations, roses, and chrysanthemums, but the potential of a biological control system for avoiding the effects of bacteria and yeasts in the vase solution has not been explored. Improper temperature management, including episodes of cooling and warming in the absence of proper precooling techniques, results in condensation and accelerated growth of pathogens on delicate petals and other floral parts, particularly when the flowers are packed under conditions that limit air movement. B. cinerea, a relatively weak pathogen, is the major pathogen of these products, and a range of chemicals have been used for postharvest protection. Wherever excessive moisture occurs the microoragnsm is found there. Although the mold can be controlled by fungicides, proper environment management is the best method (Nowak and Rudnicki, 1990). The push for organic or sustainable production, and the loss of established chemicals has led to an effort to identify alternative strategies for controlling disease. As noted above, high CO_2 levels provide effective control for species whose leaves (or petals) are not damaged by the gas. Same results have been obtained with SO_2 (Hammer et al. 1990) which provides good control of the pathogen, but damage to the host. Recently, Macnish et al. (2010d) reported the efficacy of a simple dip in a solution of $NaHClO_4$, which performed as well as the commercial fungicides under commercial conditions. Other strategies, including the use of ClO_2 (Macnish et al. 2008) and ozone generators have been tested, but with inconsistent results (Reid and Kofranek,1980). Methyl jasmonate (MJ), a natural plant growth regulator, has been tested for postharvest control of *B. cinerea* in cut flowers of a range of rose cultivars (Meir *et al.* 1998). Pulse applications of 200–400 mM MJ following either natural or artificial infection seemed to provide systemic protection. MJ applications significantly reduced lesion size and appearance of the infection apparently due to inhibition of *B. cinerea* spore germination and germ-tube elongation. Effective concentrations of MJ caused no loss of flower quality or longevity.

Flowers and foliage packed moist after harvest are very susceptible to a no of disease causing organisms. Condensation of water on flower or foliage encourages diseases. Avoid moving flowers directly from cool to warm rooms which otherwise will result in droplet formation (Reid and Lukaszewski, 1988).

10.7.6 PLANT GROWTH REGULATORS

10.7.6.1 ETHYLENE

Ethylene is certainly principal among the hormones affecting flower longevity, but other hormones can affect sensitivity to ethylene, and a large group of flowers is insensitive to ethylene. Contamination of storage and display areas with ethylene gas (e.g. in supermarkets) can cause rapid senescence of several flower crops, particularly carnations (Reid and Kofranek, 1980). The role of endogenous ethylene in triggering senescence has been well documented by a range of studies reporting the dynamics of ethylene production, changes in activity of the biosynthetic enzymes (Bufler 1984, 1986) and up-regulation of the genes encoding these enzymes (Woodson et al. 1992). The key role of ethylene has been corroborated by studies with long-lived carnation cultivars (Wu et al. 1991) and with transgenic or VIGS constructs silencing the biosynthetic pathway (Chen et al. 2004). The discovery that the action of ethylene could be inhibited by Agþ (Beyer 1976) and the subsequent development of the stable, nontoxic, yet effective silver thiosulfate complex (Veen and van de Geijn 1978) has provided an important commercial tool, still in widespread use, for preventing ethylene-mediated senescence and abscission in cut flowers and potted plants. Other inhibitors of ethylene synthesis (aminoethox-yvinyl glycine, aminooxyacetic acid, Coþþ) and action (2,5-norborna-dienehave also been found to be effective to varying degrees. The essential steps of the ethylene biosynthesis pathway have been completely identified in cut flowers. ACC (1-aminocyclopropane-1-carboxylic acid) content, ACC synthase, activity of the ethylene forming enzyme (EFE) and of /I-cyanoalanine synthase have all been shown to rise coordinately with the onset of the ethylene climacteric burst in various flower parts (Figure10.6). In a study of the control of ethylene biosynthesis during the climacteric rise, Whitehead et al. (1984) showed that, unlike some fruit tissues (where EFE activity is always high), the activity of the EFE in senescing carnation petals rises coordinately with the increase in synthesis of ACC.

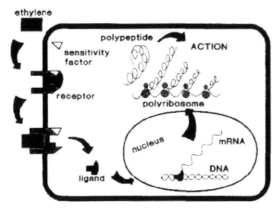

FIGURE 10.6 Hypothetical scheme for action of ethylene in induction of flower senescence (Reid and Wu, 1992).

The end of the relatively short life of carnations held in air is associated with climacteric rises in ethylene production and respiration, and coordinate rises in activity of the enzymes of the ethylene biosynthetic pathway. Carnation senescence is associated with depression of specific genes, increased polyribosome activity and major changes in patterns of protein synthesis. Isotopic competition assays indicate the presence ethylene-binding activity in carnation petals with the expected characteristics of the physiological ethylene receptor. Inhibition of ethylene production and/or ethylene-binding results in longer-lived carnations (Reid and Wu, 1992).

Diazo-cyclopentadiene (DACP), a cyclic diolefin with an attached reactive diazo group has been found very effective in inhibiting ethylene action when dissociated with UV light after being applied to the tissue (Sisler and Lallu, 1994). This material has now become a standard treatment for ethylene-sensitive flowers and potted plants (Serek et al., 1994) applied either as a gas in an enclosed space, or through the use of sachets or nanosponges (Seglie et al. 2011) that are placed in boxes prior to transportation. A nonvolatile 1-MCP formulation, N,N-dipropyl (1-cyclopropenylmethyl) amine (DPCA), has recently been successfully tested for improvement of postharvest quality of ornamental crops (Seglie et al. 2010). Spray application of this new formulation could provide a major advantage for handling ornamental crops, since they could be treated prior to harvest in the field or green-house. However, the volatile nature of 1-MCP restricts its application to an airtight environment. The effects of temperature and perhaps differences among species in the persistence of inhibition may reflect differences in the rate of turnover of the ethylene-binding site. In a study on carnation (*Dianthus caryophyllus* L. "White Sim") petals, Reid and Celikel (2008) noted some aspects of the inhibition response that were not consistent with the competitive inhibition model of 1-MCP action. They suggested an alternative model in which 1-MCP binds to a site that is exposed during the allosteric changes that accompany the enzymatic activities of the binding site in the absence of ethylene.

On the basis of their response to exogenous ethylene, pollination, and 1-MCP, flowers have been broadly classified into two groups—ethylene-sensitive and ethylene-insensitive (Table 10.4). However, this classification is undoubtedly too simplistic, since some flowers show an intermediate behavior. For example, in daf-fodil, pollinated flowers or flowers exposed to ethylene senesce rapidly, indicating an ethylene-sensitive senescence pattern (Hunter et al. 2004a). However inhibitors of ethylene action have minimal effect on the senescence of daffodil flowers held in ethylene-free air indicating that natural senescence is initiated by regulators other than ethylene. There is still considerable need for research to identify the role of other hormones in floral senescence.

TABLE 10.4   List of flowers sensitive to ethylene (Reid, 1989).

Agapanthus umbellatus.	*Freesia hybrids*
Antirrhinum majus	*Solidago spp*
Anemone spp	*Lathyrus odoratus*
Matthiola incana	*Delphinium spp.*
Dendrobium spp.	*Centaurea cyanus*
Alstroemeria hybrids	*Dianthus spp.*
Eremerus robustus	*Scabiosa spp*
Rosa spp.	*Gypsophila spp.*
Campanula spp.	*Lilium spp.*
Phlox paniculata	*Astilbe spp.*
Bouvardia hybrids	*Kniphofia uvaria*
Aconitum napellus	

10.7.6.2   CYTOKININS

Cytokinins (CK) have also been reported to extend the vase life of some cut flowers and are very effective in delaying the senescence of leafy tissue (Reid and Kofranek, 1980). Commonly used cytokinin is kinetin (o-pentenyl adenosine iPA) and the application can be spray or dipping method. Dipping the cut stems, in 10–50 mg/l solution of cytokinin for 2 min, can slow the aging process in carnations, roses, irises, tulips, and other flowers. In addition, cytokinin can also be used during storage and transportation of cut flowers, which can reduce the loss of chlorophyll in the dark. The striking effects of CK in delaying senescence of leaves were known long before the first isolation of zeatin. Given the homology between leaves and petals, it is perhaps not surprising that CKs were also found to delay petal senescence (Eisinger 1977) responsible for reducing the sensitivity of the corolla to ethylene (Mayak and Kofranek 1976) and with delaying the onset of ethylene biosynthesis (Mor et al. 1984). Endogenous CK content shows a pattern consistent with its putative role in delaying senescence. Buds and young flowers contain high CK levels, which fall as the flower ages and commences senescence (Mayak and Halevy 1970; Van Staden and Dimalla 1980; Van Staden et al. 1990). The interplay between CK content and senescence in ethylene-sensitive flowers was elegantly demonstrated by Chang et al. (2003), who transformed petunia with a SAG12-IPT construct designed to increase CK synthesis at the onset of senescence in leaves (Gan and Amasino 1995). CK content of corollas in the transformed plants increased after pollination, ethylene synthesis was delayed which resulted in flower senescence delayed by 6–10 days. As in flowers treated with exogenous CKs, the flowers from the IPT-transformed plants were less sensitive to exogenous ethylene and required longer treatment times to induce endogenous ethylene production, and the symptoms of floral senescence.

Leaf senescence is also an important component of loss of quality in floricultural crops, particularly in members of the Liliaceae and commercial pretreatments containing CKs and/or gibberellins are recommended as a prophylaxis in sensitive genera such as *Alstroemeria* and *Lilium*.

Thidiazuron is a phenyl urea compound with cytokinin-like activity which has been discovered as a good substitute for cytokinins (Reid and King, 2003; Bagheri and Sedaghathour, 2013). The chemical delayed aging and yellowing leaves of alstoemeria cut flowers by 60days at concentration above 10nM for 24 hrs (Ferrant et al., 2002). It also delayed the yellowing leaves in cut lilies, chrysanthemums, and tulips (Ferrant *et al.*, 2002 and Ferrant et al.,2003). Pulse treatment of cut *Alstroemeria* stems with as little as 5 mM TDZ essentially prevented leaf yellowing in flowers of the cultivar "Diamond", where yellowing normally starts after 4–5 days (Ferrante et al. 2001). The flowers of *Alstroemeria* are ethylene-insensitive, yet the TDZ treatment had only a minor effect on *Alstroemeria* flower life, although CKs have been shown to increase the life of *Iris*, whose natural senescence is ethylene-independent (Wang and Baker 1979; Mutui et al. 2003). In *Iris*, TDZ treatment at considerably higher concentrations (200–500 mM) significantly improved flower opening and flower life (Macnish et al. 2010b). The treatment was of particular value in that it reduced the loss of vase life that results from cool storage. Most experiments with TDZ have been conducted with flowers that are insensitive to ethylene, but in lupins and phlox, TDZ has been shown to improve flower opening and reduce ethylene-mediated flower abscis-sion and senescence (Sankhla *et al.* 2003, 2005), indicating that TDZ acts like other CKs in decreasing ethylene sensitivity and should be tested on a broader range of ornamentals.

7.6.3 ABSCISIC ACID

ABA has always been reported in the regulation of perianth senescence. Not only have researchers shown a close association between petal senescence and increased petal ABA concentrations (Onoue et al. 2000), but exogenously applied ABA has also been shown to accelerate the senescence of a number of flowers (Arditti 1971; Mayak and Dilley 1976). Such application results in many of the same physiological, biochemical, and molecular events that occur during normal senescence (Panavas et al. 1998).

In ethylene-sensitive flowers such as carnation flowers and roses, ABA-accelerated senescence appears to be mediated through induction of ethylene synthesis, since it is not seen in flowers that are pretreated with ethylene (Mayak and Dilley 1976; Ronen and Mayak 1981; Muller *et al.* 1999). This is consistent with the pattern of endogenous ABA content in rose petals, where the increase in ABA concentration occurs 2 days after the surge in ethylene production (Mayak and Halevy 1972).

ABA presumably induces senescence independently of ethylene (Panavas et al. 1998) in daylilies which are ethylene-insensitive (Lay-Yee et al. 1992). The fact that ABA accumulates in daylily tepals before any increase in activities of hydrolytic enzymes and even before the flowers have opened was an evidence that the hormone may coordinate early events in the transduction of the senescence signal. Application of ABA to pre senescent daylily tepals resulted in a loss of differential membrane permeability, an increase in lipid peroxidation, increase in the activities of proteases and nucleases and the accumulation of senescence-associated mRNAs (Panavas et al. 1998).

During senescence of daffodil flowers, however, Hunter et al. (2002) reported that although ABA accumulated in the tepals as they senesced, it did not appear to play a signaling role in natural senescence. The increase in ABA concentrations in the tepals occurred after the induction of senescence-associated genes. They concluded that the increase in ABA content is therefore most likely a consequence of the cellular stresses that occur during senescence and suggested that the hormone does not trigger senescence, but may help drive the process to completion.

10.7.6.4   OTHER HORMONES AND GROWTH REGULATORS

Gibberellins, auxins, and other plant hormones and regulators have also been shown to have positive and negative effects on floral longevity. For years, auxin was considered an important component of the rapid senescence response of orchids and other flowers to pollination (Arditti 1975), although this is more likely to be a response to auxin-induced ethylene biosynthesis. Gibberellins can delay the senescence of cut flowers thereby extending their vase life. GA3 applied at lower concentrations (25 mg L−1) renders greater beneficial effects on vase life quality, membrane stability and antioxidant activities in gladiolus cut spike, and further higher application rates cause no improvement in the flower longevity. GA3 treatment significantly influences the vase quality attributes and antioxidants capacity of gladiolus cut flowers (Saeed et al., 2013). Saks and Staden (1993) showed an increase in longevity of carnation flowers treated with 0.1 mM gibberellic acid (GA); Eason (2002) found a modest increase in life of *Sandersonia* pulsed with 1 mM GA and Hunter et al. (2004) demonstrated a similar effect on natural senescence of daffodils. Commercially, GA (sometimes in combination with BA) is used in solutions to prevent leaf yellowing in cut bulb flowers and potted flowering plants. GA treatments may have the undesirable side effect of increased stem or scape length. The anti-ethylene effects of silver ion could be utilized in the cut flower industry be preparing a stable complex between silver and thiosulfate (STS) which was mobile in the stem of cut carnations (Veen and Van De Geijin, 1978). The vase life of carnations can be greatly extended by 'pulsing the cut flowers with STS either as a 10 min treatment using 4mM STS at room temperature or as an overnight treatment using 1mM STS at 20°C (Reid and Kofranek, 1980). Naphthelein acetic acid (NAA) as a synthetic auxin can also delay the yellowing of leaves in tulips (Kawa-Miszczak et al., 2005).

10.8   POSTHARVEST MANAGEMENT TECHNIQUES FOR FRESH CUT FLOWERS

There are a series of tasks required after flowers are harvested, commonly referred to as handling, that are performed in order to prepare the flowers for market. These steps include: immediate hydration, grading, leaf removal, bunching, recutting, hydration, special treatments, packing, precooling, cold storage, and delivery to market. The specific steps required for each flower depend on the flower market being sold to. The location and exact procedures for each step depends on the market and the grower's facilities (Gast, 1997).

10.8.1   HARVESTING

Harvesting of the cut flowers should be done early in the morning and harvested flowers should be cooled immediately. In morning hours temperature is generally low which less affect the quality of the flowers. Just before harvesting of cut flowers, plants should be carefully inspected for the proper picking stage.

Flowers should be harvested when the proper stem length and inflorescence required for sale in the wholesale market are reached. Minimum harvest maturity for most of the cut flowers is the stage at which harvested buds can be opened fully and have satisfactory display life after distribution. Many flowers are best cut in the bud stage and opened after storage, transport or distribution. This technique has many advantages, including reduced growing time for single-harvest crops, increased product packing density, simplified temperature management, reduced susceptibility to mechanical damage and reduced desiccation. Many flowers are harvested when the buds started to open (rose, gladiolus), although others are normally fully open or nearly so (chrysanthemum, carnation). Flowers for local markets are generally harvested at more open stage than those intended for storage and/or long-distance transport.Stems with more than one flower are usually harvested with less than one-third of the flowers fully opened. Harvesting stage of some important cut flowers has been given in Table 10.6. The flower stem is cut at the appropriate length by hand with a sharp knife or pruning shears. Since flower condition will not improve after picking, farmers must calculate the cutting precisely so that the flower will not be past its prime when it reaches the consumer.There are automated systems that can strip leaves, trim stems, and uniformly bunch flowers that may then be transported on speciallydesigned monorails suspended on tracks from the ceiling of the greenhouse. With automated systems, surfaces are padded to minimize damage to the flowers. In the packing warehouse, stems are cut for a second time while submerged in water to allow the water to move up the stems. They are immediately placed in tepid water (110 degrees) with added floral preservative for at least 2–3 hrs to allow for a maximum amount of water uptake. They can be left temporarily in a cool (less than 60 degrees) location or stored in a 40 degree cooler overnight for subsequent grading. Flowers that do not retain water and are not kept at low temperatures will lose water and wilt quickly (Table 10.5)

TABLE 10.5   Optimal harvesting stage of cut flowers for direct sale

Species	Stage of Development	Species	Stage of Development
Freesia hybrids	First bud beginning to open	*Gaillardia pulchella*	Fully open flowers
Agapanthus umbellatus	1/4 florets open	*Gerbera jamesonii*	Outer row of flowers showing
Alstroemeria hybrids	4-5 florets open	*Helianthus annuus*	Fully open flowers

TABLE 10.5 *(Continued)*

Species	Stage of Development	Species	Stage of Development
Anthurium spp.	Spadix almost fully developed	*Hippeastrum hybrids*	Colored buds
Gladiolus cultivars	1–5 buds showing color	*Lathyrus odoratus*	1/2 florets open
Gloriosa superb	Almost fully open flowers	*Lilium spp.*	Colored buds
Campanula spp.	1/2 florets open	*Cyclamen persicum*	Fully open flowers
Chrysanthemum spp.	Fully open flowers	*Cymbidium spp.*	3–4 days after opening
Chrysanthemum morifolium		*Dahlia variabilis*	Fully open flowers
Standard cultivars	Outer petals fully elongated	*Delphinium spp.*	1/2 florets open
Spray cultivars		*Dendrobium spp.*	Almost fully open flowers
Singles	Open but before anthesis	*Dianthus barbatus*	1/2 florets open
	Open but before disk flowers	*Dianthus caryophyllus*	
Anemones	Start to elongate	Standard cultivars	Half-open flowers

Species	**Stage of Development**	**Species**	**Stage of Development**
Pompons and	Center of the oldest flower	Spray cultivars	2 fully open flowers
Decorative	Fully open	*Digitalis purpurea*	1/2 florets open
Ornithogalum spp.	Colored buds	*Narcissus spp.*	"goose neck" stage
Papaver spp.	Colored buds	**Rosa hybrids**	
Phlox paniculata	1/2 florets open		First 2 petals beginning to
Polianthes tuberose	Majority of florets open		unfold, calyx reflexed below a
		Red and pink cv.	horizontal position

TABLE 10.5   *(Continued)*

Ranunculus asi-aticus	Buds beginning to open		
Strelitzia reginae	First floret open		
Zantedeschia spp.	Just before the spathe begins to turn down-ward	Yellow cultivars	Slightly earlier than red and pink
Tulipa gesneriana	Half-colored buds	White cultivars	Slightly later than red and pink

10.8.2   PRE COOLING

The most important part of maintaining the quality of harvested flowers is ensuring that they are cooled as soon as possible after harvest and that optimum temperatures are maintained during distribution. Most flowers should be held at 0 to 1 °C (32 to 33.8 °F). Chilling-sensitive flowers for e.g. *Anthurium*, bird of paradise, ginger, tropical orchids etc should be held at temperatures above 10 °C (50 °F). These are commonly stored at temperatures between 10 and 12.5°C. Once packed, cut flowers are difficult to cool. Their high rate of respiration and the high temperatures of most greenhouses and packing areas result in heat buildup in closed containers unless measures are taken to ensure temperature reduction. It is therefore necessary to cool the flowers as soon as possible after packing (Reid, 2001).

Individually, flowers cool and warm rather rapidly. However, individual flowers brought out of cool storage into a warmer packing area will warm quickly and water will condense on the flower and inside the packages. The simplest method of ensuring that packed flowers are adequately cooled and dry is to pack them in the cool room. Although this method is not always popular with packers, and may increase labor cost and slow down packing somewhat, it will ensure a cooled, dry product. Forced-air cooling of boxes with end holes or closeable flaps is the most common and effective method for precooling cut flowers. Cool air is sucked or blown through the boxes. Care must be taken to pack them so that air can flow through the box and not be blocked by the packing material or flowers. In general, packers use less paper when packing flowers for precooling. The half-cooling time for forced-air cooling ranges from 10 to 40 min, depending on product and packaging. Flowers should be cooled for three half-cooling times (by which time they are 7/8 cool) (See section on temperature). If the packages are to remain in a cool environment after precooling, vents may be left open to assist removal of the heat of respiration. Flowers that are to be transported at ambient temperatures can be packed in polyethylene caskets, foam-sprayed boxes or boxes with the vents resealed. Ice that is used after precooling is only effective if placed to intercept heat entering the carton (ie., it must surround the product), and care must be taken to ensure that the ice does not melt onto flowers or cause chilling damage. Precooling of vertical hampers or Proconas presents a particular challenge, but can be achieved using a 'tunnel' forced-air cooling system

10.8.3   GRADING

Grading of cut flowers is done to ensure consistent standards. For many years uniform standard grades have facilitated pricing and made crop reporting and market information more meaningful for fresh produce. Flowers have been sorted into "best,""next best," and "culls" almost since the beginning of wholesale flower marketing. Such sorting became a necessity when it was found that including flowers of inferior quality in the sales unit decreased the value of those of superior quality. Stems are generally graded by stem length (18–24 inches for most flower types) and are downgraded for short or broken stems, poor flower condition, poor foliage condition, or old flowers. Sorting machines are able to grade flowers by length of stem, however, all other factors are still determined by human decision-making. The flower buds are wrapped in cone-shaped plastic sleeves to prevent damage. Stems are then tied together with string and parchment, or waxed paper is wrapped around the heads for protection. Flowers that are destined for nationwide sale are carefully prepared to maintain flower quality. Boxes are often packaged using "wet packs" to allow the flowers to remain in water throughout their transport. Packed boxes can be precooled by units that fill them with 98 percent humid, cool air for added protection. The boxes are then transported by refrigerated truck or by air to customers (Reid, 2001).

TABLE 10.6   Minimum guidelines and standards for cut flowers (Anonymous, 2009)

Flower Type	Grade	Length (cm)	Flower/ stem	Stem/ Bunch	Bloom Size (cm)	Bunch Weight (g)	Min. Diameter at Center (cm)
Alstroemeria	Super select	75	≥4	10	-	-	-
	Select	65	≥3	10	-	-	-
	Fancy	60	≥3	10	-	-	-
Aster	-	70	-	7-15	-	-	-
Strelitzia	-	100	-	-	≥16	-	-
	-	80	-	-	≥13	-	-
	-	60	-	-	≥10	-	-

TABLE 10.6 *(Continued)*

Zentedeschia	-	80	-	-	-	-	-
	-	70	-	-	-	-	-
	-	60	-	-	-	-	-
Carnations	Select	65	-	25	-	650	-
	Fancy	55	-	25	-	550	-
	Standard	50	-	25	-	450	-
	Short	50	-	25	-	<450	-
Gerbera	Extra	50+	-	-	≥ 9	-	-
	Select	40-49	-	-	≥ 8	-	-
Gypsophilla	Perfecta	65	-	6-10	-	≥ 280	-
	Million Star	70	-	5-10	-	≥ 250	-
	New Love	60	-	5-10	-	≥ 250	-
Hydrangea	Jumbo	50	-	-	≥ 20	-	-
	Select	50	-	-	≥ 12	-	-
	Mini	50	-	-	≥ 5	-	-
Asiatic lily	5+Blom	60	-	10	-	-	-
	3-5 bloom	60	-	10	-	-	-
	2 blooms	60	-	10	-	-	-
Sunflower	Extra	≥ 55	-	-	-	-	10
	Select	≥ 55	-	-	-	-	8-10
	Fancy	≥ 55	-	-	-	-	5-8
	Petite	≥ 55	-	-	-	-	<5

The designation of grade standards for cut flowers is one of the most controversial areas in their care and handling. For different type of cut flowers different standards have been made (Table 10.6) which differs in according to country (Table 10.7). Objective standards such as stem length, which is still the major quality standard for many flowers, may bear little relationship to flower quality, vase-life or usefulness. Weight of the bunch for a given length is a method that has been shown to strongly reflect flower quality. Straightness of stems, stem strength, flower size, vase-life, freedom from defects, maturity, uniformity, and foliage quality are among the factors that should also be used in cut flower grading. If used, mechanical grading systems should be carefully designed to ensure efficiency and to avoid damaging the flowers.

TABLE 10.7 International standard for stalk length and flower size in carnation, gerbera andalstroemeria as per recommendation of Aalsmeer Flower Auction Association (Ranjan et al., 2013)

Flower character	Grade	Carnation	Gerbera	Alstroemeria
Stalk length (cm)	Fancy	>55	>50	>60
	Standard	>42	>40	–
Flower diameter (cm)	Fancy	>7.5	>9.0	> 2 flowers /stem
	Standard	> 6.0	>7.0	–

10.8.4 BUNCHING

Flowers are normally bunched, except for anthuriums, orchids, and some other specialty flowers. The number of flowers in the bunch varies according to growing area, market, and flower species. Groups of 10, 12, and 25 are common for single-stemmed flowers (Table 10.11). Spray-type flowers are bunched by the number of open flowers, by weight or by bunch size. Bunches are held together by string, paper-covered wire or elastic bands and are frequently sleeved soon after harvest to unitize the bunch, protect the flower heads, prevent tangling, and identify the grower or shipper. Materials used for sleeving include paper (waxed or unwaxed), corrugated card (smooth side toward the flowers) and polyethylene (perforated, unperforated and blister). Sleeves can be preformed (although variable bunch size can be a problem), or they can be formed around each bunch using tape, heat-sealing (polyethylene), or staples (Table 10.8)

TABLE 10.8 Bunching practices in cut flowers (Reid, 2009)

Name of Flower	Bunching
Gladiolus	Bunched in 10 stems
Alstroemeria	Bunched in groups of 10
Anemone sp.	Bunched in groups of 10
Asiatic lily	bunched in groups of 5 or 10
Lisianthus	Bunched in group of 10 flower stems
Roses	The number of stems per bunch and bunch pattern depends on market preferences.
Strelitzia spp.	% stems are tied together at two points with inflorescence facing in same direction.
Bouvardia spp	Bunched in 10 stems

10.8.5 TREATMENT WITH FLORAL PRESERVATIVES

Unlike fruits and vegetables, flowers can be cut in the "bud" stage. In some flowers this is the normal commercial practice (e.g. roses, gladiolus) in others, flowers are normally cut near fully open. The high respiration of flowers and the energy required for flower growth, bud opening, and floral display requires substantial energy reserves in harvested cut flowers. The carbohydrates have a potential role in control of petal senescence. The fact that the primary component in floral "preservatives" (sometimes termed "fresh flower foods") is a simple sugar (e.g. fructose, glucose, or sometimes sucrose) which reflects the profound effects of added carbohydrates on flower development, opening, and display life. Fortunately it is possible to supply the requisite additional carbohydrate by adding it to the solutions in which flowers are held in order to inhibit the growth of microorganisms. Three types of treatments are used commercially;

(a) Bud opening. Tight cut buds are held until the flowers open in a solution containing sucrose.

b) Pulsing. The term "pulsing" means placing freshly harvested flowers for a relatively short time (a few seconds to several hours) in a solution specially formulated to extend their storage and vase life. In this case generally, buds or flowers are treated for 16–20 hr in a vase solution containing a relatively high concentration of sucrose. Pulsing solutions are specific to the individual crop. Now a days, they are used to provide additional sugar (gladiolus, tuberose, hybrid statice, lisianthus), to extend the life of ethylene-sensitive flowers (carnation, *Delphinium, Gypsophila*) and to prevent leaf yellowing (*Alstroemeria*).

- Sucrose is the main ingredient of pulsing solutions providing additional sugar, and the proper concentration ranges from 2 to 20 percent, depending on the crop. The solution should always contain a biocide appropriate for the crop being treated.
- Ethylene-sensitive flowers are pulsed with silver thiosulfate (STS). Treatments can be for short periods at warm temperatures (e.g. 10 min at 20°C) or for long periods at cool temperatures (e.g., 20 h at 2°C).
- Alstroemeria and lilies can be pulsed in a solution containing gibberellic acid to prevent leaf yellowing and this is often a useful pretreatment.
- Short pulses (10 s) in solutions of silver nitrate have been proved valuable for some crops. The function of the silver nitrate is not fully understood. In some cases it seems to function strictlyas a germicide (e.g. chrysanthemums). In all cases, residual silver nitrate solutionshould be rinsed from the stems before packing.

(c) Vase solutions. Cut flowers are often held in vase solutions containing a combination of sucrose and a biocide. Simple formulated preservative solutions are as effective as bud opening or pulsing solutions for a range of cut flowers are shown in Table 10.9.

TABLE 10.9   Preservative Solutions for cut flowers (Source: Reid and Kofranek, 1980)

Crop	Bud Opening	Pulsing (for 16-20 hrs)
Roses	None required unless buds cut very tight, then use 1.5% sucrose, 250 ppm 8-hydroxyquinoline citrate, 100 ppm 6-benzyladenine	3% sucrose, 320 ppm citric acid at 4°C and high RH
Carnations	10% sucrose, 200 ppm Physan-20 R	10–20% sucrose, 200 ppm Physan-20R at 20°C
Chrysanthemum	2% sucrose, 75 ppm citric acid, 25 ppm AgNo₃	5% sucrose, 200 ppm Physan-20R at 20°C
Gladiolus	None required but very tight buds can be opened with the pulsing solution	20% sucrose, 200 ppm Physan-20R, 20°C
Gypsophila	5% sucrose, 200 ppm Physan-20 R	10% sucrose, 200 ppm Phusan-20R, 20°C

The problem with proprietary preservatives as in vase solutions is that the optimum sucrose content of the vase solution for different flowers varies. Concentrations above 1.5 percenthave been reported to cause severe foliage burn in cut roses but have little effect on the vase life of carnations. In practice, most formulation use relatively low concentration of sucrose which avoids the danger of phytotoxicity when bud fails to provide adequate carbohydrate for maximum benefit.

Responses to sugar in the vase solution include improved floral opening (Doi and Reid 1995), improved pigmentation and size of the opening flowers (Cho et al. 2001), improved water relations (Acock and Nichols 1979) and even reduced sensitivity to ethylene (Nichols 1973). On spike-type flowers, such as gladiolus, senescing flowers appear to supply carbohydrate to those still developing. Removal of senescing florets on gladiolus species significantly reduced opening and size of florets further up the spike (Serek et al. 1994a). Perhaps the most striking effects of carbohydrate stress in harvested cut flowers is the blackening of leaves of cut flower *Proteas* (Reid et al. 1989). These bird-pollinated flowers produce copious nectar; in the postharvest environment there is insufficient photosynthate to meet the demands of the flower, resulting in necrotic death of the leaves. Girdling the stem just below the flower (Newman et al. 1989), holding the flowers in high light conditions (Bieleski et al. 1992), or providing supplementary carbohydrate (Newman et al. 1989) prevents the blackening symptoms. The study highlighted the importance of the leaves in supplying carbohydrate to the flower. It has been reported that sugar in the flower preservative is transported in the xylem to the leaves, where it enters the symplast and is transported to the flowers via the phloem (Halevy and Mayak 1979). Recent research into the effects of added carbohydrates on the life of cut flowers has focused on the potential benefits of trehalose, which has been reported to mitigate the damaging effects of ionizing radiation and to extend the life of gladiolus flowers (Otsubo and Iwaya-Inoue 2000). In a study, Yamada et al. (2003) found that trehalose, but not sucrose, delayed symptoms of senescence, and associated programmed cell-death events, including nuclear frag-mentation. These data suggest that trehalose is exerting a protective effect, perhaps on membranes (Crowe *et al*. 1984), rather than supplying the needed carbohydrate.

A continued improvement in postharvest life of lisianthus flowers at 3 percent or even 6 percent sugar has been recorded by Cho et al. (2001). Both sucrose and glucose were found effective but slightly superior performance with glucose was observed by the workers (Figure 10.7). The opening of lisianthus flowers is associated with substantially increased concentrations of glucose in the corolla. The glucose probably serves to provide osmotic potential for the expansion of the petal cells and the availability of soluble carbohydrate for that purpose is probably partly responsible for the improved opening of the flowers in preservative solution. Similarly, the availability of carbohydrate may be important for the synthesis of the lignin that is the likely reason that pedicels of sugar-treated flowers are stronger and for the anthocyanins that are the basis for color in these flowers. Kawabata *et al*. (1996) found that flowers of lisianthus grown under reduced light intensity had paler color, and also showed that the inclusion of 0.5M sucrose in the vase solution improved color.

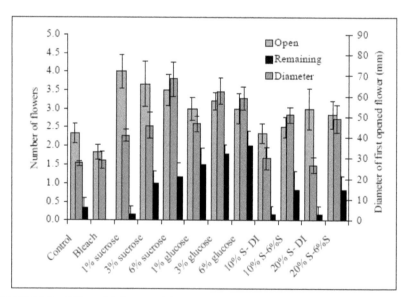

FIGURE 10.7   Effect of different pulsing and vase solutions on the postharvest performance of lisianthus flowers (Cho et al., 2001).

One of the remarkable technologies which have been successful in improving the opening and vase life of cut flowers is the provision of additional carbohydrate in high concentration or "pulse" pretreatments (Halevy and Mayak 1979). In addition to the well-known effects in gladiolus (Mayak *et al.* 1973), pulsing treatment has been successful in improving the opening of *Strelitzia* (Halevy *et al.* 1978) and *Polianthes tuberosa* (Naidu and Reid 1989; Waithaka *et al.* 2001). In *Lisianthus*, the pretreatment greatly improves the color of the newly opened blooms, and in tuberose, it ensures satisfactory bud opening which normally is inhibited by even brief periods of cool storage. The application of 25 M BAP as a pulse treatment for 24 h can be recommended to prolong the postharvest life through delayed leaf senescence and thus enhance the marketability of cut leaves of *Dracaena marginata* 'bi color', *Dracaena sanderiana* 'white' and *Dracaena deremensis* (Subhashini *et al.*, 2011).

Sucrose at concentration of 0.5 g l^{-1}, significantly increased time to stem bending and vase life of cut flower, while ascorbic acid at 150 g l^{-1} significantly increased vase life, fresh weight and percentage of total carbohydrates percentage in snapdragon cut spike flowers (Abdulrahman et al., 2012). In another study, 8-HQS + 2 percent sucrose treatment showed best water uptake, water balance, percentage of maximum increase in fresh weight of the cut flower stem and vase life which was extended up to 18 days in snapdragon (Abdul-Wasea, 2012).

Preservative solution of fresh cut flowers is not only made up of water, sugar, fungicides, ethylene inhibitor or antagonist, inorganic salts, but also needs to add some plant growth hormone which can effectively delay senescence and improve the quality of cut flowers. *Epidendrum* orchid cut flowers treated with the combination of lemon juice, Sprite and Ritebrand bleach showed the lowest petal drop (5.7), wilted leaves (3.8) and wilted florets (4.9) were obtained from those treated with. The longest vase life (21.0 days) of orchids was obtained from the treatment combinations of lemon juice, Sprite and Rite brand bleach and lime, sugar, and Listerine (Thwala *et al.*, 2013). Measurements of floral longevity are largely absent from studies of the powerful effects of jamonic acid, brassinosteroids, and salicylic acid on plant growth, development, and responses to biotic and abiotic stress (Ashraf *et al.* 2010). Similar to auxins, brassinosteroids stimulate ethylene biosynthesis and their effects on ethylene-sensitive flowers would be expected to be negative. Jasmonic acid reduced life of petunias and dendrobiums through stimulation of ethylene production (Porat *et al.* 1993). The salicylic acid signaling pathway has shown to be required for up-regulation of genes required for leaf senescence (Morris *et al.* 2000), but the effects of down-regulating this pathway on flower senescence have not been studied. *Nikkhah Bahrami*et al. (2011) evaluated effect of different concentrations of salicylic acid on vase life of cut *Lisianthus (Eustoma grandiflora)* and reported that 100 mg l-1 of SA enhanced solution uptake and vase life compared the control. Alaey et al. (2011) investigated the pre and postharvest effects of salicylic acid on cut rose and found that different levels of salicylic acid increased stem fresh and dry weight and also leaf area compared to control. Abdul-Wasea (2011) in his study on snapdragon cut flower (*Antirrhinum majus*) found that treatment with anti-ethylene compounds and antimicrobial compounds increased the amount of carbohydrates in comparison to control which is in agreement with our results. Pretreatment for 6 h with low concentrations of I-MCP (1-Methylcyclopropene, formerly designated as SIS-X), a cyclic ethylene analog, inhibited the normal wilting response of cut carnations exposed continuously to 0.4 $\mu l.l^{-1}$ ethylene (Serek et al., 1995).

Biocides. The major cause of vase life reduction in cut flowers is water relation interruption which ismostly due to vase solution microbial proliferation and consequently vascular occlusion resulting in solution uptake reduction. In order to control microbial proliferation, biocides are usually integrated in vase solution preservatives. Beside microbial proliferation control, biocides could affect cut flower's quality and physiology in various aspects (Jowkar et al., 2012). Biocides are important in the preservative solution to control microorganisms such as bacteria, yeasts, and molds. Microorganisms in the vase water can physically plug the stem, produce toxic metabolites and enzymes, release damaging levels of ethylene, and cause a hypersensitive response to low temperatures. Bacterial counts of 10–100 million per milliliter can reduce water and nutrient uptake, while counts of 3 billion per milliliter causes wilting (Teixeira da Silva, 2003). Vase solutions normally also contain a biocide such as AgNo3, 8-hydroxy quinoline citrate or Physan-2R (a quaternary ammonium compound). Some of the commonly used biocides and their doses has been listed in Table 10.10.

TABLE 10.10 Commonly used biocides and their recommended concentration for cut flowers (Reid, 2009)

Germicide Types	Common Name	Recommended Dose (ppm)
Silver Nitrate	AgNO$_3$	10–200
Silver thiosulfate	STS	0.2–4
8-hydroxyquinoline sulfate	8-HQS	200–600
8-hydroxyquinoline citrate	8-HQC	200–600
Thiobendazole	TBZ	5–300
Aluminium sulfate	A$_{12}$ (SO$_4$)$_3$	200–300
Slow release chlorine compounds	---	50–400
Quaternary ammonium salts	QAS	5–300

Flower longevity of strelitzia flowers can be substantially increased by pulsing buds or flowers for 24 h with a solution containing 10 percent sucrose, 250 ppm 8-HQC and 150 ppm citrate (Halevy et al., 1978). Fariman and Tehranifer (2011) suggested the application of ethanol and methanol as a biocide in preservative solutions for carnation flowers.

10.8.6 PACKING

There are many shapes of packing containers for cut flowers, but most are long and flat. This design restricts the depth to which the flowers can be packed in the box and this may reduce physical damage. In addition, flower heads can be placed at both ends of the container for better use of space. With this kind of flower placement, whole layers of newspaper are often used to prevent the layers of flowers from injuring each other. The use of small pieces of newspaper to protect only the flower heads, however, is probably a better practice, since it allows for more efficient cooling of flowers after packing. It is critically important that containers be packed in such a way that transport damage is minimized. To avoid longitudinal slip, packers in many flower-producing countries use one or more "cleats". These are normally foam- or newspaper-covered wood pieces that are placed over the product, pushed down, and stapled into each side of the box. Padded metal straps, elastic bands, high density polyethylene blocks, and cardboard tubes can also be used as cleats. The heads of the flowers should be placed 6 to 10 cm (2.4–4 in) from the end of the box to allow effective precooling and to eliminate the danger of petal bruising should the contents of the box shift. Gladioli, snapdragons, and some other species are often packed in vertical hampers to prevent geotropic curvature that reduces their acceptability. Cubic hampers are used for upright storage of daisies and other flowers.

Frequently, flower heads are individually protected by paper or polyethylene sleeves (Figure 10.8). Cushioning materials such as shredded paper and paper or wood wool may be placed between packed flowers to further reduce damage.

FIGURE 10.8 Corrugated fiber board sleeves for cushioning and protection of roses during transport.

In recent years, the industry has increasingly utilized the "Procona" and "Aqua-pac" systems for transportation of cut flowers. Flowers are transported in boxes with water or a flower preservative in the bottom. "Rocona" system, uses plastic bases and a cardboard sleeve to allow transport of flowers upright in water. It has widely been suggested that this transportation method reduces the need for proper temperature control and that flowers have a longer vase life after transport in water. This system is more expensive than traditional boxes and less no. of products can be packed in it, but the presence of water, may improve flower quality when they are not transported under proper temperature conditions. Specialty flowers such as Anthurium, orchid, ginger, and bird of paradise are packed in various ways to minimize friction damage during transport (Reid, 2001) (Table 10.11)

TABLE 10.11　Packaging requirements of cut flowers (Reid, 2009)

Name of Flower	PackagingRequirements
Gladiolus	Spikes wrapped in 50 gauge polyethylene sheets and Packed in fb boxes (90-100 spikes/box of dimension 33 × 18 × 122 cm.)
Alstroemeria	Sleeved and packed in Horizontal boxes
Anemone sp.	packed in standard horizontal Corrugated fiber board (CFB) boxes
Anthurium	Individually packed in moist shredded news paper Or in plastic bags and then in boxes
Asiatic lily	Packed in horizontal fiberboard boxes
Lisianthus	Flowers are sensitive to gravity so their stems will bend upwards if the flowers are held horizontal at ambient temperatures. Therefore flowers are often packed and transported in vertical hampers.
Orchids	Wrapped as individual flowers or spikes, frequently in shredded paper and then packed 12–24 flowers in each carton. Box inserts hold individual water tubes stationary. Shredded wax paper is tucked around and between the flowers for additional protection.
Roses	Sleeved in plastic, waxed paper or soft corrugated card sleeves.
Strelitzia spp.	Stem ends are evenly trimmed and a paper wrapped is placed around the bunched inflorescence. Waxed paper bag on each inflorescence and paper wrapper on each bunch is placed. Packed in shallow cartons.
Bouvardia spp	Sleeved and packed in horizontal boxes or proconas

A proprietary package, the "Procona™ system uses plastic bases and a cardboard sleeve to allow transport of flowers upright in water (Figure 10.9). This system is more expensive than traditional boxes, and less volume of product can be packed in it, but the presence of water improves out-turn when flowers are not transported under proper temperature conditions. Because of their weight, low packing density, and propensity to spill water, systems of this type are seldom used in air shipments. Specialty flowers such as anthurium, orchid, ginger, and bird-of-paradise are packed in various ways to minimize friction damage during transport. Frequently the flower heads are individually protected by paper or polyethylene sleeves.

FIGURE 10.9   Procona boxes for cut flowers.

10.8.7   COOLING

Cooling is one of the most important steps in bringing cut flowers from grower to market place. Low temperature provides many advantages to extend the vase life by:
- Reducing the respiration rate
- Reducing the internal breakdown by enzymes
- Reducing water loss and wilting
- Slowing the growth of pathogenic organisms
- Reducing the production of ethylene

Once packed, cut flowers are difficult to cool. Their high rate of respiration and the high temperatures of most greenhouses and packing areas result in heat production in packed flower containers unless measures are taken to ensure temperature reduction. It is therefore necessary to cool the flowers as soon as possible after packing. Individually, flowers cool and warm rather rapidly. However, individual flowers brought out of cool storage into a warmer packing area will warm quickly and water will condense on the flower which may result in occurrence of botrytis.

The simplest method of ensuring that packed flowers are adequately cooled and dry is to pack them in the cool room. Although this method is not always popular with packers and may increase labor cost and slow down packing somewhat, it will ensure a cooled, dry product (Reid, 2009).

Built-in-place cold storage can be constructed out of wood pole and post, steel and/or concrete block. The construction cost will depend on labor cost and on the type of materials used for the frame, walls, floor, ceiling, and insulation. The design should consider where handling equipment is located and where handling procedures will be done so product flows smoothly in and out of cold storage (Nowak and Rudnicki, 1990). Low cost cool chambers e.g. Zero Energy Cool Chambers (ZECCs) can be constructed out of sand, bricks, dry grass and wood pole.

Forced-air cooling of boxes with end holes or closeable flaps is the most common and effective method for precooling of cut flowers. Cool air is sucked or blown through the boxes. Care must be taken to pack them so that air can flow through the box and not be blocked by the packing material or flowers. In general, packers use less paper when packing flowers for precooling. The half-cooling time for forced-air cooling ranges from 10 to 40 min, depending on product and packaging. Flowers should be cooled for three half-cooling times (by which time they are 7/8 cool) (Thomson et al., 2004)).

If the packages are to remain in a cool environment after precooling, vents may be left open to assist removal of the heat of respiration. Flowers that are to be transported at ambient temperatures can be packed in polyethylene caskets, foam-sprayed boxes or boxes with the vents resealed. Ice that is used after precooling is only effective if placed to intercept heat entering the carton (ie., it must surround the product), and care must be taken to ensure that the ice does not melt onto flowers or cause chilling damage. Precooling of vertical hampers or Proconas presents a particular challenge, but can be achieved using a "tunnel" forced-air cooling system.

10.8.8   QUALITY CONTROL

Any form of reduction in the postharvest quality of cut flowers before their arrival in market lowers the potential price for growers. At the same time it is a period when flowers can be subjected to severe maltreatment in form of grading, bunching, packaging, storage, and transport and unpacking. Transportation conditions particularly can lead to quality problems e.g. damaged flowers, bent stems, uneven opening (Figure 10.10) and botrytis with the temperature often being the most important factor for optimum quality control.

(a)   (b)

FIGURE 10.10 Common quality problems in roses (a) bent stem (b) uneven flower opening.

There are few official grade standards for cut flowers. Some marketing channels, for example the British mass market chains and the Dutch auctions, have internal quality control systems that provide a check on quality of flowers. The most important quality parameter is "freshness" or vase life. This parameter is difficult to assess visually, but because of its importance, producers and receivers should set up a "quality control" program that would involve evaluation of the vase life of representative flowers on a continuing basis.

10.8.8.1 FREQUENT PROBLEMS

Common quality problems seen in bunch during transport or storage for 2–3 days are uneven bud opening, bent stems and damaged flowers. Quality controllers at the flower auctions put a remark on these flowers when they are sold. A minor quality remark will lead to a decrease in price of 5–10 per cent. Two minor remarks or one major remark can lead to a decrease in price by 20–50 per cent.

High temperatures are not always the reason for common quality problems. Bent stems for example mainly result from horizontal transport while diseases are caused from germination and infection of botrytis spores. However, higher temperature will tend to encourage these problems to develop faster.

There are the obvious reasons for the temperature increase in flowers. It can simply be due to incorrect settings in cold stores or warm conditions during transport in the aero planes. Perhaps there are high temperatures when loading and uploading of the trucks or aero planes takes place. The flowers may also be packed, distributed or unpacked in areas that are too warm. Meanwhile consideration should be given to the heat produced by the flowers themselves. Flowers are living organisms and breathe oxygen and burn their carbohydrate reserves to stay alive. This process generates heat as waste products (van der Hulst, 2004).

10.9   TRANSPORTATION OF CUT FLOWERS

Transportation of cut flowers from the field to the consumer is a process that takes particular care and planning. It is very important to decide whether flowers have to be transported in water or dry-packed. Dry-packing leaves out the weight of water in buckets, but allows for greater risk of water stress. This method is used often by commercial growers that ship flowers all over the world. Because it is important to keep low temperatures within the dry-pack boxes, ice chips are packed along with the flowers. Climate controlled trucks are also useful in maintaining low temperatures. Dry packed cut flowers should be fully hydrated and pulsed. Special care should be taken when handling flowers to avoid the bruising of petals. When transporting flowers short distances, it may be possible to ship them in buckets of water. Unlike dry-packing, there is little risk of water stress when this transport method is used. Buckets of water are more difficult to move, which would increase shipping costs (Laschkewitsch and Smith, 2000).

In less developed countries where refrigeration may be far from the production area, an efficient method is needed to preserve cut flowers while in transit. In Brazil, for example, sanitation, packaging, and logistical issues are some of the many difficulties faced by producers (Van Den Broek et al., 2004). Even in developed countries where handling methods are more modern, losses in the marketing chain can be high. Colombian cut roses en route to the U.S. are exposed to varying temperatures during transport and a prolonged transport time. It is often 4 to 6 days before flowers reach the consumer (Leonard et al., 2001, 2011). Despite the handling procedures outlined above, little major advancement has been made in extending the vase life of cut flowers in many years. Because of the remarkable effects of transport temperatures on subsequent vase life and the tendency for overheating of packed flowers, they should be transported at close to the optimal temperature ($0°C$ for most of the species). Some systems already exist for aircraft transportation of flowers under controlled temperatures. There are passive and active refrigeration systems that could usefully be applied to the transportation of cut flowers. For e.g. Envirotainers™ provide a dry-ice refrigeration system that could provide controlled temperatures suitable for use with cut flowers and would allow transport for a considerable distance at costs that would be more than recompensed by the improved quality of the flowers on arrival. Passively refrigerated and insulated containers provide alternative means for providing some control of temperature during transport. If product is properly cooled before being palletized or packed in an LD-3 container, insulation alone certainly will improve temperature management during the transportation chain. Due to the lack of temperature control in the majority of aircraft carrying cut flowers and the extreme response of cut flowers to temperature abuse, the logistics for air transport of cut flowers must focus on doing everything possible to maintain the cold chain. Flowers should be properly cooled at the grower's operation and transported to the airport in refrigerated (or at least well insulated) trucks. In some airports vacuum coolers has been installed to reduce temperature of flowers before they are air-freighted. This expensive equipment certainly provides a means of protecting product that is warm on arrival at the airport, but this is not an optimal procedure. Prior heat exposure and the additional water loss from the flowers during the vacuum cooling process undoubtedly compromise flower quality and vase life. Perhaps the weakest link in the flower postharvest chain is at the airport. Freight forwarders may be busy with late arrivals just prior to aircraft leaving. Pallets are assembled hastily, there is no time for precooling of warm flowers and boxes are handled roughly. Pallets may be at ambient temperature for up to 4 hrs while the aircraft is being loaded. The producing and transportation industries need to work together to set standards for temperature management, pallet construction, and temperature maintenance during loading.

10.9.1   TEMPERATURE MANAGEMENT BEFORE LOADING

Ideally, flowers would be packed and cooled at the production location and loaded into LD-3 containers or made into pallets on LD-9 sheets before being transported in a refrigerated vehicle to the freight forwarder or air cargo facility. Whenever active refrigeration is not available, flowers should be kept cold by the following measures:
- Consolidation into pallets to reduce the surface/volume ratio
- Reduction in radiative heat gain by holding in a shaded location
- Reduction of air infiltration by closing precooling vents (essential), and using pallet covers.
- Pallet covers should preferably be white or foil backed and has the additional benefit of reducing water loss during air transport.

Flowers cooled to 0°C and packed in a large pallet will gain about 1°C per hour due to the heat generated by the process of respiration. As the temperature rises, respiration rises and the heat gain increases too, so that by the time the flowers are at 10°C they may gain 3°C per hour and once they are at 20°C, the heat gain can be as much as 10°C per hour. Flower boxes unloaded from aircraft pallets have been probed at temperatures as high as 55°C, indicating the runaway danger of inadequate temperature management and delays in transport (Thompson et al., 2004).

10.9.1.1   PALLET CONSTRUCTION

Flowers are delicate, and box strength is compromised if corners don't line up. It's important that the pallet be constructed so that the corners are square and line up above each other. One of the most vexing problems of the flower industry is the wide array of box sizes, which may make construction of a square and well aligned papallet difficult. Walking or kneeling on flower boxes should not be permitted. If it's necessaryto provide trestles and planks to allow construction of pallets without workers kneeling or walking on the boxes, that would be money well spent. Preferably, the pallets should be constructed from the side (base to top for each successive layer. Standardization of boxes so that they easily fit the standard LD-9 and LD-3 footprint would help a gooddeal. Immediately after pallet construction, the palletized flowers should be covered with an insulated cover (or at least with shrink- or cling-wrap polyethylene film. This will reduce heat gain as a result of air movement through the pallet (Rij et al., 1979)

Problems in floriculture sector

Creating enabling environment and achieving competitiveness in the floriculture subsector require addressing of many constraints faced by the growers and exporters. These constraints are the limiting factors to attract domestic as well as foreign direct investments. These constraints are related to:

- Absence of government policies and strategies,
- Lack of information on technology and technical experts,
- Inadequate export management system,
- Lack of adequate finance,
- Inadequate infrastructures,
- High air-freight charges, etc.
- Lack of knowledge about postharvest technologies

10.10   FUTURE PROSPECTS

Research on horticultural and physiological aspects of ornamentals has given us a good understanding of the factors that affect the life of cut flowers, potted plants, and other ornamentals. These findings will be the key to future strategies for improving the postharvest life of cut flowers and potted plants. In addition to improved application of tools such as proper temperature management, existing ornamental taxa can be improved by new and emerging chemical treatments, including water-soluble ethylene inhibitors that may replace the gaseous 1-MCP treatment and registration and use of growth regulators such as TDZ and NPA for reducing leaf yellowing, improving flower opening and vase life, and preventing gravitropic bending. Chemicals for controlling vase-solution microbes and inhibiting postharvest diseases will be more environmentally friendly. In the longer term, the industry will be improved by a focus on breeding ornamentals with better postharvest characteristics, using the huge genetic variability in the wild populations of most ornamentals. This may be accomplished using conventional breeding, supplemented by modern breeding technologies.

The remarkable effects of transgenic manipulation of genes involved in petal senescence point to the potential for such strategies not only to dramatically improve the postharvest performance of commercial floricultural crops, but also to expand the palette of ornamentals in the trade. Many beautiful ephemeral flowers are seldom seen in the vase because of their short display life, but application of the transgenic techniques described above could enable the rapid commercialization of spectacular flowers. Although transgenic ornamentals are presently limited to blue carnations and roses, the rapid acceptance of transgenic technologies in agronomic crops suggests that when the political, regulatory, and social environment is ready, transgenic ornamentals will quickly play an important role in the ornamental market place.

KEYWORDS

- **Cut flowers**
- **Quality loss**
- **Flower senescence**
- **Water relation**
- **Pulsing**
- **Postharvest technology**

REFERENCES

Abdulrahman Y. A.; Ali S. F.; and Faizi, H. S.; Effect of sucrose and ascorbic acid concentrations on vase life of snapdragon (*Antirrhinum majus* L.) Cut Flowers. *Int. J. Pure Appl. Sci. Technol.* **2012**, *13*(2), 32–41.

Abdul-Wasea, A.; *J. Saudi. Soc. Agric. Sci.* **2011.** (Abstract).

Abdul-Wasea A. A.; Effects of some preservative solutions on vase life and keeping quality of snapdragon (*Antirrhinum majus* L.) cut flowers. *J Saudi. Soc. Agric. Sci.* **2012,** *11*(1), 29–35.

Acock, B.; and Nichols, R.; Effects of sucrose on water relations of cut, senescing, carnation flowers. *Am. Bot.* **1979,** *44*, 221–230.

Alaey, M.; Babalar, M.; Naderi, R.; and Kafi, M.; *Postharvest. Biol. Technol.* **2011,** *61*, 91–94.

Anonymous.; Cut flowers minimum guidelines and standards. Association of Floral Importers of Florida; **2009**, pp 52

Arditti J.; The place of the orchid pollen "poison" and pollenhormon in the history of plant hormones. *Am. J. Bot.* **1971,** *58*, 480–481.

Arditti, J.; Orchids, pollen poison, pollenhormon and plant hormones. *Orchid. Rev.* **1975,** *83*, 127–129.

Ashraf, M.; Akram, N. A.; Arteca, R. N.; and Foolad, M. R.; The physiological, biochemical and molecular roles of brassinosteroids and salicylic acid in plant processes and salt tolerance. *Crit. Rev. Plant Sci.* **2010,** *29*, 162–190.

Bagheri, H.; and Sedaghathour, S.; Effect of thidiazuron and naphthalene acetic acid (NAA) on the zase life and quality of cut Alestroemeria hybrid. *J. Ornamen. Hortic. Plant.* **2013,** *3* (2), 111–116.

Beyer, E.; A potent inhibitor of ethylene action in plants. *Plant Physiol.* **1976,** *58*, 268–271.

Bieleski, R. L.; Ripperda, J.; Newman, J. P.; and Reid, M. S.; Carbohydrate changes and leaf blackening in cut flower stems of *Protea eximia*. *J. Am. Soc. Hortic. Sci.* **1992,** *117*, 124–127.

Blessington, T. M.; Post harvest handling of cut flowers. University of Maryland Cooperative Extension Service; **2004**.

Borda, A. M.; Clark, D. G.; Huber, D. J.; Welt, B. A.; and Nell, T. A.; Effects of ethylene on volatile emission and fragrance in cut roses: the relationship between fragrance and vase life. *Postharvest Biol. Technol.* **2011**, *59*, 245–252.

Brodersen, C. R.; McElrone, A. J.; Choat, B.; Matthews, M. A.; and Shackel, K. A.; The dynamics of embolism repair in xylem: In vivo visualizations using high-resolution computed tomography. *Plant Physiol.* **2010**, *154*, 1088–1095.

Broek Van Den L.; Haydu, J. J.; Hodges, A. W.; and Neves, E. M.; Production, marketing and distribution of cut flowers in US and Brazil. In: Annual report of Florida Agricultural Experiment Station, University of Florida, Gainesville; **2004**, pp 1–19.

Bufler, G.; Ethylene-enhanced 1-aminocyclopropane-1-carboxylic acid synthase activity in ripening apples. *Plant Physiol.* **1984**, *75*, 192–195.

Bufler G.; Ethylene-promoted conversion of 1-aminocyclopropane-1-carboxylic acid to ethylene in peel of apple at various stages of fruit development. *Plant Physiol.* **1986**, *80*, 539–543.

Cevallos, J. C.; and Reid, M. S.; Effect of dry and wet storage at different temperatures on the vase life of cut flowers. *Hortic. Technol.* **2001**, *11*, 199–202.

Cevallos, J. C.; Temperature and the postharvest biology of cut flowers. M.S. Dissertation, Univ. Calif., Davis; **1998**, 92 pp.

Chandra Sekhar, R.; Kasturi, A.; Manohar Rao, A.; and Narender Reddy, S.; Effect of harvesting at different heights on growth and flower yield of carnation (Dianthus caryophyllus L.) in second season crop. *J. Hortic.* **2013**, *1*, 101. doi:10.4172/horticulture.1000101

Chang, H.; Jones, M. L.; Banowetz, G. M.; and Clark, D. G.; Overproduction of cytokinins in petunia flowers transformed with PSAG12-IPT delays corolla senescence and decreases sensitivity to ethylene. *Plant Physiol.* **2003**, *132*, 2174–2183.

Chen, J. C.; Jiang, C. Z.; Gookin, T. E.; Hunter, D. A.; Clark, D. G.; and Reid, M. S.; Chalcone synthase as a reporter in virus-induced gene silencing studies of flower senescence. *Plant Mol. Biol.* **2004**, *55*, 521–530.

Cho, M. C.; Celikel, F. G.; Dodge, L.; and Reid, M. S.; Sucrose enhances the postharvest quality of cut flowers of *Eustoma grandiflorum* (Raf.) Shinn. *Acta Hort.* **2001**. *543*, 305–315.

Clements, J.; and Atkins, C.; Characterization of a non-abscission mutant in *Lupinus angustifolius*. I. Genetic and structural aspects. *Am. J. Bot.* **2001**, *88*, 31–42.

Cochard, H.; Nardini, A.; and Coll, L.; Hydraulic architecture of leaf blades: where is the main resistance? *Plant Cell. Environ.* 27, 1257–1267.

Crowe, J. H.; Crowe, L. M.; and Chapman, D.; Preservation of membranes in anhydrobiotic organisms: The role of trehalose. *Science.* **1984**, *223*, 701–703.

Darandeh, N.; and Hadavi, E. Effect of pre-harvest foliar application of citric acid and malic acid on chlorophyll content and post-harvest vase life of Lilium cv. Brunello. *Front. Plant Sci. Crop. Sci. Hortic.* **2012**, *2* (106), 1–3.

Doi, M.; and Reid, M. S.; Sucrose improves the postharvest life of cut flowers of a hybrid *Limonium. Hortic. Sci.* **1995**, *30*, 1058–1060.

Dole, J. M.; Viloria, Z.; Fanelli, F. L.; and Fonteno, W.; Postharvest evaluation of cut dahlia, linaria, lupine, poppy, rudbeckia, trachelium, and zinnia. *Hortic. Technol.* **2009**, *19*, 593–600.

Eason, J. R.; *Sandersonia aurantiaca*: an evaluation of postharvest pulsing solutions to maximise cut flower quality. *New Zealand J. Crop. Hortic. Sci.* **2002**, *30*, 273–279.

Eidyan, B.; Effect of iron and citric acid foliar applications in combination with nitrogen fertigation on tuberose (*Polianthes tuberose* L.). Horticulture Karaj: Islamic Azad University-Karaj Branch; **2010**, 75.

Eisinger, W.; Role of cytokinins in carnation flower senescence. *Plant Physiol.* **1977**, *59*, 707–709.

Elibox, W.; and Umaharan, P.; Morphophysiological characteristics associated with vase life of cut flowers of anthurium. *Hortic. Sci.* **2008**, *43*, 825–831.

Evans, R. Y.; and Reid, M. S.; Changes in carbohydrates and osmotic potential during rhythmic expansion of rose petals. *J. Am. Soc. Hortic. Sci.* **1988**, *113*, 884–888.

Fanourakis, D.; Carvalho, S. M. P.; Almeida, D. P. F.; van Kooten, O.; van Doorn, W. G.; and Heuvelink, E.; Postharvest water relations in cut rose cultivars with contrasting sensitivity to high relative air humidity during growth. *Postharvest Biol. Technol.* **2012**, *64*, 64–73.

Fariman, J. K.; and Tehranifer, Ali; Effect of essential oils ethanol and methanol to extend the vase life of carnation (*Dianthus caryophyllus* L.) flowers. *J. Biol. Environ. Sci.* **2011**, 5 (14), 91–94.

Ferrant, A.; Hunter, D. A.; Hackett, W. P.; and Reid, M.; Thidiazuron- a pottent inhibitor of leaf senescence in Alstroemeria. *Postharvest Biol. Technol.* **2002**, *25*, 333–338.

Ferrant, A.; Tognoni, F.; Mensuali-Sadi, A.; and Serra, G.; Treatment with thidiazuron for preventing leaf yellowing in cut tulips and chrysanthemum. *Acta Hortic.* **2003**, *624*, 357–363.

Ferrante, A.; Hunter, D.; Hackett, W.; and Reid, M. S.; TDZ: A novel tool for preventing leaf yellowing in Alstroemeria flowers. *Hortic Sci.* **2001**, *36*, 599.

Friedman, H.; and Rot I.; Transportation of unrooted cuttings. Evaluation of external containers. *Adv. Hortic. Sci.* **2005**, *19*, 58–61.

Gan, S.; and Amasino, M. R.; Inhibition of leaf senescence by autoregulated production of cytokinin. *Science.* **1995**, *270*, 1986–1988.

Gast, K. L.; Post-harvest handling of fresh cut flowers and plant material. Kansas St. Coop. Ext. Serv. 2261; **1997**.

Halevy, A. H.; and Mayak, S. Senescence and postharvest physiology of cut flowers— Part 1. *Hortic. Rev.* **1979**, *1*, 204–236.

Halevy, A. H.; and Mayak, S.; Senescence and post-harvest physiology of cut flowers, Part 2. *Hortic. Rev.* **1981**, *3*, 59–143.

Halevy, A. H.; Kofranek, A. M.; and Besemer, S. T.; Postharvest handling methods for bird of paradise flowers (*Sterlitzia reginae* Ait.). *J. Am. Soc. Hortic. Sci.* **1978**, *103*, 165–169.

Hammer, P. E.; Yang, S. F.; Reid, M. S.; and Marois, J. J.; Postharvest control of Botrytis cinerea infections on cut roses using fungistatic storage atmospheres. *J. Am. Soc. Hortic. Sci.* **1990**, *115*, 102–107.

Hemming, S.; Dueck, T. A.; Janse, J.; and van Noort, F. R.; The effect of diffuse light on crops. *Acta Hortic.* **2008**, *801*, 1293–1300.

Hemming, S.; Use of natural and artificial light in horticulture: interaction of plant and technology. *Acta Hortic.* **2011**, *907*, 25–36.

Heuvelink, E.; Bakker, M. J.; Hogendonk, L.; Janse, J.; Kaarsemaker, R.; and Maaswinkel, R.; Horticultural lighting in The Netherlands: new developments. *Acta Hortic.* **2006**, *711*, 25–34.

Hogewoning, S. W.; Douwstra, P.; Trouwborst, G.; van Ieperen, W.; and Harbinson, J.; An artificial solar spectrum substantially alters plant development compared with usual climate room irradiance spectra. *J. Exp. Bot.* **2010**, *61*, 1267–1276.

Huang, L. C.; Lai, U. L.; Yang, S. F.; Chu, M. J.; Kuo, C. I.; Tsai, M. F.; and Wen Sun, C. H.; Delayed flower senescence of Petunia hybrida plants transformed with antisense broccoli ACC synthase and ACC oxidase genes. *Postharvest Biol. Technol.* **2007**, *46*, 47–53.

Hunter, D. A.; Lange, N. E.; and Reid, M. S.; Physiology of flower senescence. In: *Plant Cell Death Processes,* Nooden, L. D., Ed.; Elsevier; **2004a,** pp 307–318.

Hunter, D. A.; Ferrante, A.; Vernieri, P.; and Reid, M. S.; Role of abscisic acid in perianth senescence of daffodil (*Narcissus pseudonarcissus* "Dutch Master"). *Physiol. Plantarum.* **2004b,** *121,* 313–321.

Jones, R. B.; Serek, M.; and Reid, M. S.; Pulsing with Triton X-100 Improves hydration and vase life of cut sunflowers (*Helianthus annuus* L.). *Hortic. Sci.* **1993,** *28,* 1178–1179.

Jowkar, M. M.; Kafi, M.; Khalighi, A.; and Hasanzadeh, N.; Evaluation of aluminum sulfate as vase solution biocide on postharvest microbial and physiological properties of 'Cherry Brandy' rose. *Ann. Biol. Res.* **2012,** *3*(2), 1132–1144.

Kawabata, S.; Ohta, M.; Kusuhara, Y.; Sakiyama, R.; and Fjeld, T.; Influences of low light intensities on the pigmentation of *Eustoma grandiflorum* flowers. *Acta Hortic.* **1995,** *405,* 173–178.

Kawa-Miszczak, L.; Wegrzynowicz-Lesiak, E.; and Saniewski, M.; Retardation of tulip shoot senescence by auxin. *Acta Hortic.* **2005,** *669,* 90–183.

Kittas, C.; and Bartzanas, T.; Greenhouse microclimate and dehumidification effectiveness under different ventilators configuration. *Energ. Build.* **2007,** *42,* 3774–3784.

Laschkewitsch, B.; and Smith, R.; Growing cut flowers for market. NDSU Ext. Serv. Circ. H-1200; **2000.**

Lay-Yee, M.; Stead, A. D.; and Reid, M. S.; Flower senescence in daylily (*Hemerocallis*). *Physiologia Plantarum.* **1992,** *86,* 308–314.

Leonard, R. T.; Alexander, A. M.; and Nell, T. A.; Postharvest performance of selected Colombian cut flowers after three tansport systems to the United States. *Hortic. Technol.* **2011,** *21*(4), 435–442.

Leonard, R. T.; Nell, T. A.; Suzuki, A.; Barrett, J. E.; and Clark, D. G.; Evaluation of long term transport of Colombian grown cut roses. *Acta Hortic.* **2001,** *543,* 293–297.

Macnish, A. J.; Leonard, R. T.; and Nell, T. A.; Treatment with chlorine dioxide extends the vase life of selected cut flowers. *Postharvest Biol. Technol.* **2008,** *50,* 197–207.

Macnish, A.; deTheije, A.; Reid, M. S.; and Jiang, C. Z.; An alternative postharvest handling strategy for cut flowers: dry handling after harvest. *Acta Hortic.* **2009,** *847,* 215–221.

Macnish, A. J.; Leonard, R. T.; Borda, A. M.; and Nell, T. A.; Genotypic variation in the postharvest performance and ethylene sensitivity of cut rose flowers. *Hortic. Sci.* **2010a.** *45,* 790–796.

Macnish, A.; Jiang, C. Z.; and Reid, M. S.; Treatment with thidiazuron improves opening and vase life of iris flowers. *Postharvest Biol. Technol.* **2010b,** 56, 77–84.

Macnish, A.; Leonard, R. T.; Borda, A. M.; and Nell, T. A.; Genotypic variation in the postharvest performance and ethylene sensitivity of cut rose flowers. *Hortic. Sci.* **2010c.** 45, 790–796.

Macnish, A.; Morris, K. L.; de Theije, A.; Mensink, M. G. J.; Boerrigter, H. A. M.; Reid, M. S.; Jiang, C. Z.; and Woltering, E. J.; Sodium hypochlorite: A promising agent for reducing Botrytis cinerea infection on rose flowers. *Postharvest Biol. Technol.* **2010d,** *58,* 262– 267.

Marissen, N.; and Benninga, J.; Nursery comparison on the vase life of the rose 'First Red': Effects of growth circumstances. *Acta Hortic.* **2001,** *543,* 285–291.

Markvart, J.; Rosenqvist, E.; Aaslyng, J. M.; and Ottosen, C. O.; How is canopy photosynthesis and growth of chrysanthemums affected by diffuse and direct light? *Eur. J. Hortic. Sci.* **2010,** *75,* 253–258.

Max, J. F. J.; Horst, W. J.; Mutwiwa, U. N.; and Tantau, H. J.; Effects of greenhouse cooling method on growth, fruit yield and quality of tomato (*Solanum lycopersicum* L.) in a tropical climate. *Sci. Hortic.* **2009,** *122,* 179–186.

Max, J. F. J.; Schurr, U.; Tantau, H. J.; Hofmann, T.; and Ulbrich A.; Greenhouse cover technology. *Hortic. Rev.* **2012**, 40, 259–396.

Maxie E.; Farnham, D.; Mitchell, F.; Sommer, N.; Parsons, R.; Snyder, R.; and Rae, H.; Temperature and ethylene effects on cut flowers of carnation (*Dianthus caryophyllus*). *J. Am. Soc. Hortic. Sci.* **1973**, *98*, 568–572.

Mayak, S.; and Dilley, D.; Regulation of senescence in carnation (Dianthus caryophyllus): effect of abscisic acid and carbon dioxide on ethylene production. *Plant Physiol.* **1976**, *58*, 663–665.

Mayak, S.; and Halevy, A.; Interrelationships of ethylene and abscisic acid in the control of rose petal senescence. *Plant Physiol.* **1972**, *50*, 341–346.

Mayak, S.; and Halevy, A. H.; Cytokinin activity in rose petals and its relation to senescence. *Plant Physiol.* **1970**, *46*, 497–499.

Mayak, S.; and Kofranek, A. M.; Altering the sensitivity of carnation flowers (*Dianthus caryophyllus* L.) to ethylene. *J. Am. Soc. Hortic. Sci.* **1976**, *101*, 503–506.

Mayak, S.; Bravdo, B.; Gvilli, A.; and Halevy, A. H.; Improvement of opening of cut gladioli flowers by pretreatment with high sugar concentrations. *Sci. Hortic.* **1973**, *1*, 357–365.

Meir, S.; Droby, S.; Davidson, H.; Alsevia, S.; Cohen, L.; Horev, B.; and Philosoph-Hadas, S.; Suppression of Botrytis rot in cut rose flowers by postharvest application of methyl jasmonate. *Postharvest Biol. Technol.* **1998**, *13*, 235–243.

Mokhtari, M.; and Reid, M. S.; Effects of postharvest desiccation on hydric status of cut roses.. In: *Postharvest physiology, pathology and technologies for horticultural commodities: Recent advances*, Ait-Oubahou, A., and El-Otmani, M., Eds.; Institut Agronomique et Veterinaire Hassan II, Agadir, Morocco; **1995**, pp 489–495.

Mor, Y.; Halevy, A.; Kofranek, A. M.; and Reid, M. S.; Posthavest handling of lily of the Nile flowers. *J. Am. Soc. Hortic. Sci.* **1984**, *109*, 494–497.

Morris, K.; Mackerness, S. A. H.; Page, T.; John, C. F.; Murphy, A. M.; Carr, J. P.; and Buchanan-Wollaston V.; Salicylic acid has a role in regulating gene expression during leaf senescence. *Plant J.* **2000**, *23*, 677–685.

Mortensen, L. M.; and Fjeld, T.; Effects of air humidity, lighting period and lamp type on growth and vase life of roses. *Sci. Hortic.* **1998**, *73*, 229–237.

Mortensen, L. M.; and Gislerod, R. H.; The effect of root temperature on growth, flowering, and vase life of greenhouse roses grown at different air temperatures and CO2 concentrations. *Gartenbauwissenschaft.* **1996**, *61*, 211–214.

Muller, R.; Stummann, B. M.; Andersen, A. S.; and Serek. M.; Involvement of ABA in postharvest life of miniature potted roses. *Plant. Growth. Regul.* **1999**, *29*, 143–150.

Mutui, M.; Emongor, V. N.; and Hutchinson, M. J.; Effect of benzyladenine on the vase life and keeping quality of *Alstroemeria* cut flowers. *J. Agric. Sci. Technol.* **2003**, *5*, 91–105.

Naidu, S. N.; and Reid, M. S.; Postharvest handling of tuberose (*Polianthes tuberosa* L.). *Acta Hortic.* **1989**, *261*, 313–318.

Nardini, A.; Peda, G.; and La Rocca, N.; Trade-offs between leaf hydraulic capacity and drought vulnerability: morpho-anatomical bases, carbon costs and ecological consequences. *New Phytol.* **2012**, *196*, 788–798.

Newman, J. P.;. van Doorn, W.; and Reid, M. S.; Carbohydrate stress causes leaf blackening. *J. Intl. Protea Assoc.* **1989**, *18*, 44–46.

NHB Database; National Horticulture Board, Government of India, **2013**.

Nichols, R.; Senescence and sugar status of the cut flower. *Acta Hortic.* **1973**, *41*, 21–27.

Nikkhah Bahrami, S.; Zakizadeh, H.; and Ghasem Nejad, M.; Proceedings of the Seventh International Congress of Iranian Horticultural Science, **2011**; pp 496.

Nowak, J.; and Rudnicki, M. R.; Postharvest handling and storage of cut flowers. In: *Florist Greens and Potted Plants.* Timber Press Inc.: Portland, Oregon; **1990**.

Onoue, T.; Mikami, M.; Yoshioka, T.; Hashiba, T.; and Satoh, S.; Characteristics of the inhibitory action of 1,1-dimethyl-4-(phenylsulfonyl) semicarbazide (DPSS) on ethylene production in carnation (*Dianthus caryophyllus* L.) flowers. *Plant Growth Reg.* **2000**, *30*, 201–207.

Otsubo, M.; and Iwaya-Inoue, M.; Trehalose delays senescence in cut gladiolus spikes. *Hortic. Sci.* **2000**, *35*, 1107–1110.

Panavas, T.; Walker, E. L.; and Rubinstein, B.; Possible involvement of abscisic acid in senescence of daylily petals. *J. Exp. Bot.* **1998**, *49*, 1987–1997.

Philosoph, H. S.; Meir, S.; Rosenberger, I.; and Halevy, A. H.; Regulation of the gravitropic response and ethylene biosynthesis in gravistimulated snapdragon spikes by calcium chelators and ethylene inhibitors. *Plant Physiol.* **1996**, *110*, 301–310.

Philosoph, H. S.; Meir, S.; Rosenberger, I.; and Halevy, A. H.; Control and regulation of the gravitropic response of cut flowering stems during storage and horizontal transport. *Acta Hortic.* **1995**, *405*, 343–350.

Ranjan, P.; Ranjan, J. K.; Das, B.; and Ahmed, N.; High value flower cultivation under low cost greenhouse in NW Himalayas. *Int. J. Chem. Tech. Res.* **2013**, *5*(2), 789–794.

Reid, M. S.; and Jiang, C. Z.; Postharvest biology and technology of cut flowers and potted plants. *Hortic. Rev.* **2012**, *40*, 1–54.

Reid, M. S.; and Jiang, C. Z.; Florigen unmasked: Exciting prospects for horticulture. *Chronica. Hortic.* **2011**, *51*, 7–9.

Reid, M . S.; and King, A. I.; Non-metabolized cytokinines for improved postharvest performance of ornamentals progress. *Am. Floral. Endowm. Ann. Prog. Rep.* **2003**, 3.

Reid, M. S.; and Kofranek, A. M.; Postharvest physiology of cut flowers. *Chronica Horticulturae.* **1980**, *20*(2), 25–27.

Reid, M. S.; and Lukaszeniski, T. A.; Postharvest care and handling of cut flowers. University of California; **1988**.

Reid, M. S.; and Men-Jen Wu. Ethylene and flower senescence. *Plant Growth Regul.* **1992**, *11*, 31–43.

Reid, M. S.; and Scelikel, M. J.; Use of 1-Methylcyclopropene in ornamentals: Carnations as a model system for understanding mode of action. *Hortic. Sci.* **2008**, *43*, 95–98.

Reid, M. S.; Paul, J. L.; Farhoomand, M. R.; Kofranek, A. M.; and Staby, G. L.; Pulse treatments with the silver thiosulphate complex extend the vase life of carnations. *J. Am. Soc. Hortic. Sci.* **1980**, *105*, 2–27.

Reid, M. S.; van Doorn, W.; and Newman, J. P.; Leaf blackening in proteas. *Acta Hortic.* **1989**, *261*, 81–84.

Reid, M. S.; Advances in shipping and handling of ornamentals. *Acta Hortic.* **2001**, *543*, 277–284.

Reid, M. S.; Handling of cut flowers for air transport. IATA Perishable Corgo Manual: flowers, **2009**, p. 24.

Reid, M. S.; and Wu, M. J.; Ethylene and flower senescence. *Plant Growth Reg.* **1992**, *11*, 37–43.

Rij, R. E.; Thompson, J. F.; and Farnham, D. S.; Handling, pre-cooling and temperature management of cut flower crops for truck transportation. USDA AAT-W- 5/June 1979; **1979**, 26 p.

Rivero, R. M.; Kojima, M.; Gepstein, A.; Sakakibara, H.; Mittler, R.; Gepstein, S.; and Blumwald, E.; Delayed leaf senescence induces extreme drought tolerance in a flowering plant. *Proc. Nat. Acad. Sci.* (USA). **2007**, *104*, 19631–19636.

Riz, R. E.; Thomson, J. F.; and Farnham; Handling, precooling and temperature management of cut flower crops for truck transportation. USDA-AAT-W-5-June; **1979**.

Ronen, M.; and Mayak, S.; Interrelationship between abscisic-acid and ethylene in the control of senescence processes in carnation flowers *Dianthus caryophyllus* cultivar White Sim. *J Exp. Bot.* **1981**, *32*, 759–766.

Rot, I.; and Friedman, H.; Desiccation-induced reduction in water uptake of gypsophila florets and its amelioration. *Postharv. Biol. Technol.* **2010**, *57*, 189–195.

Sack, L.; and Holbrook, M.; Leaf hydraulics. *Annu. Rev. Plant Biol.* **2006**, *57*, 361–381.

Saeed, T.; Hassan, I.; Abbasi, N. A.; and Jilani, G.; Effect of gibberellic acid on the vase life and oxidative activities in senescing cut gladiolus flowers. *Plant Growth Regul.* **2013**. doi:10.1007/s10725-013-9839-y.

Saks, Y.; and Staden, J.; Evidence for the involvement of gibberellins in developmental phenomena associated with carnation flower senescence. *Plant Growth Regul.* **1993**, *12*, 105–110.

Sankhla, N.; Mackay, W. A.; and Davis, T. D.; Reduction of flower abscission and leaf senescence in cut phlox inflorescence by thidiazuron. *Acta Hortic.* **2003**, *628*, 837–841.

Sankhla, N.; Mackay, W. A.; and Davis, T. D.; Effect of thidiazuron on senescence of flowers in cut inflorescences *of Lupinus densiflorus* Benth. *Acta. Hortic.* **2005**, *669*, 239–243.

Savin, K. W.; Baudinette, S. C.; Graham, M. W.; Michael, M. Z.; and Nugent, N. D.; Lu, C. Y.; Chandler, S. F.; and Cornish, E. D.; Antisense ACC oxidase RNA delays carnation petal senescence. *Hortic. Sci.* **1995**, *30*, 970–972.

Savvides, A.; Fanourakis, D.; and van Ieperen, W.; Coordination of hydraulic and stomatal conductances across light qualities in cucumber leaves. *J. Exp. Bot.* **2012**, *63*, 1135–1143.

Seglie, L.; Martina, K.; Devecchi, M.; Roggero, C.; Trotta, F.; and Scariot, V.; The effects of 1-MCP in cyclodextrin-based nanosponges to improve the vase life of *Dianthus caryophyllus* cut flowers. *Postharvest Biol. Technol.* **2011**, *59*, 200–205.

Seglie, L.; Sisler, E. C.; Mibus, H.; and Serek, M.; Use of a non-volatile 1-MCP formulation, N,N-dipropyl(1-cyclopropenylmethyl)amine, for improvement of postharvest quality of ornamental crops. *Postharvest Biol. Technol.* **2010**, *56*, 117–122.

Serek, M.; Sisler, E. C.; and Reid, M. S.; Novel gaseous ethylene binding inhibitor prevents ethylene effects in potted flowering plants. *J. Am. Soc. Hortic. Sci.* **1994a**, *119*, 1230–1233.

Serek, M.; Sisler, E. C.; and Reid, M. S.; Effects of l-MCP on the vase life and ethylene response of cut flowers. *Plant Growth Reg.* **1995**, *16*, 93–91.

Sisler E. C.; and Lallu, N.; Effect of diazocyclopentadiene (DACP) on tomato fruits harvested at different ripening stages. *Postharvest Biol. Technol.* **1994**, *4*, 245–254

Subhashini R. M. B.; Amarathunga, N. L. K.; Krishnarajah, S. A.; and Eeswara, J. P.; Effect of benzylaminopurine, gibberellic acid, silver nitrate and silver thiosulphate, on postharvest longevity of cut leaves of *Dracaena. Ceylon. J. Sci. (Bio. Sci.).* **2011**, *40*(2), 157–162.

Tanase, K. K.; Amano, T. M.; and Ichimura, K.; Ethylene sensitivity and changes in ethylene production during senescence in long-lived Delphinium flowers without sepal abscission. *Postharvest Biol. Technol.* **2009**, *52*, 310–312.

Teixeira da Silva, J. A.; The cut flower: Post-harvest considerations. *Online. J. Biol. Sci.* **2003**, *3*(4), 406–442.

Thompson, J. F.; Bishop, C. F. H.; and Brecht, P. E. P.; Air transport of perishable products. University of California ANR publication #21618. 22 pp; **2004**.

Thwala, M.; Wahome, P. K.; Oseni, T. O.; and Masarirambi, M. T.; Effects of floral preservatives on the vase Life of orchid (*Epidendrum radicans* L.) Cut Flowers. *J. Hortic. Sci. Ornamen. Plant.* **2013**, *5*(1), 22–29.

Torre, S.; and Fjeld, T.; Water loss and postharvest characteristics of cut roses grown at high or moderate relative humidity. *Sci. Hortic.* **2001**, *89*, 217–226.

Torre, S.; Fjeld, T.; Gislerod, H. R.; and Moe, R.; Leaf anatomy and stomatal morphology of greenhouse roses grown at moderate or high air humidity. *J. Am. Soc. Hortic. Sci.* **2003**, *128*, 598–602.

van der Hulst, J.; Cool chain management for cut flowers. *Flower. TECH.* **2004**, *7*(6), 49–51.

van der Meulen-Muisers, J. J.; van Oeveren, J. C.; and van Tuyl, J. M.; Breeding as a tool for improving postharvest quality characters of lily and tulip flowers. *Acta Hortic.* **1997**. 430, 569–575.

Van Doorn, W. G.; and Reid, M. S.; Vascular occlusion in stems of cut rose flowers exposed to air. Role of xylem anatomy and rates of transpiration. *Physiol. Plant.* **1995**, *93*, 624–629.

Van Doorn, W. G.; Perik, R. R. J.; Abadie, P.; and Harkema, H.; A treatment to improve the vase life of cut tulips: effects on tepal senescence, tepal abscission, leaf yellowing and stem elongation. *Postharvest Biol. Technol.* **2011**, *61*, 56–63.

Van Ieperen, W.; and Trouwborst, G.; The application of LEDs as assimilation light source in greenhouse horticulture: a simulation study. *Acta Hortic.* **2008**, *801*, 1407–1414.

Van Ieperen, W.; Plant morphological and developmental responses to light quality in a horticultural context. *Acta Hortic.* **2012**, *956*, 131–140.

van Meeteren, U.; van Gelder, H.; and van Ieperen, W.; Reconsideration of the use of deionized water as vase water in postharvest experiments on cut flowers. *Postharvest Biol. Technol.* **1999**, *17*, 175–187.

Van Meeteren, U.; Why do we treat flowers the way we do? A system analysis approach of the cut flower postharvest chain. *Acta Hortic.* **2007**, *755*, 61–74.

Van Stade, J.; and Dimalla, G. G.; Endogenous cytokinins in bougainvillea 'San Diego Red': I. Occurrence of cytokinin glucosides in the root sap. *Plant Physiolgy.* **1980**, *65*, 852–854.

Van Staden, J.; Upfold, S. J.; Bayley, A. D.; and Drewes, F. E.; Cytokinins in cut carnation flowers ix transport and metabolism of isopentenyladenine and the effect of its derivatives on flower longevity. *Plant Growth Regul.* **1990**. *9*, 255–262.

VBN; *Evaluation Cards For Cut Flowers*. VBN: Leiden, The Netherlands; **2005**.

Veen, H.; and van de Geijn, S. C.; Mobility and ionic form of silver as related to longevity of cut carnations. *Plantae.* **1978**, *140*, 93–96.

Veen, H.; and Van De Geijn, S. C.; Mobility and ionic form of silver as related to longevity of cut carnation. *Plantae.* **1978**, *140*, 93–96.

Velez-Ramirez, A. I.; van Ieperen, W.; Vreugdenhil, D.; and Millenaar, F. F.; Plants under continuous light. *Trend. Plant Sci.* **2011**, *16*, 310–318.

Victoria, N. G.; Kempkes, F. L. K.; van Weel, P. A.; Stanghellini, C.; Dueck, T. A.; and Bruins, M. A.; Effect of a diffuse glass greenhouse cover on rose production and quality. *Acta Hortic.* **2012**, *952*, 241–248.

Waithaka, K.; Reid, M. S.; and Dodge, L.; Cold storage and flower keeping quality of cut tuberosa (*Polianthes tuberosa* L.). *J. Hortic. Sci. Biotechnol.* **2001**, *76*, 271–275.

Wang, C. Y.; and Baker, J. E.; Vase life of cut flowers treated with rhizobitoxine analogs, sodium benzoate, and isopentenyl adenosine. *Hortic. Sci.* **1979**, *14*, 59–60.

Whitehead, C. S.; Halevy, A. H.; and Reid, M. S.; Control of ethylene synthesis during development and senescence of carnation petals. *J. Am. Soc. Hortic. Sci. 109*, 473–475.

Woltering, E. J.; and van Doorn, W. G.; Role of ethylene in senescence of petals— morphological and taxonomical relationships. *J. Exp. Bot.* **1988**, *39*, 1605–1616.

Woltering, E. J.; The effects of leakage of substances from mechanically wounded rose stems on bacterial growth and flower quality. *Sci. Hortic.* **1987**, *33*, 129–136.

Woodson, W. R.; Park, K. Y.; Drory, A.; Larsen, P. B.; and Wang, H.; Expression of ethylene biosynthetic pathway transcripts in senescing carnation flowers. *Plant. Physiol.* **1992**, *99*, 526–532.

Wu, M. J.; van Doorn, W. G.; and Reid, M. S.; Variation in the senescence of carnation (*Dianthus caryophyllus* L.) cultivars. I. Comparison of flower life, respiration and ethylene biosynthesis. *Sci. Hortic.* **1991**, *48*, 99–107.

Yamada, T.; Takatsu, Y.; Manabe, T.; Kasumi, M.; and Wataru, M.; Suppressive effect of trehalose on apoptotic cell death leading to petal senescence in ethylene-insensitive flowers of gladiolus. *Plant Sci.* **2003**, *164*, 213–221.

Zagory, D.; and Reid, M. S.; Evaluation of the role of vase micro-organisms in the postharvest life of cut flowers. *Acta Hortic.* **1986**, *181*, 207–216.

CHAPTER 11

POSTHARVEST MANAGEMENT OF MEDICINAL AND AROMATIC PLANTS

S. DAS[1], S. SULTANA, and A. B. SHARANGI[*]

[1]Department of Floriculture and Landscape Gardening, Faculty of Horticulture, Bidhan Chandra Krishi Viswavidyalaya, Mohanpur-741252, Nadia, West Bengal, India

Department of Spices and Plantation Crops, Faculty of Horticulture, Bidhan Chandra Krishi Viswavidyalaya, Mohanpur-741252, Nadia, West Bengal, India

[*]Email: dr_absharangi@yahoo.co.in

CONTENTS

11.1 INTRODUCTION

In recent decades there has been a gradual global rise in demand for herbal medicines and healthcare products. However, the quality systems and validation procedures governing the sector are not commensurate with such growth. The issues concerning quality, efficacy, and safety vis-a vis "minimum therapeutic guarantee" must be addressed in an ingenuous way as the very purpose of any healthcare product is centered on an ethical assurance to the end-user. But poor quality produce are marketed due to inefficient postharvest handling, the demand-supply disparity and other technical flaws. Since ancient times, the physician or the traditional healer used to manage the whole supply chain, right from collection of herbs up to preparation and dispensing of the medicine which is now a multistake holders activity. The consequence gave rise to a grave situation by means of over-exploitation, habitat loss, threat to plant species and even extinction of some valuable species.

Now, the burning issue is how to come out of this vicious cycle. No doubt medicinal plants and mankind will be having the same intricate dependence as it was in the dawn of civilization, if not more. The common solution to this problem may be the easiest answer of cultivating the medicinal crops. But will the pharmacochemical complexity of the botanicals, which is the outcome of biotic and abiotic stresses the plant undergoes in a competitive environment prevailing in the wild, be mimicked in the farming conditions? Moreover, the natural survival and multiplication of many herbaceous species in forest scenario is linked to the associated flora and fauna. Diversion of fertile crop lands to nonfood crops will possibly raise huge uproar with respect to food security, ethnic issues and farm income. The possible positive outlook could be skill up-gradation of collectors, fair trade and market links for small-scale producers, natural regeneration, conservation of biodiversity and protection endangered species, efficient documentation of the species in commercial cultivation, clonal micropropagation and their subsequent herbal uses with Good Agricultural Practices (GAP).

CULTIVATION OF MEDICINAL PLANTS

Most of medicinal plants, even today, are collected from wild that caused a steady erosion of these valuable resources in their natural habitat. A scarcity is probable in the supply of raw materials to the pharmaceutical industry as well as the traditional practitioners. Immediate measure is a must to protect, conserve, and cultivate them before they may be lost from the natural vegetation forever. In situ conservation of these resources alone cannot meet the ever increasing demand of pharmaceutical industry. It is, therefore, inevitable to develop cultural practices and propagate these plants in suitable agroclimatic regions. Commercial cultivation will put a check on the continued exploitation from wild sources and serve as an effective means to conserve the rare floristic wealth and genetic diversity.

Efforts are also required to suggest appropriate cropping patterns for the incorporation of these plants into the conventional agricultural and forestry cropping systems and to formulate their proper cultivation techniques, harvesting methods, safe use of fertilizers and pesticides and waste disposal.

There are at least 35 major medicinal plants that can be cultivated in India and have established demand for their raw material or active principles in the international trade.

New varieties of medicinal plants developed in India

Crop	Variety	Characters (Institution where developed)
Psyllium *Plantagoovata*	Gujarat Isabgol-1 Gujarat Isabgol-2(GI-2)	High seed yield (1t/ha) with synchronous maturing of seed (GAU, Anand) Seed yield of 1t/ha, moderately resistant to downy mildew disease (GAU, Anand)
Opium poppy *Papaversomniferum*	Jawahar Aphim-16 (JA-16)	White flowered with serrated petals, produces oval capsules maturing early at 105-110 days for lancing. Yield 66kg of latex averaging 10% of morphine (JNKVV, Mandsur)
	Trishna (IC-42)	Medium dwarf, pink flowered, serrated petals. Produces large bumble-shaped capsules, high latex and morphine content.(over JA-16) (NBPGR, Delhi)
	Udaipur Opium (UO-285)	High latex yield in Rajasthan tract (58kg/ha) with high morphine content (12.3%) and high seed yield (1.2t/ha) (RU, Udaipur)
	NRBI-3	High latex yield in central and eastern UP. Latexyield 47-57.54kg/ha. (NBRI, Lucknow)
	Kirtiman (NOP-4)	Latex yield 45.84kg/ha, morphine content 11.94% in eastern U.P. Moderately resistant to downymildew (NDUA & T, Faizabad) With pale white peduncle, produces 66.5kglatex/ha with 18% morphine (CIMAP)
	Sweta (GS-24) Shyama (IS-34)	Foliage erect and incised, bears black flowering stalk. Produces 78.1% latex with 15.5% morphine (CIMAP).
Sarpagandha *Rauvolfiaserpentina*	RS-1	High seed germination (50%). Root yields 2.5t/ha in 18 months. Roots carry 1.45-1.80% of total alkaloids; half of it yields reserpine + serpentine combined (JNKVV, Indore)

Dioscorea flori-bunda	FB(C)-1	A composite culture, produces fast growing vines relatively free from diseases and pest attack; produces 50t/ha of fresh tubers in 2 years containing 3.5% diosgenin (IIHR, Bangalore)
	Arka-Upkar	Selection through hybridization, producing 60t of fresh tubers containing 3.5-4.0% diosgenin (IIHR, Bangalore)
Khasi-kateri *Solanumviarum*	Glaxo	Plants devoid of spines, produces high berry yield at high density planting containing 2.5-3.0% solasodine (Glaxo, India).
	IIHR 2n-11	Completely devoid of spines, produces high berry yield at high density planting containing 2.5-3% solasodine (IIHR, Bangalore)
Kangarokateri *Solanumlaciniatum*	EC-113465	Long duration crop (300 days) suitable for temperate regions. High solasodine content in leaves (1.8%) and mature berries (4%) (YSPHU,Solan)
Henbane *Hyocyamusniger*	IC-66	Short duration (100 days), early *rabi*crop inplains. Yields 2.5t/ha of dry herb with minimum0.05% total alkaloids (NBPGR, Delhi)
	Aela	A mutant characterized by yellow flower petals, produces 7.5t/ha dry herb or 23kg total alkaloids/ha (CIMAP)
Egyptian Henbane *Hyocyamusmuticus*	Auto-tetraploid	Vigorously growing and high seed fertile mutant, produces 4.5t/ha of dry herb or 23kg total alkaloids/ha (CIMAP)
Senna *Cassia*	ALFT-2	Late flowering type, tailored to produce purely leaf crop in one harvest at 100 days. Foliage sennoside *angustifolia* content (6.0%) (GAU, Anand)

Japanese mint *Menthaarvensis* var. *piperascens*	MAS-1	Yields fresh herb of 37.2t/ha in 2 cuttings; containing 0.8-1.0% oil with high leaf/stem ratio. Matures 10-15 days early. Oil yield 290kg/ha
	MAS-2	containing 83% menthol (CIMAP)
	Hyb-77	Fresh herb yield 69t/ha, oil 348kg/ha (CI-MAP)
		A tall vigorous, compact growing type, cross ofMAS-2 x MA-2. Produces 78.2t/ha fresh herb, oilyield 486kg/ha with 81.5% menthol. Highly
	Siwalik	resistant to leaf spot and rust diseases (CI-MAP)
		Introduced from China, produces compact bushy growth with thick leathery leaves, high herb and oilyield.
	EC-41911	
		A progeny selection of interspecific cross between *M. arvensis* and *M. piperita*in USSR. High herbage yield with high oil content (0.8-1%); oil contains 70-80% menthol (YSPHU, Solan).
Ocimumgratissi-mum Thymol basil- *O. viride* Sacredbasil-*O. sanctum*	Clocimum	High herbage yield with high oil content, 75%eugenol (RRL, Jammu).
	Thymol type	Herb yield - 3t/ha, 59kg/ha oil/annum.(CIMAP, Lucknow)
	EC-1828893	Superior selection with high oil yield 55l/ha in 110 days containing 53% eugenol and 19% caryophylline (NBPGR, Delhi)

11.1.2 ORGANIC FARMING

Organic farming in recent times is becoming a consumer preferred production system worldwide, for, products produced through organic farming are without inorganic fertilizers and pesticides and thus are safe to consume. Organic farming has its root in Nature. Maintenance of soil fertility, the flora and fauna of the soil, the biological cycles and the eco-system are primary concerns in organic farming. The main objective of organic farming includes mulching, crop rotation, cover cropping, green manuring, use of animal waste, composting, bio-gas slurry, bio-fertilizers and organic recycling.

As this method of crop culture requires a good and satisfactory understanding of the soil-water-crop- pathogen—insect- microbe continuum to grow safe food without chemicals, farmers need to be trained in all aspects of organic farming including the process of certification. This will enhance environment safety, biodiversity conservation and will reduce environmental pollution.

11.2　KEY POINTS IN MEDICINAL PLANT FARMING AND TRADE

- The comprehensive botanical identity of the species should be confirmed with respect to the descriptors as detailed in Ayurvedic, Homoeopathic, and Herbal Pharmacopoeia of India. Proper documentation with maximum information recorded during the collection of the produce helps in tracing the product origin, quality aspects and agreement-permission status of the collections.
- Collection of infested plant produce should be avoided and plants may be harvested at such rates that the species perpetuates indefinitely in its natural and wild habitats. The over exploitation of any herb species threatens the existence of the species in natural habitat. Lack of knowledge of the adverse environmental impact of over- exploitation may adversely affect the ecological balance and loss of genetic diversity of the surrounding habitats. Herbal crop species should be harvested within the limits of their capacity for regeneration within their geographic domain. Collection practices ought to confirm the long- term survival of wild populations, particularly the endemic plant species and their associated natural flora and fauna as per the International Standard for Sustainable Collection of wild medicinal plants.
- To have highly potent herbal formulations, medicinal plants should be harvested at the right physiological growth stage, for, the concentration of biologically active substances varies with the stages of growth and development of plants. Discard immature or over matured samples and products with foreign matters and toxic weeds. Harvested products may be sorted out just after reaping or drying, as the case may be.
- Collection procedure must adhere to the legal regulations pertaining to the certification norms by maintain sustainable frequency. Collection or harvesting tools must be thoroughly cleaned after use to avoid cross contamination with the remaining residues.
- Avoid collection during early hours to avoid dew, unless needed. Also avoid rains, foggy, and highly humid weather conditions and contaminated sources like industrial areas, sewage lines, crematoria, hospitals, mining sites, public utilities, automobile workshops, dumping grounds, anthills etc.
- Provisions should be laid down for a fair price mechanism though the supply chain and benefit sharing of the local people.
- Due respect should be given to the indigenous local value systems and ethical codes pertaining to sacred groves, water bodies and holy species.

- Timely and right processing of medicinal plant produce to preserve the quality and enhance shelf life is mandatory. Sun drying or shade drying as well as artificial methods may be used.
- Each container, of medicinal plant produce be it glass, metal, plastic, or elastomeric materials, should be labeled properly. Flammable materials should be clearly labeled. The label should contain all the required information of medicinal plant produce. Under any circumstance, already used bags/packets for food articles, construction articles such as cement, sand, or that of fertilizers or other chemicals should not be used for packing medicinal produce.
- Inappropriate storage conditions may render the produce unusable. Properly sealed and labeled containers should be kept preferably on wooden pallets, at cool and dry places. Storage house should have provisions to avoid entry of rodents, birds, and other animals and should be free from dampness, dirt, and dust.
- Training and capacity building of the collectors, particularly with respect to botanical aspects and discriminating the morphologically similar looking but botanically different species in the nature to avoid collection from wrong plant species is a must.
- Demand and supply mismatch is the primary reason for unsustainable harvesting and quality degradation. Reliable data should be generated through baseline survey on availability of medicinal plants in their wild habitat, sustainable harvesting quality and potency aspects to enable regular monitoring and their subsequent management plan.

11.3 QUALITY CONTROL

The critical issues in herbal trade are the lack of drug standardization, information, and quality control. Adulteration in the market samples and consequent lack of authentication of the botanical identity and genuineness of drugs are real threats to the stakeholders globally. As most of the Ayurvedic medicines are in the form of crude extracts which are a mixture of several ingredients and the active principles when isolated individually fail to give desired activity. Isolation and standardization of active constituents and validation of their therapeutic potency are much needed thrust areas which may be achieved by estimating their active principles, recording the anatomical features under microscope and confirming their curative effects by clinical trials.

General Scheme for Quality Assurance of Crude Drugs and Raw Materials

The general scheme for quality assurance of crude drugs and raw materials are given below.

1. Importance of quality assurance of crude drugs and raw material:
   (a) Guarantee the best final pharmaceutical products, (b) Environmental protection

(c) Sustainable utilization and development of natural resource.
2. Criteria of good quality:
    (a) Good efficacy- high active ingredient, high yield, (b) Good safety-less toxicity and side effects, minimum pesticide residues, minimum heavy metals, (c) Purity, (d) Stability
3. Gene bank conservation:
    (a) Biodiversity conservation, (b) To store plant germplasm for future uses, (c) To make germplasm available to create new cultivars, (d) *In situ* conservation—gene banks of medicinal and aromatic plants in Asia, (Table 11.1) (e) *In vitro* conservation, (f) Breeding
4. Biotechnology
    (a) Plant cell culture (eg., *Digitalis, Catharanthus*)
    (b) Hairy root culture (eg., *Salvia, Glycyrrhizauralensis, Daturastramoniun, Artemesiaannua*)
    (c) Tissue culture (eg. *Aloe, Crocus sativa, Mentha*)
    (d) Genetic engineering: Isolation and purification of an antifungal protein from *Phyto laccaamericana* against American ginseng pathogens and synthesis of its gene and expression in *E. coli.*
5. Suitable growth region: In order to get higher quality of crude drugs and raw materials selection of the most suitable growth region for relevant medicinal plant is quite important. According to the ecological conditions, flora, and other criteria, several regions of crude drug development have been identified.
6. GAP: Good agrotechnological Practices. Strong and continuing research augments large scale cultivation of medicinal plants. Plant varieties with an abundance of desired constituents can be reproduced and improved upon under cultivation even in an entirely different area. For example, cultivation of American ginseng (*Panaxquinquefolia*) in China. Attempt should be made to select appropriate region based on similar ecological conditions to introduce good cultivated variety, improve yield of the desired secondary metabolite and reduce the undesirable constituents.
7. Nonpolluted cultivation: In order to protect the environment, to sustainably utilize there sources and to get a good quality of crude drug, nonpolluted agro technology is rapidly developed in recent years. These products are commonly called as "Green crude drugs." This involves biological control of insects and pathogens and use of botanical pesticides for the control of pest and diseases.
8. Postharvest technology: Right time harvesting, good processing, good storage, extraction or distillation, quality control.

TABLE 11.1   Genebanks of **medicinal and aromatic plants in Asia**

Country	Collections	Institutions
China	2,500	IMPLAD Beijing and its 3 stations
India	1,400	NBPGR, New Delhi; CIMAP Lucknow; AMPRS, Odakkali
Korea	850	Medicinal plants gardens
Malaysia	450	National Research council and Kuala Lumpur citycouncil gardens
Nepal	340	Royal Botanic gardens
Philippines	220	University Herbal garden, Los Banos
Sri Lanka	200	Royal Botanic garden, Kandy
Thailand	100	Botanic gardens

Quality Control Requirement of New Preparation of Traditional Medicines

1. Prescription and its basis
2. Literature and research data of physicochemical characteristic concerned with quality
3. Preparation technology and its research references
4. The draft of the quality standard and explanation of medicinal material, and medicament.
5. Literature and test data of initial stability for clinical research
6. The reports of quality detection and hygiene standard detection of the preparation for clinical research
7. Property and specification of the packing material of the medicament, design draft of the label and applied instructions

General Requirement of Quality Control Standard of Medicament

1. Quantitative determination of the effective compound or indicative component of 1-2species of main medicinal materials in the prescription
2. Qualitative identification of several to half of the medicinal materials in the prescription
3. Determination of content of Pb, Cd, Hg, As, and limit test of heavy metals in medicines
4. Hygienic standard: bacteria < 1000/1gm, mold< 100/1gm, coli bacillosis-nil
5. Determination of pesticide residues (organic Cl and P) in the medicament

The General Scheme for Quality Assessment of Botanicals

The general scheme for quality assessment of botanicals is given here.

I.   **Assessment of crude plant materials**
1.   General description of the plant
2.   Parts used
3.   Production of crude drugs-cultivation, harvesting, postharvest handling, packing, storage.
4.   Quality specification: Chemical or chromatographic identification, foreign organic matter limit, ash content, acid insoluble ash content, water soluble extractive, alcohol soluble extract, moisture content, active constituent content, microbial limit, pesticide residue limit, heavy metal limit, likely contaminants and adulterants.

II. **Assessment of finished products**
1.   Tablets: Weight variation, disintegration time, identification of preservatives and active ingredients, determination of extractives in various solvents, microbial limit, heavy metals.
2.   Solutions: pH, identification of preservatives and active ingredients, alcohol content, microbial limit, Sodium Saccharic content.
3.   Infusions: Weight variations, identification of preservatives and active ingredients ,determination of extractives in various solvents, microbial limit, heavy metals, Borax.

III. **Chemical Standardization methods***:* TLC/HPTLC, HPLC, GLC, FTIR

IV. **Chemical Markers***:* Specification for raw materials, quality assurance in process control, standardization of product, obtaining stability profiles, single marker vs. *fingerprint.*

V.   **Parameters of assay validation***:* Linearity, limits of quantification and detection, precision, robustness, recovery. Complex and variable mixtures, choice of compounds to quantify, difficult sample preparation, lack of pure reference standards, lack of methods with adequate tolerances by analytical chemistry standards are some of the challenges in Chemical Standardization of plant drugs.

International Scheme for Quality Assurance of Pharmaceutics

International scheme for quality assurance of pharmaceutics involves the following standard practices.

GAP: Good Agricultural Practice

GLP: Good Laboratory Practice

GMP: Good Manufacturing Practice

GCP: Good Clinical Practice

GALP: Good Analytical/Automated Laboratory Practice

Environmentally safe GAP, GMP, and GLP have to be introduced better global market coverage. Organic farming reduces contamination of products and the environment with synthetic chemicals. Buyers now insist on fro eco-friendly products with suitable levels. These are also the requirements for ISO 9000 certifications and eco-audit procedures for safeguarding environmental damage.

Aremu et al (2012) Studies on anthelmintic activity of medicinal plants have received insufficient interest and attention from researchers despite the high incidence of helminth infections in the poorer communities of South Africa. There are only a few anthelmintic remedies available which are inadequate in terms of accessibility, affordability, and probably efficacy. In this review, we reappraised the various anthelmintic studies on South African medicinal plants to highlight how much and/or how little is known. The rich botanical and medicinal plant knowledge in South Africa is an indication of the potential of discovering potent treatments against helminth diseases in both humans and livestock. A total of 115 plant species encompassing 43 families screened for their anthelmintic potential (mainly nematodes) are listed in the current review. Combretaceae and Fabaceae were the most commonly used families. Tetradenia riparia, Hypoxis colchicifolia, Apodytes dimiata, and Leucosidea sericea are a few examples of the South African species that have demonstrated promising anthelmintic activity. Even though other species such as Dicerocaryum eriocarpum, Berchemia zeyheri, and Acorus calamus are potent anthelmintics, caution must be exercised in administering these plant extracts because of their potential toxic effects. Besides the benefit of validating the efficacy of traditional medicines, compounds from South African medicinal plants could also provide a template for novel synthetic anthelmintics. However, the major bottleneck in exploring more South African medicinal plants for anthelmintic properties is probably the lack of a robust, rapid, and reliable screening technique as well as adequate funding schemes. Perhaps, there could be more success stories via increased research outputs with the development and availability of high throughput screening methods for assessing the activity of these medicinal plants as well as their resultant bioactive compounds. In addition, accessibility of funds to acquire latest technology and motivate researchers will inevitably stimulate more studies geared toward alleviating the problems posed by helminth infections.

According to Joymati Devi et al. (2012) Among 31 essential oils, *Xylosma longifolia* oil extract was found to have the most inhibitory effect, egg hatching and larval mortality on followed by *Acorus calamus* and *Vitex negundo* extracts within 12 hrs of exposing period in *X. longifolia* at 10%. Egg hatching rate was found inversely proportional with concentration and directly with the exposure periods. Whereas mortality rate was directly proportional with concentration of extracts as well as exposure periods.

INDIAN GOOSEBERRY (*Phyllanthus emblica*); FAMILY: Euphorbiaceae

Indian gooseberry is useful in hemorrhage, leucorrhaea, menorrhagia, diarrhea, dysentery, and in combination with iron, it is useful for anemia jaundice and dyspepsia The green fruits are made into pickles, preserves to stimulate appetite and the seed is used in asthma bronchitis and biliousness. Tender shoots taken with butter milk cures indigestion and diarrhea and the leaves are also useful in conjunctivitis, inflammation, dyspepsia, and dysentery. The bark is useful in gonorrhea jaundice, diarrhea, and myalgia while the root bark is astringent and is useful in ulcerative stomatitis and gastrohelcosis. Liquor fermented from fruit is good for indigestion, anemia jaundice, heart complaints, cold to the nose and for promoting urination while the dried fruits have good effect on hair hygiene and used as ingredient in shampoo and hair oil. The fruit is a very rich source of Vitamin C (600mg/100g) and is used in preserves as a nutritive tonic in general weakness .Indian gooseberry is found throughout tropical and subtropical India Sri Lanka, Malaca, and also in Madhya Pradesh, West Bengal(Darjeeling), Sikkim and Kashmir.

It is a small to medium-sized deciduous tree of 18 m height with thin light gray bark exfoliating in small thin irregular flakes. Leaves are simple, many subsessile, closely set along the branchlets, distichous light green having the appearance of pinnate leaves and the flowers are greenish yellow in axillary fascicles, unisexual. The globose fruits are l–5cm in diameter, fleshy, pale yellow with 6 obscure vertical furrows enclosing 6 trigonous seeds in 2-seeded 3 crustaceous cocci. Two forms Amla are generally distinguished, the wild ones with smaller fruits and the cultivated ones with larger fruits and the latter are called 'itowaras'.

BRIEF AGROTECHNIQUES:

Gooseberry prefers a warm dry climate, good sunlight and rainfall. It can be grown in almost all types of soils, except very sandy type. *Amla* is usually propagated by seeds and rarely by root suckers and grafts. Seeds are soaked in water for 3–4 hrs before sowing, sown on previously prepared seed beds and irrigated. One month old seedlings can be transplanted to polythene bags and one year old seedlings can be planted in the main field with the onset of monsoon. 50 cm^3 pits are dug at 6–8m spacing and filled with a mixture of top soil and well rotten FYM before planting. During the first year irrigation and weeding are required. Application of organic manure and mulching every year are highly beneficial but chemical fertilizers are not usually applied.

11.4   HARVESTING AND YIELD

Bearing commences from the 10th year, while grafts after 3–4 years. From April to July the vegetative growth of the tree continues. Along with the new growth in the spring, flowering also commences and the fruits will mature by December-February. Fruit yield ranges from 30–50 kg/tree/year when full grown.

POSTHARVEST TECHNOLIGY:

Amla are collected from healthy plants situated near sacred sites (like temples, *chorten*, *nge*) during the first 15 days of the lunar month. Over-matured fruits are not collected. Damaged (diseased, rotten, bruised, with insect bites) amlas are den sorted out and discarded. Sorted amlas are then put in a plastic or similar container and wash thoroughly so that all dirt and other elements such as sand, stones, etc. are washed away. Fresh amlas are then boiled until it is sufficiently cooked to ease deseeding and drying. Amla fruits are often steamed (instead of boiling), which can avoid water accumulation in the fruits, better assist the deseeding work and also save water. Deseeding is the process of separating the seeds from amla pulp and needs to be done as quickly as possible after boiling/steaming, but letting the boiled amla cool down could pose difficulties in deseeding. This process can be time-consuming and laborious, and must be done with full attention. The deseeded amla fruits need to be dried at a safe, clean, dust, and smoke-free area where there is no smell of fat and meat of animals and the area which is inaccessible to wandering (domestic) animals until they become reddish in color (the color and quality required by the market). The deseeded amla are spread thinly on the mat for drying in the sun for the first 2–3 days only. Drying continuously in the sun for many days could destroy the quality of the amla (amla could turn blackish). 2–3 days of sun-drying is needed to reduce the water content to avoid fungal growth on the amla. Drying is continued in a shaded but airy place until moisture-free and hard-dried, after 2 days of drying in the sun. Dried amlas are then collected in PP, plastic bags or containers and stored inside a room until grading, packaging, and labeling works start. Unbroken pieces, fully dried and reddish in color are graded as good quality while dried amla of lesser quality are derived from pulp due to over cooking which stick to one another, broken bits and other small odd shapes, pieces with fungal attacks (not dried properly) and blackish (or other than reddish) in color. After grading and separating dried amla into two groups of good quality and lesser quality, it is packed and labeled for selling. For packaging the dried amla stock are put on a plastic/tarpaulin sheet or bamboo mat for packaging, the dirt and/or foreign materials (if any) are discarded, dried, and cleaned amlas are put in plastic bags of one or five kilograms or any other quantity (depending on the requirement of the buyer), sealed and weighed to check that they contain the correct quantity. The small plastic bags are then put in larger PP bags or Jute bags, which will protect the smaller plastic bags and provide ease for transportation and sealed. The larger PP bags are then labeled including the information's such as: month of collection and processing, date of packaging, number of small packages inside, weight of each small package and the total weight of the bag.

ALOE (*Aloe Vera*)   FAMILY: Liliaceae

Aloe Vera has been used for over 5,000 years. In Ayurvedic medicine aloe is used internally as a laxative, antihelminthic, hemorrhoid remedy, and uterine stimulant (menstrual regulator). The medicinal species of aloe are *Aloe vera, A. barbadensis* (Curacao or Barbados aloe), *A. vulgaris, A. arborescens, A. ferox* (Cape aloe), *A. perryi* (Socotrine or Zanzibar aloe). There are over 300 aloe species, most of which are native to South Africa, Madagascar, and Arabia. The aloe plant has long (up to 20 inches long and 5 inches wide), triangular, fleshy leaves that have spikes along the edges. The flowers (not used medicinally) are yellow in color. These are indigenous to South Africa and South America, but are now cultivated worldwide except in tundra, deserts, and rain forests. In the US it is commercially cultivated in southern Texas.

BRIEF AGROTECHNIQUES:

The crop grows well in tropical and subtropical regions with protection against frost and low winter temperature. It can also be grown successfully in marginal to sub marginal soils having low fertility with pH of 6–7. It flourishes well on dry sandy soils at localities with lower annual rainfall of 500 to 700 mm. It is growth is faster under medium fertile heavier soils as black cotton soils of central India, though, well-drained loam to coarse sandy-loam soils with moderate fertility and pH up to 8.5 are preferred for commercial cultivation. 3-4 months old suckers having 4–5 leaves and about 20–25 cm in length are planted in July- August during monsoon season to get better field survival and subsequent growth, in ridges (Ridge and furrow method), which are 90cm apart. Farm yard manure is applied at 15 t ha^{-1} or at 10 t + Neem cake 1.50 t ha^{-1} during planting. First irrigation is required just after planting of suckers followed by 2–3 irrigations till plant get established. However, 4–6 irrigations per year may be enough for proper growth of the plant. The field should be kept weed free throughout the growing period of crop. Beans, black gram, groundnut, sesame, coriander, cumin etc. can be grown successfully in the interspaces under arid and semiarid condition.

HARVESTING AND POSTHARVEST TECHNOLIGY:

After 10 months the leaves are removed with a sharp sickle/knife leaving 3–4 leaves from the base and the rhizomes left in the field which produces new sprouts to raise the succeeding crop. It gives commercial yield up to an age of 3 years. A precise, quick, and clean cut made at the base of the leaf ensures that the precious gel is not exposed to the open air for too long, causing irreparable oxidation before it reaches the stabilization process. The leaves are delivered to processing center (situated a hundred yards from the cultivation itself) in lots of no more than one ton at once. Harvesting of the smaller and thinner varieties (Aloe arborescens Miller, Aloe ferox, and Aloe chinensis) needs more attention. Around 55 tons of fresh leaves per hectare can be obtained per year.

Aloe leaves are gathered and the outer surface is thoroughly washed with a detergent solution based on quarternary salts, which eliminates any bacteria. The washing process of the smaller varieties requires more attention and a manual procedure.

The aloe leaf is made up of an internal gel called "parenchyma", encased in a green covering cuticle or skin, which is tough and leathery. Preparation of the leaf is usually done by finely mincing the whole leaf, resulted in a dense, greenish pulp with a very bitter taste and a highly laxative effect because of the various anthraquinones present, like aloin and aloetic acid, which are found mainly in the external part of the leaf. The elimination or extraction of such substances results in a less bitter and more palatable taste, achieved via a carbon filtering process which absorbs the undesired substances. Manual decortication of leaf is heavy and labor-intensive work, but is necessary in order to achieve a higher quality product. By hand processing, the industrial filtering process can be avoided and the active, natural ingredients of this wonderful plant are kept intact. The outer casing of leaf is completely peeled off, leaving the internal gel filet free to be minced into a pulp which is both rich in nutrients and then of a more fluid consistency. The manufacturing process for the minor varieties(Aloe arborescens Miller, ferox, and chinensis) avoids the decortication process and consists mainly of chopping the whole leaf.

The pulp obtained through the previously described extraction process remains too thick and unstable at this point. Heat (which can often reach temperatures as high as 158°F, typical of the pasteurization processes) is ueed to liquify the gel by breaking down the bonds in the mucopolysaccharides. Sometimes the product is purposely pasteurized to ensure further stabilization and even better preservation of the produce; however, the original characteristics and nutrients are either diminished or destroyed.

The last phase of production is the transportation of the raw material to the bottling plant via large containers (with a one-ton capacity), that protect the product from UVA and UVB rays. Light alter some of the qualities of the Aloe product; therefore these protected containers ensure that Aloe will never again come into contact with light. Refrigerated transportation to the manufacturing plant protect the raw material. Normal transportation could prove deleterious to the primary product. This problem increases especially in the summer (when temperatures can reach up to 140°F) transportation and potentially cause undesired effects similar to pasteurization.

ATIS (*Aconitum heterophyllum* Linn.); FAMILY: Ranunculaceae

CLIMATIC AND SOIL REQUIREMENTS:

The species is found in grassy slopes of alpine Himalayan region, between 3000 m and 4200 m altitude, sometimes descending up to 2,200 m. Although it generally prefers subalpine and alpine climate, cultivation up to 2,000 m altitude has been recommended in sandy (10 cm deep) soils with rich organic matter. Sandy loam and slightly acidic soil, with 6 pH is best for seed germination, survival, better growth, and yield. The plant prefers open, sunny sites, and abundant air and soil moisture during summer months.

BRIEF AGROTECHNIQUES:

Seeds, tuber segments or young leafy stems are used as propagules. Seeds are sown in sand and FYM (1:2) at a depth of 0.5 cm in styrofoam trays in a mist house. For raising seedlings inside the polyhouses nursery beds of 2 × 2 m or even smaller size are better. Seeds are sown during October–November or March–April in polyhouses at middle altitudes (1,800–2,200 m). Tuber segments are also planted during the same period as mentioned for seedlings. Approximately 1.5 kg seeds per hectare are required. However, about 111 000 tuber segment are needed for plantation on 1 hectare. The seedlings are transplanted after three months of the first true leaf initiation during March–April at middle altitudes and during May–August at alpine sites. 30 × 30 cm spacing in field is considered optimum for better vegetative growth and resultant high tuber yield. It is grown as a mono crop. Weeding is required every week during the rainy season. And in other seasons it may be done as and when required. During early summer planted beds need irrigation. Hence, well-drained beds are recommended for cultivation of the crop

HARVESTING AND YIELD:

Harvesting of tubers is recommended after the completion of reproductive phase and ripening of seeds during October–November, when maximum quantitative tuber yield is recorded. However, when tubers/plants are harvested in July–August at the budding stage, the active content (atisine) and other alkaloid contents have been found to be maximum. Percentage of active contents decreases slightly with the maturity of the plant. *A. heterophyllum* tubers harvested in May–June contain lower quantity of atisine (0.35%) as compared to those harvested in November and December (0.43%), which also contain a little amount of aconitine. The tubers harvested in May show higher quantity of aconitine and hypoaconitine as compared to those harvested in other seasons. Plants must be harvested during October at lower altitudes and in first week of November at higher altitudes to get maximum yield of tubers as well as seeds for multiplication.

POSTHARVEST TECHNOLIGY:

Usually whole tuber is harvested simply by digging the field. However, topmost segments of tubers can be used for further multiplication, as they have better survival and growth rates. Then the whole tubers or the tubers without the top segment are dried in partial shade or at room temperature. After complete drying, slices of tubers can be stored in wooden boxes or airtight polythene bags.

INDIAN GINSENGOR ASWAGANDHA (Withania somnifera); FAMILY: Solanaceae

Indian ginseng or **Winter cherry** is an erect branching perennial undershrub which is used in Ayurvedic and Unani medicines, to combat diseases ranging from tuberculosis to arthritis. Roots are prescribed in medicines for hiccup, several female disorders, bronchitis, rheumatism, dropsy, stomach, and lung inflammations and skin diseases while its roots and paste of green leaves are used to relieve joint pains, inflammation, and sexual weakness in male. Leaves are used in eye diseases, seeds are diuretic and it is a constituent of the herbal drug *'Lactare'* which is a galactagogue. Aswagandha is believed to have oriental origin and it is found wild in the forests of Mandsaur and Bastar in Madhya Pradesh, the foot hills of Punjab, Himachal Pradesh, Uttar Pradesh and western Himalayas and in Madhya Pradesh, Rajasthan, and other drier parts in India. It is also found wild in the Mediterranean region in North America. Aswagandha is erect, evergreen, tomentose shrub, 30–75 cm in height. The roots are stout, fleshy, cylindrical, 1–2 cm in diameter and whitish brown in color and the leaves are simple, ovate, glabrous, and opposite. Flowers, comprising 5 sepals are bisexual, inconspicuous, greenish, or dull yellow in color born on axillary umbellate cymes and the fruit is a small berry, globose, orange red when mature and is enclosed in persistent calyx. The seeds are small, flat, yellow, and reniform in shape and very light in weight.

BRIEF AGROTECHNIQUES:

Aswagandha grows well under dry climate, areas receiving 600–750mm rainfall and in sandy-loam or light red soils having a pH of 7.5- 8.0 with good drainage. *Seeds are* treated with Dithane M-45 at 3g/kg and *sown* in August at the rate of 10–12 kg/ha for broadcasting and 5kg/ha for transplanting. Seeds are sown in the nursery just before the onset of rainy season and covered with light soil. Germination takes place in 6–7 days. Six weeks old seedlings are transplanted at 60cm in furrows taken 60cm apart. It is not a fertilizer responsive crop. 25–30 days after sowing one hand weeding helps to control weeds effectively.

HARVESTING AND POSTHARVEST TECHNOLOGY:

Aswagandha is a crop of 150–170 days duration and the maturity of the crop is judged by the drying of the leaves and reddening of berries. From January harvesting usually starts and continues till March. The entire plant is uprooted for roots, which are separated from the aerial parts and the berries are plucked from dried plants, threshed to obtain the seeds. The yield is 400–500 kg of dry roots and 50-75kg seeds per hectare. The roots are separated from the plant by cutting the stem 1-2 cm above the crown and the roots are then cut into small pieces of 7–10cm to facilitate drying. Occasionally, the roots are dried as a whole. The dried roots are then cleaned, trimmed, graded, packed, and marketed. Roots are carefully hand sorted into four grades : Grade A (Root pieces 7cm long, 1–1.5cm diameter, brittle, solid, and pure white from outside), Grade B (Root pieces 5cm long, 1cm diameter, brittle, solid, and white from outside), Grade C (Root pieces 3–4cm long, less than 1cm diameter and solid), and Lower grade (Root pieces smaller, hollow, and yellowish from outside).

According to Kumar et al. (2012) Withania somnifera (L.) Dunal is a subtropical crop which is widely used in traditional medicine owing to the presence of root bioactives called withanolides. In the present study, phenological variation in withanolide content in roots and leaves was determined at various days after plantation (DAP). Results showed that during the full vegetative state, the amount of marker secondary metabolites viz., withanolide A (Wd-A) and withanone (Wn), increased significantly (p B 0.05) more than early vegetative state (8.16 and 3.2 times, respectively) and maturity stage (54 and 1.33 times, respectively). To qualify the role of temperature per se in enhancing secondary metabolite content, plants during full vegetative stage were exposed to temperature similar to lowest minimum temperature found during the growing period (8^0C). During recovery, the metabolic content of secondary metabolites again fell to the level of plants growing at 25^0C. This indicated that seasonal temperature played a key role in increasing secondary metabolites rather than the phenological stage of the plant. The physiological importance of this increase in harvesting and its balance with biomass yield has further been discussed.

BRAHMI (Bacopa monnieri)   FAMILY: Scrophulariaceae

Brahmi or **Thyme leaved gratiola** is an important drug in Ayurveda. It is good for the improvement of intelligence and memory and revitalization of sense organs, clears voice and improves digestion and it is also suggested against dermatosis, anemia, diabetes, cough, dropsy, fever, arthritis, anorexia, dyspepsia, emaciatton, and insanity. It is useful in vitiated conditions of *kapha* and *vata,* biliousness, neuralgia, flatulence, ascites, leucoderma, syphilis, sterility, leprosy, and general debility. In unani *Majun Brahmi* is considered as a brain tonic. The plant grows wild on damp places and marshy lands in the major part of the plains of India, Afghanistan, Pakistan, Nepal, Sri Lanka and other tropical countries. It is a prostrate, juicy, succulent, glabrous annual herb rooting at the nodes with numerous ascending branches with simple, opposite, decussate, sessile, obovate-oblong or spatulate shaped leaves and pale blue or whitish , axillary, solitary flowers, arranged on long slender pedicels. Fruits are ovoid, acute, 2-celled, 2-valved capsules and tipped with style base. Seeds are minute and numerous.

BRIEF AGROTECHNIQUES:

The plant grows throughout the warm humid tropics up to 1,200 m elevation. Vegetative propagation is done by 10 cm long stem cuttings. Land is prepared by plowing 2 or 3 times and 2–3 tons/ha of cow dung or compost is applied, the field is again plowed and leveled. Stem cuttings are placed at a spacing of 20 cm. Rooting starts within 15–20 days. Regular application of organic manure is beneficial. Weeding is done once in a month. Care should be taken to maintain water level at a height of 30cm during the growth period.

HARVESTING AND POSTHARVEST TECHNOLIGY

During September–October, 75–90 days after planting when plants attain a length of 20–30 cm it can be harvested by pulling out the whole plant, uprooting, or scraping off manually. In North India, where severe cold prevails, ratoon crop is not possible since the aerial parts of the plant die almost completely even after the application of irrigation and fertilizers, and the field is invaded by winter weeds. Fresh yield is 22.5 tons per hectare as a pure crop and reduced to approximately 5.5 tons per hectare on drying. As an intercrop with paddy, dry matter production of *Bacopa* is estimated to be about 3.75 tons per hectare.

The produce is dried by spreading it on clean area or sheets in the sun for four to five days, followed by shade drying for next 7–10 days. The dried material are then stored in clean containers. After six months of storage bacoside content starts reducing. Therefore, long storage should be avoided.

CURCUMA (Curcuma spp.); FAMILY: Zingiberaceae

The genus *Curcuma* belonging to the family *Zingiberaceae* comprises of a number of species which are medicinally very important. Among them, the most important species are *C. amada **Roxb.,** C. aromatica **Salisb.,** C. longa Linn. syn. C domestica Valeton., C. zedoaria **(Berg.) Rose. syn.** C. zerumbet **Roxb;** Amomum* *zedoaria* **Christm.** Turmeric is cultivated all over India, particularly in W. Bengal. T. N and Maharashtra.Mango ginger is cultivated in Gujarat and found wild in parts of West Bengal, U. P, Karnataka, and Tamil Nadu. Wild turmeric or Cochin turmeric or Yellow zeodoary is found wild throughout India and cultivated in Bengal and Kerala. The round zedoary or Zerumbet is mostly found in India and S. E. Asia.

BRIEF AGROTECHNIQUES:

Curcuma species are tropical herbs and can be grown on different types of soils both under irrigated and rain fed conditions and in rich loamy soils having good drainage. The plant is propagated by well developed, healthy, and disease free whole or split mother rhizomes. Rhizomes are to be treated with fungicides like copper oxychloride and stored in cool and dry place. April with the receipt of premonsoon showers is the best season of planting. During February-March the land is prepared to a fine tilth. On receipt of premonsoon showers in April, beds of size 3 × 1.2m with a spacing of 40 cm between, are prepared. Small pits are dug in the beds in rows with a spacing of 25–40 cm.Finger rhizomes are planted flat with buds facing upwards and covered with soil or powdered cattle manure. First mulching is done immediately after planting and second is done 50 days after first mulching. At the time of land preparation cattle manure or compost is to be applied as basal dose at 20–40t/ha. Application of NPK fertilizers increase the yield considerably. Weeding is done twice at 60 and 120 days after planting, depending upon weed intensity. Earthing-up is done after 60 days.

HARVESTING AND POSTHARVEST TECHNOLIGY:

Harvesting is done during January-March, at about 7–10 months after planting depending upon the species and variety. Harvested rhizomes are cleaned of mud and other materials adhering to them. Good fingers separated are to be used for curing (KAU, 1996).

Processing:

The whole plant is removed from the ground carefully to prevent the rhizomes being cut or bruised.

- **Sweating:** The leaves are cut off and the roots washed carefully in water and the lateral branches of the rhizomes ("fingers") are cut off from the central "bulb" ("mother"). The "'fingers" and "mothers" are then heaped separately and covered in leaves to sweat for a day.
- **Boiling:** The rhizomes are boiled or steamed to remove the raw odor, reduce the drying time, gelatinize the starch and produce a more uniformly colored product. In India, traditionally the rhizomes were placed in pans or earthenware pots filled with water and covered with leaves with cow dung over the top. The ammonia in the cow dung reacted with the turmeric and produce the final product. For hygienic reasons this method is being discouraged.

The present recommended practices are:
- The rhizomes are kept in shallow pans in large iron vats.

- Water is added 5–7cm above the rhizomes.
- 05–1 percent alkali is added (eg sodium bicarbonate).
- The rhizomes are boiled for 45 min (as is done in India) and 6 hrs (as is done in Hazare in Pakistan) depending on the variety.
- **Drying:** The boiled rhizomes are removed and dried in the sun immediately to prevent over cooking. The final moisture content should be between 8 and 10% (wet basis). When a finger snaps with a clear metallic sound it is sufficiently dry.
- **Polishing:** The dried rhizomes are polished by hand or by shaking the rhizomes in a gunny bag filled with stones to remove the rough surface. Polishing drums (simple power-driven drums) are in use in many places.
- **Adulteration:** Lead chromate is sometimes used to produce a better finish. This is actively discouraged.

Standards:

US Government Standards and the American Spice Trade Association Standards

Curcumin %	5–6.6
Moisture content %	<9
Mould %	<3
Extraneous matter % (by weight)	<0.5
Volatile oil ml/100g	<3.5

CINCHONA (Cinchona spp.); FAMILY: Rubiaceae

Cinchona, known as **Quinine, Peruvian,** or **Crown bark tree** is famous for the antimalarial drug *quinine"* obtained from the bark of the plant. Over 35 alkaloids have been isolated from the plant including quinine, quinidine, cinchonine, and cinchonidine, which exist mainly as salts of quinic, quinovic, and cinchotannic acids and are useful for the treatment of malarial fever, pneumonia, influenza, cold, whooping couphs, septicemia, typhoid, amebic dysentery, pin worms, lumbago, sciatica, intercostal neuralgia, bronchial neuritis and internal hemorrhoids. The cultivated bark contains 7–10 percent total alkaloids of which about 70 percent is quinine. Similarly 60 percent of the total alkaloids of root bark are quinine. Cinchona is native to tropical South America and grown in Bolivia, Peru, Costa Rica, Ecuador, Columbia, Indonesia, Tanzania, Kenya, Zaire, and Sri Lanka. It was introduced in 1808 in Guatemala, 1860 in India, 1918 in Uganda, 1927 in Philippines and in 1942 in Costa Rica. The first plantation was raised in Nilgiris and later on in West Bengal (Darjeeling). It first appeared in London pharmacopeia in 1677. The commonly cultivated species are C. *calisaya* Wedd., C. *ledgeriana* Moens, C. *officinalis* Linn., C. *succirubra* Pav. ex Kl., C. *lancifolia* and C. *pubescens.* It is an evergreen tree of 10-15m height. The Leaves are opposite, elliptical, ovate- lanceolate, entire, and glabrous and the flowers are reddish-brown in short cymbiform, compound cymes, terminal, and axillary. The ovoid-oblong shaped fruit is capsule; seeds elliptic, winged margin octraceous, crinulate-dentate.

BRIEF AGROTECHNIQUES:

The plant widely grows in tropical regions having an average minimum temperature of 14°C and in the mountain slopes of humid tropical areas with well distributed annual rainfall of 1,500–1,950mm. Well drained virgin and fertile forest soils with pH 4.5–6.5 are best suited for its growth. The commercial plantations are raised by seeds, but vegetative techniques such as grafting, budding, and softwood cuttings are employed in countries like India, Sri Lanka, Java, and Guatemala. *Cinchona succirubra* is commonly used as root stock in the case of grafting and budding. Seedlings are first raised in nursery beds of convenient size under shade and well decomposed compost or manure is applied. Seeds are broadcasted uniformly at 2g/m², covered with a thin layer of sand and irrigated. 10-20 days after sowing the seeds germinate and are transplanted into polythene bags after 3 months and then into the field after 1 year at l–2m spacing. Trees are thinned after third year for extracting bark , leaving 50 percent of the trees at the end of the fifth year.

HARVESTING AND POSTHARVEST TECHNOLIGY:

Harvesting can be done in one or two phases. Either the complete tree is uprooted, after 8–10 years when the alkaloid yield is maximum or the tree is cut about 30cm from the ground for bark after 6–7 years so that fresh sprouts come up from the stem to yield a second crop which is harvested with the under ground roots after 6–7 years. Stem and root both are cut into convenient pieces, bark is separated, dried in shade, graded, packed, and traded. Bark yield is 9,000–16,000kg/ha. Extraction of quinine from ground cinchona is done by extraction with supercritical CO_2 at a pressure of 100–500 bar and a temperature of 80–120°C. The extracted quinine precipitate as quinine sulfate in a separation vessel which is partly filled with sulfuric acid. Quinine is then extracted from the ground product obtained from the alkaline maceration of cinchona bark by treating it with at least one hydrocarbon and a chlorinated hydrocarbon, ketone, or alcohol solvent mixture (5–22 parts by volume of the latter to 100 parts of the hydrocarbon). These processes are simple, high in yield, low in production cost and suitable for use in industrial production.

SANDAL WOOD (*Santalum sp.*); FAMILY - SANTALACEAE

Sandal wood is distributed in the dry scrub forest of Salem, Mysore, Coorg, Coimbatore, Nilgiris up to 900 m. altitude and also found in Andhra Pradesh, Bihar, Gujarat, Karnataka, Madhya Pradesh, Maharashtra, and Tamil Nadu. Sandal wood is a small evergreen tree, a partial root parasite, attaining a height of 12–13m. and girth of 1–2.4 m. with slender druping as well as erect branching. The fruit is drupe, purplish when fully matured and single seeded. Other species of sandalwood are : red sandalwood, Australian sandalwood (*S. spicatum*) and New Caledonian Sandalwood (*S. austrocaledonicum*), these are quite different from true *Santalum album* and have very different properties and fragrances.

BRIEF AGROTECHNIQUES:

It grows well in red sandy-loam soil and requires humid and hot climate. Two type of seed beds: sunken and raised beds are used to raise sandal seedlings. Seed beds are formed with 3:1 sand and red earth and are thoroughly mixed with nematicides (Ekalux or Theimet at 500 gm. per bed of 10 × 1m.) Around 2.5 kg seed is spread uniformly over the bed, covered with straw, which are removed when leaves start appearing. Seedlings having 4–6 leaf are transplanted to poly bags of 30 × 14cm size containing soil mixture of ratio 2:1:1 (Sand: Red earth: Farmyard manure) along with a seed of "tur dal" (*Cajanus cajan*), the primary host for better growth of sandal. Shade is provided for a week immediately after transplantation. Watering is done once a day, but excess moisture is avoided. A well-branched seedling with a brown stem is ideal for planting in the field. Weeding is done at regular intervals. Organic manures like Farm Yard Manure (FYM), Vermi-Compost, Green Manure, etc. may be used as per requirement.

HARVESTING AND POSTHARVEST TECHNOLIGY:

30–60 year old sandal wood trees are harvested. It is a slow growing tree as it grows at the rate of 5 cm. of girth or more per year under favorable soil and moisture conditions. The heartwood formation starts around 10 years of age. The quality of the final sandalwood oil will depend upon the quality of the wood, the length of distillation time, and the experience of the distiller. Sandalwood essential oil is extracted primarily by steam distillation, a process in which superheated steam is passed through the powdered wood. The steam helps to release and carry away the essential oil that is locked in the cellular structure of the wood and the steam is then cooled and the result is sandalwood hydrosol and sandalwood essential oil.

Hydrodistillation is the traditional method of extraction, but rare these days, yet it is said that this method yields an oil with a superior aroma. Instead of having steam pass through the powdered wood, in a hydrodistiller the powder is allowed to soak in water and a fire from below the vessel heats the water, then carries off the steam which is allowed to cool. The Sandalwood oil is then removed from the top of the hydrosol.

The CO_2 extraction method does not use water or steam, instead CO_2 (carbon dioxide) is used as a solvent. The CO_2 is used under high pressure in which it exhibits a likeness to both a gas and a liquid (called a supercritical state), which allows the aromatic constituents of sandalwood to be extracted without heat. CO_2 is then removed from the resulting extract which is then refined and filtered. The oil produced from this method has a different look, feel, as well as an odor profile from the oil obtained by steam or hydrodistillation. CO_2 extracted Sandalwood oil is more viscous and darker in color (it is a beautiful golden color), more resinous and is deeper in the woody aroma characteristics than steam distilled Sandalwood oil. It is not as sweet smelling as the steamed distilled oil and is slightly less spicy.

HEARTWOOD FORMATION AND OIL CONTENT

Heartwood formation in sandal trees generally starts around 10–13 years of age. Certain factors, such as gravelly dry soil, insolation, and range of elevation (500–700 m), seem to provide the right environment for the formation of heartwood, irrespective of the size of the stem after 10 years of age. The occurrence of heartwood varies, like it can range from 90 percent of the stem wood to a negligible amount, or be absent. The value of heartwood is due to its oil containing high percentage of santalol. The oil content is highest in the root, next highest in the stem at ground level, and gradually tapers off toward the tip of the stem. There is also a gradient in oil content from the core to the periphery of the heartwood in a stem.

Depending upon their age, trees can be classified as : (i) Young trees (height less than 10 m, girth less than 50 cm, and heartwood diameter 0.5–2 cm) have heartwood with 0.2–2 percent oil content, which has 85 percent santalol, 5 percent acetate, and 5 percent santalenes, and (ii) Mature trees (height 15–20 m, girth 0.5–1 m, and heart-wood diameter 10–20 cm) have heartwood with oil content of 2–6.2 percent, which has over 90 percent santalol, 3–5 percent acetate, and 3 percent santalenes.

The heartwood is yellowish to dark brown in color. The yellowish heartwood has 3–4 percent oil and 90 percent santalol; light brown heart-wood contains 3–6 percent oil and 90-94 percent santalol; brown and dark brown wood has only 2–5 percent oil and 85–90 percent santalol; hence, lighter heartwood is better and superior.

CITRONELLA *(Cymbopogon winterianus Jowitt.)*; Family-Poaceae

BRIEF AGROTECHNIQUES:

Like most grasses, citronella prefers a moist loamy soil and can tolerate full sunlight but prefers a slightly shaded area that receives 6–8 hrs of sunlight per day. While citronella is a resilient and adaptable plant, it can not handle long periods of cold very well. These are vegetatively propagated with clums at the middle of rainy season i.e. June—July . The seedlings or saplings or slips can be transplanted on ridges and furrows with a planting distance of 60 × 90 sq. cm. Weeding is necessary for the first 23 months. Usually before plantation 8–10 tons of compost mixed with 40 kg P205 and 40 kg K20 per hectare and after plantation 80–1.00 kg N/ha per year are applied in 4 splits. For good yield irrigation interval of 15 days in winter and 10 days in summer is required.

HARVESTING AND POSTHARVEST TECHNOLIGY:

Harvesting (a partial harvest) is done before 12 O'clock by giving a cut by sickle at 15 cm above the ground after 3 months to induce tillering. In the first year only 3 cuts besides the partial harvest can be taken, viz., 5 months after planting and 3 months after the previous harvest and from second year onwards, 5–6 harvests can be taken per year at 2 months intervals. Cutting close to the ground results into mortality of the plant. The partially wilted herb yield is in the range of 20–30 t/ha and the average oil recovery is 1 percent. Hence the oil yield is about 200–300 kg/ha/year. A crop under average management yields not less than 200 kg oil/ha/year.

11.4.1 WITHERING OF THE FRESH HERB

The herb is allowed to wilt for 12–24 hrs after cutting to remove the excess moisture which allows better packing in the vessel and saving of steam and fuel. Wilting more than 24 hrs results loss of essential oil. Cutting the grass into shorter length gives 10–15 percent higher recovery. Dead leaves and sheaths should be removed from the harvested grass before packing into the vessel.

Distillation of oil

For better oil recovery the grass is steam distilled. The distillation equipment consists of a boiler, a distillation vessel, a condenser and two receivers / separators and the economic capacity of the unit is 1.0 ton/batch. In hilly areas small size (6–8 q) direct-fired field units are preferred and the unit should be of stainless still for durability and oil quality. Small scale growers can use properly designed direct-fired field distillation units. Distillation is completed within 3 hrs under normal pressure starting from the initial condensation of the oil but prolong distillation deteriorates oil quality.

The trend of oil recovery is as under-

Distillation time	total oil recovery %
1st hour	80 %
2nd hour	19 %
3rd hour	1 %

Larger percentage of the major components about 80percent in the total oil, such as citronellal, geraniol, citronellol, and geranyl acetate is recovered in the first hour of distillation.

GREEN CHIRETTA (Andrographis paniculata) FAMILY: Acanthaceae

Kalmegh, the Great or **Green Chiretta** is a branched annual herb. This is useful in wounds, ulcers, chronic fever, hyperdipsia, burning sensation, malarial and intermittent fevers, inflammations, cough, pruritis, intestinal worms, dyspepsia, flatulence, colic, diarrhea, dysentery, bronchitis, skin diseases, leprosy, hemorrhoids, and vitiated conditions of *pitta*. It is also a rich source of minerals. The plant is distributed throughout the tropics and found in the plains of India from UP to Assam, MP, AP, Tamil Nadu and Kerala.

It is an erect branched annual herb of 0.3–0.9m in height with quadrangular branches. Leaves are os different shapes like simple, lanceolate, acute at both ends, glabrous, with 4-6 pairs of main nerves. The small flowers are pale but blotched and spotted with brown and purple distant in lax spreading axillary and terminal racemes or panicles. Fruits are linear capsules and acute at both ends with numerous, yellowish brown and subquadrate seeds. Another species of Andrographis *A. echioides*

(Linn.) Nees. is a febrifuge and diuretic and is found in the warmer parts of India. It contains flavone echiodinin and its glucoside-echioidin.

BRIEF AGROTECHNIQUES:

May-June is the best season of planting Andrographis. The field is plowed well, mixed with compost or dried cowdung. Seedbeds of 3m length, 1/2m breadth and 15cm height are then prepared at a distance of 3m. The plant is seed propagated. Before sowing (at a spacing of 20cm) seeds are soaked in water for 6 hrs. Germination may takes place within 15–20 days of sowing. Two weedings, first at 30d ays after planting and the second at 60 days after planting are done. During summer months frequent irrigation is beneficial.

HARVESTING AND POSTHARVEST TECHNOLIGY:

From third month onwards, when flowering commences, plant are collected, tied into small bundles and sun-dried for 4–5 days. Whole plant is the economic part and the yield is about 1.25 t dried plants/ha. After uprooting the whole plant it is dried in the sun for two days and afterwards in the shade. Properly dried materials are then packed in laminated gunny bags and the harvested dry materials are stored in a dark, airy, and moisture-free place.

LEMON GRASS (*Cymbopogan flexuosus*); FAMILY: Poaceae

Most species of lemon grass are native to South Asia, South-East Asia and Australia. It grows in many parts of tropical and subtropical South East Asia and Africa. C. flexuosus also known as East Indian lemon grass or Malabar or Cochin grass, has its origin in Indo-Burma region and is native to India, Sri Lanka, Burma, and Thailand. A related species is *C. citratus* called the West Indian lemon grass, has its origin in the Malaysian region. Both species are today cultivated throughout tropical Asia. The crop is under commercial cultivation in India, along Western Ghats (Maharashtra, Kerala), Karnataka and Tamil Nadu, UP, Assam, and foothills of Arunachal Pradesh and Sikkim. It is a tall, perennial hedge throwing up dense fascicles of leaves from a short rhizome. The leaves are long, glaucous, green, linear, tapering upwards and along the margins; ligule very short; sheaths teethe, those of the barren shoots widened and tightly clasping at the base, others narrow and separating. It is a short day plant having flowers in about one meter long inflorescence. Flower panicles are 30 to over 60 cm long.

BRIEF AGROTECHNIQUES:

The crop grows well in both tropical and subtropical climates upto an elevation of 900 m. (above MSL) with sufficient sunshine, temperature ranging from 20 to 300 C and 250 to 330 cm rainfall per annum, evenly distributed over most part of the year. Lemon grass can also be grown in semiarid regions receiving low to moderate rainfall. It flourishes on a wide variety of soils ranging from loam to poor laterite, calcareous soils, but well-drained sandy-loam soils are ideally suited for better growth, yield, and oil content. Soils with poor drainage and with prolonged water logging should be avoided. The transplanting of nursery raised seedlings is found to be superior to direct sowing of seeds. The seeds are sown on well prepared raised beds of 1–1.5 m width at the onset of monsoon, covered with a thin layer of soil and watered. The recommended seed rate is 4–5 kg/ha. Seeds germinate in 5–6 days and they are ready for transplanting after a period of 60 days. For better quality and yield of oil it is propagated through slips obtained by dividing well-grown clumps. Planting is done at the beginning of the rainy season at a distance of 40 × 40 cm, 45 × 35 cm, 60 × 40 cm depending on the soil fertility status and varieties. Planting on the ridges is suggested especially in high rainfall areas. Depending upon the rainfall and its distribution the field is irrigated at an interval of 3 days during the first month and 7–10 day intervals subsequently. At the time of final land preparation FYM at 10 MT/ha is added. The normal recommended doses of fertilizers include N, P205 & K20 at 150: 60: 60 kg /ha/year. The field is kept free of weeds for the first 3–4 months after plating and weeding cum hoeing is done up to 1 month, after every harvest.

HARVESTING AND POSTHARVEST TECHNOLIGY:

Lemon grass flowers in winter. The first harvest is obtained 4–6 months after transplanting seedlings and the subsequent harvests are done at intervals of 60–70 days depending upon the fertility of the soil and other seasonal factors. Under normal conditions, 3 harvests are possible during first year, and 3-4 in subsequent years, depending on the management practices followed. The plants are cut with sickles 10 cm above ground-level and allowed to wilt in the field, before transporting to the distillation site. During first year the yield of oil is less but it increases in the second year and reaches a maximum in the third year; after this, the yield declines. 20 to 30 tons of fresh herbage is harvested per hectare per annum from 3-4 cuttings. The yield in terms of oil varies from 0.5 to 0.8 percent depending on the variety, season/ month of harvest and age of the crop, with an average oil yield of 0.65 percent. The grass is allowed to wilt for 24 hrs before distillation to reduce the moisture content by 30 percent and improving oil yield. Oil is extracted from the wilted herb by steam distillation in stainless steel unit (500 kg per charge). The factors influencing the oil production during distillation are: storage of the harvested herbage, treatment (wilting and cutting into pieces) of the material and, the method of distillation. The major cause of loss is due to oxidation and resinification of the essential oil, so the materials should be kept in a dry atmosphere with limited air circulation. The essential oils are present in the oil glands, oil sacks and glandular hairs of the plant. Before distillation, the wilted plant material is cut into small pieces to enable them to expose directly as many oil glands as possible and distilled (takes about 3–4 hrs) immediately to avoid oil loss. Dipping the chopped lemon grass in sodium chloride solution at 1 to 2 percent concentration for 24 hrs before distillation, increase the citral content. The recovery of oil from the grass ranges from 0.5–0.8 percent.

Currently the world production of Lemon grass oil is around 1,300 MT/ annum. However, another 600 MT of a substitute oil viz., Litsea cubeba (rich in citral) and synthetic citral is also available which competes with this oil. Cochin and Mumbai are the major trading centers for lemon grass oils. From India the essential oil of lemon grass is being exported to West Europe, U.S.A. and Japan.

LONG PEPPER(Piper longum)   FAMILY : Piperaceae

Long pepper is a slender aromatic climber. The spike of long pepper is widely used in Ayurvedic and Unani systems of medicine particularly for diseases of respiratory tract. Its roots also have several medicinal uses. The root is useful in bronchitis, stomach ache, diseases of spleen and tumors; fruit is useful in *vata* and *kapha,* asthma, bronchitis, abdominal complaints, fever, leukoderma, urinary discharges, tumors, piles, insomnia, and tuberculosis; root and fruit both are used in gout and lumbago and root-fruit decoction are used in acute and chronic bronchitis and cough. After parturition, the infusion of root is prescribed to induce the expulsion of placenta. Long pepper, containing the alkaloid **piperine**, has diverse pharmacological activities, including nerve depressant and antagonistic effect on electro- shock and chemo-shock seizures as well as muscular incoordination.

Long pepper is a native of Indo-Malaya region. It was introduced to Europe and was highly regarded as a flavor ingredient by the Romans. It grows wild in the tropical rain forests of India, Nepal, Malaysia, Indonesia, Rhio, Sri Lanka, Timor, and the Philippines. In India, it is seen in Assam, West Bengal, Madhya Pradesh, Uttar Pradesh, Kerala, Maharashtra, Karnataka, and Tamil Nadu. It is also cultivated in Bengal, Anamalai hills of Tamil Nadu, Akola-Amravati region of Maharashtra, Orissa, Chirapunchi area of Assam, Uduppi, and Mangalore regions of Karnataka.

Piper longum Linn. belongs to Piperaceae family. It is a glabrous perennial undershrub with an erect or subscandent nodose stem and slender branches. The latter are often creeping or trailing and rooting below or rarely scandent reaching a few meters height. Leaves have different shapes like simple, alternate, stipulate, and petiolate or nearly sessile. The lower ones are broadly ovate, cordate while upper ones are oblong, oval, all entire, smooth, thin with reticulate venation; veins raised beneath. It blooms almost throughout the year. Inflorescence is spike with unisexual small achlamydeous densely packed flowers and form very close clusters of small grayish green or darker gray berries. Female spikes with short thick stalk vary in length (1.5–2.5 cm) and in thickness (0.5 to 0.7 cm).

BRIEF AGROTECHNIQUES:

This is a tropical plant adapted to high rainfall areas. The ideal elevation is 100–1,000 m. It needs partial shade of 20–30 percent for best growth. The borders of streams are the natural habitat of this plant. In well-drained forest soils, rich in organic matter, it can be successfully cultivated. Lateritic soils with high organic matter content and moisture holding capacity are also suitable for long pepper cultivation. This is propagated by suckers or rooted vine cuttings. 3–5 nodded rooted vine cuttings of 15–20 cm length establishes very well in polybags. March-April is the best time for raising nursery. On alternate days normal irrigation is given at early stages. In two months the rooted cuttings will be ready for transplanting. The field is plowed well and brought to good tilth with the onset of monsoon in June. Raised bed of 15–20 cm height is convenient. Pits are dug on these beds at 60 x 60 cm spacing and well decomposed organic manure at 100 g/pit is applied and mixed with soil. Rooted vine cuttings from polybags are transplanted to these pits. After one month of planting, gap filling can be done. This crop needs heavy manuring of 20 l FYM/ ha every year. Once a week the crop needs irrigation. Sprinkler irrigation is ideal. *It* can also be cultivated as an intercrop in the plantations of coconut, subabul, and eucalyptus.

HARVESTING POSTHARVEST TECHNOLIGY:

Six months after planting the vines start flowering and flowers are produced almost throughout the year. The spikes mature within 2 months. When the spikes are blackish green and the pungency is highest, it is the optimum stage of harvesting. When the spikes become mature, they are handpicked and then dried. During first year the yield of dry spike is 400 kg /ha, increases to 1,000 kg during third year and thereafter it decreases. Therefore, the whole plant is harvested after 3 years. Close to the ground the stem is cut and roots are dug up. Average yield is 500 kg dry roots/ha.

The harvested spikes are dried for 4–5 days in sun until they are perfectly dry. The green to dry spike ratio is 10:1.5 by weight. In moisture proof containers the dried spikes are stored. Stem and roots are cleaned, cut into pieces of 2.5–5 cm length, dried in shade and marketed as piplamool. Based on the thickness there are three grades of piplamool. The commercial drug consists 0.5–2.5 cm long, 0.5–2.5 mm thick, cylindrical pieces with dirty light brown color, and peculiar odor with a pungent bitter taste, producing numbness to tongue.

MEDICINAL YAMS (Dioscorea spp.)   FAMILY: Dioscoreaceae

Dioscorea is the largest genus of the family having 600 species of predominant- ly twining herbs. Among the twining species, some species twine clockwise while others counterclockwise. All the species are dioceous and rhizomatous. Some of the species like *D. alata* and *D. esculenta have* been under cultivation for a long time as edible tubers. 15 species of this genus contain diosgenin, among them, *D. flori- bunda, D. composita* and *D. deltoidea* are widely grown for diosgenin production.

BRIEF AGROTECHNIQUES:

This prefer tropical climate without extreme temperature. It is adapted to mod- erate to heavy rainfall area and in a variety of soils, but light soil (with 5.5–6.5 pH) is good for easy harvesting of tubers. Propagation can be done by tuber pieces, single node stem cuttings or seed, but commercially by tuber pieces only. Propaga- tion through seed take longer time to obtain tuber yields. Three types of tuber pieces that can be distinguished for propagation purpose are crown, median and tip, of which crowns produce new shoots within 30 days and are therefore preferred. For checking the tuber rot tuber pieces are dipped for 5 min in 0.3 percent solution of Benlate followed by dusting the cut ends with 0.3 percent Benlate in talcum powder in moist sand beds. End of April is the best time of planting, so that new sprouts will grow vigorously during the rainy season commencing in June in India. Land is to be prepared thoroughly for obtaining a fine tilth. At 60 cm distance deep furrows are made with the help of a plow. The stored tuber pieces are planted in furrows with 30cm between the plants for one year crop and 45cm between the plants for 2 year crop at about 0.5 cm below soil level. The new sprouts are staked immediately. After sprouting, the plants are earthed up. For good tuber formation *Dioscorea* requires high organic matter. Besides a basal doze of 18–20t of FYM/ha and a complete fertilizer dose of 300kg N, 150kg P_2O_5 and K_2O each are to be applied per hectare. P and K are to be applied in two equal doses one during May-June and the other during August- September. At initial stages irrigation is given at weekly intervals and afterwards at about 10 days interval. *Dioscorea* vines are trailed over *pandal* system or trellis. For the first few months periodic hand weeding is essential. Intercropping with legumes decrease weeds and provide extra income.

HARVESTING AND POSTHARVEST TECHNOLIGY:

Harvesting is done by hand using sticks, spades, or diggers preferably wooden ones, as they are less likely to damage the fragile tubers; however, tools need regular replacement. It is a labor-intensive operation that involves standing, bending, squatting, and sometimes sitting on the ground depending on the size of mound, tuber size or depth of tuber penetration. Aerial tubers or bulbils are harvested by manual plucking from vine. Mechanical yam harvesting is effective, but are still limited to research and demonstration purposes.

Single harvesting or double harvesting can be done to obtain an early as well as late harvest. The first harvest is also known as "topping", "beheading", and "milking", all of which have been considered inadequate and obsolete. In single harvesting, each plant is harvested by digging around the tuber to loosen it from the soil, lifting it, and cutting from the vine with the corm attached to the tuber, once when the crop is mature. The period from planting or emergence to maturity varies from about 6–7 months or even 6–10 months depending on the cultivar.

Periods of 8–10 months and 4–5 months from planting or emergence to maturity have been recommended for double harvesting. First harvest is done by removing the soil around the tuber and cutting the lower portion, leaving the upper part of tuber or "head" to heal and continue to grow. The soil is then returned and the plant is left to grow to the end of the season for second harvest. Double harvesting is most applicable to short-term varieties like *D. rotundata* and to some extents *D. Cayenensis* and *D. alata*.

50–60t/ha of fresh tubers can be obtained in 2 years duration on an average. Diosgenin content increases with age, 2.5% in first year and 3–3.5% in the second year. Hence, 2 year crop is economical (Kumar *et al.,* 1997).

- **Transport and Packaging**

Yam tubers are traditionally placed into woven baskets made from parts of the palm tree or coconut fronds after harvest for transporting small quantity of tubers over short walking distances. For large quantity of tuber transporte over long distances tubers are packed in full telescopic fiberboard cartons with paper wrapping. Tubers can be contained in loose packs, or units of 11 kg and 23 kg. Storing yams in modified atmosphere packaging (MAP) is beneficial, particularly using appropriate packaging material with suitable size and number of holes for gas permeation. Sealing yam tubers in polyethylene film bags reduced storage losses and development of necrotic tissue.

- **Curing of Yam Tuber**

Curing of yams is recommended before storage to "heal" any physical injury, which may have occurred during harvesting and handling. Traditionally, yams tubers are cured by drying in the sun for few days. The optimum conditions for curing are 29–32°C at 90–96 percent RH for 4–8 days. Tubers cured for 24 hrs at higher temperature (40°C) or treated with gamma radiation at 12.5 krads were free of mold and had least losses during storage. Storing at 15°C with prompt removal of sprouts was found to improve the eating quality of tubers.

- **Storage**

The three main conditions: aeration, reduction of temperature, and regular inspection of produce are necessary for successful yam storage. Ventilation prevents moisture condensation on tuber surface and removes the heat of respiration. Low temperature (12–15°C) is necessary to reduce losses from respiration, sprouting, and rotting. Regular inspection of tubers is important to remove sprouts, rotted tubers, and to monitor the presence of rodents and other pests. The storage environment must also inhibit the onset of sprouting (breakage of dormancy). Both ware yam and seed yam have similar storage requirements. Fresh yam tuber can be stored in ambient and refrigerated conditions. The recommended storage temperature range is 12°–16°C. Optimum conditions of 15°C or 16°C at 70–80 percent RH or 70 percent RH are recommended for cured tubers. Under these conditions transit and storage life of 6–7 months can be achieved. During storage at ambient conditions (20°–29°C, 46–62% RH), *D. trifida* began to sprout within 3 weeks. Yam tuber decay occurs at higher humidity, and they are susceptible to chilling injury (CI) at low storage temperatures. There are several traditional storage structures for yam storage like: leaving the tubers in the ground until required, the yam barn, and Underground structures. Leaving the tubers in the ground is the simplest storage technique practiced by rural small-scale farmers. Until required the harvested yams can also be put in ashes and covered with soil, with or without grass mulch. In the major producing areas the yam barn is the principal traditional yam storage structures. Barns are usually located in shaded areas and consist of a vertical wooden framework to which the tubers are individually attached to facilitate adequate ventilation while protecting tubers from flooding and insect attack. Two tubers are tied to a fibrous rope at each end hung on horizontal 1–2 m high poles. Growing trees are used as vertical posts, which are trimmed periodically to remove excessive leaves and branches. The vegetative growth on the vertical trees also shades the tubers from excessive solar heat and rain. In barn storage is most suited for long-term varieties and yams have a maximum storage life of 6 months. Storage losses can be up to 10–15% in 3 months and 30–50% after 6 months if tubers are not treated with fungicides such as Benlate, Captan, or Thiabendazole. A typical improved yam barn has sidewall 1.2 m high and wire mesh to ward off rodents and birds with double thatched roof and with only one entry door to guard against entry of rodents. Tubers were stored on platforms or shelves. Tubers stored in such improved structures had only 10% spoilage after 5-6 months.

• **Processing**

Industrial uses of yam includes starch, poultry, and livestock feed, and production of yam flour. In many rural areas residues from sifting and peels are used as animal feed. One of the major disadvantages of industrial processing of yam for food is high nutrient losses, particularly minerals and vitamins. In products obtained from secondary processing like biscuits and fufu, the amount of loss depends on the amount of edible surface exposed during processing operations. The thiamine and riboflavin contents of *D. rotundata*, with average losses of 22 percent and 37 percent, respectively is affected by primary unit operations such as milling. Sun drying, pounding yam flour in a traditional wooden mortar or grinding in an electric mixer results in high losses of B vitamins with little change in mineral content.

PERIWINKLE (*Catharanthgus roseus*);   FAMILY: Apocynaceae

Periwinkle or Vinca is an erect handsome herbaceous perennial plant which contains the alkaloids **vinblastine** and **vincristine** present in the leaves that are recognized as anticancerous drugs. Its roots are a major source of the alkaloids, **raubasine (ajmalicine), reserpine** and **serpentine** used in the preparation of antifibrillic and hypertension-relieving drugs and it is also useful in the treatment of choriocarcinoma and *Hodgkin's* disease-a cancer affecting lymph glands, spleen, and liver. The leaves are used for curing diabetes, menorrhagia, and wasp stings. Root is tonic, stomachic, hypotensive, sedative, and tranquilizer. It is a native of Madagascar, hence the name **Madagascar Periwinkle and i**t is distributed in West Indies, Mozambique, South Vietnam, Sri Lanka , Philippines and Australia . In India it is commercially cultivated in the states of Tamil Nadu, Karnataka, Gujarat, Madhya Pradesh and Assam. USA, Hungary, West Germany, Italy, Netherlands, and UK are the major consumers. It is an erect highly branched lactiferous perennial herb growing up to a height of one meter. The leaves are oblong or ovate, opposite, short-petioled, smooth with entire margin and the flowers (**pink, pink-eyed** and **white)** are borne on axils in pairs . The dehiscent fruit consists of a pair of follicles each measuring about 25 mm in length and 2.3 mm in diameter, containing up to thirty linearly arranged seeds with a thin black tegumen.

BRIEF AGROTECHNIQUES:

Periwinkle grows well under tropical and subtropical climate with well distributed rainfall of 1,000 mm or more and on any type of soil except those which are highly saline, alkaline, or waterlogged. Light soils, rich in humus are preferable for large scale cultivation since harvesting of the roots become easy. Catharanthus is propagated by seeds(short-viable). Fresh seeds are either sown directly in the field or in a nursery and then transplanted. Seed rate is 2.5 kg/ha for direct sowing, while for transplanted crop the seed rate is 500gm/ha. Application of FYM at the rate of 15 t/ha and 80:40:40 kg $N:P_2O_5:K_2O$/ha for irrigated crop and 60:30:30 kg/ha for rainfed crop are recomended. 4 or 5 irrigations will be needed to optimize yield when rainfall is restricted while fortnightly irrigations support good crop growth when the crop is grown exclusively as an irrigated crop. Weeding is done before each topdressing and detopping of plants by 2cm at 50% flowering stage improves root yield and alkaloid contents.

HARVESTING AND POSTHARVEST TECHNOLIGY:

Beginning from 6 months the crop allows 3–4 clippings of foliage. For collection of roots with high alkaloid content the flowering stage is ideal. The crop is cut about 7 cm above the ground and dried for stem, leaf, and seed. The field is then irrigated, plowed, and roots are collected. The average yields of leaf, stem, and root are 3.6, 1.5and 1.5 t/ha, respectively under irrigated conditions and 2.0, 1.0 and 0.75t/ha respectively under rain fed conditions on air dry basis. After harvesting the stem and roots loose 80 percent and 70 percent of their weight, respectively. The whole plant is dried in shade after harvesting. At this stage, light threshing will separate the seeds, which can be used for the next sowing and the then leaves and stems are also separately collected. It is therefore advisable that only mature pods should be collected during two or three months before the crop is harvested for good percentage of germination.

According to Baskaran et al (2013) Periwinkle (Catharanthus roseus), an important medicinal plant, is the source of highly valued anticancer and antihypertension alkaloids. Seeds of variety, "Dhawal" were treated with ethyl methanesulfonate to induce mutants with increased contents of alkaloids. Two macromutants, "necrotic leaf' and "nerium leaf", were isolated from the M2 generation. The "necrotic leaf' mutant exhibited irregular light green to yellowish green and brownish lesions on seedling leaves up to about 5th leaf stage. Later, it resembled its parental variety. The 'nerium leaf' mutant had leaves distinctly different from its parental variety. Both the mutants, which were found to be true breeding in the M3 generation, were advanced to the M6 generation through manual selfing in the glasshouse and evaluated in comparison with their parental variety. Both mutants exhibited significantly higher contents of total leaf and root alkaloids and anticancer leaf alkaloids, vincristine, and vinblastine than the parental variety. Their leaf and root yields were, however, significantly lower than their parental variety. Both the macro-mutant traits were inherited as monogenic recessive traits. Gene symbols nl and ne are proposed for recessive alleles responsible for the production of 'necrotic leaf' and 'nerium leaf' traits, respectively. The "necrotic leaf" mutant appears to be the first lesion mimic mutant to be reported in any medicinal or aromatic plant

TULSI (*Ocimum sanctum*); FAMILY : Lamiaceae

Tulsi or Holy Basil, which is a species worshipped by the Hindus, is used medicinally in the treatment of heart and blood diseases, leucoderma, strangury, asthma, bronchitis, lumbago, and purulent discharge of the ear. The leaf juice possesses diaphoretic, antiperiodic, stimulant, and expectorant properties and it is used to treat infantile cough, cold, bronchitis, diarrhea, and dysentery, ringworm and other skin diseases. The oil (containing mono-terpenes, sesquiterpenes and phenols, alcohols, esters, aldehydes, ketones) extracted from the leaves is reported to possess antibacterial and insecticidal properties, and is effective as a mosquito repellent.Other compounds are terpenoid methyl chavicol, linalool, eugenol, ocimene, elemene, caryophyllene, camphor, limonene, bisabolene, 1,8-cineole, linalool, methyl eugenol, citral, nerol, geraniol type.

Among different species of genus *Ocimum*, the species *sanctum* occupies wide range of habitats. Among Indian species, *Ocimum basilicum* and *Ocimum sanctum* have the widest distribution, which cover the entire Indian subcontinent. The plants of *Ocimum sanctum* are perennial in habit and are predominantly shrubs and herbs. The plants are usually much branched. Stems and twigs are usually quadrangular and the young twigs are greenish, purplish, or brownish in color. The leaves are simple, petiolate and ovate, possess glandular hairs or stalked and sessile glands which secrete volatile oils and they exhibit racemose type of inflorescence. Flowers are hermaphrodite, zygomorphic, and complete. Seeds are mostly brownish, globose, or subglobose and are shining or nonmucilaginous.

CLIMATIC AND SOIL REQUIREMENTS:

Ocimum sanctum thrives well on a variety of soils and climatic conditions. Rich loam to poor laterite, alkaline to moderately acidic, well-drained soils are well suited for cultivation of *Ocimum sanctum*. For favorable plant growth and higher oil production fair to high rainfall, long days and high temperatures have been found to be good. Tropical and subtropical climate is suited for its cultivation but waterlogged conditions can cause root rot and result in stunted growth.

BRIEF AGROTECHNIQUES:

The seeds should be sown in the nursery, a mixture of FYM/Vermi compost and soil is thinly spread over the seeds and irrigated with a sprinkler hose. About 20–30g seeds are enough to raise the seedlings for planting one hectare land. The seeds germinate in about 8–12 days and the seedlings (5-6 cm long) are ready for transplanting (at a spacing of 60 × 60 cm) in about 6 weeks. In summer 3-4 irrigations per month are necessary whereas during the remaining period, irrigation is given as and when required. 4 or 5 weedings are enough under well managed conditions.

HARVESTING AND POSTHARVEST TECHNOLIGY:

90–95 days after planting *Ocimum sanctum* is ready for first harvest. Subsequent harvests are taken approximately once in three months. The crop has to be harvested at flower initiation for leaf production,. The crop is cut at 15–20 cm above the ground level while harvesting, in such a way that most of the tender shoots are cut leaving the woody stem portions for regeneration. From well managed organically grown *Ocimum sanctum* crop 12–15 t / ha fresh herb (equivalent to 2.4 to 3 t / ha dry matter)can be obtained in a year. *Ocimum sanctum* herb are shade dried for about 8–10 days by thinly spreading on gunny bags, preferably in well aerated drying sheds. The materials are turned over frequently to prevent fungal attack. The moisture content in the dried herb should be less than 10%.

Mangesh et al. (2012) UA has shown diverse pharmacological activities and significant medicinal potential. The optimization of the stirred batch extraction of UA is carried out. Process parameters which affect the extraction yield like extraction time, solvents, solid to solvent ratio, extraction temperature are investigated. In stirred batch extraction with final optimized conditions i.e. solid to solvent ratio of 1:120, 40°C temperature and speed of agitation of 1,000 rpm, 11.21 mg of UA per gram of OS is obtained. Spiro and siddique kinetic model shows good agreement with the experimental results. The diffusion coefficient determined ranged from 2×10^{-11} to $6.10 \times 10^{-11} m^2/s$. The activation energy for the extraction kinetics of OS is found to be $Ea = 10.45$ kJ/mol.

SERPENTWOOD (Rauvolfia serpentina); FAMILY: Apocynaceae

Serpentwood is an erect, evergreen , perennial undershrub whose medicinal use has been known since 3,000 years, dried root of which contains a number of alkaloids including reserpine, rescinnamine, deserpidine, ajamalacine, ajmaline, neoajmalin, serpentine, a-yohimbine. The root is a sedative, control high blood pressure and in ayurveda it is also used for the treatment of insomnia, epilepsy, asthma, acute stomach ache and painful delivery. The juice of the leaves is used as a remedy for the removal of opacities of the cornea. The alkaloids found in root and leaf fractions of Sarpagandha *(rauwolfia serpentina).* and the amount of alkaloids were higher in root fraction as compared to leaf.

Rauvolfia serpentina is native to India. Several species are observed growing under varying edaphoclimatic conditions in the humid tropics of India, Nepal, Burma, Thailand, Bangladesh, Indonesia, Cambodia Philippines and Sri Lanka. In India it is cultivated in Uttar Pradesh, Bihar, Tamil Nadu, Orissa Kerala Assam, West Bengal and Madhya Pradesh. Thailand is the chief exporter of *Rauvolfia* alkaloids followed by Zaire, Bangladesh, Sri Lanka, Indonesia, and Nepal. The genus *Rauvolfia* of Apocynaceae family comprises over 170 species distributed in the tropical and subtropical parts of the world including 5 species native to India The common species of the genus *Rauvolfia* .This are *R. serpentina* Benth. ex Kurz.(India, Bangladesh, Burma, Sri Lanka, Malaya, Indonesia), *R. vomitoria* Afz. (West Africa, Zaire, Rwanda Tanzania), *R. canescens* Linn. syn. *R. tetraphylla* (America, India),

*R. Mombasina (*East Africa, Kenya, Mozambique), *R. beddomei* (Western ghats and hilly tracts of Kerala), *R. densiflora* (Maymyo, India), R. microcarpa *(Thandaung),* R. verticillata *syn.* R. chinensis *(Hemsl),* R. peguana *(Rangoon-Burma hills),* R. *caffra* (Nigeria Zaire, South Africa), *R. riularis* (Nmai valley), *R. obscura* (Nigeria Zaire).R. *serpentina* is an erect perennial shrub(15-45 cm high) with roots(greenish yellow externally and pale yellow inside, extremely bitter in taste) nearly verticle, tapering up to 15 cm thick at the crown and long giving a **serpent-like** appearance, occasionally branched or tortuous developing small fibrous roots. The leaves born in whorls of 3-4 elliptic-lanceolate orobovate, pointed, and the flowers are numerous, borne on terminal or axillary cymose inflorscence. The fruit is a drupe, obliquely ovoid and purplish black in color at maturity with stone containing 1-2 ovoid wrinkled seeds. The plant is cross-pollinated, mainly due to the protogynous flowers.

CLIMATIC AND SOIL REQUIREMENTS:

Tropical humid climate an annual rainfall of 1,500–3,500 mm and the annual mean temperature is 10–38 °C is most ideal. It grows up to an elevation of 1,300–1,400m from msl. It can be grown in open as well as under partial shade conditions and on a wide range of soils. Medium to deep well drained, slightly acidic to neutral, fertile soils and clay-loam to silt-loam soils rich in organic matter are suitable for its cultivation.

BRIEF AGROTECHNIQUES:

The plant can be propagated vegetatively by root cuttings, stem cuttings or root stumps and by seeds. Seed propagation is the best method for raising commercial plantation. Fresh seeds are collected during September to November and soak in 10 percent sodium chloride solution. Those seeds which sink to the bottom should only be used. Before planting in nursery seeds are treated with ceresan or captan to avoid damping off. 5–6 kg seeds are required for 1 hectare land. Nursery beds are prepared in shade, well rotten FYM is applied at lkg/m" and seeds are dibbled 6–7 cm apart in May-June and irrigated. Seedlings of 2 months of age with 4–6 leaves are transplanted at 45-60 × 30 cm spacing in July -August in the main field. Alternatively, rooted cuttings of 2.5–5cm long roots or 12-20cm long woody stems can also be used for transplanting. In the main field 10–15 t/ha of FYM is applied as basal dose. Fertilizers are applied at 40:30:30kg N: P_2O_s :$K_2$0/ha every year. Monthly irrigation helps to increase the yield. The nursery and the main field should be kept weed free by frequent weeding and hoeing. Intercropping of soybean, brinjal, cabbage, okra, or chilly is followed in certain regions.

HARVESTING AND POSTHARVEST TECHNOLIGY:

During November -December when the plants shed leaves, become dormant and the roots contain maximum alkaloid content is the optimum time of harvesting. Harvesting is done by digging up the roots by deeply penetrating implements . The yield is 1.5-2.5 t/ha of dry roots. After harvesting the roots are cleaned washed cut into 12-15cm pieces and dried to 8-10% moisture. To protect it from mold the dried roots are then stored in polythene lined gunny bags in cool dry place. The root bark constitutes 40–45 percent of the total weight of root and contributes 90% of the total alkaloids yield.

SOLANUMS (Solanum spp.)   FAMILY: Solanaceae

A number of species of this genus are reported to be medicinal which are : *S. anguivi* Lam. syn. *S. indicum* auct. non Linn., *S. dulcamara* **Linn.**, *S. erianthum* D. Don, *syn.* *S. verbascifolium* auct. non Linn., *S. melongena* **Linn.**, *S. melongena var. incanum* **(Linn.) Prain** *syn.* *S. incanum* **Linn.**, *S. coagulens* **Forsk.**, **S. nigrum** Linn. *syn.* *S. rubrum* Mill., *S. spirale* **Roxb.**, *S. stramoniifolium* Jacq., *syn.* *S. ferox* auct. non Linn., *S. surattense* Burm. F. *syn.* *S. xanthocarpum* schrad. & Wendl., **S. jacquinii** Willd., *S. torvum* **Sw.**, *S. trilobatum* **Linn.**, *S. viarum* Dunal, *syn.* *S. Khasianum* **C. B. Clarke**.

BRIEF AGROTECHNIQUES:

The plants of solanaceous group come up very well in tropical and subtropical climate upto 2000m altitude and can be raised on a variety of soils, reach in organic matter. Seed propagation is done. The seedlings are raised in the nursery and transplanted to the main field 30–45 days after sowing(when the plants attain 8–10cm height). Planting is done on ridges during rainy season while in furrows during summer, at a spacing ranging from 30–90cm depending upon the stature and spreading habit of the plant. During summer the transplanted seedlings need temporary shade for 2–4 days. At the time of land preparation FYM or compost at 20–25t/ha is applied. A moderate fertilizer dose of 75:40:40 N, P_2O_5, K_2O/ha may be given, P as basal dose and N and K in 2–3 split doses. Weeds can be controlled by one or two intercultural operations. Earthing-up is done after weeding and topdressing. During summer irrigation is needed at 3-4 days interval and on alternate days during fruiting period. Staking is done to avoid lodging due to heavy bearing.

HARVESTING AND POSTHARVEST TECHNOLIGY:

Harvested is done at different stages of fruit development. The time from flowering to harvest may be 10–40 days depending on cultivar and temperature. Immature fruits (before seeds begin to significantly enlarge and harden) are generally harvested. Firmness and external glossiness are also indicators of a prematurity condition. Over mature fruits become pithy and bitter.

- **Grades, Sizes, and Packing:** Grades include Fancy, U.S. No. 1, U.S. No. 2, and Unclassified, based on size, external appearances, and firmness. Sizes are defined as: Small, Medium, Large, and Extra Large, and they are defined as 32 fruit/box with fruit length 12 to 14 cm; 24 fruit/box with fruit length 19–21 cm; 18 fruit/box with fruit length 21–24 cm; and 16 fruit/box with fruit length 24–26 cm respectively. Packages (0.39 m) are one-piece waxed fiberboard boxes or wire-bound crates, containing 15 kg. Fruit are individually wrapped with paper.

- **Precooling conditions:** Immediately after harvest rapid cooling to 10 °C is necessary to retard discoloration, weight loss, drying of calyx, and decay. The most effective processes are hydrocooling and forced-air cooling, but room-cooling after washing or hydrocooling is common.

- **Optimum Storage Conditions:** Fruit are stored at 10–12 °C with 90–95 percent RH. As visual and sensory qualities deteriorate rapidly the storage is generally less than 14 days. Short-term storage is often used to reduce weight loss, but result in chilling injury after transfer to retail conditions.

- **Controlled Atmosphere (CA) Considerations:** Low O_2 levels of 3–5 percent delay deterioration. This plant tolerates up to 10 percent CO_2 but storage life is not extended beyond that under reduced O_2. Wrapping fruits with plastic film to create modified atmosphere reduces weight loss and maintains firmness, due to the high RH. It has been found especially on Japanese eggplant types, which have a high transpiration rate. Wrapped fruit in high density polyethylene (HDPE) maintain a fresher flavor, firmness, and quality for a longer period.

CHEBULIC MYROBALAN (Terminalia chebula *Retz. Syn.* Myrobalanus chebula *(Retz.) Gaertner)* FAMILY: Combretaceae

Chebulic myrobalan fruit is a common constituent of "*Triphala*" capable of imparting youthful vitality and receptivity of mind and sense. This is a major constituent of several Ayurvedic preparations. It is used in astringent ointments in allopathy. It is used as a blood purifier in Unani system. The fruit pulp is given in piles, flatulence, asthma, urinary disorders, chronic diarrhea, dysentery, costiveness, vomiting, hiccup, intestinal worms, ascites, and enlarged spleen and liver. Powder of the fruit is used in chronic ulcers and wounds, carious teeth and bleeding ulceration of the gums.

The plant is found throughout India mainly in deciduous forests, on dry slopes upto 900m especially in Bengal, Tamil Nadu, West Coast and Western Ghats. This is also reported in Sri Lanka, Nepal, and Burma. This is a medium sized deciduous tree with a cylindrical bole, rounded crown, spreading branches with dark brown bark and brownish gray heartwood. Leaves are of different shapes like simple, alternate, or subopposite, ovate or elliptic ovate with short petioles bearing two glands below the blades. Flowers are pale yellow or white in 4–10cm long axillary spikes. Yellow to orange brown color fruit is a drupe, ovoid glossy, glabrous, faintly angled. The pale yellow colored seeds are hard.

BRIEF AGROTECHNIQUES:

It requires direct overhead light and cannot tolerate shade. But the young plants appreciate a certain amount of shade. It is frost hardy and drought-resistant to some extent. It can be grown on loam as well as on lateritic soils with moderate fertility.

The plant attains its best development on loose, well-drained soil and an average temperature ranging from 10 to 48°C.

- **Natural regeneration:**

  The tree propagates by natural regeneration in some localities. In the rainy season the seeds germinate better if it is covered up with the earth or debris than, if it is lying in the open. Good drainage and shelter from the side is desirable for natural regeneration. Growth is poor during rains and the seedling is often killed by heavy and continuous rain. Regeneration is facilitated by manipulation of canopy by creating small gaps, and this is supplemented by sowing seeds in gaps.

- **Artificial regeneration:**

  The tree can be successfully raised in the field by:
  1. Direct sowing of seeds.
  2. Transplanting seedlings.
  3. Planting root and shoot cuttings.

HARVESTING POSTHARVEST TECHNOLIGY:

Fruit collection is done during January to March. Fruit should be collected from the ground as soon as they have fallen. January is the best time for collection of the fruit for optimum tannin content. Collection prior to or after January will yield inferior quality of Harra, containing 32 % tannin, range of which usually varies from 12 to 49 percent. Freshly collected and immediately dried, yellowish color. Harra fetch a better price. The fruits, lie on the ground have darker color with sometimes mold attack with very low tannin content. Mold attack also is a major cause of poor quality of myrobalans.

- **Collection and processing:**

  The collection of fruits is generally done by shaking the trees and picking up from the grounds. For avoiding contamination the fruits are then dried in the sun with arrangements for 3–4 weeks. For this purpose contractors generally erect temporary sheds to store myrobalans in the event of rain. The raw myrobalan is graded based upon their solidness, color, and freedom from insect attack. Grading generally consists of separating inferior fruits, which constitute a second grade, the remainder being the first grade. And these were also graded by appearance for the export market and for some tanneries within the country. Myrobalans of fair quality from any area were marketed without grading as FAQ (Fair avg. Quality), which consists of 75 percent solid and 25 percent of hollow and decayed nuts.

- **Grading:**

  The different grades of myrobalans are at present known by the names of their exported area. Following four grades are known: (1) Jabalpore from MP and partly from Orissa, (2) Bimilipatnam from Andhra Pradesh and partly from Tamil Nadu, (3) Rajpores or Bombay variety mostly from Kolhapur and other parts of Maharashtra, and (4) Salem or Madras variety mostly from Tamil

Nadu. The fruits fall on the ground soon after ripening. In trade parlance, Harra is divided into three categories.

Bal/Choti/Jawa Harra–This is primarily used for Ayurvedic medicines and harvesting is usually done in Januar. Price is around Rs. 40 per kg. The fruit is collected before maturity as small Harra has more medicinal value. Though harvesting at such an early stage is not considered sustainable, forest dwellers are forced to do it as it fetches more prices. When Harra becomes mature, it loses medicinal value as well as fetch less money.

Badi Harra–It is not very useful for Ayurvedic medicines as it has lesser medicinal values, but it is used in Tanneries. Average price is about Rs. 3 per kg. February is the best month for collection of this variety. Badi Harra is losing its ground rapidly as tannin production companies have developed a substitute.

Kacheria -It is the crushed pulp of Badi Harra as the astringent quality is the same and can be used as substitute of bal Harra. Whole fruits are preferred to avoid adulterations while the crushed Myrobalans are preferred as it reduces bulk and weight of the material. Transport difficulties forces exporters to send Myrobalans in crushed form. 60 kg of Kacheria comes out from 100 kgs of Badi Harra which fetches about Rs.10 per kg.

- **Value addition:**

Myrobalan extracts (spray dried powder form) are manufactured in an open vat made of wood or concrete with copper or brass fittings (iron not being used in order to avoid chemical reaction of the tannin solution) by leaching with water at 70 degree C. The extract is cooled to 15 degree C, as it causes separation loss of tannin. The manufacture of spray dried tannin powder consists in principle in projection under pressure of finely dried spray of the liquid in a closed chamber where it is so arranged that water vapor that is given off is continually withdrawn. The evaporation is also helped by the introduction of hot air, the circulation of which can be so regulated as to control at will the moisture content of the dry powder formed. Before being pressed through the chamber, the hot air is completely dried. The dry powder extract so obtained falls to the bottom of the chamber on an inclined base so as to facilitate easy removal.

KEYWORDS

- **Herbs**
- **Medicinal plants**
- **Quality**
- **Biochemical compounds**
- **Postharvest technology**

REFERENCES

Aremu, A. O.; Finnie, J. F.; Finnie and Van Staden J.; Potential of South African medicinal plants used as anthelmintics – Their efficacy, safety concerns and reappraisal of current screening methods. *S. Afr. J. Bot.* **2012**, *82*, 134–150.

Baskaran, K.; Srinivas, K. V. N. S.; and Kulkarni, R. N.; Two induced macro-mutants of periwinkle with enhanced contents of leaf and root alkaloids and their inheritance. *Indian Crops. Prod.* **2013**, *43*, 701–703.

Joymati Devi, L.; Christina Devi, Kh.; and Ronibala Devi, Kh.; *Effect of oil Extracts of Medicinal Plants Against Root Knot Nematodes;* Department of Zoology Nematology laboratory, D. M. College of Science: Meghalaya – 795 001, India, **2012**.

Kumar, A.; Abrol, E.; Koul, S.; and Vyas, D.; Seasonal low temperature plays an important role in increasing metabolic content of secondary metabolites in *Withania somnifera* (L.) Dunal and affects the time of harvesting. *Acta Phys. Plant.* **2012**, *34*, 2027–2031.

Mangesh, D. V.; Vikesh, G. L.; and Rathod, V. K.; Extraction of ursolic acid from *Ocimum sanctum* leaves: Kinetics and modeling. *Food Bio. Prod. Proc.* **2012**, *90*, 793–798.

CHAPTER 12

POSTHARVEST MANAGEMENT OF SPICE CROPS

S. SULTANA, S. DAS[1] and A. B. SHARANGI[*]

[1]Department of Floriculture and Landscape Gardening, Faculty of Horticulture, Bidhan Chandra Krishi Viswavidyalaya, Mohanpur-741252, Nadia, West Bengal, India

Department of Spices and Plantation Crops, Faculty of Horticulture, Bidhan Chandra Krishi Viswavidyalaya, Mohanpur-741252, Nadia, West Bengal, India

[*]Email: dr_absharangi@yahoo.co.in

CONTENTS

12.1   INTRODUCTION

India is often referred to as the "Home of spices" mainly because most of the 70 spices grown in the world are native to India. Further, since antiquity, India pioneered in growing spices and export. India has enjoyed virtual monopoly in the international spice trade since ancient times. Out of the 70 spices, several of them can be grown in India, whereas in other countries a few spices are only grown. This is because; India has a great extent of diversity in the climate and soils, which enables to grow a variety of spices. Spices are always export oriented crops not only in India, but also in other spice producing countries. India produces annually about 5.1 million tons of different spices valued at about Rs. 10,200 crores, contributing to 25–30 percent of the world production. India is the biggest exporter of spices and annually exporting about 2 million tons of different spices and spice products earning a foreign of about Rs. 5,240 crores. Spices are the aromatic substances of vegetable origin obtained from various plants used in food, beverages, cosmetics and confectionary and also as preservatives and food flavorants. The plant parts that are used vary from fruits to seeds, flowers and bark. Spices and condiments contain essential oils and which provide the flavor and taste. They are of little nutritive value. They are used as whole, ground, paste or liquid form, mainly for flavoring and seasoning food. Most spices increase the shelf-life of food.

Spices are not perishable like that of vegetable but most crop loss occurred during postharvest management. Postharvest management do not related only after harvest, it start before harvest. Good panting material, proper agronomic practices, proper harvesting all have certain influences in postharvest management practices. Postharvest management practices that reduce product loss to spoilage or shrinkage will reduce microbial risks. These include: cleaning the product, sorting, packaging, quick cooling, good refrigerated storage. Good transportation and distribution. However each spices is processed in a specific manner depending on its usage. Postharvest management of spices has great scope considering the present International trade scenario. We expect a huge lump in the export of curry powder and other value-added products in the coming decade. The total export in value-added products may cross US $600 million in the coming 10 years. The research programs should orient for this demand by focusing more attention on better agrotechniques in product diversification, varieties suitable for such products and following GAP. Considering the huge potential of spices as source of eco-friendly nutraceuticals, agrotechniques which release such compounds needs to be formulated (Parthasarathy et al. 2011).

Sagoo et al, (2008) worked on Assessment of the microbiological safety of dried spices and herbs from production and retail premises in the United Kingdom. *Salmonella* spp were detected in 1.5 percent and 1.1 percent of dried spices and herbs sampled at production and retail, respectively. Overall, 3.0% of herbs and spices contained high counts of *Bacillus cereus* (1%, 105 cfu g1), Clostridium perfringens (0.4%, 103 cfu g1) and/or Escherichia coli (2.1%, 102 cfu g 1). Ninety percent of samples examined were recorded as being "ready- to-use", 96 percent of which was of satisfactory/acceptable quality. The potential public health risk of using spices and herbs as an addition to ready-to-eat foods that potentially undergo no further processing is therefore highlighted in this study. Prevention of microbial contamination in dried herbs and spices lies in the application of good hygiene practices during growing, harvesting and processing from farm to fork, and effective decontamination. In addition, the importance of correct food handling practices and usage of herbs and spices by end users cannot be overemphasized. Kant et al, (2013) studied on postharvest storage losses by cigarette beetle (*Lasioderma serricorne* Fab.) in seed spices and estimated the damage and reproductive potential of *Lasioderma serricorne* Fab on some seed spices, viz., cumin, coriander, fennel, ajowan and dill at different storage conditions. The result showed that the beetle causes huge storage losses which were maximum in fennel seed (58.02%) and minimum in dill seed (39.0%). Population growth was also related to damaging potential on different seed spices. Maximum population of insect was recorded in fennel seed and minimum in dill seed. Maximum damage and reproduction was noticed in July to September and minimum in the month of January to March in coriander, ajowan and cumin and April to June in case of fennel and dill. Pramila et al (2013) studied on assess seed quality of some important seed spices, supplied in the local markets of Bengaluru. The results revealed that the highest physical purity was recorded in fenugreek seed lots with an overall mean of 98.76 percent followed by fennel (97.99%) and coriander (97.68%). The moisture content of all the seed spices ranged from 6.89% to 7.15%. The highest seed germination (89.42%) was recorded in fenugreek followed by fennel (76.82%) and lowest (64.33%) in coriander. The seedling vigor index was highest in fenugreek (range 1116-1819; mean 1532) and lowest in cumin (range 621–832; mean 737). A 12.0 percent increase in germination and improvement in the vigor index (997–1226) were also noticed in treated seed samples compared to control.

12.2   FENUGREEK (*TRIGONELLA FOENUM-GRAECUM*, FAMILY: FABACEAE)

It is one of the most common spices grown throughout the country. It has a multitude of uses including using the leaves fresh or dried, using the seeds whole or ground as a spice or the whole plant as an effective green manure to improve the soil. Good soil of medium texture is preferred. Tolerated pH range is 5.3–8.2. A sunny, well-drained position and adequate water is required for its proper growth and development.

Fenugreek does not like to be transplanted, so should be sown directly into a well- drained sunny spot. Sow 0.5 cm deep into drills 20 cm apart, and 5 cm spacing within row. It can be sown during "rabi" (winter) season and will withstand some frost. Germination will normally take place within a week. The plant grows rapidly and is vigorous enough to compete against most weeds. The small leafed variety will continue to grow slowly throughout the winter whereas the larger one will generally die off. Although fenugreek is a legume, it doesn't always fix nitrogen. For this to happen, the right bacteria (*Rhizobium meliloti*) need to be present in the soil. To see if the plants are fixing nitrogen, carefully dig (don't pull!) up a plant and look for pink colored nodules (2mm diameter) on the roots. Plants that are fixing nitrogen should be able to produce lush green growth on a low fertility soil with few problems, whereas those that can't tend to develop pale colored leaves and smaller plants. Poor soil nutrient status can affect the flavor of the crop.

HARVESTING

Harvesting the plants for leaves is done by cutting the stem a few cm above the base when the plants are up to 25 cm tall. The larger white flowered variety will not regrow after flowering so needs successional sowing whereas the yellow variety can be cut a number of times and should be cut regularly to prevent it seeding and keep it productive. They will generally be ready by 6 weeks after sowing depending on the weather. The quality of leaves will decline once flower buds start to appear so try and harvest before then.

POSTHARVEST PROCESSING

Fenugreek will flower and produce thin seed pods. The pods are allowed to ripen and turn yellow on the plant then are harvested shortly before the seed pods pop open. The plants are uprooted and dried in the sun on threshing floor for 2-3 days. Pods are threshed. Seeds are separated by winnowing, cleaned and sun dried. Seeds are stored in gunny bags lines with white alkathene paper. Seeds are ground if required. Whole dried seeds or ground fenugreek powder are stored in airtight containers.

Yield: Leaf yield is 700–800 kg per ha. Grain yield is 750–800 kg per ha under rain fed conditions. 1,200 to 1,300 kg per ha under irrigated conditions. Depending on the variety and the season it produces 7–8 tn. of per hectare.

12.3 CORIANDER (CORIANDRUM SATIVUM, FAMILY: APIACEAE)

Coriander is a hardy annual plant and has been cultivated for thousands of years. It has slender sparse branches with a highly aromatic stem and leaves that can be harvested and used fresh as well as a dried spice. Coriander can be grown successfully in most soil types provided that there is sufficient moisture.

Coriander is normally sown from April to September and needs to be sown in succession to ensure continual production. It is normally drilled directly 1 inch apart in 15 inch rows in an open system, 8.5kg – 14kg depending on the seed count. Coriander takes from 4 to 8 weeks from sowing to the start of harvesting and the length of harvest will vary throughout the year depending on the conditions. The coriander has a tendency to bolt rapidly when subjected to any stress, causing the plant to stop producing the leaves that the plant is grown for.

HARVESTING AND PROCESSING

Coriander is normally sown from April to September and needs to be sown in succession to ensure continual production. It is normally drilled directly 1 inch apart in 15 inch rows in an open system, the approximate seed rate would 8.5–14 kg depending on the seed count. Coriander takes about 4–8 weeks from sowing to the start of harvesting and the length of harvest will vary throughout the year depending on the conditions. The coriander has a tendency to bolt rapidly when subjected to any stress, causing the plant to stop producing the leaves that the plant is grown for. If harvesting is delayed – seeds shattered fruits splitted.

When the green color turn to straw colored, plants are cut or pulled, tied in bundles, piled in shade for drying to avoid grain shattering and loss of essential oil. After 2–3 days of shade drying, grain is threshed, winnowed and sundried. The moisture content is reduced from 2 to 6 percent. The cleaned, dried produce is stored in gunny bags lined with white polythene

POSTHARVEST TREATMENT

The coriander leaves need to be cooled as quickly as possible after harvest and stored in a cool area to slow deterioration. Coriander is currently sold by weight; the seed count is typically between 75 – 120 seeds per gram.

QUALITY ISSUES

The quality of coriander relates to size, shape, appearance, color, odor and aroma characteristics. The produce must be safe and free from any health hazard substances and contaminants. Moisture, volatile oil, oleoresin content and major chemical constituents present in coriander determine the intrinsic quality. As quality is the most urgent challenge facing the industry, there is a need to ensure that the product for the market, either for domestic or export purposes, is completely free from pesticide residue, aflatoxin, other mycotoxin and unfavorable microbial contamination. After processing, coriander should be graded according to the International Organization for Standardization (ISO) or according to the requirement of the importing country. Most of the importing countries have their own grades. Therefore, grading and standardization become the essential prerequisites ensuring quality.

12.4 CUMIN (CUMINUM CYMINUM, FAMILY: *APIACEAE*, ORIGIN: MEDITERRANEAN REGION)

It have aromatic fragrance is due to an alcohol 'cuminol'. Used as a spice in curry powder. It is used as Carminative, stomachic and astringent).Volatile oil content ranges from 2.8 to 4.7 percent.

Tropical plant and it can be cultivated as Rabi crop. Its required well drained, medium to heavy textured soils. On light textured soils wilt is more and can be grown in slightly alkaline soils having a pH of 8.9. It is sown in lines or broadcasting is done. Soaking of seed for 24 – 36 hours is suggested to enhance the germination percentage. In line sowing, lines are spaced at 20 cm apart. Seeds are covered by fine soil. Irrigate lightly. At 5 cm height, thin the population to 15 cm spacing. A light irrigation immediately after sowing and subsequent irrigation is given when required. At 5 cm tall; first hoeing and weeding. Later one or two hoeing or weeding is done.

HARVESTING

The crop matures in 100–120 days. The plants are uprooted when leaves become yellow. Then after sun drying, threshing is done by beating. Cleaning is basically done by winnowing and the clean and dried seeds are stored in polythene line gunny bags in cool and dry places. Yield varies from 500–800 kg per ha grains and can be up to 1,000 kg per ha under good management.

PROCESSING AND STORAGE

Cumin oil and oleoresin are obtained from well crushed seeds by steam distillation method. The ripe seeds are used for oil production, both as whole seeds or coarsely ground seeds. Hydro distillation is used for essential oil extraction, producing a color less or pale yellow oily liquid with a strong odor. The yield for oil production varies from 2.5 to 4.5%, depending on whether the entire seed or the coarsely ground seed is distilled. Cumin oil can be readily converted artificially into thymol. The volatile oil should be kept in well-sealed bottles or aluminum containers.

QUALITY SPECIFICATIONS

Seeds – Oblong in shape, thicker in the middle, compressed laterally, about 3–6 mm long, resembling caraway seeds, but lighter in color and ,almost straight. They have nine fine ridges. The odor and taste are somewhat like caraway, but less agreeable. The powdered seeds are yellowish brown with an aromatic, slightly camphoraceous odor and taste.

Specific quality indices are:
- Seed moisture: less than 6 percent
- Total ash: 7 percent
- Acid insoluble ash: 1.5 percent
- Volatile oil: minimum 2 percent
- Foreign organic matter: 2 percent (US maximum for harmless foreign matter: 5%)

12.5   AJWAN (*TRACHYSTERMUM AMMI.* FAMILY: APIACEAE)

It is an annual herbaceous plant grown for its herb and grayish brown seeds. It is known as 'Bishop's weed' and native of Egypt. It is mainly grown as winter crop in subtropical and temperate climate. It prefers humus rich loam or clay loam soils. Seeds are sown directly in the field in line sowing method during September-October. Spacing is 45 × 30 cm apart. Seeds germinate in 7–14 days. Irrigation is given immediately after sowing and subsequent irrigation is given as when required. Weeding is generally done twice.

HARVESTING AND PROCESSING

The plants start flowering 2 months after flowering. The crop is harvested at about 120–140 days after planting when the flower heads turn brown in color. The mature umbels are cut with sickle manually. After cutting, the crop is tied in small bundles and kept for drying. Then threshed on a cleaned floor and winnowed to separate the clean seeds.

Indian standards have 3 grades i.e Special, Good, and Fair .Cleaned seeds packed in clean gunny bags of 40–45 kg capacity and stored in hygienic warehouse or godowns. Seeds are normally stored for 6–8 months.

The value-added products are ajowan oil, thymol, dethymolised oil, thyme, and fatty oil.

PROCESSING OF SPICE ESSENTIAL OIL

The dried seeds are crushed and distilled to obtain essential oil. Hydro or steam distillation use of subcritical liquid carbon-dioxide is used. The seeds lose the essential oil when stored for long time.

12.6   BLACK CUMIN (*NIGELLA SATIVA,* FAMILY: RANUNCULACEAE)

Nigella is classified as a mild spice and, on the basis of plant organs used. Nigella seeds are aromatic and contain a disagreeable odor. It is known to grow wild and cultivated in India, Egypt and the Middle East. It is primarily exported from India and Egypt.

Fairly warm weather during sowing with a temperature of 20–25°C is desirable. Cold weather is congenial for the early growth period and the crop requires warm sunny weather during seed formation. Plants are frost sensitive at any growth stage. It can thrive on a wide range of soils, which are rich in organic matter and free from water logging. However, loamy, medium to heavy soils with a better fertility level are most suitable. It is propagated by seed. Seeds are sown at row spacing of 30 cm and plant spacing of 15–20 cm and a seed rate of 8 kg/ha is required during the month of October has been found appropriate. The seeds germinate within 12 days. Light irrigation should be given immediately after sowing if initial moisture is low. Irrigation should be given at 5–6 day intervals initially and thereafter at 10–15 days.

HARVESTING AND PROCESSING

The crop is harvested after 110–140 days after planting. At the mature stage the crop (capsules) turns yellow in color. The capsules when split the fingers discharge deep black colored seeds with pleasant aroma. After harvesting the whole plants are cut and dried in shade. The seeds are separated by threshing and winnowing. After this the seeds are dried under sun until it reached moisture 8–10%. Later the seeds are stored in air tight container or poly bags in cool and dry place.

DISTILLATION OF OIL

The seeds are steam distilled to obtain yellowish brown volatile oil with pleasant aroma. The oil contains 45–60 percent carvone, d- limonene, and cymene. Carbonyl compound, nygellone, is also present.

QUALITY SPECIFICATIONS

The Indian Agmark grade specifications for Nigella seeds with minimum specific quality indices. Seed moisture = not more than 11 percent by weight.
- Total ash = not more than 6 percent by weight.
- Ash insoluble in acid = not more than 1 percent by weight.
- Organic extraneous matters = not more than 3 percent by weight.
- Inorganic extraneous matters = not more than 2 percent by weight.
- Volatile oil = not less than 1% (v/w).
- Ether extract (crude oil) = not less than 35% (v/w).
- Alcoholic acidity as oleic acid = not more than 7% (v/w).

Nigella powder is produced by grinding dried, cleaned and sterilized seed. After sieving through the required mesh size, the powder should be packed in airtight containers

12.7   FENNEL (*FOENICULUM VULGARE*), FAMILY – APIACEAE

Fennel grow under dry climate and susceptible to frost. Well-drained loamy soil and black cotton soils are suitable for cultivation of this crop. Rain fall during maturity spoil color and reduce quality of fennel seeds. Generally, fennel is cultivated as Kharif as well as Rabi crop. Best sowing time is 15th October. 5–6 kg of seed/ha is required for sowing drilled fennel crop in the nursery beds . Beds should be covered with plant waste for up to 12 days. Tender seedlings should be protected from direct sunlight, till they are under danger, by erecting temporary shed over the bed. When seedlings are free from the danger of sunlight, temporary shed should be lifted. 1–2 hand weeding should be done depending on weed growth. First light irrigation should be given just after sowing and second after 3–4 days after sowing and then light irrigations are given as and when required to keep the soil moist.

HARVESTING AND PROCESSING

The crop is harvested 5–6 month after sowing before the fruits are fully ripe to avoid shattering. For the chewing type, the crop can be harvested when the grain is half length size. During harvesting plants are cut and spread out in loose bundles for drying in the sun. The dried fruits are threshed and cleaned by winnowing. The dried and cleaned seeds are packed in jute bags. The oil obtained from well crushed seeds by steam distillation method which gives pale yellow colored liquid with characteristic taste and aroma.

12.8   SMALL CARDAMOM (*ELETTARIA CARDAMOMUM*, FAMILY: ZINGIBERACEAE, ORIGIN: WESTERN GHATS, SOUTH INDIA

Cardamom is popularly known as the Queen of Spices and also Green Gold. It is one of the ancient species of India and is also one of the most valued spices of the world. It is next only to black pepper as the largest foreign exchange earner among various Indian spices.

Small cardamom is a humid tropical plant. It is grown under natural conditions of ever green forests and grows best on well drained humus rich forest soils with a soil pH 5.0–6.5. Cardamom can be propagated by seeds, rhizomes and suckers. Seedlings are normally raised in primary and secondary nurseries. Sowing may be taken up during November – January and is done in rows. Seeds are mulched to a thickness of 2 cm with paddy straw or any locally available material and are watered regularly. The germination commences in about 30 days and may continue to a month or two. After germination the mulch is to be removed. Seedlings are transplanted in the secondary nursery in March—May at a spacing of 20 × 20 cm and mulched. Immediately beds are to be covered with an overhead pandal and should be watered regularly. Seedlings are also raised in poly bags containing rich forest soil. Some of the common shade trees in cardamom estates are *Diospyros ebenum, D. elongi , Mimusops elengi, Artocarpus fraxinifolius*, Jack, *Cedrella toona*. Cardamom is generally raised as rain fed crop. It is necessary to irrigate the crop during dry periods to get increased yields. 2–3 weeding per year may be necessary. Fallen leaves of the shade trees and up rooted weeds are utilized for mulching

HARVESTING

Cardamom plants start bearing in about 3 years after planting. Flowering starts in April—May and continues up to August—September. Peak flowering will be in the month of May- June. From flowering to maturity the fruit takes 5-6 months. Only ripe capsules are harvested at 25-30 days interval, the harvesting is completed in 5– 6 pickings. In most of the areas the peak period of harvest is during October—November.

Yield: Although the Cardamom plant start bearing from 2nd or 3 rd year of planting, an economic crop can be obtained only from 4th or 5th year. Yield varies with variety and age. Optimum average yield is 50–70 kg of dry capsule per ha. Yields decline from 10th year to 12th year.

1st year of bearing – 25–50 kg per ha (dry capsules)
2nd year of flowering 50–70 kg per ha (dry capsules)
3rd year of flowering 70–100 kg per ha (dry capsules)

Processing: The commercial product of Cardamom is the dried capsules. At the time of harvesting the capsules are juicy and fleshy, so they must be cured before sending them to the market.

Bleaching: Green color of the cardamom capsules plays a vital role in the market. Green color of the capsules can be preserved by alkali treatment. So freshly harvested cardamom capsules are soaked in 2 percent washing soda (Na 2 CO 3) solution for 10 min.

Drying: After bleaching, the capsules are dried either by sun drying or in fuel kilns and electric driers. The capsules are sun dried for 3–5 days. These capsules get bleached and does not store well. Hence, now a days capsules are dried artificially in which drying is complete and the green color remains. In electrical drier in capsules are dried at 45–50°C for 18 hrs.

Fuel kilns: Temperature is set at 50 – 60°C overnight.

The capsules kept for drying are spread thinly and stirred frequently to ensure uniform drying. The dried capsules are rubbed with hands or coir mat or wire mesh and winnowed to remove any foreign matter.

Storage: Then they are stored according to size and color and stored in black polythene lined gunny bags to retain green color during storage. These bags are then kept in wooden chamber.

Sorting: The dried capsules are stored according to their size, color and stored in black polythene lined gunny bags to retain green color during storage.

12.9   LARGE CARDAMOM

HARVESTING

The indication of the time of harvest is when the seeds of the topmost capsules turn brown. To enhance maturity, bearing tillers are cut to a height of 30–45 cm and left for another 10–15 days for full maturity. The spikes are harvested using special knives. The harvested spikes are heaped and capsules are separated and dried. The cured capsules are rubbed on a wire mesh for clearing and removal of the calyx (tail). Leela et al, (2008) studied on Essential oil composition of selected cardamom genotypes at different maturity levels and indicated that the oil composition did not vary much over a period of 95–140 DAF. This might be because the biosynthesis of the secondary metabolites would have been completed by 95 DAF in cardamom capsules. The study also revealed the close chemical similarity between Malabar and Vazhukka types.

CURING

Traditionally large cardamom is cured in a bhatti where the capsules are dried by direct heating. Under this system the cardamom comes in direct contact with smoke which turns the capsules to a darker browner black color with a smoky smell. Improved curing techniques are available by which cardamom is processed to give better quality and appearance. One such method is the ICRI Spices Board improved bhatti system of curing in which the cardamom is dried by indirect heating at 45–50 degrees C. After studying the traditional bhatti system used for curing of large cardamom in Sikkim , It was observed that apart from energy efficiency, emphasis for technological development should be to improve the quality of dried cardamom. Accordingly, under an ISPS (Indo Swiss Project Sikkim) sponsored project TERI has developed an advanced gasifier based system for improving the energy efficiency and quality of dried cardamom. To make available the system to the cardamom growers a fabrication unit of the gasifier system is required to be set up.

Flue pipe curing -In Flue pipe curing houses, flue pipes are laid inside a room (curing house) and connected to a furnace installed outside. Fresh cardamom is spread over wire meshes fixed above the flue pipes. This is an indirect system of drying and smoke does not come into contact with the produce at any stage. This type of drier resulted in early drying and gave better quality capsules, including a better color.

CFTRI system-The Central Food Technological Research Institute, (CFTRI), Mysore, has designed and developed a low cost natural convection dryer. In this system the flue ducts are arranged in double-deck fashion and connected in series to the furnace. The convection current passes upward through the bed of capsules. Thermal efficiency is much better, the cost of drying cheaper, the quality of the product superior and the annual product output higher, than in the case of a curing house or any other existing system.

Treatment with diluted HCl solution (0.025%) of the freshly harvested capsules, improved the color after drying as revealed by better retention of the anthocyanin content of 461.43 mg/100 g as compared to freshly harvested ones that contain 1,159 mg/100 g.

PACKAGING

The properly dried capsules should be allowed to cool and then packed in polythene lined jute bags. The bags may be stored on a wooden platform to avoid absorption of moisture, which may result in fungus growth damaging the stored produce. Large cardamom is usually stored in bulk on bamboo matting spread on the ground or packed immediately into gunny bags which may then be stored in plywood tea-chests. A key issue in storage is maintaining the right level of moisture. The moisture content of capsules has to be brought down to 12–14 percent to achieve a longer shelf-life.

QUALITY ISSUES

The quality of large cardamom depends mainly on external appearance (which provides visual perception of quality as influenced by color, uniformity of size, shape, consistency and texture) and flavor (which is influenced by composition of aromatic compounds. Cineole contributes to pungency while terpinyl acetate toward pleasant).

A draft International Standards Organization (ISO) proposal on large cardamom was prepared by Spices Board, India in conjunction with CFTRI, Mysore and submitted to the Bureau of Indian Standards (BIS). The draft proposal for BIS adoption reads as follows:

CAPSULES

1. Extraneous matter – Not more than 5 percent by weight
2. Insect damaged capsules – Not more than 5 percent by weight

3. Moisture – Not more than 14 percent by weight
4. Volatile oil (%) ml/100 g – Not less than 1.5 percent
5. Colour should be natural and capsules free from added colors

SEEDS

1. Moisture – Not more than 13 percent by weight
2. Volatile oil – Not less than 2 percent by weight
3. Total ash – Not more than 5 percent by weight
4. Acid insoluble ash – Not more than 2percentby weight
5. Extraneous matter – Not more than 2 percent by weight
6. The seeds should be free from molds and insects
7. Insect damaged seeds – Not more than 2 percent by weight
8. Colour and flavor should be natural and characteristic

Leela et al. (2008) evaluated essential oil composition of selected cardamom genotypes at different maturity levels. It is desirable toharvest capsules at physiologically mature stage (110–124 days after flowering) as it results in high oil yield. Besides this, green color retention of the capsules and dry recovery were also reported to be better at physiologically mature stage.

12.10 GINGER (*ZINGIBER OFFICINALE*, FAMILY: ZINGIBERACEAE, ORIGIN: SOUTH EAST ASIA)

Ginger is one of the five most important major spices of India. India is the largest producer of dry zinger in the world, accounting for more than 60 percent of world production. One third of the production of Ginger in the country is exported. Kerala is the largest producer of ginger accounting for more than 40 percent of the total production of India. In Andhra Pradesh, it is cultivated in Nellore, East and West Godavari, Medak, Visakhapatnam, and Srikakulam districts. Dried rhizomes scraped or peeled are greatly esteemed for their aroma, flavor and pungency.

It requires warm humid climate and deep, well drained, humus rich soil. Ginger is always propagated through rhizomes, known as seed rhizomes. Carefully preserved seed rhizomes are cut into small bits (sets) of 2.5–5.0cm long, weighing 15–20 g each having one or two buds. It is planted in beds or ridges at a spacing 45 × 15 cm and The crop is mulched after sowing. Sprouting starts within a week and continues for another 3–4 weeks. Irrigate the crop at 4–10 day interval and 3–4 weedings are necessary.

HARVESTING AND YIELD

For fresh ginger, the crop should be harvested before attaining the full maturity means when rhizomes are still tender, low in pungency and fiber content, usually from fifth month onwards after planting. Harvesting for preserved ginger should be done after 5–7 months of planting while harvest for dried spices and oil is best at full maturity, i.e. between 8 and 9 months after planting when leaves start yellowing. Rhizomes to be used for planting material should be harvested until the leaves become completely dry. After digging out, these rhizomes should be treated with fungicide Dithane M-45 at 3 g/ liter of water, dried in the shade, and stored in pits covered with 20 cm layer of sand alternating every 30 cm layer of rhizome. These pits should be dug under a thatched roof to protect the rhizomes from rain, water and direct sun. Average yield varies from 12 to 15 t/ha. However, recovery of dry ginger varies from 20 to 22 percent.

Kizhakkayil and Sasikumar (2009) evaluated on variabilty for quality traits in a global germplasm collection of ginger (Zingiber officinale R.) about 46 ginger accessions originating from India, China, Pakistan, Brazil, Jamaica, Oman, Nepal, Nigeria, and Queensland analyzed for the quality attributes revealed that the primi- tive types/land races are rich in oleoresin and essential oil and low in crude fiber content as compared to the improved varieties. Juliani et al., (2007) evaluated on Chemistry and Quality of Fresh Ginger Varieties (Zingiber officinale) from Ghana and find that Ginger, Zingiber officinale, one of the most important and oldest of spices, consists of rhizomes with a warm pungent taste and a pleasant odor. The essential oils are responsible for the aroma while the nonvolatile components are responsible for the pungency with gingerol the most pungent component in fresh ginger. Ghana has a long history of producing ginger rhizomes for local markets. In this study, we evaluated the quality and essential oils and pungent principles of two different types of Ghanaian ginger, each from three regions, compared to ginger rhizomes found in the US Market. The essential oils composition of variety 2 exhibited the typical ginger oil composition with geranial, neral and zingiberene as the main components. Variety 1 exhibited a distinctly different essential oils composition dominated by zerumbone (85–87%). Variety 2 possessed a warm pungent taste similar to the commercial samples whereas variety 1 of gingers possessed a bitter taste. Variety 2 and the US commercial gingers contain 6-gingerol as the main component. The samples from Oframanse exhibited the highest amount of total gingerols (0.66%) (6-, 8-, and 10 gingerols), being almost 4 times higher than the commercial US ginger (0.17). Ginger variety 2 appears promising for the export market since these gingers could be offered both as a dried and/or ground spice as well as a source of gingerols for the nutraceutical industry.

WASHING AND DRYING

After harvest, the fibrous roots attached to the rhizomes should be trimmed off and soil is removed by washing. Rhizomes should be soaked in water overnight and then cleaned. The skin can be removed by scrapping with sharp bamboo splits or wooden splice. Use of metallic knives should be avoided since they will discolor the rhizomes. Peeling or scraping reduces drying time, thus minimize mold growth and fermentation. However scraping process tends to remove some of the oils constituents which are more concentrated in the peel. By removing the outside corky skin the fiber content also decreases. After scrapping, the rhizomes should be sun-dried for a week with frequent turnings and well rubbed by hand to remove any outer skin. This is called as the unbleached ginger. The peeled rhizomes should be repeatedly immersed in 2 percent lime solution for 6 hrs and allowed to dry in the sun for 10 days while rhizomes receive a uniform coating of lime and moisture content should be 8–10 percent. This is called as the bleached ginger which has improved appearance with light bright color. Mechanical drying is rapid, gives more homogenous and cleaner product over sun-drying method where peeled ginger takes 8–9 days to reach a moisture content of 8–9 percent. To reduce losses in quality cleaning and drying should be done as fast as possible after harvesting. To avoid discoloration the temperature should not exceed 60°C during mechanical drying.

Jayashree and Visvanathan (2010) Studied on Mechanical Washing of Ginger Rhizomes in a rotary mechanical vegetable washer at three peripheral washing drum speeds of 2.83,3.45 and 4.08 m.s^{-1} for three varying washing durations of 5, 10 and 15 min. The performance of ginger washer was evaluated based on bruise index, mechanical washing efficiency, microbial washing efficiency and color of washed ginger. For microbial washing efficiency to be above 80 percent, washing speed of 3.45 m.s^{-1} for 5 min duration corresponding to microbial washing efficiency of 92% was considered as optimum. At the optimum washing condition, the bruise index and mechanical washing efficiency were 7.5 percent and 97.8 percent, respectively. The color values of ginger washed at optimized conditions were 46.34, 7.44, and 18.71, respectively. The capacity of the mechanical washer was found to be 60 kg h^{-1} of fresh ginger.

GRADING, PACKAGING AND STORAGE

Proper care should be taken during the grading and packaging to supply quality ginger. Dried gingers can be categorized or graded as Unpeeled, Peeled, Rough scraped, Bleached, Splits and slices and Ratoons based on drying process. These different forms of rhizomes should be packed in jute sacks, wooden boxes or lined corrugated cardboard boxes depending upon distance for transportation and type of market. Dry slices or powder should be packed in multiwall laminated bags or polyethylene film pouches. Fresh ginger should be stored at 10–12°C and 90 percent relative humidity in cold room. A "zero energy" cool chamber which maintains the temperature 6 to 7°C below than outside temperature can be used in the producing areas where cold storage are not available. Gamma-irradiation at doses of 0.05–0.06 KGy inhibits sprouting of fresh ginger. Besides this storage of fresh ginger in polyethylene bags with *2 percent* ventilation prevents dehydration and mold development. Dried rhizomes, slices, or splits should be stored at 10-15°C in cold room. If cold storage facility is not available, extraction or distillation of dried ginger should be done rapidly because after three months at room temperature storage the oil content decreses considerably. Gamma-irradiation of dried rhizomes at doses of 5–10 KGy prevents mold and bacteria growth. Ethylene oxide @ 50 ppm can be used as a fumigation treatment of rhizomes.

Yamaguchi et al., (2010) worked on the Effects of High Hydrostatic Pressure (HHP) Treatment on the Flavor and Color of Grated Ginger. He applied HHP to grated ginger in order to inactivate quality-degrading enzymes in a nonthermal manner. The effects of HHP treatment on the flavor and the color of the grated ginger were investigated just after treatment and during storage. After HHP treatment (400 MPa, 5 min), gera- niol dehydrogenase (GeDH) was inactivated to less than 5 percent, but the activity of polyphenol oxidase (PPO) was reduced only to 37 percent. Heat treatment (100 C, 10 min) inactivated GeDH to 43 percent and PPO to about 10 percent. In storage, the reduction of geranial, neral, and citronellal to the corresponding alcohols was observed in the untreated and the heat-treated ginger, while it was not in the HHP-treated grated ginger. In the HHP-treated sample, terpene aldehydes almost disappeared without the formation of the corresponding alcohols. Browning was not observed immediately after HHP treatment, while it was complete in the heat-treated sample. The color change during storage appeared to reflect the residual activity of PPO. Sagar and Kumar (2009) worked on Effect of packaging and storage on the quality of ginger, onion and garlic (GOG) mix powder these could be prepared successfully by mixing them in the ratio of 1:4:1 and treated with spraying of 1 percent potassium meta bisulfite (KMS) solution with the help of hand spryer, and drying in a cabinet dryer at 60°C for 10 h, followed by grinding in a laboratory powder mill and sieving with 30 mesh sieve. The powder could be stored up to 6 months in 200 gauge high density polyethylene (HDPE) pouches at low temperature (7°C) and 3 months at ambient condition with better pungency and other nutrient contents. Retention of color and pungency as pyruvic acid was found to be more in the samples packed in 20 gauge HDPE pouches stored at 7°C, followed by 200 gauge low density polyethylene (LDPE) pouches. For storage of the product having 3.85percent moisture, the optimum RH was found to be 47.5 percent and the critical and danger point as 9.14 and 6.75 percent moisture, respectively. The product was stable at low temperature as compared to high temperature in respect of color, flavor and texture. Juliani et al. (2007) studied on chemistry and quality of fresh ginger varieties from Ghana ginger, *Zingiber officinale*, one of the most important and oldest of spices, consisting of rhizomes with a warm pungent taste and a pleasant odor. The essential oils are responsible for the aroma while the nonvolatile components are responsible for the pungency with gingerol the most pungent component in fresh ginger. Ghana has a long history of producing ginger rhizomes for local markets. In this study, they evaluated the quality and essential oils and pungent principles of two different types of Ghanaian ginger, each from three regions, compared to ginger rhizomes found in the US Market. The essential oils composition of variety 2 exhibited the typical ginger oil composition with geranial, neral and zingiberene as the main components. Variety 1 exhibited a distinctly different essential oils composition dominated by zerumbone (85–87%). Variety 2 possessed a warm pungent taste similar to the commercial samples whereas variety 1 of gingers possessed a bitter taste. Variety 2 and the US commercial gingers contain 6-gingerol as the main component. The samples from Oframanse exhibited the highest amount of total gingerols (0.66%) (6-, 8-, and 10 gingerols), being almost 4 times higher than the commercial US ginger (0.17). Ginger variety 2 appears promising for the export market since these gingers could be offered both as a dried and/or ground spice as well as a source of gingerols for the nutraceutical industry.

PROCESSING

Ginger oil may be produced from fresh or dried rhizomes. Oil from dried rhizomes will have less of the low boiling point volatile compounds since they tend to evaporate during the drying processes. The difference between oils produced from fresh and dried rhizomes can be seen in the citral content, usually lower in the oil from dried plant material. Additionally, unpeeled or coated rhizomes are preferably used for oil or oleoresin extraction to improve yield. For steam distillation, dried rhizomes are ground to a coarse powder and loaded into a still. Live steam is passed through the powder, thus entraining the volatile components, which are then condensed with cold water. Upon cooling, the oil separates from the water. Cohobation, or redistillation, is practiced in India to increase oil yield. Oil yield from dried rhizomes is generally from 1.5 percent to 3.0 percent 42. Indian (Cochin) ginger yields 1.5% to 2.2 % of an oil rich in citral. The rhizome powder stripped from its oil (marc) is made of about 50% starch and may be used as livestock feed. It may also be further dried and powdered to produce an inferior spice. Major components in ginger essential oil are zingeberene (20–37%), ar-curcumene (5–20%), and farnesene, bisabolene and -sesquiphellandrene. The low boiling point monoterpenes α-pinene, cineole, borneol, geraniol, geranial, and neral are less abundant and present in various proportions, and they impart aromas characteristic to the products. For instance, citral with its two isomers geranial and neral, is especially high in the Brazilian-grown cultivars 'Capira' (6.6–7.0 percent citral) and "Gigante" (14.3–20.7% citral), while it is only 1.9–4.3 percent in some Chinese oils. Australian oils also have a high citral content, up to 27 percent, averaging 19 percent, imparting a lemony aroma to the final product.

Jayashree et al (2012) worked on quality of dry ginger (*Zingiber officinale*) by different drying methods viz., ginger rhizomes sliced to various lengths of 5, 10, 15, 20, 30, 40 and 50 mm and dried by various drying methods like sun drying, solar tunnel drying and cabinet drying at temperatures of 50, 55, 60 and 65°C. The study indicated that slicing (5mm) significantly reduced the drying time (4 days) compared to drying whole rhizomes (9 days) under sun. In case of mechanical drying, with the increase in temperature from 50 to 65°C, reduction in drying time from 8 to 6 days for drying whole rhizomes was observed. Gas chromatographic analysis of volatile constituents of ginger essential oil like zingiberene, limonene, linalool, geraniol and nerolidol showed significant reduction in its content as the slice length decreased when compared to whole rhizome. Nonvolatile constituents of dry ginger like total gingerols and total shogoals decreased as the sliced length reduced. Sun drying and solar tunnel drying retained the maximum essential oil content (13.9 mg/g of dry ginger) and oleoresin content (45.2 mg/g of dry ginger). In case of mechanical drying, the optimum drying temperature was considered as 60 °C and whole rhizomes lost about 12.2 percent of essential oil and 5.3 percent oleoresin when dried at this temperature.

EXTRACTION: OLEORESIN PRODUCTION

Gingerols (6-, 8-, and 10-gingerol) are the compounds responsible for ginger pungency; however, because they are readily decomposed to the less pungent shoagols and zingerones upon heating, oleoresins obtained by solvent extraction are preferred when pungency is desired. Commercial solvents include ethanol, acetone, trichloroethane or dichloroethane, although the latter two are known carcinogenic and ethyl acetate or hexane is preferred. Dried powdered rhizomes are extracted by percolation, and the extract is then cold-distilled at 45–55 °C to remove all the solvent, while assuring integrity of gingerols by not overheating. Hydrophilic solvents such as ethanol, and acetone also extract water-soluble gums, which may need to be further separated by centrifugation. However, water-soluble solvents may be preferred to prepare extractive to be used by the beverage industry to assure water solubility. Supercritical fluid extraction uses carbon dioxide (CO_2) under high pressure and cold temperature. This extraction technique is preferred for higher quality extracts because there is no thermal degradation, and the aromatic profile is therefore closer to the profile in the plant. Zingerone, shogaol and gingerol were present in cold pressed oil and supercritical extract of Chinese ginger, and were absent from the steam distilled oil. Geranial and citral were 10.9 and 2.0 percent in supercritical extract, as compared to 0.63 and 1.31 percent in steam distilled oil, respectively. The use of steam or CO_2 is environmentally preferred over hydrocarbon or halohydro carbon solvents since they generate little or no hazardous wastes. Kim et al. (1992) reported extracts yield of 6.9 percent with CO_2. For certifiable organic production, synthetic solvents are not allowed. Therefore, solvents derived from petrochemicals such as hexane, pentane, di- and tri-chloroethanes, acetone, cannot be used in organic production. The International Federation of Organic Movement (IFOAM) specifies that only ethanol, water, edible oils or carbon dioxide are allowed.

GINGER OIL

Ginger oil can be prepared by steam distillation of grind paste or dried powdered ginger which is used as a flavoring agent for soft drinks, ginger beer and in food preparation. For oil extraction, dried rhizomes are ground to a coarse slurry, paste or powder, loaded into a still for distillation and steam is passed through the slurry/paste/powder. This steam containing the volatile components is condensed with cold water and collected in separate container. The oil can be separated from the water upon cooling by the separatory funnel. Redistillation can be done to increase oil yield. Usually oil yield obtained from dried rhizomes is 1.5 percent to 3.5 percent on dry weight basis and 0.4% on green weight basis depending upon variety of ginger used.

GINGER OLEORESIN

It is a blend of oil and resinoids. Oleoresin is obtained by extraction of dried ginger, pulverized to coarse powder, with organic solvents like ethanol or acetone. Oleoresin content ranges from 3.5 to 9.5 percent.

GINGER CANDY

Ginger candy can be prepared by selecting big sized rhizomes of low fiber content. These rhizomes should be washed with water to remove the adhered dirt and debris. The peel should be removed with the help of wooden splinters or knives and wash thoroughly with water. After this, rhizomes should be pricked properly with the help of forks so that sugar can penetrate deep in the tissues. The pricked rhizomes should be cut into pieces of 1–2 cm thickness. The pieces should be boiled for almost one hour until they become soft. After boiling sufficiently, the ginger pieces should be removed from water and kept in shade for drying. After this, the ginger pieces should be spread in a stainless steel utensil having alternate layer of sugar and ginger pieces (1 Kg ginger pieces:1 Kg sugar) and kept for 24 hrs. On second day the ginger pieces are removed from sugar syrup, 2 g citric acid is added and the syrup is boiled until sugar strength is reached up to 60° Brix. The syrup is then allowed to cool, the ginger pieces are added into the syrup and kept it for 24 hrs. On third day the ginger pieces are removed, 1 g citric acid is added and the syrup is boiled until sugar strength is reached up to 65° Brix. The syrup is then allowed to cool, the ginger pieces are added into the syrup and again kept it for 24 hrs. On next day the same process should be repeated by addition of 1 gm citric acid and boiling the syrup until the sugar strength reached up to 75° Brix. The syrup is then allowed to cool and the ginger pieces are added into the syrup. It is kept for 4 days in syrup. Then the well soaked pieces are remove d from syrup and dried in oven at 60°C for 6–8 hrs. These dried pieces can be coated with powdered sugar or confectioner's sugar or glucose powder by sprinkling the powder over the pieces and mixed thoroughly. The coated ginger candy is filled in glass jars or packed in polyethylene pouches and stored in cool and dry place.

GINGER SOFT DRINK (RTS)

Ginger ready to serve (RTS) soft drink can be prepared by selecting healthy and blemish free rhizomes. The rhizomes are washed with water and peeled with the help of wooden splinters or knives. Then cut into small pieces and pulp is made by passing through mixer-grinder by addition of little water to facilitate easy pulping. After pulping the pulp is strained and kept for 1 hr to settle down the sediments at bottom. Then the clear juice is siphoned off and mixed with sugar syrup solution which can be prepared by addition of sugar + citric acid + water at 120 g + 3 g citric acid + 850 ml water. The sugar syrup is strained with muslin cloth to remove the impurities from dissolved sugar and the ginger juice or pulp at 30 ml are mixed and then the preservative potassium metabisulphite at 40 mg/ liter of RTS is added add. All the ingredients are mix thoroughly, filled into the bottles then crown corked. The sealed bottles should be pasteurized at 85°C for 15 min and then air cooled and can be kept for storage in cool and dry place.

GINGER SHREDS

Ginger shreds can be prepared by washing and peeling of the rhizomes. After peeling rhizomes should be grated in to small pieces. Then grated small pieces of ginger should be kept in the muslin cloths and squeezed slightly to remove excess juice content. Then black salt and common salt at 4% are added and should be kept in oven for drying at 60°C for 2 days. Final product should be packed in polyethylene pouches and kept in cool and dry place for use.

GINGER PICKLE

Ginger pickle can be prepared by washing and peeling of the rhizomes. Then peeled rhizomes should be cut into small rectangular pieces. These pieces should be dried in shade for removal of outer moisture. The mixture of spices (ajowain + black pepper + cumin seed + chilli powder and citric acid at 10 g each for 250 gm of ginger pieces) should be prepared and mixed together. After this, all material is filled into glass jar and kept for sun drying up to two weeks with occasional stirring. Finally it can be stored in cool and dry place.

GINGER CHUTNEY

Chutney of good quality and taste can be prepared by using ginger. For this ginger rhizome (250 g) should be washed, peeled and grind in mixer. Tamarind (250 g) and garlic (100 g) should also be grind in mixer and then grinded paste of ginger, tamarind and garlic should be mixed. This mixture should be heated to a little and add salt (100 g). Then frying of another garlic paste (100 g), fenugreek powder (20 g) should be done in little mustard oil (100 ml). This fried mixture of spices should be mixed with ginger paste, sugar (500 g) and fill into glass jar. Final product should be stored in cool and dry place.

12.11   TURMERIC (*CURCUMA LONGA*, FAMILY: ZINGIBERACEAE, ORIGIN: SOUTH EAST ASIA)

Underground rhizome is used as condiment, dye stuff, drug and cosmetic. India is the largest producer of Turmeric which is one of the traditional items of export. In India, A.P. leads in area and production .Turmeric ranks 4th in foreign exchange earner among the spices after pepper, cardamom and ginger. Three categories of turmeric are there: long duration types (9 months duration; Duggirala, Tekurpeta, Armoor and Mydukur, CLL 324, 325, 326 and 327), medium duration types (8 months duration; Kothapet, Krishna, Kesari, CL), short duration types (6–7 months duration; Amalapuram, Dindigram, PCT – 13[Suguna], PCT – 14 [Sudarshan]). Armoor is the popular type in Nizamabad district. Kasturi and Kesari are good in curcumin content, but poorer in curing percentage.

It requires warm and moist climate, rainfall ranging 100–200 cm and thrives best in well drained, friable, rich sandy, or clay loam soils. Crop stands neither water logging nor alkalinity. Temperature range preferable is 20°–30°C it is propagated through mother rhizomes. Bed system gives higher yield by 54 to 80 percent. Beds of 1 m width and convenient length with a spacing of 40–50 cm between beds where natural drainage does not exist, ridges and furrows are prepared at 45–60 cm spacing during May-July. Germination starts in 10–20 days and will be over by 60 days. Mulching is done with dry leaves thickly on which a layer of cow dung is spread. Second It is usually done after weeding and application of fertilizers, after 50–60 days of sowing. A good soaking irrigation is given immediately after sowing. Thereafter, irrigate at weekly interval, 3–4 weedings are required at 60, 90, 120 and 150 days of planting and hoeings are done simultaneously.

HARVEST

Turmeric readiness for harvest is indicated by the drying of the plant and stem, approximately 7– 10 months after planting, depending on cultivar, soil and growing conditions. The rhizome bunches are carefully dug out manually with a spade, or the soil is first loosen with a small digger, and clumps manually lifted. It is better to cut the leaves before lifting the rhizomes. Rhizomes are cleaned from adhering soil by soaking in water, and long roots are removed as well as leaf scales. Rhizomes are then further cured and processed, or stored for the next year's planting. Rhizomes for seed purposes must be stored in well-ventilated rooms to minimize rot, but covered with the plant dry leaves to prevent dehydration. They can also be stored in pits covered with sawdust, sand, or panal (*Glycosmis pentaphylla*) leaves that may act as insect repellent. The Indian Institute of Spice Research recommends the following fungicides as a prestorage dip treatment for rhizome seeds: quinalphos at 0.075 percent, and mancozeb at 0.3 percent. It indicates that bulbs (mother rhizomes) are preferred to fingers as a seed stock.

Extent of losses in turmeric rhizomes when stored under different storage methods were studied. The methods tried were storing rhizomes with lining materials viz., sawdust, sand, *Glycosmls pentaphyla* (pannel) leaves, *Olea dioica* (vettey) leaves, hyderated lime, coir dust, storing rhizomes after hot water treatment, cow dung treatment, storing by providing smoke, storing in pit without any lining material, keeping in heap, storing by treating with pesticide solution (Control). Parameters studied were moisture loss in weight or rhizomes, percentage of decayed and healthy rhizomes obtained after storage. Healthy rhizomes were significantly higher in all the storage methods tried except in heap, cow dung and smoke treatment. The storage of rhizomes in heap without any lining material is not advocated due to heavy losses in weight of rhizomes. Sprouting of stored healthy rhizomes was significantly influenced by different storage methods. Germination percentage was the maximum for rhizomes stored with saw dust, which was on par with rhizomes stored with sand, *Glycomis pentaphylla* leaves, *Olea dioica* leaves, hot water treated rhizomes and rhizomes soaked in pesticide solution. The paper outlines a brief attempt to assess the efficacy of nonchemical methods including indigenous methods of control of decay and moisture losses during storage of turmeric (Thankamani, 2002).

Thomas et al., (2010) worked on curcuminoid profiling of Indian turmeric and found that curcumin, an orange yellow pigment of turmeric (Curcuma longa L.) consists of three curcuminoids namely curcumin (curcumin I), demethoxy curcumin (curcumin II), and bis demethoxy curcumin (curcumin III). The curcuminoid profile of three export grade Indian turmerics and three popular varieties was estimated using HPLC. HPLC was performed on an amino column using chloroform and ethanol. Highest levels of curcumin I and curcumin III were recorded in the popular variety "Prathibha" (3.34 and 1.30%, respectively) and least level was recorded in the traded variety "Wynadan" turmeric (1.31 and 0.19%).

POSTHARVEST HANDLING: CURING, DRYING, AND POLISHING

Turmeric rhizomes are cured before drying. Curing involves boiling the rhizomes until soft. It is performed to gelatinize the starch for a more uniform drying, and to remove the fresh earthy odor. During this process, the coloring material is diffused uniformly through the rhizome. Recommendations as to the acidity or alkalinity of the boiling water vary by author. The Indian Institute of Spice Research, Calicut, Kerala, and the Agricultural Technology Information Center simply recommend boiling in water for 45 min to one hour, until froth appears at the surface and the typical turmeric aroma is released. They report the color deteriorates as a result of overcooking, but that the rhizome becomes brittle when undercooked. Optimum cooking is attained when the rhizome yields to finger pressure and can be perforated by a blunt piece of wood. Boiling in alkaline water by adding 0.05 percent to 1 percent sodium carbonate, or lime, may improve the color. For the curing process, it is important to boil batches of equal size rhizomes since different size material would require different cooking times. Practically, fingers and bulbs are cured in separate batches, and bulbs are cut in halves. Cooking may vary from one to four or six hours, depending on the batch size. Curing is more uniform when done with small batches at a time. It is recommended to use perforated containers that allow smaller batches of 50–75 kg, which are immersed in the boiling water; by using this method, the same water may be used for cooking several batches. Curing should be done two or three days after harvest, and should not be delayed to avoid rhizome spoilage. The quality of cured rhizomes is negatively affected for material with higher initial moisture content. Benefits of curing turmeric include reduction of the drying time, and a more attractive product (not wrinkled) that lends itself to easier polishing. However, it was reported that while the total volatile oil and color remained unchanged, curcuminoid extractability might be reduced. The curing by boiling process has the advantage of sterilizing the rhizomes before drying. Slicing the rhizomes reduces drying time and yield turmeric with lower moisture content as well as better curcuminoid extractability. In rural Bolivia, slicing the boiled rhizomes is done by women. The "Fundación Poscosecha", with the support of FAO has developed a slicing machine in order to ease the women's work. The slicing machine has a simple design, is easy to use, and can be made at a low cost. It consists of a metallic structure (rack), a transmission system, and a metallic box containing a disc and two stainless steel circular blades. The transmission system is made of a pinion, an escape wheel, and a chain, and has a transmission report of 3:1. The slicing machine is all metallic, and when well maintained (lubrication of axles and bearings), it can be used up to eight years. The advantages of this machine are ease of use and installation, and ease of transport (it weighs about 40 kg). It has a high capacity (up to 120 kg/hr), and considerable reduces the traditional cutting by hand. Cooked fingers or bulbs are dried to a moisture level of 5 percent–10 percent. Sun drying may take 10–15 days, and the rhizomes should be spread in 5–7 cm thick layers to minimize direct sunlight that results in surface discoloration. Turmeric is one of the spices for which it is more advantageous to use mechanical driers because of the sensitivity to light. Those can be drums, trays, or continuous parallel or cross-flow hot air tunnels. Like with ginger rhizomes, the optimum drying temperature is 60°C. An example of solar drier was developed in Bolivia, the "Secadora Solar". It was designed to dry turmeric and ginger under the hot and humid conditions of tropical Bolivia, but it can also be used to dry other foods. The maximum temperature achieved by the drier depends on the outside climatic conditions. The body of the machine is quite simple; it consists of a metallic rack supporting the rest of the components, two parallel inserted plastic trays where the products are put on to dry, and a plastic cover that should be designed to assure major protection from ultraviolet radiation. Ideally, the plastic should be black, or contain a UV protector. Because the sun is serving as energy source, satisfying outputs cannot be achieved in regions where cloudiness and humidity is high. Approximately 4 KWH/m² of solar energy is needed to use these techniques successfully. The best outputs are obtained in regions with a humidity of 40 percent to 60 percent and average temperatures of 14° to 18°C. The advantages of this type of drier are simple construction with appropriate technology, ease of dismantling and transport, and versatility of use. Dried fingers are polished to remove scales and rootlets from the rhizomes by using rotating drums lined with a metallic mesh that abrades the rhizome's surface. Turmeric powder suspended in water is sprinkled over the rhizomes at the final stage of polishing to give an attractive color

GRADING, PACKING AND STORAGE

Quality specifications are imposed by the importing country, and pertain to cleanliness specifications rather than quality of the spice. Proper care must be taken to meet minimum requirements, otherwise a lot may be rejected and need further cleaning and/or disinfection with ethylene oxide or irradiation. Bulk rhizomes are graded into fingers, bulbs and splits.

The Indian Standards for turmeric follow the Agmark specifications (Agricultural Directorate of Marketing), to insure quality and purity.

Agmark Standards for Turmeric Rhizomes

Grade	Flexibility	Broken pieces, fingers<15 mm No more than (% by weight)	Foreign Matter No more than (% by weight)	Defectives No more than (% by weight)	Percentage of bulbs by weight, max. No more than (% by weight)
Alleppey fingers					
Good	Hard to touch	5	1	3	4
Fair	Hard	7	1.5	5	5

Fingers other than Alleppey					
Special	Hard to touch, metallic twang on break	2	1	0.5	2
Good	Same	3	1.5	1	3
Fair	Hard	5	2	1.5	5
Rajapore" fingers					
Special	Hard to touch, metallic twang on break	3	1	3	2
Good	same	5	1.5	5	3
Fair	Hard	7	2	7	5
Nonspeci-fied	-	-	4	-	-
Bulb's					
Special	-	-	1	1	-
Good	-	-	1.5	3	-
Fair	-	-	2	5	-

Fingers shall be of secondary rhizomes of Curcuma longa L.; shall be well set and close grained; free from bulbs; be perfectly dry and free from weevil damage and fungus attack; not be artificially colored with chemicals. Bulbs shall be primary rhizomes of Curcuma longa L.; shall be well developed, smooth and free from rootlets; have the characteristics of variety; be perfectly dry and free from weevil damage and fungus attack; not be artificially colored with chemicals. Rhizomes may be packed in jute sacks, wooden boxes or lined corrugated cardboard boxes for shipping. Storage of bulk rhizomes should always be in a cool and dry environment, to prevent moisture absorption and chemical degradation. Turmeric has traditionally been adulterated with related Curcuma species, specifically *C. xanthorrhiza* Roxburg, *C. aromatica*, and *C. zedoaria*. However, due to strong competition between spice processors, the quality of turmeric destined to the Western markets is usually guaranteed by the exporter in contracts negotiated between the buyer and the seller

GRINDING AND MILLING

Grinding is a simple process involving cutting and crushing the rhizomes into small particles, then sifting through a series of several screens. Depending on the type of mill, and the speed of crushing, the spice may heat up and volatiles be lost. In the case of turmeric, heat and oxygen during the process may contribute to curcumin degradation. Cryogenic milling under liquid nitrogen prevents oxidation and volatile loss, but it is expensive and not widespread in the industry. Ground spices are size sorted through screens, and the larger particles can be further ground. Most quality control laboratories use the U.S. Standards (U.S.S.) screen size system. However, there are other systems that use a different numbering, and comparisons between specifications may be difficult. For instance, the U.S.S. screen numbering goes from 4 to 80 mesh screens (i.e. 4 to 80 openings per inch), while the Mill screen system goes from 4 to 55 mesh with different increments than the U.S.S. system.

Grade	Moisture (% w/w) max	Total ash (% w/w) max	Acid insoluble ash max (% w/w)	Lead max (ppm)	Starch max (% w/w)	Chromate Test
Turmeric powder						
Standard	10	7	1.5	2.5	60	Negative
Coarse ground powder						
Standard	10	9	1.6	2.5	60	Negative

AGMARK STANDARDS FOR TURMERIC POWDER

The Indian Agmark standards for turmeric powder refer to a grain size that would pass through a 300 micron sieve (turmeric powder), and 500 micron sieve for coarse powder

EXTRACTION: OLEORESIN PRODUCTION

Since curcumin is the compound of interest in turmeric rhizome, it is important to know the solubility of curcumin in different solvents in order to choose the appropriate solvent. Curcumin is soluble in polar solvents (acetone, ethyl acetate, methanol, ethanol), and quite insoluble in nonpolar solvents such as hexane, and insoluble in water. Dried powdered rhizomes are extracted by percolation with the polar solvent. The particle size, uniform packing in the extractor, temperature and percolation rate of the solvent are all important parameters for optimum extraction. If the oleoresin is the desired product, the solvent is completely evaporated by distillation at 45–55°C. If curcumin is the final product, the solvent is only partially removed, and the color material is separated from the solvent by freezing, then centrifugation or vacuum-filtration. At this step, curcumin is further purified with a wash with hexane. Hexane will extract all the gummy matter, oils, fats, and volatile essential oils that would otherwise impart a turmeric flavor. The yield of curcumin from dried turmeric root is about 5 percent.Oleoresin composition will vary greatly with the type of solvent, temperature and extraction methods, in addition to the effect due to quality of the raw material. The commercial methods of extraction will vary by manufacturer and are proprietary information. The yield of oleoresin from dried root is typically 10–12 percent.For organic production, synthetic solvents are not allowed. Solvents derived from petrochemicals such as hexane, pentane, di- and tri-chloroethanes, acetone, cannot be used in organic production. The International Federation of Organic Movement (IFOAM) specifies that only ethanol, water, edible oils or carbon dioxide are allowed. Therefore, possible solvents for curcumin extraction would be ethanol in the first step, and wash with vegetable oils and water for purification. There is no published study by using these restricted solvents, and it would be worth pursuing experimentally.

12.12   NUTMEG AND MACE (*MYRISTICA FRAGRANS* FAMILY – MYRISTICACEAE, ORIGIN – MOLUCCAS ISLAND

Nutmeg and mace are two different parts of the same fruit of the nutmeg tree commercially considered as spice. The major nutmeg growing areas are Indonesia and Grenada (West Indies) Sri Lanka, India, China, Malaysia, Western Sumatra, Zanzibar, Mauritius and the Solomon Islands. Nutmeg is the dried kernel of the seed and mace is the dried aril surrounding the seed. Both the spices have similar flavor. It grows well in warm humid conditions in locations with an annual rainfall of 150 cm. Soils of clay loam, sandy loam and red lateritic are ideal for its growth. Both dry climate and water-logged conditions are not good for nutmeg. Nutmeg trees are usually propagated through *seed.* The fleshy rind of the fruit as well as the mace covering the seed is removed before sowing. The seeds should be sown in prepared nursery beds immediately after collection. Seedlings are transplanted in the main field when they are 12–18 months old at 8 × 8 m spacing. in 60 cm cube pit. Young plants are provided with artificial shade and irrigated during summer months

HARVESTING AND PROCESSING

The ripe or mature fruit splits open at the groove while still on the tree and the seed surrounded by the red aril falls to the ground after two days. Harvesting involves collecting the seed or seed with aril from the ground. Sometimes fruits with partially opened pods may be picked from the tree using a long pole "rodding". The latter method affords a better quality aril, and pods that could be used in agro processing. This procedure may also lead to excessive dropping of flowers and young fruits. The frequency with which nutmegs are harvested is dependent on the location of the field, the availability of labor, the level of production, and the price offered to farmers. Most farmers collect the fallen seeds daily during the two peak production periods – January to March and June to August, and every two to three days during the rest of the year. Once the field is readily accessible nutmegs are harvested with a higher frequency. In the cases where farmers are part-time, fields located in distant areas, or when the farm is comprised of several plots of land at different locations, then the collection rate may be as low as once per week. Observations show that a larger proportion of women are usually involved in harvesting.

POSTHARVEST HANDLING OPERATIONS IN THE FIELD— PREPARATION FOR MARKETING

In the case where rodding was used, open fruits may fall to the ground intact. The seed with the surrounding red aril is removed from the pod which oftentimes is discarded. The collected seeds, and seeds with mace are transported from the field by workers, on their heads or assisted by animal (donkey) or vehicle to the farmer's residence where the mace is carefully separated from the seed, graded and allowed to dry directly in the sun. Care is taken so that drying mace does not get wet. Wetting will encourage mold growth and such mace will have to be discarded. The seeds are usually delivered green (fresh), within 24 hrs after harvesting to the receiving station. However, depending on the distance from the receiving station and the quantities of nutmegs involved, deliveries may be made once weekly or at a much later period if the nutmegs are being delivered in the dry state. This is usually the situation with large estates with adequate drying facilities. Mace is always delivered to receiving station dried.

YIELD

A tree from seedling usually starts to yield within five to eight years. Trees propagated vegetatively by marcots may fruit as early as in three years. Yields increase gradually and at 25–30 years the plant may have peaked to its maximum production level. It continues to bear up to 100 years and over. However, after age 70–15 yields tend to decline. A good producing tree may give on an average a yearly production in the vicinity of 14–22 kg green nutmegs (7–11 kg of shelled, dry nutmegs). The proportion of dried shelled nutmeg to dried mace is approximately 20:3. During drying nutmeg loses about 25 percent of its weight. Yields vary greatly between trees, and between plantations or field locations.

HANDLING OPERATIONS

(a) Nutmeg

Sorting

On delivery at the receiving station the green nutmegs are emptied into sorting trays. These trays are wooden and four sided, 152 cm long by 76 cm wide and 23 cm deep with a slightly perforated wooden base and angled gently toward an emptying hole at one end, and standing on legs about 91 cm off the floor.

The inspector or inspectors spread out the seeds with a wooden pallet, and hand select out broken seeds, slightly discolored seeds, water-logged seeds, empty or rotten seeds, moldy seeds, very light seeds and germinating seeds. These are usually returned to the delivering farmer.

The remaining seeds are scooped into a receiving bag. When all of that particular farmer's consignment is sorted, the bag or bags are weighed and the weight entered into the farmer's assigned book and recorded at the station. The farmer is given a bill and is paid at whatever is the then rate per pound. At the end of the day all collected nutmegs are reweighed and checked against the weight paid out for, and this weight is entered in the "reweighed book" at the station.

Drying

Bags with about 45.5 kg of seeds are hand sewn at the top and carted to the hoisting area for lifting to the drying floors. Smaller quantities may be handled at the smaller receiving stations. Also at the peak periods or times of very high deliveries, nutmegs are transported from the receiving stations to the processing stations daily.

For drying, the fresh seeds are spread on large trays to a depth of 5–7.5 cm. The seeds are turned daily with wooden rakes or spades. Seeds are shade dried in the buildings at a temperature of 29–32°C for 6–8 weeks. Drying trays have wooden bottoms and wooden sides with periodic trap doors. Trays vary in size according to the station, but are usually arranged in tiers with about 4–8 trays above each other separated by 30.5 cm between trays. Trays are arranged so that there is always easy passage between stacks of trays to afford ready access to all seeds. However, for the higher trays workers have to climb up to gain access.

The energy for drying comes from the sun on the large galvanized roof of the station and the warm circulating air. Seeds closer to the roof tend to dry much faster. Drying completion is indicated by simple inspection. Usually after 6 weeks a sample of seeds is taken, these are cracked, cut with a knife and inspected for moisture.

The characteristics looked for are rattling in shells, difficulty or ease to cut, degree of oiliness and intensity of the aromatic smell. After the drying period and satisfaction with the inspection, the seeds are heaped in the trays close to a trap door which on raising allows easy scooping into bags. Each bag weighing 68 kg. Bags are sewn and dried nutmeg in shells is stored until an order is received.

Grading and Flotation

The first grading of shelled kernels is effected by flotation in water using the principle of varying density. The procedure is to place 914 kg kernels in a wicker basket. The wicker basket is then immersed in water held in a concrete trough to a level just about 2.54 cm below its rim. The kernels are then agitated by hand. Once stirring stops, some kernels are seen to remain at the bottom of the basket while others float. All "floats" are removed as defectives along with any kernels seen to be moving or in suspension (doubtfuls). Workers (female) try to affect this in as short a time as possible. The kernels remaining at the bottom are classified as Sounds. The detectives (floats) are grouped, basketed and spread on trays to dry for 48–72 furs. The sounds are spread usually on the upper trays to dry for 24 furs. Both grades are turned twice daily while drying. When detectives "floaters" are inspected by cutting in half, they usually show incomplete kernels (large airspaces, or whitish cork tissue with reduced brownish endosperm).

Metal Sieve Grading

Using large metal sieves with uniform regular circular perforations and sieves of different sizes, workers pour on hand-graded sounds and gently massage them. The appropriate kernels fall through the appropriate holes into collecting bags.

The grades collected are:

110 S	(242 to the kg)
80 S	(176 to the kg)
60/65 S	(132/143 to the kg)

Inspection of Grades

As an additional quality control measure, before putting into new labeled bags, sound graded nutmegs are further inspected visually. The worker will spread the nutmegs of a particular grade in a small wooden tray and hand remove any broken pieces, cracked nutmegs or shriveled and discolored. Such a worker is expected to handle a minimum of 2 (68 kg) daily.

Fumigation

The final processing step for nutmegs before export is fumigation of the bagged nutmegs with methyl bromide in a special fumigation chamber overnight.

Defectives

Sometimes the defective "heavies" are sorted and exported separately. The "floats" after drying are heaped, then packed in bags and stored until an order for that class of product is received. For export they are then packed in labeled bags and fumigated.

(b) Mace

The mace delivered at the receiving station though preclassified by the farmer as No. I or No. 2 is still carefully inspected on delivery and classified as No. 1 or No. 2. The mace is weighed and the weight entered in the farmer's book and the book at the receiving station and the farmer is paid.

The mace is then bagged according to grades and at the end of the day the separated grades are reweighed, the weights noted, and the mace placed in separate wooden curing bins—1.83 × 1.22 × 1.22 m. Each bin may be loaded to the level of 727–772 kg and left for 3 months. Into each bin is suspended a bottle with carbon disulfide (CS2) to keep away any insect pest. After the three-month curing period, the mace is now ready for export. The cured graded mace is bagged accordingly and fumigated.

PACKAGING AND DISTRIBUTION

MACE	No. 1 - PLASTIC BAGS - 25.5 kg
	No. 2 - PLASTIC BAGS - 36.5/40 kg
	No. 3 - JUTE BAGS - 50/51 kg
FORMERLY:	No. 1 - IN PLY BOXES - 72.7 kg
	No. 2 - IN PLY BOXES - 91.0 kg
	No. 3 - IN JUTE BAGS - 51 kg
NUTMEGS	SOUND UNASSORTED - JUTE BAGS - 63.5 kg
	SOUND SELECTED - JUTE BAGS - 63.5 kg
	DEFECTIVES - JUTE BAGS - 63.5 kg
	DRY IN SHELL - JUTE BAGS - 50 Or 51 kg
ESSENTIAL OIL	The nutmeg oil distilled in Belgium is packaged in 45 gal. drums and is sold to Switzerland by the company Puressence.

For shipping to foreign ports containers (20 ft) are utilized and for nutmegs average capacities are 12 tons (240 bags) for dry nutmegs in shells, 16 tons (240 bags) for unasserted and 14 tons (224 bags) for detectives.

The principal constituents of the spices nutmeg and mace are steam volatile oil (essential oil), fixed (fatty) oil, proteins, cellulose, pentosans, starch, resin and mineral elements. Thus the fixed oil content of sound nutmegs varies from 25 to 40 percent while that of mace is 20–30 percent. Worm eaten nutmegs have a higher content of volatile oils than sound nutmegs since in the former the starches and fixed oil have been selectively eaten by insects.

1. Fixed Oil

There are two general methods by which the fixed oil of nutmeg is extracted. In one method, sound, ground nutmeg is subjected to intense hydraulic pressure and heat (heated plates in the presence of steam) while, in the other the ground nutmeg is extracted by refluxing with a solvent like diethyl ether. Both processes will result in the crude fixed oil containing significant quantities of essential oil in the average of 10–12 percent. Prior steam distillation will lead to a significant reduction of essential oil in the prepared fixed oil.

The extracted or expressed fixed oil is a semisolid aromatic (smell and taste of nutmeg), orange colored fat, known as concrete, expressed oil or nutmeg butter which melts at 45–51°C and has a density of 0-990–0.995. It is completely soluble in hot alcohol, but sparingly so in cold. However, it is freely soluble in ether and chloroform.

The major component of the fixed oil is Trimyristin and Power and Salways (1908) gave the following components and their relative abundance for the analysis of a fixed oil for which there was no prior distillation to remove essential oil from the nutmeg raw material:

Trimyristin	73.09%
Essential oil	12.5%
Oleic acid (as glyceride)	3.0%
Linolenic acid "	0.5%
Unsaponifiable constituents	8.5%
Resinous material	2.0%
Formic, acetate and cerotic acid	(traces)

If essential oil is previously extracted, then the relative abundance of trimyristin in the fixed oil will increase.

2. Essential oil

The essential oil is usually obtained by steam distillation of dried kernels. It is a color less or yellow liquid with the characteristic odor and taste of nutmeg. The oil is insoluble in water but soluble in alcohol and has a density at 25°C of 0.859–0.924, refractive index at 20°C, 1.470– 1.488 and optical rotation at 20°C of + 10° - +45°. This oil keeps best in the cool in tightly closed containers protected from light.

Extensive analyses have been carried out on the volatile oil of nutmeg and these have provided the major classes of compounds constituting the oil as: monoterpene hydrocarbons, 61 - 88 percent; oxygenated monoterpenes (simple and others) ie. monoterpene alcohols, monoterpene esters; aromatic ethers; sesquiterpenes, aromatic monoterpenes, alkenes, organic acids and some miscellaneous.

The compounds identified in the volatile oils of nutmeg and mace, A-and b-pinene and sabinene constitutes the major components of the monoterpene hydrocarbon fraction where as myristicin is the major constituent of the aromatic-ether fraction.

It must be noted that the composition of distilled volatile oil is not identical to the natural oil in the kernel or oleoresin extract. Thus about 30–55 percent of the kernel consist of oil and 45–60 percent of solid matter. The essential or volatile oil accounts for 5–15 percent of the nutmeg kernel while the fixed oil accounts for 24 to 40 percent of the nutmeg kernel. Fixed oils are virtually absent from mace and volatile oil accounts for 4–17 percent of the composition of mace. There is always in the distilled volatile oil a higher percentage of monoterpenes, especially a and b-pinene and sabinene since there is incomplete distillation of the oxygenated components which possess higher boiling points.

On organoleptic grounds it has been stated that West Indian oils are weaker in odor and less spicy than East Indian oils. From studying and comparing East and West Indian oil, Baldry et al. (1976) have suggested that the composition differences are more a reflection of the proportion and the constituting compounds more so than absence of constituents. The major quantitative differences were lower proportions of a-pinene, safrole and myristicin and higher proportions of sabinene in the West Indian oils.

Zachariah et al. (2007) experimented on Chemical composition of leaf oils of Myristica beddomeii (King), Myristica fragrans (Houtt.) and *Myristica malabarica* (Lamk.) and reported on essential oil constituents of leaves of three *Myristica* species namely, *Myristica beddomeii*, M. fragrans and *M. malabarica* as determined by gas chromatography and gas chromatography-mass spectrometry. *M. fragrans* was dominated by monoterpenes (91%), M. beddomeii contained mono- (48%) and sesquiterpenes (35%) whereas M. malabarica was dominated by sesquiterpenes (73%). The leaf oil of *M. beddomeii* was dominated by α-pinene (19.59%), t-caryophyllene (14.63%) and β-pinene (12.46%). The leaf oil of *M. fragrans* contained sabinene (19.07%), α-pinene (18.04%), 4- terpineol (11.83%), limonene (8.32%) and β-pinene (7.92%) as major compounds, while t- caryophyllene (20.15%), α-humulene (10.17%), nerolidol (9.25%) and δ-cadinene (6.72%) were predominant in the oil of *M. malabarica*. Linalool, α-terpineol, t-caryophyllene, β-elemene and γ- elemenet were present in all the three species. This is the first report on the essential oil composition of *M. beddomeii* leaves.

12.13 CLOVE (*EUGENIA CARYOPHYLLATA*), FAMILY-MYRTACEAE

It is a perennial plant Cloves come from unopened dried flower buds. The plants grow well in rich loamy soils in the wet tropics. It can also grow in heavier red soils, but in either case, it needs good drainage and prefers partial shade and even rainfall. Cloves are propagated by seeds or by cuttings. The seeds can be directly planted, or soaked in water overnight to remove the outer lining.

HARVESTING AND PROCESSING

The trees begin to flower in 6 years. Full bearing is achieved by about 20 years and the production continues for 80 years or more. Bearing between years shows much variation. Clove clusters are handpicked, when the buds reach full size and turn pink but before they open. Clove buds are harvested when they have reached their full size and the color has turned reddish. At this stage, they are less than 2 cm long. Harvesting has to be done without damaging the branches, as it adversely affects the subsequent growth of the trees. On an average, a clove tree yields 3.5–7.0 kg/year, depending upon the age, size and condition of the tree.

After being harvested, the buds are separated from the stems, by hand or thresher machine. They are spread thinly on mats and stirred frequently for uniform drying. Well dried cloves will snap cleanly with a sharp click across the thumb nail and weigh about one third of the green weight. The opened flowers are not valued as a spice. Clove yields different types of volatile oil [oil extracted from. leaves, stem, buds and fruit. These oils differ considerably in yield and quality. The yield and composition of the oil obtained are influenced by its origin, season, variety and quality of raw material, maturity at harvest, pre- and postdistillation treatments and method of distillation. The chief component of all the types of oil is eugenol. The stem remaining after the separation of the bud from the fleshy harvested clusters are dried similarly and are used to distill clove oil by steam distillation method. The duration of distillation ranges 8–25 hrs. The leaves and small twigs yield clove leaf oil. The oil yield is 17–19 percent from clove bud, 6 percent from stem and 2–3 percent from leaves. Value added products are i.e clove oil, ground clove, oleoresin, clove stem oil, clove leaf oil, clove bud oil, oil of mother cloves, clove root oil.

QUALITY

The quality required for clove products depends on each country, for use in domestic or export and nature of the product. The specific requirement for clove ground is 12percent for minimum quercitrinnic content, clove stem content maximum 5 percent, and volatile ether extract minimum 15 percent (Reineccius 1994).

12.14   BLACK PEPPER (*PIPER NIGRUM*, FAMILY-PIPERACEAE, ORIGIN: WESTERN GHATS OF SOUTH INDIA)

Pepper is the most important of all spices and popularly known as the „king of spices'. It is one of the most important earliest known spices produced and exported from India. It is the most valuable and important foreign exchange earner among the important spices earning nearly 50% of the total export earnings from all the spices. Because of its importance in the spices and unique position in trade and large share in export earnings, it is popularly referred as king of spices and black gold in trade. The alkaloid "piperine" is considered to be the major constituent responsible for the bitter taste of black pepper.

Pepper is a tropical plant it requires warm humid climate for commercial production and grown in a wide range of soils such as clay loam, red loam, sandy loam and lateritic soils with a pH of 4.5–6.0. It thrives better in soils rich in organic matters. Pepper is invariably propagated by vegetatively (stem cuttings). Pepper cuttings are generally planted with onset of South West monsoon. When pepper is grown as pure crop, pits of 0.5 m3 are dug at a spacing of 2.5 × 2.5 m. Erythrina stem cuttings of 2 m length from 2 year old seedlings are planted on receipt of early monsoon showers. Pepper cuttings are to be planted 100 to 120 cm away from the tree trunk. The pepper vines are tied firmly as and when they grow. The vines are trimmed at the top and prevented from growing too tall for convenience of picking. In pepper plantation, shade is given to the pepper vines, especially during the hot weather to keep the soil cool and moist and to allow sun light during cool weather to encourage production of flowers and fruits. Judicious and regular manuring is necessary to get good yields. Manures are applied around the vines at a distance of 30 cm and forked in to the soil.

HARVESTING

Pepper vines start yielding usually from the 3rd or 4th year. The vines flower in May-June. It takes 6–8 months from flowering to ripening stage. Harvesting is done from November to February in plains and January to March in hills. When one or two berries on spike turn red in early the whole spike is plucked. Yields vary with the variety and season. A full bearing vine yields one kg of dry pepper. However, individual vines recorded yields up to 3–5 kg of dried pepper. Harvesting of pepper is carried out according to the purpose for which it is harvested. For preparation of white pepper the berries are harvested at a slightly advanced stage of ripeness i.e. when the berries turn red (bright orange). To get black pepper the berries are gathered at younger stages.

Yield: Pepper vine attain full bearing stage in the 7th or 8th year after planting and yield starts decline after 20–25 years and replanting has to be done thereafter. 7th or 8th year old pepper vine gives 800 to 1,000 kg of Black Pepper per ha.

PROCESSING OF PEPPER

Almost all the produce in India is processed in to black pepper and only a very limited quantity is converted in to white pepper. Black pepper: It consists of fully developed, but unripe dried berries of Pepper. The harvested spikes are sun dried for 7–10 days on cement floor or mats, until the outer skin becomes tough black, shrink and wrinkled. Drying is carried till the moisture content gets reduced to 10–15 percent. Then the dried berries are separated from the spikes by beating or rubbing between hands or trampling them under the feet. For making good quality of Black pepper of uniform color, the separated berries are collected in a perforated bamboo basket or vessel and the basket with the berries is dipped in boiling water for 1 minute. The basket is then taken out and drained. The treated berries are sun dried on a clean bamboo mat or cement floor. The recovery of black berry is about 33 % (26– 36percent depending upon the variety).

White Pepper: This consists of dried ripe fruits without pericarp (skin). It is prepared by removing the outer skin along with the pulp before drying. White pepper is prepared by one of the two methods

    I. Water steeping technique (traditional method)

    II. Steaming or boiling technique (improved method)

I. Water steeping technique: It is a traditional and slow method. It involves five steps.

1. Steeping: Spikes with fully ripe berries are filled in gunny bags and are steeped in flowing water for about 7 –8 days. During this steeping process, the skin gets loosened from the seed.
2. Depulping: At the end of steeping, the berries are taken out and the skin with the pulp is removed either by rubbing between hands or by trampling under feet.
3. Washing: These depulped seeds are then washed and cleaned with fresh water repeatedly (3–4 times)
4. Drying: The cleaned seeds are sun dried for 3–5 days on cement floor or mats till they become white and the moisture gets reduced to 10–15 percent.

5. Polishing: The dried seeds are now dull white with color. They are further cleaned by winnowing or by rubbing with a cloth. The percentage of recovery of white pepper is about 25 percent of ripe berries.

II. Steaming or boiling technique: This is an improved and quick method developed at CFTRI, Mysore. It involves four steps.
   1. Boiling: Freshly harvested spikes or berries are boiled for about 15 min.
   2. Depulping: The boiled berries are then pulped mechanically. Boiled berries pass through motorized fruit pulping machine.
   3. Bleaching: The depulped berries are washed thoroughly by using bleaching powder or any bleaching agent.
   4. Drying: The cleaned berries are sun dried for 3-5 days on cement floor or mats till they become white and the moisture gets reduced to 10–15 percent.

Jayashree et al. (2009) experimented on physicochemical properties of black pepper from selected varieties in relation to market grades. The varieties "Panniyur-1", "Panniyur-2", "Panniyur-5", "Sreekara", and "Subhakara" were graded in a hand operated rotary sieve cleaner-cum-grader. The grader was provided with 3 sets of sieves with pore size of 3.5, 3.8 and 4.8 mm placed one after the other in the increasing order. The variety 'Subhakara' had highest amount of berries of size between 3.8 and 4.8 mm (33.3%) which belongs to TGSEB (Tellicherry Garbled Special Extra Bold) grade. 'Panniyur-1', 'Panniyur-2' and "Panniyur-5" had more than 60 % of its berries under the grade TG (Tellicherry Garbled). Bulk density ranged from 450 to 571 g/l. Bulk density increased with increase in size. However, bulk density decreased when the berry size was > 4.8 mm. The starch content increased as the grade size increased whereas protein and crude fiber did not show any consistent trend. The oleoresin and piperine contents were highest for the lower grade (<3.5 mm) but the oil content was not related to grade in all the varieties.

Value added product are White pepper, pepper oil, pepper oleoresin, piperine, green pepper in brine, Dehydrated green pepper, frozen green pepper.

12.15 CHILLI (*CAPSICUM SP*, FAMILY- SOLANACEAE, ORIGIN - SOUTH CENTRAL AMERICA)

A pepper is an annual or perennial herb. It is used as a colorant, flavourant, or as a source of pungency, depending on the processed product. Peppers can be used fresh, dried, fermented, or as an oleoresin extract Peppers are used as a colorant, flavourant, and source of pungency. The main source of pungency in peppers is the chemical group of alkaloid compounds called capsaicinoids (CAPS), which are produced in the fruit. It preferred deep, loamy, fertile soils rich in organic matter. Also need well drained soils with adequate soil moisture for the growth of the crop. Chilli grows well in the dry weather condition. Planta produced in nursery bed and transplanted in raised bed are preferred by the crop for satisfactory growth, Planting is done during September to February at proper spacing. Land should be keep free and hand weeding done when necessary.

HARVESTING

Time and stage of harvesting chilli is governed by the purpose for which it is grown. The large part of the crop is producing dry chilli fruits. The crop is ready for harvesting green chilli in about a month after transplanting. One or two pickings of green fruits can be taken and the produce is disposed of in local market to be used as green salad, vegetable or condiment. This practice not only supplements net returns to farmer but also enhances growth of plants and induces them to produce more flowers and fruits. For dry chillies the fruits should not underripened or overripened. Crop is ready for harvesting ripe fruits in about three and a half months. Picking of fruits continues for about 2 months and 6 pickings are taken annually. While harvesting fruits, care should be taken to hold stalks firmly, and fruit should be pulled upward gently, breaking the base of the stalk. If it is rainfed crop 2–4 pickings and for irrigated crop 6–8 pickings are generally taken. Farmers are under the impression that frequent pickings in irrigated chilli cause breaking of twigs and more labor requirement and thus they will reduce the number of pickings which may not be correct. Delayed harvesting of fruits gives poor quality produce. Chilli is harvested by hand picking and extends up to two months. Yield of fresh green chilli is 3-4 time more than that of fresh red ripe chilli and 6–10 times than that of dry chilli.

POSTHARVEST TECHNOLOGY

Pungency, initial color, and color retention properties of fruits are closely related to maturity. Pods left to ripen and partially withered on plant are superior in the above said three qualities. However, it should be noted that care must be taken over the extent of withering permitted prior to harvesting, since if prolonged, it can sometimes results in a product with gray color.

TRADITIONAL SUN DRYING

Chillies on harvesting have moisture content of 65–80 percent depending on whether partially dried on the plant or harvested while still succulent, this must be reduced to 10 percent to prepare dried spice. Traditionally, this has been achieved by sun—drying of fruits immediately after harvesting without any special form of treatment. Sun drying even today the most widely used method in the world. Immediately after harvesting of fresh fruits heaped indoors for 2 or 3 days, so that the partially ripe fruits if any ripen fully and whole produce develops a uniform red color. The best temperature for ripening is 22–25°C and direct sun light is to be avoided since this can result in the development of white patches. Heaped fruits then spread out in the sun on hard dry ground or on concrete floors or even on the flat roofs of houses, frequent stirrings are given during day time in order to get uniform drying and thereby no discoloration or mold growth. The drying fruits are heaped and covered by tarpaulins or gunny bags during nights and spread during day time. After 2 or 3 days, the larger types are flattened by trampling or rolling to facilitate subsequent packing into bags for storage and transport. Drying by this procedure takes 5–15 days depending on prevailing weather. Out of 100 kg of fresh fruits 25–35kg of dried fruits may be obtained. Recently in majority of areas the fresh produce dried on open spaces like roadsides and remains exposed to weather for the entire drying period (5–15 days) may cause contamination with dust and dirt, damaged by rainfall animals, birds and insects. The losses may range 70–80 percent of total quantity due to this method. Traditional method of harvesting and sun drying involved poor handling of fruits results in bruising and splitting. Bruising shows up as discolored spots on pods, splitting leads to an excessive amount of loose seeds in a consignment, there is a considerable loss in weight and then in price. If the harvested fruits are not properly dried and protected from rain and pests, it will lose the color, glossiness and pungency.

IMPROVED CFTRI METHOD OF SUN - DRYING:

- A four—tier system of wire—mesh trays or a single tray of perforated aluminium took 14 days in sun to dry fruits having a moisture content of 72 to 74 percent reducing it to about 6 percent, the traditional method of sun drying takes about 3 weeks to achieve a moisture level of 15–20 percent.
- The improved CFTRI technology involves the following steps.
- Dip fresh chilli in "Dipsol" emulsion for a short period (approx 5 min)

DRAIN EXCESS EMULSION

- For 100 kg of fresh chillies 15 liters of emulsion is required which costs Rs.4/- only.
- Spread the material for drying on racks having multitier wire net trays at 5-10 kg per m² of tray area depending upon ambient temperatures.
- The treated material dry to the commercial level moisture content in about a weeks' time.

ADVANTAGES OF IMPROVED CFTRI METHOD ARE:

- Rate of drying is fast and hence the drying period is only a week as compared to 15–21 days in traditional method.

- Requires less space.
- Helps in better retention of color and pungency
- Gives a more hygienic and superior quality product.
- Gives 2 percent more finished product by weight and thus more profits.
- Preparation of Emulsion "DIPSOL":
- 'Dipsol is a water-based emulsion containing Potassium carbonate (2.5%), refined groundnut oil (1%), Gum acacia (0.1%) and Butylated hydroxy anisole (BHA) (0.001). Thus, 100 kilograms of "Dipsol" contains as follows.
- Potassium carbonate 2.5 kg
- Refined groundnut oil 1.0 kg
- Gum acacia 0.1 kg
- BHA 0.001 kg
- Dissolve potassium carbonate and gum acacia in water separately. Similarly, dissolve BHA in refined groundnut oil.
- Mix water—phase solutions and add BHA dissolved in groundnut oil slowly while stirring.
- The mixture is passed through a homogenizer twice at 200 kg/sq.cm.

SOLAR DRYING

- Recently attempts have been made to develop solar equipment to improve upon the sun - drying techniques, which lead to:
- better use of available solar radiation
  (b) reduction in drying time,
  (c) cleaner and better quality product, free from dust, dirt and insect infestation.
- This equipment is called "Solar Drier".
- The RRL (Jammu) has devised a solar drier for drying chillies.
- Red chillies of Kashmir are very popular throughout the country as these impart attractive bright red color to dishes.
- Chillies are produced in substantial quantity in Kashmir valley and it is a common scene to find chillies strung together in thread and hung on walls and doors or spread on roof tops.
- Commercially, plants with fruits still unplucked are harvested and spread out on the ground for about a week for partial drying.
- Thereafter, the fruits are plucked by hand and spread in field for final drying.
- The entire operation takes about 15 days during which chillies are exposed to dirt, dust, fungus attack besides uneven drying.
- A solar drier has been made near Pampore (Jammu and Kashmir) which effects complete drying of the commodity in 4–5 days with a marked improvement in color and storage characteristics.

- The gadget is very simple and is made of mud, stone pebbles and glass panes only and is specially suited for rural areas.
- It can be conveniently constructed by village artisans.
- With the extensive use of such solar driers, sizeable quantities of red chillies and other dried vegetables of improved quality can be produced in rural areas.
- Work at Agricultural Research Station., Lam on mechanical drying has shown that the produce can be dried within a period of 18 hours with the aid of air blown drier keeping the temperature at 44–46°C.
- This method not only saves time and avoids watch for 10–15 days, but also imparts deep red color and glossy texture to the fruits which are liked in Foreign trade and fetch higher premium than that of sun drying (the moisture content of dry pods is to be kept at 8–10%).
- The cost of mechanical drying worked out to 25 paise per kg of dry fruits.
- Packing is done after the removal of defective and decolorized pods in gunny bags, Jute boras or palmyrah baskets.

GRADING

Grading of fruits based on color, size ,stage of maturity
- Pack the fruits separately according to grade before sending them to market
- The losses due to grading at farmer level is 5–8 percent. If the grading is not good at farmer level another 4–5 percent loss in produce is possible.

PACKING AND STORAGE

- The discoloration of the red pigment of chilli during storage is greatly influenced by moisture content of pods at the time of storage and temperatures at which the produce is stored.
- Storage has a marked influence on the color of the dried chillies though it has little effect on their pungency.
- Since, color is one of the main determinants of the price, which a producer receives.
- Greatest influence on color retention is not infact of the storage conditions but rather of variety of capsicum or chilli grown.
- Delaying in harvesting until pods are partially withered on the plant and then curing the sliced pods provide a product with superior color retention properties.
- Exposure of dried chillies to air and light accelerates rate of bleaching and so storage in airtight containers away from sunlight is desirable.
- Moisture content higher than 15 percent is critical with respect to mold growth.

- Chillies should be conditioned to 10% moisture and compressed at 2.5 kg / cm2 by using a baling process.
- For retail or consumer packing of chilli powder.
- Packing in 3,000 gauge low density polyethylene film pouches are suitable for 100 g consumer unit packs to give a shelf—life of 3–6 months.
- Under tropical conditions, 200—gauge low and high density polyethylene films are suitable for packing of whole chilli in units of 250 g each.
- Such packs can be stored at a cool, dark, dry place for about a year.
- Detachment of stalks from pods resulting in bleeding of seeds from within the pods, leading to loss in pungency.
- As this commodity emits strong odor; it shall be stored in separate compartment as far as possible.
- Chillies are attacked by spice beetle and cigarette beetle during storage.
- The storage temperature has a greater influence in color retention than does light, air, kind of container or when the spice is stored in the whole or ground from.
- Application of fat—soluble antioxidants has been found to improve color retention.
- Addition of antioxidants is more effective after curing than before and in the ground spice rather than whole pods.
- Rats have a great liking for chillies in spite of their pungency, and therefore care should be taken in storage to protect chillies against this noxious animal.
- Shelf life of green pepper can be prolonged by using perforated polyethylene bags of 150–200 gauge.
- Ventilation of packages should be adequate to avoid off-flavor development and moisture condensation in packages.
- The lowest temperature range recommended for storing green bell peppers is 7–10°C for up to 2–3 weeks.
- At temperatures below 7° C bell peppers are subjected to chilling injury.
- Peppers having a large surface to volume ratio are particularly susceptible to water loss.
- They must be held in high relative humidity of 90–95 percent or else they will rapidly become wilted.
- For controlled atmospheric storage of bell peppers, the recommendations are 4–8 percent oxygen, 2–4 percent of Carbondioxide at 13°C.
- Oxygen concentration below 2 percent combined with 10 percent Carbondioxide may cause injury.

Such packs can be stored at a cool, dark, dry place for about a year.

As this commodity emits strong odor; it shall be stored in separate compartment as far as possible. The storage temperature has a greater influence in color retention than does light, air, kind of container or when the spice is stored in the whole or grind form

TRANSPORTATION

In general, farmers use bullock carts or tractors for sending the produce to nearby market.

- From market yards the produce is transported to distance places in ordinary trucks and lorries.
- Perfect packing, care in loading and unloading and quick transport results in less spoilage of fruits.

12.16 ALLSPICE (*PIMENTA DIOICA*, FAMILY MYRTACEAE)

It is a polygamodioecious evergreen tree, the dried unripe fruits of which provide the culinary spice pimento of commerce. The tree is indigenous to West Indies (Jamaica). Jamaica is the largest producer and exporter of pimento, accounting for 70 percent of the world trade. The remaining 30% is produced by Honduras, Guatemala, Mexico, Brazil and Belize. The dried mature but unripe berries, berry oleoresin, berry oil and leaf oil are the products.

Allspice is planted on a wide range of soils, a well-drained, fertile, loam limestone soil with a pH of 6–8 suits the crop best. Pimento grows well in semitropical lowland forests with a mean temperature of 18–24°C, a low of 15°C and a maximum of 32°C, annual rainfall of 150–170 cm evenly spread throughout the year.

Allspice is traditionally propagated through seeds, but vegetative propagation is also adopted to get true to type plants. The spacing recommended for allspice is 6 m × 6 m. Pits of about 60 cm deep and 30 cm. Transplanting should be done at the beginning of the rainy season. The base of the young seedlings should be kept free of weeds. After three to four years of growth, slashing once or twice annually around the tree would be sufficient. The base of the young seedlings should be kept free of weeds. After 3–4 years of growth, slashing once or twice annually around the tree would be sufficient

HARVESTING AND PROCESSING

Fruit is harvested 3–4 months after flowering when fully mature but not ripe which give most strongly flavored. Small and overripe fruits should be removed as they detract from the appearance of the finished product. Harvesting is done manually by breaking the twigs of which bearing the berry clusters. After harvesting the ripe berries are separated from green berries and the berry clusters should be taken in a container. After that the seed berries are spread out in the sun and turned over with a wooden rake so that they dry evenly. The end product should be bright brown in color.

Drying is complete by judging sharp, crispy, rattling sounds when a handful is shaken close to the ear. Then berries are cleaned by winnowing and stored by removing dust. A well grown trees give BOUT 50–60 Kg of dry berries. Pimento should be stored in gunny bags lined with polyethylene and well protected from sun rain and excessive heat. The dried fruits should be stored in poly-lined corrugated cardboard containers or in airtight containers and kept in a cool, dry area with a maximum temperature of 21°C and maximum humidity of 70 percent. The essential oil is stored in sealed opaque containers. The industry standard has recommended a shelf-life of 24 months.

It is marketed as whole or ground pimento. There are four grades i.e Mexican, Guatemala, Honduras, Jamaican.

QUALITY ISSUES

As per the ISO specifications, allspice is described as the dried, fully mature but unripe, whole berry of *Pimenta dioica* (L.) Merrill, 6.5–9.5 mm in diameter, a dark brown color, the surface somewhat rough and bearing a small annulus formed by the remains of the four sepals of the calyx. Allspice may also be in the pure ground form. The odor and taste of pimento, either whole or ground, shall be fresh, aromatic and pungent and shall be practically free from dead insects, insect fragments and rodent contamination

VALUE ADDED PRODUCTS

Berry oil, leaf oil, oleoresin, bark and wood

12.17   CURRY LEAF (*MURRAYA KOENIGII* LINN.; FAMILY-RUTACEAE)

Curry leaf herb is a culinary plant whose leaves are used as an aromatic and the fruit of the plant is a component of desserts in some Eastern nations. The plant is tropical to subtropical and produces small fragrant white flowers that become small black berry-like fruits.

Red sandy loam soils with good drainage are ideal for better leaf yield. The optimum temperature requirement is 26° to 37°C. The main season of sowing is July—August. Within 3–4 days of collection of fruits, the seeds should be pulped and sown in nursery beds or poly bags. One year old seedlings are suitable for planting. One seedling is planted at the center of the pit. Pit size of 30 x30x30 cm is dug one to two months before planting at a spacing of 1.2–1.5 m. After planting the pits are irrigated. On the third day the second irrigation is given and then the irrigation is given once in a week. After each harvest 20 kg of FYM/plant is applied and mixed with soil. Plant should be protected from diseases and pest.

HARVESTING AND PROCESSING

The crop is harvested at the end of first year and gives economic yield from second year onwards. During harvesting new shoot produced from the final pruning are harvested with the Sharpe knives. The tree harvested four times in a year. The young shoot and tender leaves are harvested, packed in gunny bags or tied in bundles and transported. The leaves are dried and ground into powder and used as curry leaf powder. The value-added products of curry leaf oil are volatile oil and dehydrated curry leaf.

12.18 CINNAMON (*CINNAMOMUM VERUM* /*CINNAMOMUM ZEYLANICUM*), FAMILY: LAURACEAE

Cinnamon (sweet wood) is the earliest known spice in India. It is a native of Sri Lanka. Bark of cinnamon is used as a spice. Well-drained soil rich in humus content and Sandy loam soils is most suitable. Cinnamon requires hot and humid climate. Annual precipitation of 150–250 cm and average temperature of 27°C are ideal.

It is commonly propagated through seed, though it can be propagated by cuttings and air layers. The seeds are sown in sand beds or polythene bags containing a mixture of sand, soil and well–powdered cowdung in a 3:3:1 ratio. The seeds start germinates within 10–20 days. Proper irrigations are required for proper growth. The seedlings require artificial shading till they become 6 months old. Pits size 50 cm³are dug at a spacing of 3 × 3 m. It is planted during June-July to take advantage of monsoon for the establishment of seedlings. One-year-seedlings are planted. In each pit, in some cases, the seeds are directly dibbled in pits that are filled with compost and soil. Partial shade in the initial years is advantageous for healthy and rapid growth of plants. It is pruned when become 2–3 years old; the shoot is cut back to a height of 30 cm from ground level to produce side shoots. This is called '*coppicing*'. Weeding is done 3–4 times in a year.

HARVESTING

For the quills the plant harvested after 3 years of planting. The crop is harvested by cutting the shoot for the extraction of bark during May- November. After cutting young shoot spring up from the stump, they will be ready for removal in the subsequent seasons within 18 months. The shoot selected for the cutting having 1.5–2 m long and 2–2.5 cm thick. Peeling is done with a small knife having round edge at the end.

PROCESSING, GRADING AND STORAGE

The cut stems are given longitudinal splits from one end to another end. If the bark does not peel easily; the stems are rubbed in between hard pieces of wood which enables the easy detachment of the bark.

During rolling the barks are packed together and placed one above the other and pressed well. The length of bark slip is reduced to 20 cm. they are then covered with dry leaves or mats to preserve the moisture for the next day of operation which is known piping. After peeling and rolling the quills are bundled and taken to piping yard for piping .the sticks are driven into ground in such way that they cross each other at a height of 30 cm from te ground level and scrape off the outer skin with a small knife. The scrape slips are shorted in to different grades according their thickness. After piping they are allowed to dry.

The bark free ones of finest ,smoothest quality graded as "00000", the coarsest being grade "0" and remaining are graded as "chips", pieces, quillings, and featherings'. The outer bark possesses a lightly acidic flavor and its removal enhances the delicate aroma.

Cinnamon should be stored in a cool, dry place. Excessive heat will volatilize and dissipate its aromatic essential oils. Under good storage conditions, the qualities of aroma and flavor for which cinnamon is prized will be retained long enough to meet any normal requirements. On prolonged storage, owing to oxidation, it becomes contaminated with resin and changes cinnamic acid to cherry red.

The leaf and bark oil of cinnamon can be obtained by distilling the dried cinnamon leaves and bark. About 4 kg of bark oil is obtained pre hectare of cinnamon plantation. The value-added products are whole and ground cinnamon, essential oil of cinnamon bark and leaf, seed oil, root oil and oleoresin.

Quality Issues

The bark obtained from the central branches is superior to that from the outer shoots and that from either the base or the top. The bark of thick branches is coarse and that of young shoots is thin and straw colored with very little flavor. Plants grown under shade produce inferior quality quills. The quality of cinnamon is assessed primarily on the basis of its appearance and on the content and aroma or flavor characteristics of the volatile oil. It should be light brown color with wavy lines and a produced a fractured sound when broken. When chewed, it should become soft, melt in the mouth and sweeten the breath. Good quality cinnamon should not be thicker than thick paper. Freshly ground cinnamon bark of good quality contains 0.9 to 2.3 percent essential oil depending on the variety.

SCOPE OF RESEARCH

1. Reviewing the current status of research, development and adoption of post-harvest management and value addition of spices crops
2. Technology cluster on industrial crops covers cardamom, black pepper, chilli, ginger,turmeric, vanilla, cinnamon. While some of these crops may currently serve the fresh market, their greater potential lies in their role as raw materials for down-stream processing of value-added products.
3. Research on industrial crops has a tremendous scope involving disciplines such as breeding and selection of new improved varieties, agronomy (cultural practices and soil fertility management), pest and disease management (including weeds) and postharvest handling of the crop produce.
4. Tissue culture as a method of producing virus-free planting material (by the Biotechnology Research Centre) and product development (by the Food Technology Research Centre) is another area of intervention.
5. Trials are to be carried out at research stations on a range of soil types and agro-ecologies which are currently cultivated with the industrial crops, or have potential for future cultivation.
6. Multilocation and local verification trials are also to be carried out at stations and on farmers' fields.
7. Latest technologies on postharvest, value addition, food safety and information management are to be emphasized especially on low cost technologies for small farmers.
8. The available information on success stories and replicable models on low cost technologies are to be explored.
9. Identifying strategies for promoting linkages between farmers, markets and processors including public-private participation.
10. Development of a regional action plan for strengthening cooperation in postharvest and value   addition technologies, policy framework, and advocacy.

KEYWORDS

- **Spices and condiments**
- **Quality issues**
- **Processing technology**
- **Medicinal value**
- **Biochemical composition**
- **Essential oils**

REFERENCES

Baldry, J.; Dougan, J.; Matthews, W. S.; Nabney, J.; Pickering, G. R. and Robinson, F. V.; 'Composition and flavour of nutmeg oils.' *Flavours*. **1976**, *7*, 28–30.

Jayashree, E.; and Visanathan, R.; Studies on mechanical washing of ginger rhizomes. *J. Agric. Eng.* **2010**, *47* (4).

Jayashree, E.; Vlsvanathan, R.; and John Zachariah T.; Quality of dry ginger (*Zingiber officinale*) by different drying methods. *J. Food Sci. Technol.* **2013**, doi: 10.1007/s13197-012-0823-8.

Jayashree, E.; John Zachariah, T.; Gobinath, P.; Physico-chemical properties of black pepper from selected varieties in relation to market grades. *J. Food Sci. Technol.* **2009**, *46*(3), 263–265.

Juliani, H. R.; Koroch, A. R.; Simon, Asante-Dartey, J. E. J.; and Dan Acquaye; Chemistry and Quality of Fresh Ginger Varieties (*Zingiber officinale*) from Ghana. New Use Agriculture and Agribusiness in Sustainable Natural Plant Products Program; **2007**.

Kant, K.; Ranjan, J. K.; Mishra, B. K. Meena, Lal, S. R. G.; and Vishal, M. K.; Post harvest storage losses by cigarette beetle (*Lasioderma serricorne* Fab.) in seed spice crops. *Indian J. Hortic.* **2013**, *70*(3), 392–396.

Kim, J.; Marshall, M.; and Wei, C.; Antimicrobial activity of some essential oils components against five foodborne pathogens. *J. Agric. Food Chem.* **1995**, *43*, 2839–2845

Kizhakkayil, J.; and Sasikumar, B.; Variabilty for quality traits in a global germplasm collection of ginger (*Zingiber officinale* R.). *Curr. Trend. Biotechnol. Pharm.* **2009**, *3*(3), 254–259.

Leela, N. K.; Prasath, D.; and Venugopal, M. N.; Essential oil composition of selected cardamom genotypes at different maturity levels. *Ind. J. Hortic.* **2008**, *65*(3) , 366–36.

Parthasarathy, V. A.; Zachariah, J. T.; and Jayashree, E.; Managing spices in better way for marketing. *Indian Hortic.* **2011**, *56*(3).

Pramila, C. K.; Prasanna, K. P. R.; Balakrishna, P.; Devaraju, J.; and Siddaraju; Assessment of seed quality in seed spices. *J. Spices. Arom. Crops.* **2013**, *22*(2), 233–237.

Reineccius, G. Flavor and aroma chemistry. In *Quality Attributes and Their Measurement in Meat, Poultry and Fish Products;* Advances in Meat Research Series 9, **1994**; pp 184–201.

Sagar, V. R.; and Kumar, R.; Effect of packaging and storage on the quality of ginger, onion and garlic (GOG) mix powder. *Indian. J. Hortic.* **2009**, *66*(3), 367–373.

Sagoo, S. K.; Little, C. L.; Greenwood, M.; Mithani, V.; Grant, K. A.; McLauchlin, J.; Pinna, E. D.; and Threlfall, E. J.; Assessment of the microbiological safety of dried spices and herbs from production and retail premises in the United Kingdom. *Food Microbiol.* **2009**, *26*, 39–43.

Thomas, E. T.; John Zachariah, S.; Syamkumar and Sasikumar, B.; Curcuminoid profiling of Indian turmeric. *J. Med. Arom. Plant Sci.* **2010**, *33*(1), 36–40.

Zachariah, J. T.; Leela, N. K. M.; Maya, Rema, K. J.; Mathew, P. A.; Vipin, T. M.; and Krishnamoorthy, B.; Chemical composition of leaf oils of *Myristica beddomeii* (King), *Myristica fragrans* (Houtt.) *and Myristica malabarica* (Lamk.) *J. Spices. Arom. Crop.* **2008**, *17*(1), 10–15.

CHAPTER 13

BIOTECHNOLOGICAL APPROACHES TO IMPROVE POSTHARVEST QUALITY OF FRUITS AND VEGETABLES

BISHUN DEO PRASAD[1], SANGITA SAHNI[2], and
MOHAMMED WASIM SIDDIQUI[3]

[1]Department of Plant Breeding and Genetics, Bihar Agricultural University, Sabour, Bhagalpur, Bihar (813210) India; Email: dev.bishnu@gmail.com

[2]Department of Plant Pathology, T. C. A, Dholi, RAU, Pusa

[3]Department of Food Science and Technology, Bihar Agricultural University, Sabour, Bhagalpur, Bihar (813210) India

CONTENTS

13.1   INTODUCTION

Fruits and vegetables are considered as a commercially important and nutritionally essential food commodity due to providing not only the major dietary source of vitamins, sugars, organic acids, and minerals, but also other phytochemicals including dietary fiber and antioxidants with health-beneficial effects. In addition, fruits and vegetables provide variety in color, shape, taste, aroma, and texture to refine sensory pleasure in human's diet. There is an increasing demand for fresh produce at the consumer level, because of the raising awareness of people about the superior of fresh, natural foods than processed products resulting in the active encouragement by health agencies and public media as well as several medical researches demonstrating various health benefits of fresh produce consumption (Wills et al., 2007). Unfortunately, fruits and vegetables are highly perishable in nature and may be unacceptable for consumption if not handled properly after harvesting (Kader, 2002, 2005; Kays and Paull 2004; Wills et al., 2007). Furthermore, fresh Fruits and vegetables have great potential for foreign exchange. Longer shipments and distribution periods may eventually increase the potential of heavy losses. It is important to stress that postharvest losses occurs due to high moisture content, active metabolisms, tender nature, and rich in nutrients, etc., are immense. It is estimated that the magnitude of these losses due to inadequate postharvest handling, transportation, and storage in fresh fruits and vegetables is relatively higher, 20–50 percent, in developing countries when compared to 5–25 percent in developed countries (Kader 2005). Technically advanced countries such as the USA, Japan, Australia, and European countries can apply relatively sophisticated technologies to minimize losses whereas in developing countries postharvest losses are one of the most significant factors limiting agricultural production. Therefore, the importance of proper cares and techniques for handling fresh produce after harvest has been recognized and emphasized. The complete elimination of postharvest losses may be impossible and uneconomical, but to diminish them by 50 percent is possible and desirable (Kader 2005).

Since last decade, there has been an increasing spate of interest in postharvest biotechnology research as the standards and expectations have changed across the spectrum of products harvested for use of food. In the mist of recent global challenges such as increasing population, increasing demand for food, climate change, and water scarcity, plant biotechnology has become a necessity tool for growth and yield performance to meet the food needs of today. From a biotechnological point of view it is therefore important to establish the nature of the crop and the particular problem before establishing the approach to be attempted. Modern genetic engineering techniques allow us to cross species barriers (and even kingdom barriers) and therefore genes that would not normally be accessible by conventional breeding can now be incorporated into the plant species being targeted. Serious efforts have been put to develop fruits and vegetables plants with reduced postharvest losses by delayed ripening, reduced fruit softening, increased shelf life and resistance to postharvest pathogen and pest attacks. The understanding of the fundamental processes which influence fruit set, maturation, and ripening are required to manipulate fruits and vegetable yield and quality. Biotechnology has played a significant role in this respect. Various constraints lies in the improvement of fruits and vegetables includes long juvenile periods, the complex reproductive biology, high degree of heterozygosity, inter and intra incompatibility, and sterility of breeding of fruits and vegetables plants such as tomatoes, orange etc. (Bapat et al., 2010). Further, the ability to maintain the quality of stored fruit and vegetables during postharvest storage is highly related to the physiological, biochemical, and molecular traits of the plant from which they derive. These traits are genetically determined and can be manipulated using molecular breeding and/or biotechnology to enable plants tolerate the biotic and abiotic stresses, and plant resistances to problematic pests and disease, which may provide higher nutritional contents, and extend the shelf life of the produce. The application of biotechnological knowledge should not only lead to major improvements in postharvest storage of fresh fruits and vegetables but also improve human food supply. Detailed knowledge is now available regarding the relationship between the biology of the fresh produce and its postharvest characteristics. This knowledge has great potential to be used in biotechnology based specific postharvest related traits in fruits and vegetables. In this chapter the biotechnologically based improvements of postharvest quality of fruits and vegetables are discussed in detail.

13.1.1 EFFECT OF BIOTECHNOLOGICAL APPROACHES ON RIPENING AND PERISHABILITY (SHELF LIFE) IN FRUITS AND VEGETABLES

The postharvest life of fruits and vegetables is limited by a number of factors that include primarily ripening, postharvest diseases and senescence. The commercial value of fruit depends heavily on the ripening process, which ideally should proceed to a point that is attractive to the consumer, but not too far, or else the fruit becomes overripe and eventually rots. Overripe produce is a significant economic problem, with an average of 20 percent or more of perishable fruits and vegetables being lost postharvest and total U.S. losses estimated at several billion dollars annually. Poor postharvest characteristics such as deficient flavor development, very short shelf life, quick softening, easy spoilage, sensitivity to low temperatures (chilling injury) and easy pathogen attack (bacteria, fungi, etc.), are major constraints to profitability for the domestic market, and to the expansion of existing and new export markets. Among all fruits, tropical fruits are notorious for their poorer- than- average postharvest quality.

Conventional remedies, such as storage in a controlled atmosphere or chemical treatment, can be costly. Biotechnological tools may be an alternate choice to manipulate the biochemical pathway leading to the synthesis of ethylene, the plant hormone that controls the ripening process. Extension of shelf life and resistance to pathogen attack are the two major obvious targets to improve the postharvest losses. The ripening process involves a large number of biochemical pathways governed by a large number of genes leading to marked changes in texture, taste, and color of fruits and vegetables. In general, fruit are classified as climacteric or nonclimacteric depending upon their patterns of respiration and ethylene synthesis during ripening. Climacteric fruits are characterized by an increased respiration rate at an early stage in the ripening process accompanied by autocatalytic ethylene production whereas nonclimacteric fruits show a different respiration pattern and display a lack of autocatalytic ethylene synthesis. Many of the economically important fruit crops are climacteric; therefore a large amount of research has been developed to studying the biochemical and molecular pathways operating during the climacteric ripening of fruits.

Most of the genetic engineering approaches attempted in order to improve the shelf life and general appearance of fruits have centered on the set of genes controlling fruit firmness (membrane and cell wall properties) and the ripening rate (ethylene production or perception). These approaches have targeted endogenous genes with vital functions in the ripening process aiming to down regulate their activity by gene silencing.

13.1.2   MAINTAINING FRUIT FIRMNESS

Fruit firmness is an important quality in fruit production that can decide which fruit will be harvested, transported, stored, or marketed. Fruit softening which effects cell wall structure and their integrity are one of the primary targets of researchers. Factors influencing fruit softening includes both internal factor like cell wall metabolism and external factors like temperature, humidity, shading, etc. The temperature of a plant growth environment is critical for plant development and has been shown to affect the rate of fruit growth and time required for fruit to ripen. Temperature differences or fluctuations have not been directly linked fruit softening, but there is evidence that temperature may influence cell wall degradation resulting in softening of fruit. Several studied showed that the humidity of the postharvest storage environment also has an effect on the rate of ethylene production, thus affecting fruit softening. Shading of fruit on melon plants exhibited an increase in ethylene production leading to loss a reduction in firmness of the inner mesocarp of the fruit correlating to the increase in ethylene concentration (Nishizawa et al. 2000). Apart from above mentioned external factors Oxygen Concentration, nutrients, and salt concentration in irrigated water and soil may directly or indirectly influence fruit repining.

Over the past two decades, the manipulation of different cell wall hydrolysis enzymes genes which affect fruit softening at harvesting and during postharvest storage has become firmly entrenched in literature (Vicente et al., 2007: Goulao and

Oliveira, 2008; Li et al., 2010). Among the genes involved in firmness, the most extensively studied is the polygalturonase (PG), a cell wall-degrading enzyme that catalyzes the hydrolysis of polygalacturonic acid chains (Della-Penna et al., 1986; Grierson et al., 1986). Polygalacturonic acid is an important component of the plant cell wall that significantly contributes to the fruit firmness. Partial silencing of the PG gene has been achieved in tomato by sense and antisense techniques. Experiments using either a partial or the full length PG gene successfully reduced the levels of PG mRNA and enzyme activity (Sheehy et al., 1988; Smith et al., 1988, 1990). Low PG tomatoes were more resistant to cracking and splitting than regular fruit. They also had superior handling and transport characteristics showing a severely reduced degree of damage during those processes (Schuch et al., 1991). "FlavrSavr", the commercial name for a low PG tomato, marked an important milestone in plant biotechnology being the first genetically modified plant food to reach the market, commercialized by calgene in the USA in 1994.

Recent studies demonstrated that manipulation of cell wall hydolases can be important to the postharvest qualities of leafy vegetables. Transgenic lettuce plants (*Lactucasativa* cv. Valeria) were produced in which the production of the cell wall-modifying enzyme xyloglucanendotransglucosylase/hydrolase (XTH) was down-regulated by antisense inhibition. Consequently, xyloglucanendotransglucosylase (XET) enzyme activity and action were down-regulated in the cell walls of these leaves and it was established that leaf area and fresh weight were decreased while leaf strength was increased in the transgenic lines. Overall an extended shelf life of transgenic lines was observed relative to the nontransgenic control plants, which illustrated the potential for manipulation of cell wall related genes for improving postharvest quality of leafy crops.

13.1.3 CONTROL ON ETHYLENE BIOSYNTHESIS ORPERCEPTION

Fruit ripening and softening are major attributes that contribute to perishability in both climacteric and nonclimacteric fruits. Fruits and vegetables such as tomato, banana, mango, avocado etc. take about a few days after which it is considered inedible due to over ripening. The spoilage include excessive softening and changes in taste, aroma, and skin color. This unavoidable process brings significant losses to both farmers and consumers alike. Even though ripening in fruit and vegetables can be delayed through several external procedures, the physiological and biochemical changes associated with ripening is an irreversible process and once started cannot be stopped (Prasanna et al., 2007; Martínez-Romero et al., 2007). Ethylene has been identified as the major hormone that initiates and controls ripening in fleshy fruits and vegetables. Ethylene has a central role in the regulation of different biological processes associated with the postharvest life of fruits and vegetables. Influencing ethylene biosynthesis during ripening in fleshy commodities has been the foremost attempt for combating postharvest deterioration. Many of the manipulations of ethylene biosynthesis were achieved by inhibition of two central biosynthesis related genes, 1-aminocyclopropane-1-carboxylate oxidase (*ACC*) synthase (*ACS*) or ACC oxidase (*ACO*). Hormone perception was manipulated in most cases by introduction of a mutated form of ethylene receptor genes which found to confer insensitivity when introduced into heterologous plant systems (Wilkinson et al., 1997).

In tomato, the ACC synthase gene active during ripening (*LEACC2*) was silenced using antisense technique. This silencing effectively reduces the production of ethylene by the ripening fruit by 99.5 percent (Oeller et al., 1991). While control fruits begin to produce ethylene 48–50 days after pollination and immediately undergo a respiratory burst, genetically modified tomatoes produced minimal levels of ethylene and failed to produce the respiratory burst.

FIGURE 13.1   Ethylene biosynthesis pathway.

Instead of altering the levels of enzyme controlling the biosynthesis of ethylene, two commercial companies (Monsanto and Agritope) have opted for alternative strategies aimed at depleting the intermediate substrates of the pathway. Monsanto used a bacterial enzyme (ACC deaminase) to drain the cell of the immediate precursor of ethylene (ACC). Overexpression of an ACC deaminase gene in tomato plants led to a marked depletion of the levels of ACC and therefore reduced the availability of this precursor to be converted in to ethylene (Klee et al., 1991). Transgenic plants overexpressing ACC deaminase were indistinguishable from controls with no differences observed during development even though there was a dramatic decrease in the levels of ethylene produced in vegetative tissues. When fruits were picked from the plant at the breaker stage and stored at room temperature, control achieved fully red stage in 7 days compared with 24 days for the transgenic fruits. Softening behavior was also affected with control showing a strong incidence of softening 2 weeks after picking' in contrast transgenic fruits remained firm for 5 months. When fruits were left on the plant to ripen, transgenic fruits remained firm for much longer than controls and did not abscise for more than 40 days.

Agritope has used a bacteriophage gene encoding S-adenosyl methionine (SAM) hydrolase, in conjunction with a ripening specific promoter, to hydrolyze the first intermediate of the ethylene biosynthetic pathway (SAM) in ripening cherry tomoto fruits. The Cnr and rin mutations are recessive and dominant mutations, respectively, and effectively block the ripening process. This was attributed to failure to produce elevated ethylene or to respond to exogenous ethylene during ripening (Vrebalov et al., 2002; Manning et al., 2006). These mutant loci encode putative transcription factors were revealed to provide the first insights into dedicated fruit-specific transcriptional control of ripening. The rin was reported to encode a partially deleted MADS-box protein of the SEPELATTA clade, whiles Cnr is a genetic gene control unassociated with DNA change but alters the function of the promoter methylation of a SQAMOSA promoter binding (SPB) protein. The Nr mutation revealed an ethylene receptor gene, and Gr has been found to encode a novel component of ethylene signaling. GR was cloned by positional cloning of the gene underlying a dominant ripening mutation (Barry and Giovannoni, 2006). The biochemical nature of the GR remains unclear, but amino acid sequence suggests its membrane localization and possible copper-binding activities. Apart from studies carried out on various mutants, development of transgenic tomato fruit with different genes has provided a better insight of fruit ripening and genes involved with this process. SAMDC has been isolated from different plants and has been utilized to modify fruits with an idea that overexpression of SAMDC, might enhance the flux of SAM through the polyamine pathway, thus reducing the amount available for ethylene biosynthesis. The rate of ethylene production in transgenic tomatoes with yeast SAMDC gene under the control of E8 promoter exhibitor was reported to be lower than in the nontransgenic control fruit, suggesting that polyamine and ethylene biosynthesis pathways may act simultaneously in ripening tomato fruit (Mehta et al., 2002). Both ACO and ACS are encoded by a multigene family of five and nine members, respectively in tomato, whose expressions are differentially regulated during fruit development and ripening. In tomato, the antisense copy of one member of ACS gene family with its untranslated region was used to develop transgenic plants. Transgenic tomatoes showed 99.5% decrease in ethylene production and did not ripen without exogenous treatment of ethylene. In another attempt, anti-ACS containing transgenic tomato plants showed a 30 percent decrease in ethylene production by fruits.

13.1.4 CONTROL ONCYTOKININS BIOSYNTHESIS OR PERCEPTION

Cytokinins are essential hormones for plant growth and development. Many vegetables exhibit a very short life span after harvesting and require very elaborate measures to expand their life. Cytokinins have the capacity to delay leaf senescence (Van Staden et al., 1988). Modification of cytokinin biosynthesis using genetic manipulation in transgenic plants, and especially during the senescence phase, can significantly delay senescence. The *Agrobacterium tumefaciens IPT gene*, encoding for isopentenyl transferase (IPT), a key enzyme in cytokinin biosynthesis, was recognized to be key enzyme in cytokinin biosynthesis, was recognized to be a good candidate for manipulation.

The use of the senescence specific *SAG12* promoter for activation of the *IPT gene* resulted in an efficient autoregulatory cytokinin production system (Gan and

Amasino, 1995, 1997). At the onset of leaf senescence, the senescence specific promoter activates the expression of *IPT*, resulting in increased cytokinin levels, which in turn prevent the leaf from senescing. The inhibition of leaf senescence renders the senescence specific promoter inactive to prevent cytokinins from accumulating to very high levels, since overproduction of cytokinins may interfere with other aspects of plant development. Because cytokinin production is targeted to senescing leaves, overproduction of cytokinins before senescence should be avoided.

Transformation of lettuce with the *SAG12-IPT* gene resulted in a significant delay of development and postharvest leaf senescence in mature heads of transgenic lettuce (McCabe et al., 2001). Also in broccoli, the *PSAG12-IPT* construct was used successfully for retarding postharvest yellowing (Chen et al., 2001).

13.1.5 CONTROL ON ABSCISIC ACID (ABA BIOSYNTHESIS) OR PERCEPTION

For fruits and vegetables low temperature storage is widely used because it is considered as one of the best storage methods to delay different physiological, biochemical, and pathological processes associated with the loss of postharvest quality (McGlasson et al., 1979). Low temperature slows down the cell metabolism rate and delays senescence and ripening processes that are very important for postharvest storage. Low temperature also inhibits the rate of growth and level of pathogenicity of many postharvest pathogens.

The Plant hormone ABA, known to mediate the signal transduction of different abiotic stresses, has a central role in increasing plant cold tolerance (Kim,2007). Manipulation of the expression of genes responsible for cold tolerance can be targeted for improving plant cold tolerance.

13.2 EFFECT OF BIOTECHNOLOGICAL APPROACHES ON NUTRITIONAL QUALITY OF FRUITS AND VEGETABLES

Fruits and vegetables play a significant role in human nutrition, especially as sources of vitamins, minerals, and dietary fiber (Craig and Beck, 1999; Quebedeaux and Bliss, 1988; Quebedeaux and Eisa, 1990; Wargovich, 2000). Their contribution as a group is estimated at 91 percent of vitamin C, 48 percent of vitamin A, 30 percent of folacin, 27 percent of vitamin B , 17 percent of thiamine, and 15 percent of niacin in the U.S. diet. Fruits and vegetables also supply 16 percent of magnesium, 19percent of iron, and 9 percent of the calories. Legume vegetables, potatoes, and tree nuts (e.g., almond, filbert, pecan, pistachio, and walnut) contribute about 5 percent of the per capita availability of proteins in the U.S. diet, and their proteins are of high quality as to their content of essential amino acids. Other important nutrients supplied by fruits and vegetables include riboflavin (B$_2$), zinc, calcium, potassium, and phosphorus.

13.2.1   VITAMINS AND DIET ENRICHMENT COMPOUNDS

Vitamins are essential compounds for humans and other vertebrates and they must be obtained from the diet. In addition, some vitamins are used as functional additives in food products. Ascorbic acid (vitamin C) is used to prevent oxidation in apples, peaches, apricots, potatoes, peanut butter, potato chips, beer, fat, and oils. The carotenoids (vitamin A precursors) are used as colorants in margarine, cheese, ice cream, pasta, juices, and beverages. In tomato the transformation with a bacterial phytoenedesaturase increased up to twofold β carotene content in fruits. Another lipid-soluble vitamin whose function is linked to an antioxidant role is vitamin E (tocopherol). Daily intake of this vitamin in excess of a recommended minimum is associated with decreased incidence of several diseases. Plant oils are the main source of dietary vitamin E and they generally have a high content of the vitamin E precurs_o-trocopherol. Flavonols are another group of secondary metabolites whose inclusion in the human diet may give protection against cardiovascular diseases. The biosynthetic pathway leading to the synthesis of these compounds has been known for a long time. However, recent information regarding the pathway has allowed the design of specific strategies to increase the content of selected bioactive compounds. Thus, the transformation of tomato with a gene from Petunia encoding a *chalconeisomerase* has produced tomato fruits with a 78-fold increase in the content of flavonols in the peel (Muir et al., 2001). It was observed that more important, 65% of the flavonols were retained in the paste obtained after processing the transgenic fruits.

13.2.2   MODIFICATION OF FRUIT COLOR AND SWEETNESS

It is becoming widely accepted that plant biotechnology is entering a second phase of development that looks for modifying fruit color and sweetness. Anthocyanins are the pigments responsible for color in many fruits, such as grapes and strawberries. Deeply colored fruits are generally more desirable to consumers. Further, anthocyanins and related flavonoids have antioxidant properties that reduce the risk of cardiovascular disease and cancer. Fruits with consistently higher levels of anthocyanins, produced through genetic modification, could reach the supermarket within 15 years. These will likely be produced by altering the expression of whole biochemical pathways rather than through modulation of specific enzymes. Studies found that tomato plants transformed with yeast *SAMDC* gene under the control of E8 promoter showed improvement in tomato lycopene content, better fruit juice quality, and vine life (Bapat et al, 2010). Fruit coloration and softening were essentially unaffected, and all the seedlings from first generation seed displayed a normal triple response to ethylene. Overexpression of Nr (wild-type) gene, in tomato using constitutive 35S promoter produced plants that were less sensitive to ethylene (Ciardi et al., 2000). As ethylene receptors belong to a multigene family, antisense reduction in expression of individual receptors did not show a major effect on ethylene sensitivity possibly due to redundancy except in case of LeETR4. Antisense plants developed using LeETR4 under the control of CaMV35S promoter exhibited a constitutive ethylene response and were severely affected (Tieman et al., 2000). When antisense plants were developed, using this receptor with fruit-specific promoter, fruits showed early ripening (Kevany et al., 2008). Hackett et al. (2000) developed transgenic Nr plants by inhibition of the mutant Nr gene. In these transgenic plants, normal ripening of Nr fruit was restored and fruit achieved wild-type levels of expression of ripening-related (PSY1 and ACO1) and ethylene-responsive (E4) genes. Own-regulation of PG mRNA accumulation by constitutive expression of an antisense PG transgene driven by the cauliflower mosaic virus 35S promoter yielded transgenic fruits, retaining only 0.5–1 percent of wild-type levels of PG enzyme activity though overall fruit ripening and softening was not affected (Rose et al., 2003).

Suppression of PME activity in tomato by introducing antisense PME2/PEC2 transgenes under the control of the constitutive CaMV35S promoter modulated degree of pectin methyl esterification. In transgenic antisense PME fruit esterification was higher than controls throughout ripening, but the fruit otherwise ripened normally. In another study, antisense suppression of pectinesterase under CaMV35S promoter produced fruits with reduced PE activity and suppression in the rate of softening during ripening (Phan et al., 2007). In tomato, a large and divergent multigene family encodes EGases (cellulases), which consists of at least eight members. mRNA accumulation of the highly divergent EGases LeCel1 and LeCel2 was suppressed individually by constitutive expression of antisense transgenes (Rose et al., 2003). In both cases, most suppressed lines showed decreased mRNA accumulation in fruit pericarp by 99 percent as compared to wild-type, without affecting the expression of the other EGase and fruit softening. Galactosidases in tomato are encoded by a multigene family having seven members (TBG1–7). These members show differential expression patterns during fruit development (Smith and Gross, 2000). Transgenic plants have been developed using members of this family to reduce softening process. Sense suppression by a short gene specific region of TBG1 cDNA reduced TBG1 mRNA abundance to 10 percent of wild-type levels in ripe fruit, but did not reduce total exo-galactanase activity and did not affect cell wall galactose content or fruit softening (Carey et al., 2001). Antisense tomato beta-galactosidase4 (TBG4) and 7 (TBG7) cDNAs driven by the CaMV35S promoter resulted in transgenic tomatoes with modulated fruit firmness in comparison to control fruit (Moctezuma et al., 2003). Ethylene response factors (ERFs) play important role in modulating ethylene induced ripening in fruits. These ERFs belong to multigene family and are transcriptional regulators. These mediate ethylene-dependent gene expression by binding to the GCC motif found in the promoter region of ethylene-regulated genes. Modulation of expression of these individual ERFs in tomato has demonstrated their role in plant development and ripening. The sense and antisense LeERF1 transgenic tomato under the control of CaMV35 promoter were developed. Overexpression of LeERF1 in tomato caused the typical ethylene triple response on etiolated seedling. Antisense LeERF1 fruits showed longer shelf life compared with wild-type tomato (Li et al., 2007).

13.3 CONCLUSION

The application of biotechnological approaches to improve nutritional quality and shelf life of fruits and vegetables were analyzed. It was evident that developed biotechnological approaches have the potential to enhance the yield, quality, and shelf life of fruits and vegetables to meet the demands of the 21st century. However, the developed biotech approaches for fruits and vegetables were more of academic jargon than a commercial reality. To make sure that the current debates and complexities surrounding the registration and the commercialization of genetically modified fruits and vegetables are adequately addressed, various stakeholders in the industry (policy makers, private sectors, agriculturalists, biotechnologists, scientists, extension agents, farmers, and the general public) must be engaged in policy formulations, seed embodiments, and products development. The full benefit of the knowledge can be reaped if there are total commitment by all stakeholders regarding increased and sustained funding, increase agricultural R&D, and less cost and time for registration and commercialization of new traits.

KEYWORDS

- **Postharvest quality**
- **Fruits and vegetables**
- **Shelf Life**
- **Ripening**
- **Ethylene**
- **Enzymatic changes**
- **Genetic modification**

REFERENCES

Bapat, V. A.; Trivedi, P. K.; Ghosh, A.; Sane, V. A.; Ganapathi, T. R.; and Nath, P.; Ripening of fleshy fruit: molecular insight and the role of ethylene. *Biotechnol. Adv.* **2010**, *28*, 94–107.

Carey, A. T.; Smith, D. L.; and Harrison, E.; et al.; Down-regulation of a ripening-related β-galactosidase gene (*TBG1*) in transgenic tomato fruits, and characterization of a related cDNA clone. *Plant Physiol.* **2001**, *108*, 1099–1107.

Chen, L. F. O.; Hwang, J. Y.; Charng, Y. Y.; Sun, C. W.; and Yang, S. F.; Transformation of broccoli (*Brassica oleraceavar.* italica) with isopentenyltransferase gene via *Agrobacterium tumefaciens* for post-harvest yellowing retardation. *Mol. Breed.* **2001**, *7*, 243–257.

Ciardi, J. A.; Tieman, D. M.; Lund, S. T.; Jones, J. B.; Stall, R. E.; and Klee, H. J.; Response to *Xanthomonas campestris* pv. vesicatoria in tomato involves regulation of ethylene receptor gene expression. *Plant. Physiol.* **2000**, *123*, 81–92.

Craig, W.; and Beck. L.; Phytochemicals: Health protective effects. *Can. J. Diet. Pract. Res.* **1999**, *60*, 78–84.

Della-Penna, D.; Alexandra, D. C.; and Bennett, A. B.; Molecular cloning of tomato fruit polygalacturonase: Analysis of polygalacturonase mRNA levels during ripening. *Proc. Natl. Acad. Sci. U.S.A.* **1986**, *83*, 6420–6424.

Gan, S.; and Amasino, R. M.; Inhibition of leaf senescence by auto regulated production of cytokinin. *Science* **1995**, *270*, 1986–1988.

Gan, S.; and Amasino, R. M.; Making sense of senescence. *Plant Physiol.* **1997**, *113*, 313–319.

Goulao, L. F.; and Oliveira, C. M.; Cell wall modifications during fruit ripening: when the fruit is not the fruit. *Trends Food Sci. Technol.* **2008**, *19*, 4–25.

Grierson, D.; Maunders, M. J.; Slater, A.; Ray, J.; Bird, C. R.; Schuch, W.; Holdsworth, G. A.; and Knapp, J. E.; Gene expression during tomato ripening. *Philos Trans R Soc. Lond.-Biol. Sci.* **1986**, *314*, 399–410.

Hackett, R. M.; Ho, C. W.; Lin, Z. F.; Foote, H. C. C.; Fray, R. G.; and Grierson, D.; Antisense inhibition of the Nr gene restores normal ripening to the tomato Never-ripe mutant, consistent with the ethylene receptor-inhibition model. *Plant Physiol.* **2000**, *124*, 1079–1085.

Kader, A. A.; Increasing food availability by reducing postharvest losses of fresh produce. *Acta Hortic.* **2005**, *682*, 2169–2175.

Kader, A. A.; Postharvest biology and technology: an overview. In: *Postharvest Technology of Horticultural Crops,* 3rd ed.; Kader, A. A., Ed.; University of California, Agriculture & Natural Resources, Publication #3311; **2002**; pp 39–47.

Kays, S. J.; and Paull, R. E.; Post*harvest Biology;* Exon Press, USA; **2004**, 568 pp.

Kevany, B. M.; Tieman, D. M.; Taylor, M. G.; Cin, V. D.; and Klee, H. J.; Ethylene receptor degradation controls the timing of ripening in tomato fruit. *Plant. J. Cell. Mol. Biol.* **2007**, *51*, 458–467.

Kim, J.; Perception, transduction, and networks in cold signaling. *J. Plant. Biol.* **2007**, *50*, 139–147.

Klee, H. J.; Hayford, M. B.; Kretzmer, K. A.; Barry, G. F.; and Kishore, G. M.; Control of ethylene synthesis by expression of a bacterial enzyme in transgenic tomato plants. *Plant. Cell.* **1991**, *3*, 1187–1193.

Li, X.; Xu, C. J.; Korban, S. S.; andChen, K. S.; Regulatory mechanism of textural changes in ripening fruits. *CRC Crit. Rev. Plant Sci.* **2010**, *29*, 222–243.

Li, Y.; Zhu, B.; Xu, W.; Zhu, H.; Chen, A.; Xie, Y.; Shao, Y.; and Luo, Y.; *LeERF1* positively modulated ethylene triple response on etiolated seedling, plant development and fruit ripening and softening in tomato. *Plant Cell. Rep.* **2007**, *26*, 1999–2008.

Martinez-Romero, D.; Bailen, G.; Serrano, M.; Guillen, F.; Valverde, J. M.; Zapata, P.; Castillo, S.; and Valero, D.; Tools to maintain postharvest fruit and vegetable quality through the inhibition of ethylene action: A review. *Crit. Rev. Food Sci. Nutr.* **2007**, *47*, 543–560.

McGlasson, W. B.; Scott, K. J.; and Mendoza, J. D. B.; The refrigerated storage of tropical and subtropical products. *Int. J. Refrig.* **1979**, *2*, 199–206.

Mehta, R. A.; Cassol, T.; Li, N.; Ali, N.; Handa, A. K.; and Mattoo, A. K.; Engineered polyamine accumulation in tomato enhances phytonutrient content, juice quality and vine life. *Nat. Biotechnol.* **2002**, *20*, 613–618.

Moctezuma, E.; Smith, D. L.; and Gross, K. C.; Antisense suppression of a β-galactosidase gene (*TBG6*) in tomato increase fruit cracking. *J. Exp. Bot.* **2003**, *54*, 2025–2033.

Muir, S. R.; Collins, G. J.; Robinson, S.; Hughes, S.; Bovy, A.; De Vos, C. H. R.; van Tunen, A. J.; and Verhoeyen, M. E.; Overexpression of petunia chalcone isomerase in tomato results in fruit containing increased levels of flavonols. *Nat. Biotechnol.* **2001**, *19*, 470–474.

Nishizawa, T; Ito, A; Motomura, Y; Ito, M; and Togashi, M.; Changes in fruit quality as influenced by shading of netted melon plants (*Cucumismelo* L. 'Andesu' and 'Luster'). *J. Jap. Soc. Hortic. Sci.* **2002**, *69*(5), 563–569.

Oeller, P. W.; Min Wong, L.; Taylor, L. P.; Pike, D. A.; and Theologis, A.; Reversible inhibition of tomato fruit senescence by antisense RNA. *Science.* **1991**, *254*, 437–439.

Phan, T. D.; Bo, W.; West, G.; Lycett, G. W.; and Tucker, G. A.; Silencing of the major salt dependent isoform of pectinesterase in tomato alters fruits softening. *Plant Phys.* **2007**, *144*, 1960–1967.

Prasanna, V.; Prabha, T. N.; and Tharanathan, R. N.; Fruit ripening phenomena –An overview. *Crit. Rev. Food. Sci. Nutr.* **2007**, *47*, 1–19.

Quebedeaux, B.; and Bliss, F. A.; Horticulture and human health. Contributions of fruits and vegetables. Proceeding of 1st International Symposium on Horticulture and Human Health, Prentice Hall, Englewood, NJ, **1988**.

Quebedeaux, B.; and Eisa, H. M.; Horticulture and human health. Contributions of fruits and vegetables. Proceeding of 2nd International Symposium on Horticulture and Human Health. *Hortic. Sci.* **1990**, *25*, 1473–1532.

Rose, J. K. C.; Catalá, C.; Gonzalez-Carranza, C. Z. H.; and Roberts, J. A.; Plant Cell Wall Disassembly. *In: The Plant Cell Wall*; Rose, J. K. C., Ed., Blackwell Publishing Ltd, Oxford, **2003**; Vol. 8, pp 264–324.

Schuch, W.; Kanczler, J.; Robertson, D.; Hobson, G.; Tucker, G.; Grierson, D.; Bright, S.; and Bird, C.; Fruit quality characteristics of transgenic tomato fruit with altered polygalacturonase activity. *Hortic. Sci.* **1991**, *26*, 1517–1520.

Sheehy, R. E.; Kramer, M.; and Hiatt, W. R.; Reduction of polygalacturonase activity in tomato fruit by antisense RNA. *Proc. Natl. Acad. Sci. U.S.A.* **1988**, *85*, 8805–8809.

Smith, D. L.; and Gross, K. C.; A family of at least seven beta☐galactosidase genes is expressed during tomato fruit development. *Plant Physiol.* **2000**, *123*, 1173–1183.

Smith, C. J. S.; Watson, C. F.; Ray, J.; Bird, C. R.; Morris, P. C.; Schuch, W.; and Grierson, D.; Antisense RNA inhibition of polygalacturonase gene expression in transgenic tomatoes. Nature **1988**, *334*, 724–726.

Smith, C. J. S.; Watson, C. F.; Morris, P. C.; Bird, C. R.; Seymour, G. B.; Gray, A. J.; Arnold, C.; Tucker, G. A.; Schuch, W.; Harding, S.; and Grierson, D.; Inheritance and effect on ripeing of antisense polygalacturonase genes in transgenic tomatoes. *Plant Mol. Biol.* **1990**, *14*, 369–379.

Tieman, D. V.; Taylor, M. G.; Ciardi, J. A.; and Klee, H. J.; The tomato ethylene receptors NR and LeETR4 are negative regulators of ethylene response and exhibit functional compensation within a multigene family. *Proc. Natl. Acad. Sci. U.S.A.* **2000**, *97*, 5663–5668.

Van Staden, J.; Cook, E. L.; and Nooden, L. D.; Cytokinins and senescence. In: *Senescence and Aging in Plants*; Nooden, L. D., Leopold, A. C., Eds., Academic Press, San Diego; **1988**, pp 281–328.

Vicente, A. R.; Saladie, M.; Rose, J. K. C.; and Labavitch, J. M.; The linkage between cell wall metabolism and fruit softening: looking to the future. *J. Sci. Food. Agric.* **2007**, *87*, 1435–1448.

Vrebalov, J.; Pan, I. L.; Arroyo, A. J.; Arroyo, A. J.; McQuinn, R.; McQuinn, R.; Chung, M.; Poole, M.; Rose, J.; Seymour, G.; Grandillo, S.; and Giovannoni, J.; Fleshy fruit expansion and ripening are regulated by the tomato SHATTERPROOF gene TAGL1. *Plant Cell.* **2009**, *21*, 3041–3062.

Wargovich, M. J.; Anticancer properties of fruits and vegetables. *Hortic. Sci.* **2000**, *35*, 573–575.

Wilkinson, J. Q.; Lanahan, M. B.; Clark, D. G.; Bleecker, A. B.; Chang, C.; Meyerowitz, E.; M.; and Klee, H. J.; A dominant mutant receptor from Arabidopsis confers ethylene insensitivity in heterologous plants. *Nat. Biotechnol.* **1997**, *15*, 444–447.

Wills, R. B. H.; McGlasson, W. B.; Graha, D.; and Joyce, D. C.; *Postharvest – An Introduction to the Physiology and Handling of Fruits, Vegetables and Ornamentals,* 5th ed.; CAB International: Oxfordshire, UK; **2007**, 227 pp.

CHAPTER 14

ADVANCES IN POSTHARVEST DISEASE CONTROL IN VEGETABLES

MD. ARSHAD ANWER[1*]

[1*]Department of Plant Pathology, Bihar Agricultural University, Sabour, Bhagalpur, Bihar; E-mail: arshad_anwer@yahoo.com

CONTENTS

14.1 INTRODUCTION

Most of the vegetables are consumed fresh worldwide due to high quality. Vegetables decay is mainly caused by fungal and bacterial plant pathogens resulting considerable postharvest loses. It is well established that these losses are significant (Burchill and Maude, 1986; Pathak, 1997). In underdeveloped and tropical countries these losses have been estimated up to 50 percent (Coursey and Booth, 1972; Jeffries and Jeger, 1990).Vegetables are very susceptible to pathogenic fungi due to have low pH, high moisture content and rich in nutrient composition (Moss, 2002). Out of 100,000 species of fungi, about 10 percent are plant pathogenic and among them about 100 species of fungi are responsible for causing most of the postharvest diseases (Eckert and Ratnayake, 1983). In some tuber crops and brassicas, viral infections present before harvest can sometimes develop more rapidly after harvest. In general, however, viruses are not an important cause of postharvest disease. Postharvest diseases are often classified according to how infection is initiated. The so-called "quiescent" or "latent" infections are those where the pathogen initiates infection of the host at some point in time (usually before harvest), but then enters a period of inactivity or dormancy until the physiological status of the host tissue changes in such a way that infection can proceed. The dramatic physiological changes which occur during fruit ripening or maturity are often the trigger for reactivation of quiescent infections eg. gray mold of carrot caused by *Botrytis cinerea* (postharvest diseases arising from quiescent infections). The other major group of postharvest diseases are those which arise from infections initiated during and after harvest. Mostly these infections occur through surface wounds created by mechanical or insect injury. The size of the wounds need not be large for infection to take place and in many cases may be microscopic in size. Common postharvest diseases resulting from wound infections include blue and green mold (caused by *Penicillium* spp.) and transit rot (caused by *Rhizopus stolonifer*). Bacteria such as *Erwinia carotovora* (soft rot) are also common wound invader. Many pathogens, such as the banana crown rot fungi, also gain entry through the injury created by severing the crop from the plant.

Postharvest disease losses may occur at any time during postharvest handling, from harvest to consumption. When estimating these losses, it is important to consider reductions in quantity and quality resulting reduced product value. Apart from direct economic losses, diseased vegetable poses a potential health risk. Many fungal genera such as *Penicillium*, *Alternaria*, and *Fusarium* are known to produce mycotoxins under suitable conditions. Generally, highest mycotoxin contamination occurs when diseased produce is used in the production of processed food or animal feed. In most of the cases, fresh produce, which is diseased would not be consumed.

Losses due to postharvest disease are affected by various factors viz., commodity type, cultivar susceptibility to postharvest disease, postharvest environment (temperature, relative humidity, atmosphere composition, etc.), produce maturity, treatments used for disease control, produce handling methods and postharvest hygiene.

14.2 CAUSES OF POSTHARVEST DISEASE

Correct identification of the pathogen causing postharvest disease is central to the selection of an appropriate disease control strategy. Table 14.1 lists some common postharvest diseases and pathogens of vegetables. Many of the fungi, which cause postharvest disease belong to the phylum Ascomycota and the associated Fungi Anamorphici (Fungi Imperfecti). In the case of the Ascomycota, the asexual stage of fungus (the anamorph) is usually encountered more frequently in postharvest diseases than the sexual stage of the fungus (the teleomorph). Important genera of anamorphic postharvest pathogens include *Penicillium, Aspergillus, Geotrichum, Botrytis, Fusarium, Alternaria, Colletotrichum, Dothiorella, Lasiodiplodia,* and *Phomopsis.* Some of these fungi also form ascomycete sexual stages.

In the phylum Oomycota, the genera *Phytophthora* and *Pythium* are important postharvest pathogens, causing a number of diseases such as cottony leak of cucurbits (*Pythium* spp.). *Rhizopus* and *Mucor* are important genera of postharvest pathogens in the phylum Zygomycota. *R. stolonifer* is a common wound pathogen of a very wide range of fruit and vegetables, causing a rapidly spreading watery soft rot. Genera within the phylum Basidiomycota are generally not important causal agents of postharvest disease, although fungi such as *Sclerotium rolfsii* and *Rhizoctonia solani,* which have basidiomycete sexual stages, can cause significant postharvest losses of vegetable crops such as tomato and potato. While diseases caused by these pathogens are primarily field diseases, the development of symptoms often accelerates after harvest.

Bacterial soft rots are very important postharvest diseases of many vegetables. The major causal agents of bacterial soft rots are various species of *Erwinia, Pseudomonas, Bacillus, Lactobacillus,* and *Xanthomonas.*

TABLE 14.1 Examples of common postharvest diseases and pathogens of vegetables.

Disease	Pathogen	
	Anamorph	**Telomorph**
Cucurbits		
Bacterial soft rots	Various *Erwinia* spp.,	
	Bacillus polymyxa,	
	Pseudomonas spp.	
	Xanthomonas campestris	

TABLE 14.1   *(Continued)*

Disease	Pathogen	
	Anamorph	**Telomorph**
Grey mold	*Botrytis cinerea*	*Botryotinia fuckeli-ana*
Fusarium rot	*Fusarium* spp.	
Alternaria rot	*Alternaria* spp.	
Charcoal rot	*Macrophomina phaseolina*	
Cottony leak		*Pyhium* spp.
Rhizopus rot		*Rhizopus* spp.
Tomato, Eggplant and Capsicum		
Bacterial soft rots	Various *Erwinia* spp.	
	Bacillus polymyxa	
	Pseudomonas spp. And	
	Xanthomonas campestris	
Grey mold	*Botrytis cenerea*	*Botryotinia fuckeli-ana*
Alternaria rot	*Alternaria* spp.	
Cladosporium rot	*Cladosporium* spp.	
Rhizopus rot		*Rhizopus* spp.
Watery soft rot		*Sclerotinia* spp.
Cottony leak		*Pythium* spp.
Sclerotium rot	*Sclerotium rolfsii* (slerotial state)	*Athelia rolfsii*
Brassicas		
Bacterial soft rot	Various *Erwinia* spp.	
	Bacillus spp.	
	Pseudomonas spp. and	
	Xanthomonas campestris	
Grey mold	*Botrytis cinerea*	*Botryotinia fuckeli-ana*
Watery soft rot		*Slerotinia* spp.
Phytophthora rot		*Phytophthora porri*

TABLE 14.1   *(Continued)*

Disease	Pathogen	
	Anamorph	**Telomorph**
Leafy Vegetables		
Bacterial soft rot	Various *Erwinia* spp.	
	Pseudomonas spp. and	
	Xanthomonas campestris	
Grey mold	*Botrytis cenerea*	*Botryotinia fuckeliana*
Watery soft rot		*Sclerotinia* spp.
Onions		
Bacterial soft rot	Various *Erwinia* spp	
	Lactobacillus spp. And	
	Pseudomonas spp.	
Black mold rot	*Aspergillus niger*	
Fusarium basal rot	*Fusarium oxysporum* f. sp.	
	Cepae	
Smudge	*Colletotrtchum circinans*	
Carrots		
Bacterial soft rot	Various *Erwinia* spp. And	
	Pseudomonas spp.	
Rhizopus rot		*Rhizopus* spp.
Grey mold	*Botrytis cinerea*	*Botryotinia fuckeliana*
Watery soft rot		*Sclerotinia* spp.
Sclerotium rot	*Sclerotium rolfsii* (sclerotial state)	Athelin rolfsii
Chalara and	*Chalara thielavioides*	
Thielaviopsis rots	*Thielaviopsis basicola*	
Potatoes		
Bacterial soft rot	*Erwinia* spp.	

TABLE 14.1   *(Continued)*

Disease	Pathogen	
	Anamorph	**Telomorph**
Dry rot	*Fusarium* spp.	*Gibberella* spp.
Gangrene	*Phoma exigua* var *exigua* and var *foveata*	
Black scurf	*Rhizoctonia sonani* (sclerotial state)	*Thanatephorus cucumeris*
Silver scurf	*Helminthosporium solani*	
Skin spot	*Poyscytalum pustulans*	

14.3   PHYSIOLOGICAL STATUS OF HOST

The development of postharvest disease is intimately associated with the physiological status of the host tissue. To create the right environment for minimizing postharvest losses due to disease, it is important to understand the physiological changes that occur after produce is harvested. All plant organs undergo the physiological processes of growth, development, and senescence. Growth and development generally only occur while the organ is attached to the plant (with the exception of seed germination and sprouting of storage organs), but senescence will occur regardless of whether the organ is attached or not. When an organ such as a fruit is harvested from a plant, it continues to respire and transpire depleting both food reserves and water. Such changes ultimately lead to senescence. Treatments which slow respiration and water loss, such as cool storage, therefore help to delay senescence.

14.4   MODE OF INFECTION

Infection of vegetables by postharvest pathogens can occur before, during, or after harvest. Infections which occur before harvest and then remain quiescent until some point during ripening are particularly common amongst tropical crops. Anthracnose is an important postharvest disease of a number of vegetables such as bean etc. and temperate fruits (e.g., strawberry). Various species of *Colletotrichum* can cause anthracnose. Some species (e.g., *C. musae*) are host-specific, whereas others can attack a wide range of fruit and vegetables (e.g., *C. gloeosporioides*). The infection process begins when conidia germinate on the surface of host tissue to produce a germ tube and an appressorium. Although it is known that there is a quiescent phase in the life cycle of the fungus, it is not entirely clear whether the ungerminated or the germinated appressorium represents the quiescent stage, as different studies have reported conflicting results. It may be that the fungus behaves differently on different hosts, or perhaps some researchers have been unable to detect appressorial germination due to the limitations of the techniques used. In avocado for example, early studies reported that ungerminated appressoria were the quiescent phase of *C. gloeosporioides*. Studies conducted two decades later however showed that appressoria germinated to produce infection hyphae prior to the onset of quiescence. In any case, the fungus ceases growth soon after appressorium formation and remains in a quiescent state until fruit ripening occurs. During ripening, the fungus resumes activity and colonizes the fruit tissue, leading to the development of typical anthracnose symptoms. Symptoms may develop more rapidly after harvest, particularly if storage conditions favor pathogen development. Examples of postharvest diseases which can arise from late season infections include brown rot of peach (caused by *Monilinia fructicola*), gray mold of grape (caused by *Botrytis cinerea*), yeasty rot of tomato (caused by *Geotrichum candidatum*) and sclerotium rot of various vegetables (caused by *Sclerotium rolfsii*).

Many common postharvest pathogens are unable to directly penetrate the host cuticle. Such pathogens therefore infect through surface injuries or natural openings such as stomata and lenticels. Injuries can vary in size from microscopic to clearly visible and may arise in a number of ways. Mechanical injuries such as cuts, abrasions, pressure damage, and impact damage commonly occur during harvesting and handling. Insect injuries may occur before harvest yet remain undetected at the time of grading, providing ideal infection sites for many postharvest pathogens. Some chemical treatments used after harvest, such as fumigants used in insect disinfestation and disinfectants such as chlorine, may also injure produce if applied incorrectly.

Some physiological injury such as chilling and heat injury can predispose produce to infection by postharvest pathogens. Symptoms of chilling injury, peel and flesh discoloration, water-soaking and pitting. Chilling injury can increase the susceptibility of produce to postharvest disease considerably. For example, the incidence of Alternaria rot various vegetable crops is increased by exposure to excessive cold. High temperatures can also increase susceptibility of harvested produce to disease.

The natural resistance of vegetables to disease declines with storage duration. Weak pathogens which normally require a wound in order to infect can become a problem in produce that has been stored for long periods of time. Treatments which help to maintain the natural 'vitality' of vegetables aid in delaying the onset of disease in stored produce.

14.5 CONVENTIONAL POSTHARVEST DISEASE MANAGEMENT

14.5.1 FUNGICIDES

Fungicides are used extensively for postharvest disease control in fruit and vegetables. Time of application and type of fungicide used depend primarily on the target pathogen and when infection occurs. For postharvest pathogens which infect produce before harvest, field application of fungicides is often necessary. This may involve the repeated application of protectant fungicides during the growing season, and/or strategic application of systemic fungicides.

In the postharvest situation, fungicides are often applied to control infections already established in the surface tissues of produce or to protect against infections, which may occur during storage and handling. In the case of quiescent field infections present at the time of harvest, fungicides must be able to penetrate to the site of infection to be effective. Systemic fungicides are generally used for this purpose, although how deeply they penetrate when applied in this way is not well documented.

In the case of infections, which occur during and after harvest, fungicides can be used to interrupt pathogen development. How successful fungicides are in doing this depends largely on the extent to which infection has developed at the time of fungicide application and how effectively the fungicide penetrates the host tissue. In general, fungicides for the control of wound-invading pathogens should be applied as soon as possible after harvest. If infection is well advanced at the time of postharvest treatment, control will be difficult to achieve. The usual approach with controlling wound pathogens is to maintain a certain concentration of the fungicide at the injury site which will suppress (though not necessarily kill) pathogen development until the wound has healed. In this sense, most of the 'fungicides', which are used postharvest are actually fungistatic rather than fungicidal in their action under normal usage. Disinfectants such as sodium hypochlorite can be used to kill pathogen propagules on the surface of fruit, but are unable to control pathogens once they have gained entry to host tissue.

Postharvest fungicides can be applied as dips, sprays, fumigants, treated wraps and box liners or in waxes and coatings. Dips and sprays are very commonly used and depending on the compound, can take the form of aqueous solutions, suspensions, or emulsions. Fungicides commonly applied as dips or sprays include the benzimidazoles (e.g., benomyl and thiabendazole) and the triazoles (e.g., prochloraz and imazalil).

14.5.2 MANAGEMENT OF POSTHARVEST ENVIRONMENT FOR PRESERVATION OF HOST RESISTANCE TO INFECTION

The ability to control postharvest environment provides a tremendous opportunity to delay senescence. Temperature is one of the most important factor influencing disease development after harvest. It not only directly influences the rate of pathogen growth, but also to delay disease development. Low temperature storage of vegetables is commonly used for storage are not lethal to the pathogen. For example, many temperate fruit and vegetables (e.g., apples, peaches, and broccoli) can be stored at 0°C, whereas many tropical fruits and vegetables cannot be stored below 10°C without developing symptoms of chilling injury.

The relative humidity of the storage environment can also have a major influence on the development of postharvest disease. High humidity minimizes water loss of produce that can increase disease levels, particularly if free moisture accumulates in storage containers or on the surface of produce.

14.5.3 HYGIENE PRACTICES

Maintenance of hygiene at all stages especially during production and postharvest handling is critical in minimizing sources of inoculum for postharvest diseases. To reduce inoculum effectively, a sound knowledge of the pathogen's life cycle is essential. Sources of inoculum for postharvest diseases depend mainly on the pathogen and when infection occurs. In the case of postharvest diseases which cause from pre harvest infections, practices which make the crop environment less favorable to pathogens will help reduce the amount of infection which occurs during the growing season. For example, in tomato crop, pruning can increase ventilation within the plant canopy, making conditions less favorable for fungal and bacterial infection. In case of soil-borne diseases, irrigation can encourage pathogen spread and infection, minimizing contact of leaves and fruit with the soil is desirable and microsprinkler irrigation systems may be more appropriate.

14.5.4 PREHARVEST FACTORS

Various preharvest factors viz., weather (rainfall, temperature, etc.), cultivar, cultural practices (irrigation, planting density, fertilization, mulching, pesticide application, fruit bagging, etc. and planting material) influence the development of postharvest disease. These factors may have a direct influence on the development of disease by reducing inoculum sources or by discouraging infection. Alternatively, they may affect the physiology of the produce in a way that impacts on disease development after harvest. For example, the application of certain nutrients may improve the fruit skin so that it is less susceptible to injury after harvest and therefore less prone to invasion by pathogens.

14.5.5 PREVENTION OF INJURY

As many postharvest pathogens gain entry through wounds or infect physiologically damaged tissue, prevention of injury at all stages during production, harvest, and postharvest handling is critical. Injuries can be either mechanical (e.g., cuts, bruises, and abrasions), chemical (e.g., burns), biological (e.g., insect, bird, and rodent damage), or physiological (e.g., chilling injury, heat injury). Injuries can be minimized by careful harvesting and handling of produce, appropriate packaging of produce, controlling insect pests in the field, storing produce at the recommended temperature and applying postharvest treatments correctly. Where injuries are present, the process of wound healing can be accelerated in some instances through manipulation of the postharvest environment (e.g., temperature and humidity) or by application of certain chemical treatments. Wound healing has been shown to be associated with resistance to certain postharvest diseases such as bacterial soft rot of potatoes caused by *Erwinia* sp.

14.6 EMERGING TECHNOLOGIES FOR POSTHARVEST DISEASE MANAGEMENT

Recently consumers are highly concern over the presence of pesticide residues in food has warrant the search for nonchemical disease control measures. Use of fungicides after harvest is of particular concern and are applied close to the time of consumption. A number of noble approaches to control postharvest diseases are currently under study, including biological control, constitutive, or induced host resistance and natural fungicides.

14.6.1 BIOLOGICAL CONTROL

Use of antagonistic microorganisms for the control of postharvest diseases has been shown promising (Khan and Anwer, 2011). These organisms may be isolated from different sources viz. food products (fermented) and the surfaces of vegetables, fruits, or leaves. After the isolation of organisms, it can be screened in various ways for inhibition of selected pathogens. In most reported cases, pathogen inhibition is greater when the antagonist is applied prior to infection takes place. Preferably, an antagonist should be effective against a broad spectrum of pathogens on a wide range vegetables and fruits; it should be unique and be able to be produced on inexpensive growth media. Formulations incorporating such antagonists should have a long shelf life and be able to be manufactured at low cost. Most antagonists do not satisfy all of these criteria. Many are quite specific in their activity against pathogens and for this reason may not be particularly appealing to prospective investors.

There are two basic approaches for using the microbial antagonists for controlling the postharvest diseases of vegetables: (i) use of microorganisms which already exist on the produce itself (natural microbial antagonists), which can be promoted and managed, or (ii) those that can be artificially introduced against postharvest pathogens.

14.6.1.1 NATURAL MICROBIAL ANTAGONISTS

Natural occurring antagonists are those, which are present naturally on the surface of fruits and vegetables, and after isolation, antagonists are used for the control of postharvest diseases (Janisiewicz, 1987; Sobiczewski et al., 1996). Chalutz and Wilson (1990) found that when concentrated washings from the surface of citrus fruit were plated out on agar medium, only bacteria and yeast appeared while after dilution of these washings, several rot fungi appeared on the agar, suggesting that yeast and bacteria may be suppressing fungal growth. Thus, it indicates that when fruits and vegetables are washed, they are more susceptible to decay than those, which are not washed at all.

14.6.1.2 INTRODUCED MICROBIAL ANTAGONISTS

Although the first reported use of a microbial antagonist was the control of Botrytis rot of strawberry with *Trichoderma* spp. (Tronsmo and Denis, 1977), the first classical work was the control of brown rot of stone fruits by *Bacillus subtilis* (Pusey and Wilson, 1984). Since then, several antagonists have been identified, and used for controlling postharvest diseases of different fruits and vegetables. Artificial introduction of microbial antagonists is more effective in controlling postharvest diseases of fruits and vegetables than other means of biological control. Several microbial antagonists have been identified and artificially introduced on a variety of harvested commodities including vegetables for control of postharvest diseases (Table 14.2). For instance in several vegetables like tomatoes (Chalutz et al., 1988; Saligkarias et al., 2002; Xi and Tian, 2005), cabbage (*Brassica oleracea* var. capitata L.) (Adeline and Sijam, 1999), chillies (*Capsicum fruitsecence* L.) (Chanchaichaovivat et al., 2007) and potato (Colyer and Mount, 1984) (Table 14.2). The success of some of these microbial antagonists in laboratory studies and pilot tests conducted in packing houses have generated interest by several agrochemical companies in the development and commercialization of bioproducts containing microbial antagonists for control of postharvest diseases of fruits and vegetables. Several microbial antagonists have been patented and evaluated for commercial use, of which, ASPIRE (*Candida oleophila*), YieldPlus (*Candida oleophila*), and BIOSAVE (*Pseudomonas syringae*) are used worldwide for controlling postharvest diseases of fruits and vegetables effectively.

TABLE 14.2 Microbial antagonists used for the successful control of postharvest diseases of vegetables.

Antagonists	Disease (pathogen)	Vegetables	Reference(s)
Candida oleophila	Gray mold (*Botrytis cinerea*)	Tomato	Saligkarias et al. (2002)
Cryptococcus laurentii	Gray mold (*Botrytis cinerea*)	Tomato	Xi and Tian (2005)

Pichia guilliermondii	Alternaria rot (*Alternata alternata*)	Tomato	Chalutz et al. (1988)
	Rhizopus rot (*Rhizopus nigricans*)	Tomato	Zhao et al. (2008)
Pseudomonas aeruginosa	Bacterial soft rot (Erwinia carotovora sub sp. Carotovora	Cabbage	Adeline and Sijam (1999)
Pseudomonas putida	Soft rot (*Erwinia carotovora* sub sp. *carotovora*)	Potato	Colyer and Mount (1984)

14.6.1.3   MODE OF ACTION OF MICROBIAL ANTAGONISTS

Several researchers have been reviewed on the use of the microbial antagonists (Droby et al., 1989; Wisniewski et al., 1991; Filonow et al., 1996; Chand-Goyal and Spotts, 1997; Korsten et al., 1997; Filonow, 1998; Calvente et al., 1999; Janisiewicz et al., 2000; El-Ghaouth et al., 2004; Khan et al., 2011). However, the mechanism(s) by which microbial antagonists exert their influence on the pathogens has not yet been fully understood. It is important to understand the mode of action of the microbial antagonists because, it will help in developing some additional means and procedures for better results from the known antagonists, and it will also help in selecting more effective and desirable antagonists or strains of antagonists (Wilson and Wisniewski, 1989; Wisniewski and Wilson, 1992). Several modes of action have been suggested to explain the biological control activity of microbial antagonists (Sharma et. al., 2009). Still, competition for nutrient and space between the pathogen and the antagonist is considered as the major modes of action by which microbial agents control pathogens causing postharvest decay (Droby et al., 1992; Wilson et al., 1993; Filonow, 1998; Ippolito et al., 2000; Jijakli et al., 2001). In tomato, *Debaryomyces hansenii* controls Rhizopus rot, Gray mold and Alternaria rot (Chalutz et al., 1988) with the mechanism of competition for nutrients. In addition, production of antibiotics (antibiosis), direct parasitism, and possibly induced resistance are other modes of action of the microbial antagonists by which they suppress the activity of postharvest pathogens on vegetables (Janisiewicz et al., 2000; Barkai-Golan, 2001; El-Ghaouth et al., 2004).

COMPETITION FOR NUTRIENTS AND SPACE

Competition for nutrition and space between the microbial antagonist and the pathogen is considered as the major mode of action by which microbial antagonists suppress pathogens causing decay in harvested fruits and vegetables (Droby et al., 1989; Wilson and Wisniewski, 1989). To compete successfully with pathogen at the wound site, the microbial antagonist should be better adapted to various environmental and nutritional conditions than the pathogen (Barkai-Golan, 2001; El-Ghaouth et al., 2004).

COMPETITION FOR SPACE

Rapid colonization of fruit wound by the antagonist is critical for decay control, and manipulations leading to improved colonization enhance biocontrol (Mercier and Wilson, 1994). Thus, microbial antagonists should have the ability to grow more rapidly than the pathogen. Similarly, it should have the ability to survive even under conditions that are unfavorable to the pathogen (Droby et al., 1992). The biocontrol activity of microbial antagonists with most harvested commodities increased with the increasing concentrations of antagonists and decreasing concentrations of pathogen. For example, *Candida saitona* was effective at a concentration of 10^7–10^8 CFU/ml for controlling *Penicillium expansum* on apples (McLaughlin et al., 1990; El-Ghaouth et al.,1998). This qualitative relationship, however, is highly dependent on the ability of the antagonists to multiply and grow at the wound site. This was demonstrated by using a mutant of *Pichia guilliermondii*, which lost its biocontrol activity against *Penicillium digitatum* on grapefruit and against *Botrytis cinerea* on apples, even when applied to the wounds at concentrations as high as 10^{10} CFU/ml (Droby et al., 1991). The cell population of this mutant remained constant at the wound sites during incubation period, while that of the wild type increased 10- to 20-fold, within 24 h.

Attachment by microbial antagonist to the pathogen hyphae appears to be an important factor necessary for competition for nutrients as shown by the interactions of *Enterobacter cloacae* and *Rhizopus stolonifer* (Wisniewski et al., 1989), and *Pichia guilliermondii* and *Penicillium italicum* (Arras et al., 1998). In vitro studies conducted on such interactions revealed that due to direct attachment, antagonistic yeasts and bacteria take nutrients more rapidly than target pathogens and thereby prevent spore germination and growth of the pathogens (Droby et al., 1989, 1998; Wisniewski et al., 1989). In contrast, direct physical interaction did not appear to be required for the antagonistic activity of *Aureobasidium pullulans* against *Botrytis cinerea*, *Penicillium expansum*, *Rhizopus stolonifer*, and *Aspergillus niger* infecting table grapes (Vitis vinifera L.) and *Botrytis cinerea* and *Penicillium expansum* on apple fruit (Castoria et al., 2001). In these examples, antagonism was not the result of direct attachment of the microbial antagonist(s) with hyphae of the pathogens, but other mechanisms like antibiosis might have played a significant role for antagonism.

COMPETITION FOR NUTRIENTS

This mode of action of microbial antagonists supports the hypothesis that competition for nutrients plays a major role in the mode of action of nonpathogenic species of *Erwinia*, such as, *E. cypripedii*, showed antagonistic activity against various isolates of *Erwinia caratovora* sub sp. *caratovora*, the causal agent of soft rot of many vegetables like carrot, tomatoes, and pepper, primarily by competing for nutrients (Moline, 1991; Moline et al., 1999). It has been demonstrated through in vitro studies that microbial antagonists take up nutrients more rapidly than pathogens, get established and inhibit spore germination of the pathogens at the wound site (Wisniewski et al., 1989; Droby and Chalutz, 1994; Droby et al., 1998).

POPULATIONS OF THE MICROBIAL ANTAGONIST

The amount of control done by the microbial antagonists is also highly dependent on the initial concentration of the antagonists applied on the wound site and the ability of the antagonist to rapidly colonize the wound site (Janisiewicz and Roitman, 1988; Wisniewski et al., 1989; McLaughlin et al., 1990; Khan et al., 2011). In general, microbial antagonists are most effective in controlling postharvest decay on fruits and vegetables when applied at a concentration of 10^7–10^8 CFU/ml (McLaughlin et al., 1990; El-Ghaouth et al., 2004), and rarely, higher concentrations are required. Currently, there is only fragmented data regarding the antagonist- pathogen interaction in terms of competitions for limiting nutrients essential for pathogenesis. Once more information regarding the specificity of competition between antagonistic and pathogens in fruit wounds is available and genes responses of antagonism of biocontrol agents have been characterized, it will be possible to develop antagonistic strains with a higher rate of transport and/or metabolism of limiting nutrient essential for pathogenesis. This may allow us to circumvent some of the limitations of microbial antagonists.

PRODUCTION OF ANTIBIOTICS

Production of antibiotics is the second important mechanism by which microbial antagonists suppress the pathogens of harvested fruits and vegetables. For instance, bacterial antagonists like *Bacillus subtilis* and *Pseudomonas cepacia* are known to kill pathogens by producing the antibiotic iturin (Gueldner et al., 1988; Pusey, 1989). The antagonism so produced by *Bacillus subtilis* was effective in controlling fungal rot in citrus (Singh and Deverall, 1984). Similarly, the bacterial antagonist, *Pseudomonas syringae*, controlled green mold of citrus and gray mold of apple, by producing an antibiotic syringomycin (Bull et al., 1998). However, the production of this antibiotic was never detected on the fruit and vegetables despite extensive efforts, raising a doubt on the role of the antibiosis in postharvest diseases control and suggesting the operation of a different mechanism not dependent on the production of syringomycin (Bull et al., 1998). Although, antibiosis might be an effective tool for controlling postharvest diseases in a few fruits and vegetables, at present emphasis is being given for the development of nonantibiotic producing microbial antagonists for the control of postharvest diseases of fruits and vegetables (El-Gha-outh et al., 2004; Singh and Sharma, 2007). In transposon mutagenesis it was found that mutant did not able to produce antibiotic, but phenazine antibiotic produced by a wild strain of *Pseudomonas fluorescens* and able to suppress the take-all pathogen of wheat, *Gauemannomyces graminis* var. *tritici* whereas mutagenic isolates was not able to produce and did not suppress the pathogen. Researchers are aiming to concentrate to develop antagonistic microorganisms that control postharvest diseases by the mechanism of competition for space, nutrient, and direct parasitism or induced resistance (Droby, 2006).

DIRECT PARASITISM

To control postharvest disease of fruits and vegetables with the help of microbial antagonists has been reported comparatively less than other mechanism. However, it was observed that while *Pichia guilliermondii* cells had the ability to attach to the hyphae of *Botrytis cinerea* and *Penicillium,* the hyphal surface became concave and there was partial degradation of the cell wall of the pathogen (*Botrytis cinerea*) at the attachment sites (Wisniewski et al., 1991). Lytic enzymes are also produced by microbial antagonists such as chitinase, gluconase, and proteinases which help in the cell wall degradation of the pathogenic fungi (Kapat et al., 1998; Mortuza and Ilag, 1999; Castoria et al., 2001; Chernin and Chet, 2002). So, Wisniewski et al. (1991) reported that a very strong attachment of microbial antagonist with cell wall degradation enzymes may be responsible for microbial antagonistic agents in controlling the postharvest diseases of vegetables. This is also found that attachment of the microbial antagonists to a site increases their possible action for the utilization of nutrients at the invasion site and may affects the access of the pathogen to nutrients as well (El-Ghaouth et al., 2004).

INDUCED HOST RESISTANCE

To protect against infection, plants are having various biochemical and structural defence mechanisms. First mechanisms, plants have before arrival of the pathogen that is constitutive resistance, while others are only activated in response to infection that is induced resistance. In respect to plants, less is known related to host defence responses in harvested fruits and vegetables.

Phytoalexins are produced following pathogen invasion, where as in some cases they can be elicited by certain physical and chemical treatments, viz. in various crops nonionizing ultraviolet-C radiation is known to induce production of phytoalexins. 6-methoxymellen is produced after UV treatment of carrot slices that is inhibitory to *Sclerotinia sclerotiorum* and *Botrytis cinerea* (Arras, 1996)

Cell wall of many fungi having Chitosan, which is a natural fungicides compound and elicitor of host defence responses (Ippolito et al., 2000) and also has direct fungicidal effect against various postharvest pathogens. It can stimulate production of chitinase, accumulation of phytoalexins and increase in lignification. *Trichoderma* have potent antifungal activity and it produce antibiotics by various species against various pathogens like *Botrytis cinerea, Sclerotinia sclerotiorum, Corticium rolfsii,* and other plant pathogens. Some other natural compounds which have been isolated and having enough antifungal activity and may be more desirable than synthetic chemicals. Other compounds such as salicylic acid, methyl jasmonate and phosphonates and heat treatments can also induce host defences in harvested vegetables (El-Ghaouth et al.,1998).

14.6.2 NATURAL PRODUCTS' CONTROL

Scientists have also given a considerable attention to the potential of natural products of plant origin to control postharvest diseases. These compounds are environmentally safe, biodegradable, nontoxic and specific in their action viz., flavor compounds (e.g., acetaldehyde, benzaldehyde, hexanal, etc.), acetic acid, glucosinolates, jasmonates, propolis fusapyrone, essential oils, deoxyfusapyrone, chitosan, and some plant extracts.

5.6.2.1 FLAVOUR COMPOUNDS

Flavour compounds are secondary metabolites of plants having properties of volatility, low-water and fat solubility, and having antimicrobial properties (Culter et al., 1986). Being volatile in nature, they are not easily water soluble, easily adsorbed, and very useful in postharvest protection. Many such compounds would be harmless in mammalian systems and there would be less chance of any off-odors in treated produce. Most of the volatiles are effective at very low concentration. Wilson et al. (1987) found that a number of fruit volatiles produced by peaches as they ripen are highly fungicidal. Acetaldehyde has been used as a fumigant to control rot (Shaw, 1969), and *Botrytis cinerea, Rhizopus stolonifer* (Avissar and Pesis, 1991) of strawberries in storage. It was also reported that acetaldehyde has also been utilized to inhibit postharvest pathogen such as *Penicillium* spp. (Stalelbacher and Prasad, 1974), *Erwinia carotovora, Pseudomonas fluorescens, Monilinia fructicola* (Aharoni and Stadelbacher, 1973), and some species of yeast (Barkai-Golanand Aharoni, 1976) commonly found on vegetables. Some plant volatiles, such as acetaldehyde, benzaldehyde, cinnamaldehyde, ethanol, benzyl alcohol, nerolidol, 2-nonanone have also been found to have antagonistic activity against the vegetable pathogens, *P. digitatum, R. stolonifer, Colletotrichum musae* and *Erwinia caratovora* during in vitro trials (Utama et al., 2002).

Another volatile compound, (*E*)-2-Hexenal is strongly antagonistic with *B. cinerea* (Hamilton-Kemp et al., 1992; Hatanaka, 1993; Fallik et al., 1998; Archbold et al.,1999). In vitro, vapor of hexenel also inhibited hyphal growth of *Penicillium expansum* and *B. cinerea* on apple (Song et al., 1996). Hexenel vapors have a number of attributes that may be important in consumer demand for more natural control measures for fruit diseases with fewer toxic residues. Six carbon (C_6) aldehydes have been found to inhibit the hyphal growth of *Alternaria alternata* and *B. cinerea* (Hamilton-Kemp et al., 1992). These aldehydes released by plant material through the lipoxygenase pathway after tissue damage (Vick and Zimmerman, 1987). Use of these aldehydes also seems to be a possible future option for the postharvest disease control of vegetables.

5.6.2.2   ACETIC ACID

Acetic acid is a metabolic intermediate occurs naturally in many fruits (Nursten, 1970) is being used as fumigant for surface-sterilization of wide range of vegetables. The inhibitory effect of acetic acid on pathogen is greater than that due to pH alone and can penetrate the microbial cell to exert its toxic effect (Banwart, 1981). It has no phytotoxic effect as shown by Sholberg and Gaunce (1995) that low concentrations of acetic acid in air were extremely effective for control of *B. cinerea* conidia on apple fruit, as its vapor initially sterilizes the fruit surface by killing surface borne spores. Use of acetic acid as fumigant has several advantages as it is a natural compound found throughout the biosphere, having little or no residual hazard at the low levels required to suppress fungal spores, inexpensive, compared to other fumigants such as acetaldehyde, and can be used in low concentrations than others. It has been shown to be an effective fumigant for commercial use (Sholberg et al., 1996).

5.6.2.3   JASMONATES

Methyl jasmonates (MJ) and Jasmonic acid (JA) collectively known as jasmonates, are naturally plant growth regulators that are present in the plant, and regulate different aspect of plant development and responses to environmental stresses (Creelman and Mullet, 1997). It plays an important role as signal molecules in plant defence responses against pathogen attack and accumulates in plant tissues (Doares et al., 1995; Nojiri et al., 1996). Many jasmonates have the property to activate genes encoding antifungal proteins such as osmotin (Xu et al., 1994), thionin (Andresen et al., 1992), a novel ribosome inactive protein RIP (Chaudhry et al., 1994), and several other genes involved in phytoalexin biosynthesis (Gundlach et al., 1992). Moline et al. (1997) reported that MJ can be applied effectively as a postharvest treatment to suppress gray mold rot caused by *B. cinerea* in strawberry. MJ has a pleasant aroma and easily binds to polymeric materials that may prolong MJ presence in storage rooms or fumigation chambers. JA being highly soluble in water is suitable for use in solution as dip or drench. When applied at low concentrations, jasmonates are potential postharvest treatments to enhance natural resistance and to reduce decay in fruit. Moreover, jasmonates are naturally occurring compounds and are given in low concentration and doses; they may provide an eco-friendly means of reducing the current chemical usage.

5.6.2.4 GLUCOSINOLATES

Glucosinolates are mainly produced by Crucifereae that have highly antimicrobial activity (Fenwick et al., 1983). On hydrolysis of glucosinolates, a series of compounds such as isothiocyanate (ITC), thiocyanate, dglucose, sulfate ion and nitrile are produced. The antagonistic effect of six glucosinolates has been reported on several postharvest pathogens, both in vitro (Mari et al., 1993) and in vivo (Mari et al., 1996). Allyl-isothiocyanate (AITC), a naturally occurring volatile flavor compound in horseradish and mustard, has a antimicrobial activity (Delaquis and Mazza, 1995) and used successfully in modified atmosphere packaging or as a gaseous treatment before storage. Exposed pear fruit with AITC-enriched atmosphere resulted in good control of blue mold, and a TBZ-resistant strain on pears (Mari et al., 2002, 2003).

5.6.2.5 PROPOLIS

Propolis is obtained from leaf buds and bark of poplar and conifer trees. It is a natural resinous substance contains protein, amino acids, vitamins, minerals, and flavonoids (Walker and Crane, 1987; Stangaciu, 1997). It has high antibiotic, antibacterial, and antifungal activities (Tosi et al., 1996). Propolis has been found to inhibit the postharvest pathogens like *B. cinerea* and *P. expansum* (Lima et al., 1998).

5.6.2.6 FUSAPYRONE AND DEOXYFUSAPYRONE

Fusapyrone, an antifungal metabolite purified from cultures of *Fusarium semitectum*. The antgonistic activity of fusapyrone against growth and germination of *B. cinerea* has been tested in vitro and in vivo on grapes. Absence of phytotoxic effects of fusapyrone have promoted its use in control of *B. cinerea* on grapes and other crops (Altomare et al., 2000).

5.6.2.7 CHITOSAN

Chitosan is soluble form of chitin and its derivatives have plant protective and fungicidal properties. At very low concentrations, they can trigger defensive mechanisms in plants against pathogenic attacks. They have been used in solution, powder form or as wettable coatings of fruit and seeds (Choi et al., 2002). Chitosan has been used as an alternative control agent against blue mold in harvested apple fruit and has been shown to induce resistance in the fruit than direct inhibiting the pathogen (Capdeville et al., 2002).

5.6.2.8 ESSENTIAL OILS

Antifungal activity of essential oils is well known (Alankararao et al., 1991; Gogoi et al., 1997; Pitarokili et al., 1999; Meepagala et al., 2002) especially on postharvest pathogens (Bishop and Thornton, 1997). The benefit of essential oils is their bioactivity in the vapor phase and that makes them attractive as fumigants for the protection of stored products. These oils have role in plant defence mechanisms against plant pathogens (Mihaliak et al., 1991). It is documented that most of the essential oils inhibit postharvest fungi in in vitro conditions (Bishop and Reagan, 1998; Singh and Tripathi, 1999; Bellerbeck et al., 2001; Hidalgo et al., 2002). Whereas some of them have been tested for in vivo efficacy to protect stored commodities from biodeterioration. There are so many reports on essential oils in enhancing storage life of vegetables and fruit by managing their fungal rot. Essential oils from leaves of *Ocimum canum, Citrus medica* and *Melaleuca leucadendron*, were able to protect at 500-2000 μgml^{-1} of several stored food commodities from deterioration caused by *Aspergillus versicolor* and *A. flavus*. The essential oils have been tested by spraying or dipping to control postharvest decay in fruit and vegetables (Smid et al., 1994; Dixit et al., 1995). Thymo, an essential oil (component from thyme, *Thymus capitatus*) has been used as food preservative, and beverage ingredient (Mansour et al., 1986). In 1964, thymol was initially registered as a pesticide in the US and provided effective control at low concentration (2-4 mg l^{-1}) over brown rot on apricots without causing any phytotoxicity.

Hartmans et al. (1995) and Oosterhaven (1995) reported that carvone, a monoterpene, isolated from the essential oil of *Carum carvi* has been shown fungicidal activity in protecting the potato tubers from rotting and to inhibit sprouting of potatoes during storage without altering taste, quality, and without exhibiting mammalian toxicity (Oosterhaven, 1995). TALENT, a trade name in The Netherlands is the essential oil of *Salvia officinalis* has also shown great practical utility in enhancing the shelf life of some vegetables by protecting them from fungal pathogens (Bang, 1995).

The fungicidal property of the essential oils may be due to synergism among their components. Thus, there would be a negligible chance of development of fungal resistant races following application of essential oils to vegetables or fruits (Migheli et al., 1988; Pandey and Dubey, 1997). Whereas some researchers said that their phytotoxic effect increases due to this synergistic action. However, in vitro and in vivo more work on synergistic action of plant products is required.

5.6.2.9 PLANT EXTRACTS

Plant extracts from aromatic plants have preservatives in nature and therefore its antimicrobial properties have been taken into consideration. Some plants extracted have shown inhibitory action against different storage fungi (Hiremath et al., 1996;

Rana et al., 1999). However, phytochemically active principles of some of plants have been isolated and have shown strong inhibitory action against postharvest fungi. Irilin A, irilin B, the flavonone dihydrowogonin and sesquiterpene pygmol (four compounds) were isolated from aerial parts of *Chenopodium procerum* (dichloromethane extract). All four compounds inhibited the growth of the plant pathogenic fungus *Cladosporium cucumerinum* except irilin A (Bergeron et al., 1995). Inhibition of *Botryodiplodia theobromae* causing Java black rot in sweet potato was induced by phenolic compounds at 20µgml⁻¹ (Mohapotra et al., 2000). To recommend their formulation in control of postharvest diseases the investigation on the mode of action and practical applicability of such plant products is required.

14.6.3 INTEGRATED MANAGEMENT WITH PESTICIDES

Fungicides resistance biocontrol agents commonly being used in integrated disease management programs. Isolate *Aureobasidium pullulans* L47, suppressed efficiently better foliar disease on grapes and strawberries when applied in combination with a little dose of fungicide as compared to the biocontrol agent alone (Ippolito et al., 1998). The integration of biocontrol agents with fungicides provides the opportunity to reduce the amount of fungicide in preharvest application, thus decreasing the level of residues on marketable vegetables and fruits.

14.6.4 COMBINING PHYSICAL, CHEMICAL, AND BIOLOGICAL METHODS FOR SYNERGISTIC CONTROL

A case study was conducted by Eshel et al. (2009) to manage Black Root Rot of carrot with the combined application of physical, chemical, and biological method. He found that combined use of different control methods can improve control efficacy, increase the range of controlled pathogens and reduce the possibility of resistance development. To be effective, however, the different methods must be compatible: the first treatment should not have any adverse effect on the succeeding one; preferably, it should contribute to its efficacy. Carrot growers usually brush carrots before storage to remove the outer peel of the root which enhances the appearance of Black Root Rot during storage, a postharvest disease caused by the fungus *Thielaviopsis basicola*. The chemical fungicide iprodione is usually applied before storage to reduce the development of postharvest diseases. A combined of physical, low-residue chemical and biological control agents as an alternative to the conventional chemical control approach. In this study the precise application of steam (85 °C for 3 seconds) and combined application with 20 percent stabilized hydrogen peroxide (Tsunami® 100) or a yeast (*Metschnikowia fructicola*) commercial product (ShemerTM, 2gl⁻¹) were tested. If used alone, both the steam and Tsunami were highly effective at reducing disease decay but were phytotoxic to the roots. Application of combined treatments of sublethal steam followed by a sublethal dosage of Tsunami or Shemer improved efficacy and disease control by 80 percent and 86 percent, respectively. These combinations showed a synergistic effect as compared to each of the treatments alone. In other method up to 54 percent disease control was observed with the noncompatible combination of applying Tsunami first, washing it off with water and then applying Shemer. This study showed synergistic effects of sublethal treatments applied sequentially to control postharvest disease as a potential method to reduce the use of chemicals in fruit and vegetables.

14.6.5 ELICITORS OF INDUCED DISEASE RESISTANCE IN POSTHARVEST VEGETABLES' CROP

Conventional fungicides increases the pathogen resistance and having health and environmental risk, necessitate the introduction of integrated disease management (IDM) programmes. Recently an ideal strategy has been getting attention for disease management is induction of natural disease resistance (NDR) in vegetables or horticultural crops using physical, biological and/or chemical elicitors. Induced NDR can play a major role in achieving practical suppression of postharvest diseases as part of an IDM approach (Terry and Joyce, 2000).

14.6.5.1 INDUCED RESISTANCE WITH CHEMICAL

Many natural and synthetic substances stimulate local acquired resistance (LAR) and/or systemic acquired resistance (SAR) or induced systemic resistance (ISR) in horticultural produce including vegetables. Some chemicals modify the plant-pathogen reaction so that it shows an incompatible interaction with defence-related mechanisms induced prior to or after challenge (Sticher et al., 1997). Researchers have been concentrated on chemical activators on preharvest diseases (Joyce and Johnson, 1999). However, chemical elicitors that are applied pre- and/or postharvest have also been shown to enhance NDR in large number of harvested horticultural crops (e.g., in potato) (*Solanum tuberosum*), Acibenzolar (preharvest application) worked as a chemical elicitor against *Fusarium semitectum* pathogen in field as well as postharvest crop (postharvest application) (Bokshi et al., 2000).

14.6.5.2 NATURAL ORGANIC ELICITORS

Salicylic acid (SA) is proved to be a natural inducer of disease resistance in plants (Sticher et al., 1997; Mandal et. al., 2009; Khan et al, 2011). Applications of 2.0 mg SA ml^{-1} in pre or postharvest stage tended to suppress postharvest anthracnose (*Colletotrichu gloesporioides*) disease severity in mango (Zainuri et al., 2001). Zainuri et al. (2001) also recognized the effects of SA to inhibition of mango skin ripening. Natural disease resistance (NDR) in kiwifruit was enhanced against *B. cinerea* when dipped in 0.14 mg ml^{-1} concentration of SA before storage (Poole et al., 1998). SA also increased phenylalanine ammonia-lyase (PAL) and peroxidase activities related to untreated control fruits (Poole et al., 1998).

Preharvest and/or postharvest applications of 2.0 mg SA ml^{-1} suppressed the disease severity caused by *Alternaria* and *Epicoccum* sp. in Geraldton wax flower cv. CWA Pink (*Chamelaucium uncinatum*) flowers (Beasley et al., 1999). It was not clear whether SAR was induced but it is likely that salt effects caused by presence of SA suppressed pathogens and saprophytes on dead bracteoles of the flower.

Jasmonate has been proved to decrease storage diseases on horticultural crops, cut flowers (rose flowers, Meir et al., 1998) and fruits (grapefruit, Droby et al., 1999 and apple fruit, Saftner et al., 1999). Although, jasmonates have the antagonistic property against *B. cinerea* (Meir et al., 1998). Chitosan, a high molecular weight cationic polysaccharide (β-1,4-glucosamine), produced by the deacetylation of chitin, usually extracted from crustacean shell wastes and has been used as a surface coating on fruit and vegetables results delay ripening, prolong shelf life and limit fungal decay. These effects have been recognized to its direct antifungal activity, modified atmosphere effects and/or induction of postharvest resistance responses in plant tissues. Treatment with chitosan elicits accumulation of chitinases, proteinase inhibitors and hytoalexins, and also promotes lignification. However, chitosan treatment of strawberry fruit had only direct fungistatic effects against *B. cinerea* and *Rhizopus stolonifer* (El Ghaouth et al., 1992).

14.6.5.3   INORGANIC ELICITORS

Preharvest treatment with phosphonate induced disease resistance in cucumber and bean plants (Sticher et al., 1997). In contrast, postharvest potassium phosphonate (1.0 mg ml^{-1}) treatment after wounding reduced decay in apple (Wild et al., 1998). Phosphonate severely inhibited growth and pigmentation of antagonistic biocontrol yeasts (Holmes et al., 1998). Thus, phosphonate may not be an appropriate treatment in an IDM strategy.

14.6.5.4   SYNTHETIC ORGANIC ELICITORS

Development of novel synthetic organic chemical activators with increased efficacy has been focused and many modern synthetic plant activators having broad-spectrum efficacy against pathogens on fruits, vegetables, and other crops (Tally et al., 2000).

Metraux et al. (1990) reported 2,6-Dichloroisonicotinic acid (INA; CGA 41396) and its methyl ester as a first synthetic chemicals shown to activate SAR. In vitro, INA is a weak fungistatic, but effectively elicits expression of SAR genes in tobacco plant prior to TMV inoculation (Ward et al., 1991).

Acibenzolar (benzo-(1,2,3)-thiadiazole-7-carbothioic acid); S-methyl ester; ASM; BTH; CGA 245704; Actigard™ ; Bion™ is one of the most potent synthetic SAR activator (Friedrich et al., 1996). Acibenzolar is a functional analog of SA (Tally et al., 2000). Like INA, acibenzolar acts downstream of SA (Friedrich et al., 1996) and elicits accumulation of the same SAR genes and pathogenesis-related proteins (PRPs) as SA. Unlike INA, acibenzolar is not phytotoxic and has proven an effective SAR elicitor in cucumber; when applied as a foliar spray (Tally et al., 2000).

Bokshi et al. (2000) reported that in potato (*Solanum tuberosum)*, Acibenzolar applied as a preharvest spray suppressed the postharvest disease cause by *Fusarium semitectum.* Similarly, preharvest treatment of 2-year-old passion fruit cv. Supersweet 1 vines with 10 fortnightly applications of acibenzolar (0.05 mg AI ml^{-1}) combined with the strobilurin azoxystrobin and/or industry standard fungicides (copper oxychloride, mancozeb, iprodione) reduced disease incidence and severity of fruit scab caused by *Cladosporium oxysporum* compared to industry standard fungicides alone (Willingham et al., 2002). Acibenzolar was not effective against Alternaria spot caused by *Alternaria alternata*. In addition, a single acibenzolar foliar spray (0.05 or 0.025 mg AI ml^{-1}) prior to flowering reduced *Alternaria* spp. and *Fusarium* spp. disease incidence and severity in both cvs. Eldorado and South Cross rock melons and also suppressed Rhizopus disease incidence in cv. Early yellow hami melons (Huang et al., 2000). Thus, acibenzolar and perhaps other plant activators could prove valuable in the commercial management of various postharvest diseases. Continued research in this area is required to elucidate the precise SAR mechanism(s) involved.

Most published literature has shown positive effects of chemical elicitors in inducing SAR. The efficacy of acibenzolar (0.588 mg AI ml^{-1}) was not very much eliciting defence responses in cucumber against powdery mildew (*Sphaerotheca fuliginea*) (Wurms et al., 1999). An early treatment (0.05 mg AI ml^{-1}) of potato plants at 30 days after sprouting induced resistance in tubers against *Fusarium semitectum*. Therefore timing of acibenzolar treatment and developmental stage of the plant may be important in determining its efficacy. Thus, SAR elicitors may be commercially viable in some plant-pathogen systems. Trade-offs in plant defence has been shown using acibenzolar or JA to occur between pathogens and herbivores (Thaler et al., 1999). Herbivore resistance was compromised in acibenzolar- treated tomato plants induced against bacterial speck disease (*Pseudomonas syringae* pv. *Tomato*). Similarly, treatment of tomato plants with JA induced ISR against insects but reduced PRP expression and resistance to bacterial speck disease. Thus, successful utilization of chemical plant activators of SAR or ISR for reducing postharvest disease requires greater understanding of such potential trade-offs.

14.6.5.5 PHYSICALLY INDUCED RESISTANCE

In recent years, induction of NDR in horticultural crops using physical elicitors has received increasing attention (Wilson et al., 1994, 1997). Disinfection of commodities is the primary mode of action of many physical treatments. Thus, fungal spores and mycelial infections on and in the fruit or vegetables are removed and/ or destroyed. However, physical stress can cause induced resistance against future infection in some species. Low temperature storage, wounding (Ismail and Brown, 1979), CO_2 treatment (Prusky et al., 1993), heat treatment (Schirra et al., 2000), ionizing irradiation (McDonald et al., 2000) and UV-C irradiation (Wilson et al., 1997) can each enhance NDR.

Prestorage heat treatment is an important method for reducing disease incidence and severity (see reviews by Lurie, 1998; Schirra et al., 2000). Postharvest heat treatments increased the resistance of tomatoes against *B. cinerea* (Lurie et al., 1997) and reduced the development of *P. expansum* on apple fruit (Fallik et al., 1995). Heat treatments have also been used to control *B. cinerea* and *A. alternata* on red bell pepper (Fallik et al., 1996), *M. fructicola* on nectarines (Anthony et al., 1989) and *B. cinerea* on table grapes in storage (Lydakis and Aked, 2003). A thermal dip treatment for flower heads was found effective against gray mold in five rose cultivars and one carnation cultivar naturally infected with *B. cinerea* (Elad and Volpin, 1991). Low doses of short-wave ultraviolet light (UV-C, 190–280 nm wavelength) can control many storage rots of fruit and vegetables. UV-C irradiation at low doses (0.25–8.0 kJm^{-2}) target the DNA of microorganisms. For this reason UV-C treatment has been used as a germicidal or mutagenic agent. In addition to this direct germicidal activity, UV-C irradiation can modulate induced defence in plants. At appropriate wavelength and dose rates, UV-C irradiation can stimulate accumulation of stress-induced phenylpropanoids (Ben-Yehoshua et al., 1998) and PRPs (Porat et al., 2000). Increased UV-C dose lead to higher concentrations of the phytoalexin scoparone induced in flavedo tissue. Similar phytoalexin-mediated responses and associated increases in postharvest NDR have been observed in carrot (Mercier et al., 1993, 2000) and tomato (Charles et al., 1999). Most postharvest UV-C treatments have used the 254 nm wavelength due to its ready availability as commercial lamps.

The possibility of other wavelengths within the UV-C band (190–280 nm) enhancing NDR merits further investigation. In addition, effects of other light wavelengths (e.g., visible region of electromagnetic spectrum) on NDR are not yet fully explored.

14.6.5.6 BIOLOGICALLY INDUCED RESISTANCE

The use of microbial antagonists for the control of postharvest fruit decay has been actively pursued (Wilson and Wisniewski, 1989; Ippolito and Nigro, 2000). A number of antagonistic microorganisms are capable of inducing defence reactions in host tissue (Adikaram et al., 2002; Droby et al., 2002; Khan et al., 2006). Moreover, some biotic inducers and biological extracts, like yeast cell wall extracts (Adikaram et al., 1988), can boost general defence reactions in plants. The use of avirulent or attenuated strains of either pathogenic or saprophytic microorganisms to induce SAR in vegetative host tissue has been relatively well researched (Adikaram, 1990; Kuc, 2000), including for control of postharvest rots. A relatively new product Messanger™ (EDEN Bioscience Corp., WA, USA) is based on the 44 kDa harpin protein derived from *Erwinia amylovora* (Wei et al., 1992), the pathogen which causes fire blight of pear, apple, and related plants. Harpin (0.04–0.16 mg ml^{-1}) was shown to suppress blue mold disease incidence/severity caused by *P. expansum* of apple cvs. Empire, McIntoch, and Red Delicious fruit stored in normal atmosphere for 120 days at 0.5 °C when sprayed 4 or 8 days before harvest (de Capdeville et al., 2003). Similarly, harpin induced resistance in apple cv. Red Delicious fruit against *P. expansum* when applied as a postharvest spray (0.04–0.16 mg ml^{-1}). The level of resistance increased with harpin concentration and was affected by inoculum concentration and interval between treatment and inoculation.

Mercier and Arul (1993) reported that SAR in carrots could be achieved by pre-inoculating carrot roots with mycelial *B. cinerea* plugs. The incidence of gray mold was reduced when carrot crowns were subsequently challenged with *B. cinerea* 25 days later. Preinoculation with *B. cinerea* was shown to elicit systemic accumulation of the phy-toalexin 6-methoxymellein (Mercier et al., 2001).

14.7 CONCLUSIONS

Mainly fungal and bacterial pathogens cause postharvest disease in vegetables. Some of these infect produce before harvest and then remain quiescent until conditions are more favorable for disease development after harvest. Some pathogens infect produce during and after harvest through surface injuries. In the development of strategies for postharvest disease management, it is imperative to take a step back and consider the production and postharvest handling systems in the whole. Various preharvest factors directly and indirectly influence the development of postharvest disease, even in the case of infections initiated after harvest. In one hand traditionally fungicides have played a central role in postharvest disease control and in other hand, trends toward reduced chemical usage in horticulture are warrants the development of new strategies such as natural microbial antagonists and natural products to manage postharvest diseases of vegetables. This provides an exciting challenge for the twenty-first century.

KEYWORDS

- **Fruits and vegetables**
- **Postharvest disease**
- **Shelf life**
- **Natural antagonist**
- **Bacterial soft rot**
- **Pathogenic fungi**
- **Postharvest quality**

REFERENCES

Adeline, T. S. Y.; and Sijam, K.; Biological control of bacterial soft rot of cabbage. In: *Biological Control in the Tropics: Towards Efficient Biodiversity and Bioresource Management for Effective Biological Control*, Hong, L. W., Sastroutomo, S. S., Caunter, I. G., Ali, J., Yeang, L. K., Vijaysegaran, S., and Sen, Y. H., Eds.; Proceedings of the Symposium on Biological Control in the Tropics. CABI Publishing: Wallingford, UK; **1999**, pp 133–134.

Adikaram, N. K. B.; Possibility of control of post-harvest fungal diseases by manipulation of host defence system. In: Proceedings of the Third Conference on Plant Protection in the Tropics. Genting Highlands; **1990**, *5*, 31–36.

Adikaram, N. K. B.; Brown, A. E.; and Swinburne, T. R.; Phytoalexin induction as a factor in the protection of *Capsicum annuum* L. fruits against infection by *Botrytis cinerea* Pers. *J. Phytopathology.* **1988**, *122*, 267–273

Adikaram, N. K. B.; Joyce, D. C.; and Terry, L. A.; Biocontrol activity and induced resistance as a possible mode of action of *Aureobasidium pullulans* against grey mould of strawberry fruit. *Aust. Plant Pathol.* **2002**, *31*, 223–229.

Aharoni, Y.; and Stadelbacher, G. L.; The toxicity of acetaldehyde vapour to postharvest pathogens of fruits and vegetables. *Phytopathology.* **1973**, *63*, 544–545.

Alankararao, G. S. J. G.; Baby, P.; and Rajendra Prasad, Y.; Leaf oil of *Coleus amboinicus* Lour: the in vitro antimicrobial studies. *Perfumerie Kosmetics.* **1991**, *72*, 744–745.

Altomare, C.; Perrone, G.; Zonno, M. C.; Evidente, A.; Pengue, R.; Fanti, F.; and Polonelli, L.; Biological characterization of fusapyrone and deoxyfusapyrone, two bioactive secondary metabolites of *Fusarium semitectum*. *J. Nat. Prod.* **2000**, *63*, 1131–1135

Andresen, I.; Becker, W.; Schluter, K.; Burges, J.; Parthier, B.; and Apel, K.; The identification of leaf thionin as one of the main jasmonate induced proteins in barley (*Hordeum vulgare*). *Plant Mol. Biol.* **1992**, *19*, 193–204.

Anthony, B. R.; Phillips, D. J.; Bard, S.; and Aharoni, Y.; Decay control and quality maintenance after moist air heat treatment of individually plastic-wrapped nectarines. *J. Am. Soc. Hortic. Sci.* **1989**, *114*, 946–949.

Archbold, D. D.; Hamilton-Kemp, T. R.; Clements, A. M.; and Collins Randy, W.; Fumigating 'Crimson Seedless' table grapes with (*E*)-2-hexenal reduces mold during long-term postharvest storage. *Hortic. Sci.* **1999**, *34*, 705–707.

Arras, G.; Mode of action of an isolate of Candida famata in biological control of Penicillium digitatum in orange fruit. *Postharvest Biol. Technol.* **1996**, *8*, 191–198.

Arras, G.; de-Cicco, V.; Arru, S.; and Lima, G.; Biocontrol by yeasts of blue mold of citrus fruits and the mode of action of an isolate of Pichia guilliermondii. *J. Hortic. Sci. Biotechnol.* **1998**, *73*, 413–418.

Avissar, I.; and Pesis, E.; The control of postharvest decay in table grapes using acetaldehyde vapours. *Ann. Appl. Biol.* **1991**, *118*, 229–237.

Bang, U.; Essential oils as fungicides and sprout inhibitors in potatoes. In: Proceedings of the EAPR Pathology Section Meeting, *Phytophthora infestancs* ISO, Dublin, Sept 10–16, **1995**.

Banwart, G. J.; *Basic Food Microbiology*. AVI: Westport, CT, **1981**.

Barkai-Golan, R.; Aharoni, Y.; The sensitivity of food spoilage yeasts to acetaldehyde vapours. *J. Food Sci.* **1976**, *41*, 717–718.

Barkai-Golan, R.; *Postharvest Diseases of Fruit and Vegetables: Development and Control*. Elsevier Sciences: Amsterdam, The Netherlands, **2001**.

Barkai-Golan, R.; and Aharoni, Y.; The sensitivity of food spoilage yeasts to acetaldehyde vapours. *J. Food Sci.* **1976**, *41*, 717– 718.

Beasley, D. R.; Joyce, D. C.; Coates, L. M.; and Wearing, A. H.; Effect of salicylic acid treatment on postharvest diseases of Geraldton wax flower. In: Proceedings of the 12th Australasian Plant Pathology Conference. Canberra, **1999**; p 222.

Bellerbeck, V. G.; De Roques, C. G.; Bessiere, J. M.; Fonvieille, J. L.; and Dargent, R.; Effect of *Cymbopogon nardus* (L) W. Watson essential oil on the growth and morphogenesis of *Aspergillus niger*. *Can. J. Microbiol.* **2001**, *47*, 9–17.

Ben-Yehoshua, S.; Rodov, S.; Kim, J. J.; and Carnelli, S.; Preformed and induced antifungal materials of citrus fruit in relation to enhancement of decay resistance by heat and UV treatment. *J. Agric. Food Chem.* **1992**, *40*, 1217–1221.

Ben-Yehoshua, S.; Rodov, V.; and Peretz, J.; Constitutive and induced resistance of citrus fruit against pathogens. *ACIAR Proc.* **1998**, *80*, 78–92.

Bergeron et al.; Bergeron, C., Marston, A., Hakizamungu, E., and Hostettmann, K., Antifungal constituents of *Chenopodium procerum*. *Int. J. Pharmacognosy*. **1995**, *33*, 115–119.

Bishop, C. D.; and Reagan, J.; Control of the storage pathogen *Botrytis cinerea* on Dutch white cabbage (*Brassica oleracea* var. *capitata*) by the essential oil of *Melaleuca alternifolia*. *J. Essent. Oil Res.* **1998**, *10*, 57–60.

Bishop, C. D.; and Thornton, I. B.; Evaluation of the antifungal activity of the essential oils of *Monarda citriodora* var. *citriodora* and *Melaleuca alternifolia* on the post harvest pathogens. *J. Essent. Oil. Res.* **1997**, *9*, 77–82.

Bokshi, A.; Morris, S.; Deverall, B.; and Stephens, B., Induction of systemic acquired resistance in potato. In: Proceedings of the Australian Potato Research, Development And Technology Transfer Conference. Adelaide, Australia; **2000**.

Bull, C. T.; Wadsworth, M. L. K.; Sorenson, K. N.; Takemoto, J.; Austin, R.; Smilanick, J. L.; and Syringomycin E produced by biological agents controls green mold on lemons. Biological Control. **1998**, *12*, 89–95.

Burchill, R. T.; and Maude, R. B.; Microbial deterioration in stored fresh fruit and vegetables. *Outlook Agric.* **1986**, *15*, 160–166.

Calvente, V.; Benuzzi, D.; and de Tosetti, M. I. D.; Antagonistic action of siderophores from Rhodotorula glutinis upon the postharvest pathogen Penicillium expansum. *Int. Biodeteriorat. Biodegrad.* **1999**, *43*, 167–172.

Cantos, E.; Espin, J. C.; and Tomas-Barberan, F. A.; Postharvest induction modelling method using UV irradiation pulses for obtaining resveratrol-enriched table grapes: a new "functional" fruit. *J. Agric. Food Chem.* **2001**, *49*, 5052–5058.

Capdeville, G.; De Wilson, C. L.; Beer, S. V.; and Aist, J. R.; Alternative disease control agents induce resistance to blue mold in harvested Red Delicious apple fruit. *Phytopathology.* **2002**, *92*, 900–908.

Castoria, R.; de Curtis, F.; Lima, G.; Caputo, L.; Pacifico, S.; and de Cicco, V.; Aureobasidium pullulans (LS-30), an antagonist of postharvest pathogens of fruits: study on its mode of action. *Postharvest Biol. Technol.* **2001**, *32*, 717–724.

Chalutz, E.; Ben-Arie, R.; Droby, S.; Cohen, L.; Weiss, B.; and Wilson, C. L.; Yeasts as biocontrol agents of postharvest diseases of fruit. *Phytoparasitica*, **1988**, *16*, 69–75.

Chalutz, E.; and Wilson, C. L.; Postharvest biocontrol of green and blue mold and sour rot of citrus fruit by Debaryomyces hansenii. *Plant Dis.* **1990**, *74*, 134–137.

Chanchaichaovivat, A.; Ruenwongsa, P.; and Panijpan, B.; Screening and identification of yeast strains from fruit and vegetables: potential for biological control of postharvest chilli anthracnose (Colletotrichum capscii). *Biol. Cont.* **2007**, *42*(3), 326–335.

Chand-Goyal, T.; and Spotts, R. A.; Biological control of postharvest diseases of apple and pear under semi-commercial and commercial conditions using three saprophytic yeasts. *Biol. Cont.* **1997**, *10*(3), 199–206.

Charles, M-T.; Arul, J.; and Gosselin, C.; Induction of resistance to gray mold and accumulation of the phytoalexin rishitin in tomato fruits by UV-C. (Abstr.) *Phytopathology.* **1999**, *89* (Suppl.), S14.

Chaudhry, B.; Muller-Uri, F.; Cameron-Mills, V.; Gough, S.; Simpson, D.; Skriver, K.; and Mundy, J.; The barley 60 kDa jasmonate-induced protein (JIP60) is a novel ribosome-inactivating protein. *Plant J.* **1994**, *6*, 815–824.

Chernin, L.; and Chet, I.; Microbial enzymes in the biocontrol of plant pathogens and pests. In: *Enzymes in the Environment: Activity, Ecology, and Applications*, Burns, R. G., Dick, R. P., Eds.; Marcel Dekker Inc.: New York, USA; **2002**.

Choi, W. Y.; Park, H. J.; Ahn, D. J.; Lee, J.; and Lee, C. Y.; Wettability of chitosan coating solution on Fiji apple skin. *J. Food Sci.* **2002**, *67*, 2668–2672.

Colyer, P. D.; and Mount, M. D.; Bacterization of potatoes with Pseudomonas putida and its influence on postharvest soft rot diseases. *Plant Dis.* **1984**, *68*, 703–706.

Creelman, R. A.; and Mullet, J. E.; Biosynthesis and action of jasmonates in plants. *Annu. Rev. Plant Physiol. Plant Mol. Biol.* **1997**, *48*, 355–381.

Culter, H. G.; Steverson, R. F.; Cole, P. D.; Jackson, D. M.; and Johnson, A. W.; Secondary Metabolites from Higher Plants. Their Possible Role as Biological Control Agents. ACS Symposium Series, American Chemical Society, Washington, DC, **1986**; pp 178–196.

Coursey, D. G.; and Booth, R. H.; The postharvest phytopathology of perishable tropical produce. *Rev. Plant Pathol.* **1972**, *51*, 751–765.

D'hallewin, G.; Schirra, M.; Manuedda, E.; Piga, A.; and Ben-Yehoshua, S.; Scoparone and scopoletin accumulation and ultraviolet-C induced resistance to postharvest decay in oranges as influenced by harvest date. *J. Am. Soc. Hortic. Sci.* **1999**, *124*, 702–707.

de Capdeville, G.; Beer, S. V.; Watkins, C. B.; Wilson, C. L.; Tedeschi, L. O.; and Aist, J. R.; Pre- and post-harvest harpin treatments of apples induce resistance to blue mold. *Plant Dis.* **2003**, *87*, 39–44.

Delaquis, P. J.; and Mazza, G.; Antimicrobial properties of isothiocyanates in food preservation. *Food Technol.* **1995**, *49*, 73–84.

Dixit et al.; Dixit, S. N.; Chandra, H.; Tiwari, R.; and Dixit, V.; Development of botanical fungicide against blue mold of mandarins. *J. Stored Prod. Res.* **1995**, *31*, 165–172.

Doares, S. H.; Syrovets, T.; Weiler, E. W.; and Ryan, C. A.; Oligogalacturonides and chitosan activate plant defense genes through the octadecanoid pathway. *Proc. Natl. Acad. Sci. U.S.A.* **1995**, *92*, 4095–4098.

Droby, S.; Improving quality and safety of fresh fruit and vegetables after harvest by the use of biocontrol agents and natural materials. *Acta Hortic.* **2006**, *709*, 45–51.

Droby, S.; and Chalutz, E.; Mode of action of biological agents of postharvest diseases. In: *Biological Control of Postharvest Diseases – Theory and Practice,* Wilson, C. L., Wisniewski, M. E., Eds.; CRC Press: Boca Raton, **1994**, pp 63–75.

Droby, S.; Chalutz, E.; Horev, B.; Cohen, V.; Gaba, V.; Wilson, C. L.; and Wisniewski, M.; Factors affecting UV-induced resistance in grapefruit against the green mould decay caused by *Penicillium digitatum. Plant Pathol.* **1993**, *42*, 418–424.

Droby, S.; Chalutz, E.; Wilson, C. L.; and Wisniewski, M. E.; Characterization of the biocontrol activity of Debaryomyces hansenii in the control of *Penicillium* digitatum on grapefruit. *Canadida J. Microbiol.* **1989**, *35*, 794–800.

Droby, S.; Chalutz, E.; Wilson, C. L.; and Wisniewski, M. E.; Biological control of postharvest diseases: a promising alternative to the use of synthetic fungicides. *Phytoparasitica.* **1992**, *20*, 1495–1503.

Droby, S.; Cohen, A.; Weiss, B.; Horev, B.; Chalutz, E.; Katz, H.; Keren-Tzur, M.; and Shachnai, A.; Commercial testing of aspire: a yeast preparation for the biological control of postharvest decay of citrus. *Biol. Cont.* **1998**, *12*, 97–100.

Droby, S.; Porat, R.; Cohen, L.; Weiss, B.; Shapiro, B.; Philosoph- Hadas, S.; and Meir, S.; Suppressing green mold decay in grapefruit with postharvest jasmonate application. *J. Am. Soc. Hortic. Sci.* **1999**, *124*, 1–5.

Droby, S.; Vinokur, V.; Weiss, B.; Cohen, L.; Daus, A.; Goldschmid, E.; and Porat, R.; Induction of resistance to *Penicillium digitatum* in grapefruit with the biocontrol agent *Candida oleophila. Phytopathology.* **2002**, *92*, 393–399.

Eckert, J. W.; and Ratnayake, M.; Host–pathogen interactions in postharvest disease. In: *Post Harvest Physiology and Crop Preservation,* Lieberman, M., Ed.; Plenum Press: New York; **1983**.

El Ghaouth, A.; Arul, J.; Grenier, J.; and Asselin, A.; Antifungal activity of chitosan on two postharvest pathogens of strawberry fruit. *Phytopatholgy.* **1992**, *82*, 398–402.

Elad, Y.; and Volpin, H.; Heat treatment for the control of rose and carnation grey mold (*Botrytis cinerea*). *Plant Pathol.* **1991**, *40*, 278–286.

El-Ghaouth, A.; Wilson, C. L.; and Wisniewski, M. E.; Untrastructural and cytochemical aspects of biocontrol activity of Candida saitona in apple fruit. *Phytopathology.* **1998**, *88*, 282–291.

El-Ghaouth, A.; Wilson, C. L.; and Wisniewski, M. E.; Biologically based alternatives to synthetic fungicides for the postharvest diseases of fruit and vegetables. In: *Diseases of Fruit and Vegetables;* Naqvi, S. A. M. H., Ed.; Kluwer Academic Publishers: The Netherlands; **2004**, Vol. 2, pp 511–535.

Eshel, D.; Regev, R.; Orestein, J.; Droby, S.; and Gan-Mor, S.; Combining physical, chemical and biological methods for synergistic control of postharvest diseases: A case study of Black Root Rot of carrot. *Postharvest Biol. Technol.* **2009**, *54*, 48–52.

Fallik, E.; Archbold, D. D.; Hamilton-Kemp, T. R.; Clements, A. M.; Collins, R. W.; and Barth, M. E.; (*E*)-2-Hexenal can stimulate *Botrytis cinerea* growth in vitro and on strawberry fruit in vivo during storage. *J. Am. Soc. Hortic. Sci.* **1998**, *123*, 875–881.

Fallik, E.; Grinberg, S.; Alkalai, S.; Yekutieli, O.; Wiseblum, A., Regev, R., Beres, H., and Bar-Lev, E.,. A unique rapid hot water treatment to improve storage quality of sweet pepper. *Postharvest Biol. Technol.* **1999**, *15*, 25–32.

Fallik, E.; Grinberg, S.; Gambourg, M.; Klein, J. D.; and Lurie, S.; Prestorage heat treatment reduces pathogenicity of *Penicillium expansum* in apple fruit. *Plant Pathol.* **1995**, *45*, 92–97.

Fenwick, G. R.; Heaney, R. K.; and Mullin, W. J.; Glucosinolates and their breakdown products in food and food plants. *CRC Crit. Rev. Food Sci. Nutr.* **1983**, *18*, 123–201.

Filonow, A. B.; Role of competition for sugars by yeasts in the biocontrol of gray mold of apple. *Biocont. Sci. Technol.* **1998**, *8*, 243–256.

Filonow, A. B.; Vishniac, H. S.; Anderson, J. A.; and Janisiewicz, W. J.; Biological control of Botrytis cinerea in apple by yeasts from various habitats and their putative mechanism of antagonism. *Biol. Cont.* **1996**, *7*(2), 212–220.

Friedrich, L.; Lawton, K.; Ruess, W.; Masner, P.; Specker, N.; Gut Rella, M.; Meier, B.; Dincher, S.; Staub, T.; Uknes, S.; Métraux, J.-P.; Kessman, H.; and Ryals, J.; A benzothiadiazole derivative induces systemic acquired resistance in tobacco. *Plant J.* **1996**, *10*, 61–70.

Gogoi, R.; Baruah, P.; and Nath, S. C.; Antifungal activity of the essential oil of *Litsea cubeba* Pers. *J. Essent. Oils. Res.* **1997**, *9*, 213–215.

Gorlach, J.; Volrath, S.; Knauf-Beiter, G.; Hengy, G.; Beckhove, U.; Kogel, K-H.; Oostendorp, M.; Staub, T.; Ward, E.; Kessmann, H.; and Ryals, J.; Benzothiadiazole, a novel class of inducers of systemic acquired resistance, activates expression and disease resistance in wheat. *Plant Cell.* **1996**, *8*, 629–643.

Gueldner, R. C.; Reilly, C. C.; Pussey, P. L.; Costello, C. E.; Arrendale, R. F.; Cox, R. H.; Himmelsbach, D. S.; Crumley, F. G.; and Culter, H. G.; Isolation and identification of iturins as antifungal peptides in biological control of peach brown rot with Bacillus subtilis. *J. Agric. Food Chem.* **1988**, *36*, 366–370.

Gundlach, H.; Muller, M. J.; Kutchan, T. M.; and Zenk, M. H.; Jasmonic acid is a signal transducer in elicitor-induced plant cell cultures. *Proc. Natl. Acad. Sci. U.S.A.* **1992**, *89*, 2389–2393.

Hamilton-Kemp, T. R.; McCracken Jr., C. T.; Loughrin, J. H.; Anderson, R. A.; and Hildebrand, D. F.; Effect of some natural volatile compounds on the pathogenic fungi *Alternaria alternata* and *Botrytis cinerea*. *J. Chem. Ecol.* **1992**, *18*, 1083–1091.

Hartmans, K. J.; Diepenhorst, P.; Bakker, W.; and Gorris, L. G. M.; The use of carvone in agriculture, sprout suppression of potatoes and antifungal activity against potato tuber and other plant diseases. *Indian. Crops Prod.* **1995**, *4*, 3–13.

Hatanaka, A.; The biogeneration of green odour by green leaves. *Phytochemistry.* **1993**, *34*, 1201–1218.

Hidalgo, P. J.; Ubera, J. L.; Santos, J. A.; LaFont, F.; Castelanos, C.; Palomino, A.; and Roman, M.; Essential oils in *Culamintha sylvatica*. Bromf. ssp. ascendens (Jorden) P. W. Ball wild and cultivated productions and antifungal activity. *J. Essent. Oil. Res.* **2002**, *14*, 68–71.

Hiremath, S. P.; Swamy, H. K. S.; Badami, S.; and Meena, S.; Antibacterial and antifungal activities of *Striga densiflora* and *Striga orabanchioides*. *Indian J. Pharm. Sci.* **1996**, *58*, 174–176.

Holmes, R. J.; de Alwis, S.; Shanmuganathan, N.; Widyastuti, S.; and Keane, P. J.; Enhanced biocontrol of postharvest disease of apples and pears. *ACIAR Proc.* **1998**, 80, 162–166.

Huang, Y.; Deverall, B. J.; Tang, W. H.; Wang, W.; and Wu, F. W.; Foliar application of acibenzolar-S-methyl and protection of postharvest Rock melons and Hami melons from disease. *Eur. J. Plant Pathol.* **2000**, *106*, 651–656.

Ippolito, A.; El-Ghaouth, A.; Wilson, C. L.; and Wisniewski, M. A.; Control of postharvest decay of apple fruit by Aureobasidium pullulans and induction of defense responses. *Postharvest Biol. Technol.* **2000**, *19*, 265–272.

Ippolito, A.; and Nigro, F.; Impact of preharvest application of biocontrol agents on postharvest diseases of fresh fruits and vegetables. *Crop. Prot.* **2000**, *19*, 723–725.

Ippolito, A.; Nigro, F.; Romanazzi, G.; and Campanella, V.; Field application of *Aureobasidium pullulans* against Botrytis storage rot of strawberry. In: *Non-conventional Methods for the Control of Post- Harvest Disease and Microbiological Spoilage*, Bertolini, P., Sijmons, P. C., Guerzoni, M. E., Serra, F., Eds.; Workshop Proceedings COST 914-COST 915. Bologna, Italy; 1997, pp 127133.

Ismail, M.; and Brown, G. E.; Postharvest wound healing in citrus fruit: induction of phenylalanine ammonia-lyase in injured 'Valencia' orange flavedo. *J. Am. Soc. Hortic. Sci.* 1979, *104*, 126–129.

Janisiewicz, W. J.; Postharvest biological control of blue mold on apple. *Phytopathology.* **1987**, *77*, 481–485.

Janisiewicz, W. J.; and Roitman, J.; Biological control of blue mold and gray mold on apple and pear with Pseudomonas cepacia. *Phytopathology.* **1988**, *78*, 1697–1700.

Janisiewicz, W. J.; Tworkoski, T. J.; and Sharer, C.; Characterizing the mechanism of biological control of postharvest diseases on fruit with a simple method to study competition for nutrients. *Phytopathology.* **2000**, *90*(11), 1196–1200.

Jeffries, P.; and Jeger, M. J.; The biological control of postharvest diseases of fruits. *Postharvest News Inform.* **1990**, *1*, 365–368.

Jijakli, M. H.; Grevesse, C.; and Lepoivre, P.; Modes of action of biocontrol agents of postharvest diseases: challenges and difficulties. *Bulletin-OILB/SROP.* **2001**, *24*(3), 317–318.

Joyce, D. C.; and Johnson, G. I.; Prospects for exploitation of natural disease resistance in harvested horticultural crops. *Postharvest News Inform.* **1999**, *10*, 45N–48N.

Kelman, A.; In: *Postharvest Pathology of Fruits and Vegetables: Postharvest Losses in Perishable Crops*, Moline, H. E. Ed.; University of California Agricultural Experimental Station Bulletin, **1984**, pp 1–3.

Khan, M. R.; and Anwer, M. A.; Fungal based bioinoculants for plant disease management. In: *Microbes and Microbial Technology: Agricultural and environmental Applications*, Pichtel, J. (USA), and Ahmad, I. (India), Eds.; Springer, **2011**; pp 447–489.

Khan, M. R.; Anwer, M. A.; and Shahid, S.; Management of grey mould of chickpea, *Botrytis cinerea* with bacterial and fungal biopesticides using different modes of inoculation and application. *Biol. Cont.* **2011**, *57*(1), 13–23.

Khan, M. R.; Haque, Z.; Anwer, M. A.; and Khan, M. M.; Morphological and biochemical response of selected germplasm of tobacco to soil inoculation with *Pythium aphanidermatum*. *Archiv. Phytopathol. Plant Prot.* **2012**, *45*(1), 99–109.

Khan, M. R.; Anwer, M. A.; Khan, S. M.; and Khan, M. M.; An evaluation of isolates of *Aspergillus niger* against *Rhizoctonia solani*. *Test. Agrochem. Culti*, **2006**, *27*, 31–32.

Khan, M. R.; Haque, Z.; Anwer, M. A.; and Khan, M. M.; Morphological and biochemical responses of selected germplasm of tobacco to soil inoculation with *Pythium aphanidermatum*, Archives Of Phytopathology And Plant Protection, **2011**, pp 1–11.

Korsten, L.; de-Villiers, E. E.; Wehner, F. C.; and Kotze, J. M.; Field sprays of *Bacillus* subtilis and fungicides for control of preharvest fruit diseases of avocado in South Africa. *Plant Dis.* 1997, *81*, 455–459.

Ku´c, J.; Development and future direction of induced systemic acquired resistance in plants. *Crop Prot.* **2000**, *19*, 859–861.

Lima, G.; De Curtis, F.; Castoria, R.; Pacifica, S.; and De Cicco, V.; Additives and natural products against post harvest pathogens compatibility with antagonistic yeasts. In: *Plant Pathology and Sustainable Agriculture*. Proceedings of the Sixth SIPaV Annual Meeting, Campobasso, Sept 17–18, **1998**.

Liu, J.; Stevens, C.; Khan, V. A.; Lu, J. Y.; Wilson, C. L.; Adeyeye, O.; Kabwe, M. K.; Pusey, P. L.; Chalutz, E.; Sultana, T.; and Droby, S.; Application of ultraviolet-C light on storage rots and ripening of tomatoes. *J. Food Prot.* **1993**, *56*, 868–872.

Lurie, S.; Postharvest heat treatments of horticultural crops. *Hortic. Rev.* **1998**, *22*, 91–121.

Lurie, S.; Fallik, E.; Handros, A.; and Shapira, R.; The possible involvement of peroxidase in resistance to *B. cinerea* in heat treated tomato fruit. *Physiol. Mol. Plant Pathol.* **1997**, *50*, 141– 149.

Lydakis, D.; and Aked, J.; Vapour heat treatment of Sultanina table grapes. I. Control of *Botrytis cinerea*. *Postharvest Biol. Technol.* **2003**, *27*, 109–116.

Mandal et. al., **2009**.

Mandal S.; Mallick, N.; and Mitra, Y.; Salicylic acid-induced resistance to *Fusarium oxysporum* f. sp. *lycopersici* in tomato. *Plant Physiol. Biochem.* **2009**, *47*(7), 642–649.

Mansour, F.; Ravid, U.; and Putievsky, E.; Studies of essential oils isolated from 14 species of Labiateae on the carimine spider mint *Tetranychus cinnabarinus*. *Phytoparasitica.* **1986**, *14*, 137–142.

Mari, M.; Bertoii, P.; and Prateiia, G. C.; Non-conventional methods for the control of post harvest pear diseases. *J. Appl. Microbiol.* **2003**, *94*, 761–766.

Mari, M.; Leoni, O.; Lori, R.; and Cembali, T.; Antifungal vapour-phase activity of allyl isothiocyanate against *Penicillium expansum* on pears. *Plant Pathol.* **2002**, *51*, 231–236.

Mari, M.; Leoni, O.; Lori, R.; and Marchi, A.; Bioassay of glucosinolate derived isothiocyanates against post harvest pear pathogens. *Plant Pathol.* **1996**, *45*, 753–760.

Mari, M.; Lori, R.; Leoni, O.; and Marchi, A.; In vitro activity of glucosinolate derived isothiocyanates against post harvest pear pathogens. *Ann. Appl. Biol.* **1993**, *123*, 155–164.

Marquenie, D.; Lammertyn, J.; Geeraerd, A. H.; Soontjens, C.; Van Impe, J. F.; Nicola¨ı, B. M.; and Michiels, C. W.; Using survival analysis to investigate the effect of UV-C and heat treatment on storage rot of strawberry and sweet cherry. *Int. J. Food Microbiol.* **2002**, *73*, 187–196.

Marquenie, D.; Michiels, C. W.; Van Impe, J. F.; Schrevens, E.; and Nicola¨ı, B. M.; Pulsed white light in combination with UV-C and heat to reduce storage rot of strawberry. *Postharvest Biol. Technol.* **2003**, *28*, 455–461.

McDonald, R. E.; Miller, W. R.; and McCollum, T. G.; Canopy position and heat treatments influence gamma-irradiationinduced changes in phenylpropanoid metabolism in grapefruit. *J. Am. Soc. Hortic. Sci.* **2000**, *125*, 364–369.

McLaughlin, R. J.; Wilson, C. L.; Chalutz, E.; Kurtzman, W. F.; and Osman, S. F.; Characterization and reclassification of yeasts used for biological control of postharvest diseases of fruit and vegetables. *Appl. Environ. Microbiol.* 56, 3583–3586.

Meepagala, K. M.; Sturtz, G.; and Wedge, D. E.; Antifungal constituents of the essential oil fraction of *Artemisia dracunculus* L. var. *dracunculus*. *J. Agric. Food Chem.* **2002**, *50*, 6989–6992.

Meir, S.; Droby, S.; Davidson, H.; Alsevia, S.; Cohen, L.; Horev, B.; and Pilosophadas, S.; Suppression of Botrytis rot in cut rose flowers by postharvest application of methyl jasmonate. *Postharvest Biol. Technol.* **1998**, *13*, 235–243.

Mercier, J.; and Arul, J.; Induction of systemic disease resistance in carrot by pre- inoculation with storage pathogen. *Can. J. Plant Pathol.* **1993**, *15*, 281–283.

Mercier, J.; Arul, J.; and Julien, C.; Effect of UV-C on phytoalexin accumulation and resistance to *Botrytis cinerea* in stored carrots. *J. Phytopathol.* **1993**, *139*, 17–25.

Mercier, J.; Baka, M.; Reddy, B.; Corcuff, R.; and Arul, J.; Shortwave ultraviolet irradiation for control of decay by *Botrytis cinerea* in bell pepper: induced resistance and germicidal effects. *J. Am. Soc. Hortic. Sci.* **2001**, *126*, 128–133.

Mercier, J.; Roussel, D.; Charles, M-T.; and Arul, J.; Systemic and local responses associated with UV-induced and pathogen induced resistance to *Botrytis cinerea* in stored carrot. *Phytopathology.* **2000**, *90*, 981–986.

Mercier, J.; and Wilson, C. L.; Colonization of apple wounds by naturally occurring microflora and introduced Candida oleophila and their effect on infection by Botrytis cinerea during storage. *Biol. Cont.* **1994**, *4*, 138–144.

Métraux, J. P.; Signer, H.; Ryals, J.; Ward, E.; Wyss-Benz, M.; Gaudin, J.; Raschdorf, S. E.; Blum, W.; and Inverardi, B.; Increase in salicylic acid at the onset of systemic acquired resistance in cucumber. *Science.* **1990**, *250*, 1004–1006.

Migheli, Q.; Aloi, C.; and Gullino, M. L.; Evaluation of the in vitro activity of diethoferrocarb phenyl carbamate againstsome pathogens sensitive or resistant to benzimidazole. *Difesa Piante.* **1988**, *11*, 3–12.

Mihaliak, C. A.; Gershenzo, J.; and Croteau, R.; Lack of rapidmonoterpene turnover in rooted plants, implications for theories of plant chemical defense. *Oecologia.* **1991**, *87*, 373–376.

Mohapotra, N. P.; Pati, S. P.; and Ray, R. C.; In vitro inhibition of *Botryodiplodia theobromae* (Pat.) causing Java black rot in sweet potato by phenolic compounds. *Ann. Plant Prot. Sci.* **2000**, *8*, 106–109.

Moline, H. E.; Biocontrol of postharvest bacteria diseases of fruits and vegetables. In: *Biological Control of Postharvest Diseases of Fruits and Vegetables*, Wilson, C. L., Chalutz, E., Eds.; Workshop Proceedings, US Department of Agriculture, ARS-92, **1991**; pp 114–124.

Moline, H. E.; Buta, J. G.; Saftner, R. A.; and Maas, J. L.; Comparison of three volatile natural products for the reduction of post harvest diseases in strawberries. *Adv. Strawberry Res.* **1997**, *16*, 43–48.

Moline, H. E.; Hubbard, J. E.; Karns, J. S.; and Cohen, J. D.; Selective isolation of bacterial antagonists of Botrytis cinerea. *Eur. J. Plant Pathol.* **1999**, *105*, 95–101.

Mortuza, M. G.; and Ilag, L. L.; Potential for biocontrol of Lasiodiplodia theobromae in banana fruit by Trichoderma species. *Biol. Cont.* **1999**, *15*, 235–240.

Moss, M. O.; Mycotoxin review. 1. *Aspergillus* and *Penicillium. Mycologist.* **2002**, *16*, 116–119.

Nafussi, B.; Ben-Yehoshua, S.; Rodov, V.; Peretz, J.; Ozer, B. K.; and D'hallewin, G.; Mode of action of hot water dip in reducing decay in lemon fruit. In: Proceedings of the International Symposium on Postharvest. Jerusalem, Israel, **2000**, p 8.

Nigro, F.; Ippolito, A.; Lattanzio, V.; Di Venere, D.; and Salerno, M.; Effects of ultraviolet-C light on postharvest decay of strawberry. *J. Plant Pathol.* **2000**, *82*, 29–37.

Nojiri, H.; Sugimori, M.; Yamane, H.; Nishimura, Y.; Yamada, A.; Shibuya, N.; Kodama, O.; Murofushi, N.; and Omori, T.; Involvement of jasmonic acid in elicitor-induced phytoalexin production in suspension-culture rice cells. *Plant Physiol.* **1996**, *110*, 387–392.

Nursten, H. E.; Volatile compounds. The aroma of fruits. In: *The Biochemistry of Fruits and Their Products,* Hulme, A. C., Ed.,; Academic Press: New York, **1970**, pp 239–268.

Oosterhaven, J.; Different aspects of *S*-carvone—a natural potato sprout growth inhibitor. Thesis, Landbouwuniversiteit, Wageningen, Cip-data Konin Klije Bibliotheek Den Haag, **1995**. 152 pp, ISBN 90-5485-435-9.

Paiva, N. L.; An introduction to the biosynthesis of chemical used in plant-microbe communication. *J. Plant Growth. Reg.* **2000**, *19*, 131–143.

Pandey, V. N.; and Dubey, N. K.; Synergistic activity of extracted plant oils against *Pythium aphanidermatum* and *P. debaryanum*. *Trop. Agric.* **1997**, *74*, 164–167.

Pathak, V. N.; Postharvest fruit pathology—present status and future possibilities. *Indian Phytopathol.* **1997**, *50*, 161–185.

Pavoncello, D.; Lurie, S., Droby, S., and Porat, R.; A hot water treatment induces resistance to *Penicillium digitatumand* promotes the accumulation of heat shock proteins and pathogenesis-related proteins in grapefruit flavedo. *Physiol. Plantarum.* **2001**, *111*, 17–22.

Pitarokili, D.; Tzakou, O.; Couladis, M.; and Verykokidou, E.;. Composition and antifungal activity of the essential oil of *Salvia pomifera* subsp. *calycina* growing wild in Greece. *J. Essent. Oil. Res.* **1999**, *11*, 655–659.

Poole, P. R.; McLeod, L. C.; Whitmore, K. J.; and Whitaker, G.; Periharvest control of *Botrytis cinerea* rots in stored kiwifruit. *Acta Hortic.* **1998**, *464*, 71–76.

Porat, R.; Lers, A.; Dori, S.; Cohen, L.; Ben-Yehoshua, S.; Fallik, E.; Droby, S.; and Lurie, S.; Induction of resistance against *Penicillium digitatum* and chilling injury in star ruby grapefruit by a short hot water-brushing treatment. *J. Hortic. Sci. Biotechnol.* **2000**, *75*, 428–432.

Porat, R.; Lers, A.; Dori, S.; Cohen, L.; Ben-Yehoshua, S.; Fallik, E.; Droby, S.; and Lurie, S.; Induction of resistance against *Penicillium digitatum* and chilling injury in star ruby grapefruit by a short hot water-brushing treatment. *J. Hortic. Sci. Biotechnol.* **2000**, *75*, 428–432.

Prusky, D.; Pathogen quiescence in postharvest diseases. *Ann. Rev. Phytopathol.* **1996**, *34*, 413–434.

Prusky, D.; Mechanisms of resistance of fruits and vegetables to postharvest diseases. *ACIAR Proc.* **1998**, *80*, 19–33.

Prusky, D.; Ardi, R.; Koblier, I.; Beno-Moalem, D., and Leikin, A., Mechanism of resistance of avocado fruits to *Colletotrichum gloeosporioides* attack. *ACIAR Proc.* **1998**, *80*, 63– 71.

Prusky, D.; Freeman, S., Rodriguez, R. J.; and Keen, N. T.; Anonpathogenic mutant strain of *Colletotrichum magna* induces resistance to *Colletotrichum gleosporioides* in avocado fruit. *Mol. Plant-Microbe. Interact.* **1994**, *7*, 326–333.

Prusky, D.; Kobiler, I.; Ardi, R.; and Fishman, Y.; Induction of resistance of avocado fruit to *Colletotrichum gleosporioides* attack by CO2 treatment. *Acta Hortic.* **1993**, *343*, 325–331.

Pusey, P. L.; Use of Bacillus subtilis and related organisms as biofungicides. *Pest. Sci.* **1989**, *27*, 133–140.

Pusey, P. L.; and Wilson, C. L.; Postharvest biological control of stone fruit brown rotby Bacillus subtilis. *Plant Dis.* **1984**, *68*, 753–756.

Rana, B. K.; Taneja, V.; and Singh, U. P.; Antifungal activity of an aqueous extract of leaves of garlic creeper (*Adenocalymna alliaceum* Miers.). *Pharm. Biol.* **1999**, *37*, 13–16.

Reddy, M. V. B.; Belkacemi, K.; Corcuff, R.; Castaigne, F.; and Arul, J., Effect of pre-harvest chitosan sprays in post-harvest infections by *Botrytis cinerea* and quality of strawberry fruit. *Postharv. Biol. Technol.* **2000**, *20*, 39–51.

Reglinski, T.; Poole, P. R.; Whitaker, G.; and Hoyte, S. M.; Induced resistance against *Sclerotinia sclerotiorum* in kiwifruit leaves. *Plant Pathol.* **1997**, *46*, 716–721.

Rodov, V.; Ben-Yehoshua, S.; Kim, J. J.; Shapiro, B.; and Ittah, Y.; Ultraviolet illumination induces scoparone production in kumquat and orange fruit and improves decay resistance. *J. Am. Soc. Hortic. Sci.* **1992**, *117*, 788–792.

Saftner, R. A.; Abbott, J. A.; Conway, W. S.; and Barden, C. L.; Effects of postharvest heat, methyl jasmonate dip, and 1-methylcyclopropene vapor treatments on quality maintenance and decay development in 'Golden Delicious' apples. In: Proceedings of the 96th Annual Conference of American Society for Horticultural Science. Minneapolis, USA; **1999**, pp 27–31.

Saks, Y.; Copel, A.; and Barkai-Golan, R.; Improvement of harvested strawberry quality by illumination: color and *Botrytis* infection. *Postharvest Biol. Technol.* **1996**, *8*, 19–27.

Saligkarias, I. D.; Gravanis, F. T.; and Epton, H. A. S.; Biological control of Botrytis cinerea on tomato plants by the use of epiphytic yeasts *Candida guilliermondii* strains 101 and US 7 and *Candida oleophila* strain I-182: in vivo studies. *Biol. Contr.* **2002**, *25*(2), 143–150.

Schirra, M.; D'hallewin, G.; Ben-Yehoshua, S.; and Fallik, E.; Host-pathogen interactions modulated by heat treatment. *Postharvest Biol. Technol.* **2000**, *21*, 71–85.

Sharma R. R.; Singh, D.; and Singh, R.; Biological control of postharvest disease of fruits and begetables by microbial antagonists: a review. *Biol. Cont.* **2009**, *50*, 205–221.

Shaw, G. W.; The effect of controlled atmosphere storage on the quality and shelf life of fresh strawberries with special reference to *Botrytis cinerea* and *Rhizopus nigricans*. Ph.D. Thesis. University of Madison; **1969**, 62 pp.

Sholberg, P. L.; and Gaunce, A. P.; Fumigation of fruit with acetic acid to prevent post harvest decay. *Hortic Sci.* **1995**, *30*, 1271–1275.

Sholberg, P. L.; Reynolds, A. G.; and Gaunce, A. P.; Fumigation of table grapes with acetic acid to prevent post harvest decay. *Plant Dis.* **1996**, *80*, 1425–1428.

Singh, D.; and Sharma, R. R.; Postharvest diseases of fruit and vegetables and their management. In: *Sustainable Pest Management,* Prasad, D., Ed.; Daya Publishing House: New Delhi, India; **2007**.

Singh, J.; and Tripathi, N. N.; Inhibition of storage fungi of black gram (*Vigna mungo* L.) by some essential oils. *Flavour. Fragr.* J. **1999**, *14*, 42–44.

Singh, V.; and Deverall, B. J.; Bacillus subtilis as a control agent against fungal pathogens of citrus fruit. *Trans. Br. Mycol. Soc.* **1984**, *83*, 487– 490.

Smid, E. J.; Witte, Y.; de Vrees, O.; and Gorris, L. M. G.; Use of secondary plant metabolites for the control of post harvest fungal diseases on flower bulbs. *Acta Hortic.* **1994**, *368*, 523–530.

Sobiczewski, P.; Bryk, H.; and Berezynski, S.; Evaluation of epiphytic bacteria isolated from apple leaves in the control of postharvest diseases. *J. Fruit. Ornamen. Plant Res.* **1996**, *4*, 35–45.

Song, J.; Leepipattanawit, R.; Deng, W.; and Beaudry, R. M.; Hexenal vapor is a natural, metabolizable fungicide: inhibition of fungal activity and enhancement of aroma biosynthesis in apple slices. *J. Am. Soc. Hortic. Sci.* **1996**, *121*, 937–942.

Stalelbacher, G. J.; and Prasad, K.; Postharvest decay control of apple by acetaldehyde vapour. *J. Am. Soc. Hortic. Sci.* **1974**, *99*, 364–368.

Stangaciu, S.; *A Guide to the Composition and Properties of Propolis.* Dao Publishing House: Constanta, Romania.

Sticher, L.; Mauch-Mani, B.; and Métraux, J. P.; Systemic acquired resistance. *Ann. Rev. Phytopathol.* **1997**, *35*, 235–270.

Tosi, B.; Donini, A.; Romagnoli, C.; Bruni, A.; 1996. Antimicrobial activity of some commercial extracts of propolis prepared with different solvents. *Phytother. Res.* **1997**, *10*, 335–336.

Tsao, R.; and Zhou, T.; Interactions of monoterpenoids, methyl jasmonate, and Ca2+ in controlling postharvest brown rot of sweet cherry. *Hortic Sci.* **2000**, 35, 1304–1307.

Tally, A.; Oostendorp, M.; Lawton, K.; Staub, T.; and Bassi, B.; Commercial development of elicitors of induced resistance to pathogens. In: *Induced Plant Defenses Against Pathogens and Herbivores,* Agrawal, A. A., Tuzun, S., Bent, E., Eds.; APS Press: St. Paul, MN, **2000**, pp 357–369.

Terry, L. A.; and Joyce, D. C.; Suppression of grey mould on strawberry fruit with the chemical plant activator acibenzolar. *Pest Manage. Sci.* **2000**, *56*, 989–992.

Thaler, J. S.; Fidantsef, A. L.; Duffey, S. S.; and Bostock, R. M.; Trade-offs in plant defense against pathogens and herbivores: a field demonstration of chemical elicitors of induced resistance. *J. Chem. Ecol.* **1999**, *25*, 1597–1609.

Tronsmo, A.; and Denis, C.; The use of Trichoderma species to control strawberry fruit rots. *Netherlands J. Plant Pathol.* **1977**, *83*, 449–455.

Utama, I. M. S.; Wills, R. B. H.; Ben-Ye-Hoshua, S.; and Kuek, C.; In vitro efficacy of plant volatiles for inhibiting the growth of fruit and vegetable decay microorganisms. *J. Agric. Food Chem.* **2002**, *50*, 6371–6377.

Van Toor, R. F.; Jaspers, M. V.; and Stewart, A.; Evaluation of acibenzolar-S-methyl for induction of resistance in camellia flowers to *Ciborinia camelliae* infection. *New Zealand. Plant Prot.* **2001**, 54, 209–212.

Vick, B. A.; and Zimmerman, D. C.; Oxidative systems for modification of acetic acids: the lipoxygenase pathway. In: *The Biochemistry of Plants,* Stumpf, P. K., Ed.; Academic Press, **1987**, Vol. 9, pp 53–90.

Walker, P.; and Crane, E.; Constituents of propolis. *Apidologie.* **1987**, *18*, 327–334.

Ward, E. R.; Uknes, S. J.; Williams, S. C.; Dincher, S. S.; Weiderhold, D. L.; Coordinate gene activity in response to agents that induce systemic acquired resistance. *Plant Cell.* **1991**, *3*, 1085–1094.

Wei, Z. M.; Laby, R. J.; Zumoff, C. H.; Bauer, D. W.; He, S. Y.; Collmer, A.; and Beer, S. V.; Harpin, elicitor of the hypersensitive response produced by the plant pathogen *Erwinia amylovora. Science.* **1992**, *257*, 85–88.

Wild, B. L.; Wilson, C. L.; and Winley, E. L.; Apple host defence reactions as affected by cycloheximide, phosphonate and citrus green mould, *Penicillium digitatum. ACIAR Proc.* **1998**, 80, 155– 161; **1998**.

Willingham, S. L.; Pegg, K. G.; Langdon, P. W. B.; Cooke, A. W.; Beasley, D.; and Mclennan, R.; Combinations of strobilurin fungicides and acibenzolar (Bion) to reduce scab on passionfruit cause by *Cladosporium oxysporum. Aust. Plant Pathol.* **2002**, *31*, 333–336.

Wilson, C. L.; El Ghaouth, A.; Chalutz, E.; Droby, S.; Stevens, C.; Lu, J. Y.; Khan, V.; and Arul, J.; Potential of induced resistance to control postharvest diseases of fruits and vegetables. *Plant Dis.* **1994**, *78*, 837–844.

Wilson, C. L.; El Ghaouth, A.; Upchurch, B.; Stevens, C.; Khan, V.; Droby, S.; and Chalutz, E.; Using an on-line apparatus to treat harvest fruit for controlling postharvest decay. *Hortic. Technol.* **1997**, *7*, 278–282.

Wilson, C. L.; Franklin, J. D.; and Otto, B. E.; Fruit volatiles inhibitory to *Monilinia fructicola* and *Botrytis cinerea. Plant Dis.* **1987**, *71*, 316–319.

Wilson, C. L.; and Wisniewski, M. E.; Biological control of postharvest diseases of fruit and vegetables: an emerging technology. *Ann. Rev. Phytopathol.* **1989**, *27*, 425–441.

Wilson, C. L., Wisniewski, M. E.; Droby, E.; and Chalutz, E.; A selection strategy for microbial antagonists to control postharvest diseases of fruit and vegetables. *Sci Hortic.* **1993**, *53*, 183–189.

Wisniewski, M.; Biles, C.; and Droby, S.; The use of yeast Pichia guilliermondii as a biocontrol agent: characterization of attachment to Botrytis cinerea. In: *Biological Control of Postharvest Diseases of Fruit and Vegetables*, Wilson, C. L., and Chalutz, E., Eds,. Proceedings of Workshop, US Department of Agriculture, ARS-92,**1991**; pp 167–183.

Wisniewski, M.; Wilson, C. L.; and Hershberger, W.; Characterization of inhibition of Rhizopus stolonifer germination and growth by Enterobacter cloacae. *Plant Dis.* **1989**, *81*, 204–210.

Wisniewski, M. E.; and Wilson, C. L.; Biological control of postharvest diseases of fruit and vegetables: recent advances. *Hortic. Sci.* **1992**, *27*, 94–98.

Wurms, K.; Labbe, C.; Benhamou, N.; and Belanger, R. R.; Effects of Milsana and benzothiadiazole on the ultrastructure of powdery mildew haustoria on cucumber. *Phytopathology.* **1999**, *89*, 728–736.

Xi, L.; and Tian, S. P.; Control of postharvest diseases of tomato fruit by combining antagonistic yeast with sodium bicarbonate. *Sci. Agric Sin.* **2005**, *38*(5), 950–955.

Xu, Y.; Chang, P. L.; and Liu, D.; Narasimhan, M. L.; Raghothma, K. G.; Hasegawa, P. M.; Bressan, R. A.; Plant defense genes are synergistically induced by ethylene and methyl jasmonate. *Plant Cell.* **1994**, *6*, 1077–1085.

Zainuri, Joyce, D. C.; Wearing, A. H.; Coates, L.; and Terry, L.; Effects of phosphonate and salicylic acid treatments on anthracnose disease development and ripening of "Kensington Pride" mango fruit, *Aust. J. Exp. Agric.* **2001**, *41*, 805–813.

Zhao, Y.; Tu, K.; Shao, X.; Jing, W.; and Su, Z.; Effects of the yeast Pichia guilliermondii against Rhizopus nigricans on tomato fruit. *Postharvest Biol. Technol.* **2008**. *49*(1), 113–120.

INDEX

Q